Nucleophilicity

ADVANCES IN CHEMISTRY SERIES **215**

Nucleophilicity

J. Milton Harris, EDITOR
University of Alabama in Huntsville

Samuel P. McManus, EDITOR
University of Alabama in Huntsville

Developed from a symposium sponsored by
the Division of Organic Chemistry
at the 190th Meeting
of the American Chemical Society,
Chicago, Illinois,
September 8-13, 1985

American Chemical Society, Washington, DC 1987

Library of Congress Cataloging-in-Publication Data

Nucleophilicity.
(Advances in chemistry series, ISSN 0065-2392; 215)

"Developed from a symposium sponsored by the Division of Organic Chemistry at the 190th annual meeting of the American Chemical Society, Chicago, Illinois, September 8-13, 1985."

Bibliography: p.
Includes index.

1. Nucleophilic reactions—Congresses.

I. Harris, J. Milton. II. McManus, Samuel P. III. American Chemical Society. Meeting (190th: 1985: Chicago, Ill.) IV. American Chemical Society. Division of Organic Chemistry. V. Series.

QD1.A355 no. 215 [QD501] 540 s [547.1'39] 86-28843
ISBN 0-8412-0952-9

PRINTED IN THE UNITED STATES OF AMERICA

ABOUT THE EDITORS

J. MILTON HARRIS is Professor of Chemistry at the University of Alabama in Huntsville. He was Assistant Professor of Chemistry at California State University, Fullerton, from 1970 to 1973 and Visiting Research Fellow at Princeton University from 1969 to 1970. He received a B.S. in chemistry at Auburn University in 1963 and a Ph.D. in chemistry from the University of Texas at Austin in 1969. His research interests in physical organic chemistry include predicting transition-state structure, characterizing solvents and nucleophiles, and studying nucleophilic substitution. Additional research interests include synthesis of polyethers and applications in biological purifications, surface modification, and enzyme modification and synthesis of new organic solids for use in optics and growth of crystals of these materials in microgravity.

SAMUEL P. MCMANUS is Professor of Chemistry at the University of Alabama in Huntsville. He was graduated in 1960 with a B.S. degree in chemistry from The Citadel in Charleston, South Carolina, and received M.S. and Ph.D. degrees from Clemson University, specializing in organic chemistry. He began investigating intramolecular nucleophilic reactions while still a graduate student. While a chemist at E. I. du Pont de Nemours' Marshall Laboratory in Philadelphia and with the U.S. Army at Redstone Arsenal in Alabama, his attention turned to the chemical reactions of polymers, a field that still attracts his attention. At the University of Alabama, Dr. McManus resumed studies of nucleophilicity and has been especially active in the study of nucleophilic reactions with carbocations and ion pairs. Lately, he has explored the effects of microgravity on chemical processes. His research has led to more than 75 articles. He has also edited *Organic Reactive Intermediates* (1973) and coauthored *Neighboring Group Participation* (1976, with B. Capon).

FOREWORD

The ADVANCES IN CHEMISTRY SERIES was founded in 1949 by the American Chemical Society as an outlet for symposia and collections of data in special areas of topical interest that could not be accommodated in the Society's journals. It provides a medium for symposia that would otherwise be fragmented because their papers would be distributed among several journals or not published at all. Papers are reviewed critically according to ACS editorial standards and receive the careful attention and processing characteristic of ACS publications. Volumes in the ADVANCES IN CHEMISTRY SERIES maintain the integrity of the symposia on which they are based; however, verbatim reproductions of previously published papers are not accepted. Papers may include reports of research as well as reviews, because symposia may embrace both types of presentation.

Advances in Chemistry Series

M. Joan Comstock, *Series Editor*

CONTENTS

PREFACE

OUR UNDERSTANDING OF THE REACTIONS of popular reagents dates back to the early 1920s, when Lewis, Lowry, and Brønsted began developing their acid–base theories. Shortly thereafter, Lapworth, who had pioneered the study of carbonyl addition reaction mechanisms in the early 1900s, proposed the classification of polar reagents into the classes we know today as electrophiles and nucleophiles.

In the early 1930s, Ingold proposed that nucleophilic displacement reactions be divided into S_N1 and S_N2 types. Could he have known how broadly these classifications might apply? Consider the novel gas-phase displacement by a hydride ion of benzene from a chromium center:

$$H^- + C_6H_6Cr(CO)_3 \longrightarrow HCr(CO)_3^- + C_6H_6$$

Squires and Lane recently found that this reaction may be similar to a simple S_N2 reaction.

The present book is the outgrowth of a symposium on nucleophilicity held in Chicago in September 1985. The chapters were contributed by symposium participants. Although it is impossible to cover the topic completely in this format, we attempted to arrange the symposium and solicit chapters so as to cover the major areas of endeavor. We apologize to those whose interests or works may have been slighted.

This book and the symposium upon which it is based were made possible by the financial support of the ACS Division of Organic Chemistry, the U.S. Army Research Office, and the Petroleum Research Fund, administered by the American Chemical Society. We gratefully acknowledge this support. Also, we thank the contributors to this book for their timely submission of manuscripts.

J. MILTON HARRIS
SAMUEL P. MCMANUS
Department of Chemistry
University of Alabama in Huntsville
Huntsville, AL 35899

September 1985

1

Introduction to Nucleophilic Reactivity

J. Milton Harris and Samuel P. McManus

Department of Chemistry, University of Alabama in Huntsville, Huntsville AL 35899

The purpose of this chapter is to give an introduction to the subject of nucleophilicity. The chapters of the present volume are collected into five groups: (1) Marcus theory, methyl transfers, and gas-phase reactions; (2) Brønsted equation, hard–soft acid–base theory, and factors determining nucleophilicity; (3) linear free-energy relationships for solvent nucleophilicity; (4) complex nucleophilic reactions; and (5) enhancement of nucleophilicity. The present chapter is divided in the same way, giving an introduction to each of the five topics followed by a description of key points in each chapter as they relate to current studies of nucleophilicity and the other chapters of the book.

THE FIELD OF NUCLEOPHILICITY is briefly surveyed in this chapter, and the contents of this volume as they relate to the general subject of nucleophilicity are briefly described. First, we must ask, what is a nucleophile? In an article published 60 years ago, Lapworth (*1*) recognized that polar reagents fall into two classes that he termed "cationoid" and "anionoid." Ingold (*2*, *3*) later proposed that the two classes be designated "electrophilic," defined as electron seeking, and "nucleophilic," defined as nucleus seeking. Bunnett (*4*) has suggested that the term nucleophile be restricted to reagents that supply a pair of electrons to form a new bond with another atom. In its modern form, nucleophilicity is restricted to kinetic phenomena, referring to the kinetic ability of a nucleophile in a substitution or combination process. Basicity, on the other hand, is a thermodynamic term, referring to the ability of an electron-rich species to displace an equilibrium. The term basicity, given alone, usually refers to hydrogen basicity; that is, the thermodynamic ability of a Brønsted base to remove a proton from a proton donor (a Brønsted acid). If reference is made to reaction of a base with some element other than hydrogen, that element is named for the sake of clarity (e.g., carbon basicity). Similarly, the term nucleophilicity

0065–2393/87/0215–0001$06.50/0

is frequently used in reference to attack on carbon, although the term is certainly not so restricted. Again, for clarity, reference should be made to the element attacked (e.g., hydrogen nucleophilicity).

The following chapters of this book are concerned with elucidating those factors determining and affecting reactions of nucleophiles. Not surprisingly the various authors have common interests such as solvent effects, nucleophilicity scales, relationships between kinetics and thermodynamics, transition-state structure, Marcus theory, and the Brønsted equation. Detailed discussion of these specific subjects will be given later, but interestingly, the concerns of the present group of authors are similar to those of earlier years. In a 1963 review of nucleophilicity, Bunnett (*4*) listed 17 factors contributing to nucleophilic reactivity for which he found a sound experimental or theoretical precedent. These factors are (1) geometrical factors, including steric hindrance, steric acceleration from initial state compression, and entropic advantages of intramolecular reactions; (2) solvation; (3) ion aggregation; (4) thermodynamic affinity of the nucleophile for the electrophilic center; (5) capacity of Z in :Z–C–X to facilitate expulsion of a leaving group, X; and (6) negative potential at the nucleophilic center, Y; (7) bond strength of the new bond formed, M–Y; (8) ease of release of electrons from the nucleophilic atom; (9) recession of electrons toward the backside of the entering nucleophilic atom, Y; (10) π bonding between Y and M of the p–d π type; (11) London attraction between Y and M; (12) London attraction between Y and substrate substituents; (13) attraction between a substrate dipole and a dipole induced in Y; (14) acid catalysis or hydrogen bonding by protic sites on the nucleophile; (15) aggregation of large, nonpolar groups from nucleophile and substrate; (16) the α effect; and (17) base-catalyzed removal of a hydrogen from a neutral substrate of type Y–H, which leads to assisted expulsion of the leaving group.

As the individual chapters will show, our understanding of many of these 17 factors has advanced dramatically in the past 20 years. For example, separation of thermodynamic factors from nucleophilicity is now possible and complicating solvent effects can be avoided in gas-phase reactions and in reactions in aprotic solvents. On the other hand, many old problems remain. For example, the α effect is still not understood. Also, it is informative to recall the challenging pessimism expressed by Pearson et al. (*5*) in 1968 when they stated that "at present it is not possible to predict quantitatively the rates of nucleophilic displacement reactions when a number of substrates of widely varying properties are considered." In 1986, we still have no single equation, theoretical, empirical, or semiempirical, with which to refute this assessment, although some methods approach this goal (*6*).

Many of the leading figures in physical organic chemistry have made contributions to the study of nucleophilic reactivity. The present volume should serve to present these contributions and to summarize accomplishments and challenges for the future.

In the remainder of this chapter, we review the current study of nucleophilicity by describing the other chapters of this volume and giving a brief survey of the literature leading up to these works. These chapters are interrelated in many ways, so that any grouping of them must be somewhat arbitrary. Nonetheless, we have placed the chapters in one of five categories to facilitate discussion. Also, the chapters are not reviews, but rather are reports of ongoing investigations with substantial review sections and extensive references to earlier work.

Marcus Theory, Methyl Transfers, and Gas-Phase Reactions

As Bunnett has noted (*4*), the kinetic barrier to nucleophilic attack is affected by the thermodynamics of the reaction. If this thermodynamic contribution could be removed, then *intrinsic* nucleophilicities for substitution reactions could be obtained that would be independent of the leaving group. Pioneering work by Albery and Kreevoy (*7*), Pellerite and Brauman (*8*), and Lewis et al. (*9*) has shown that Marcus theory can be applied to methyl-transfer reactions to separate thermodynamic and kinetic contributions and provide intrinsic barriers to nucleophilic attack. One expression of Marcus theory is given in equation 1, where ΔE^* is the activation energy, ΔE° is the heat of reaction, and ΔE_0^* is the "intrinsic" activation energy or the barrier to reaction in the absence of any thermodynamic driving force.

$$\Delta E^* = (\Delta E^\circ)^2/16\Delta E_0^* + \Delta E_0^* + \tfrac{1}{2}\,\Delta E^\circ \qquad (1)$$

Brauman and his co-workers have applied Marcus theory to gas-phase reactions of anions with methyl halides (using ion cyclotron resonance) to obtain intrinsic nucleophilicities. This work is reviewed in Chapter 2 by Brauman et al. Interestingly, some supposedly good nucleophiles (such as alkoxide) are not good nucleophiles, by this definition, but owe their reactivity to thermodynamics. These workers argue that solvent effects generally will not overwhelm these basic energy factors, so the gas-phase results can be used to make solution-phase predictions (discussed later). Brauman et al. also note that linear-free energy relationships (LFERs) are only expected when the intrinsic barrier is constant, as is generally the case for family-type correlations. Finally, these workers observe a rather good correlation between carbon and proton basicity; several of the other authors in this volume have observed similar relationships, and a consensus exists that such relationships are key to correlations such as the Brønsted equation.

The contention that solvent effects will not overwhelm the basic energy factors revealed by Marcus analysis is supported in Chapter 3 by the solution-phase study of Lewis et al. of methyl-transfer reactions. These workers measure identity rates, k_{XX}, for $CH_3X + X^-$ and equilibrium constants, K_{Xr}, for reactions of the methylating agents with the reference nucleophile $C_6H_5SO_3^-$. Use of these data in the Marcus equations permits calculation of other nonidentity rates, which agree well with observed rates.

Lewis and co-workers also are concerned with the reactivity–selectivity postulate (RSP), which can also be derived from the Marcus expression. In the examples given here, selectivity does not vary with reactivity, in apparent contradiction to Marcus theory. This result can be explained on the basis that the intrinsic barriers are not constant and by assuming that the quadratic term of the Marcus equation contributes very little when the identity barriers are high (as they are when rates are well below diffusion control). Other important contributions to understanding the RSP have been made recently (*9a, 9b*).

Assuming that the Marcus quadratic term can be ignored, Lewis and co-workers then show the Marcus equation reduces to a simple equation for the rate of substitution k_{YX} of X^- on CH_3Y:

$$\log k_{YX} = M_Y + N_X$$

where M_Y is a property of CH_3Y and N_X is a property of the nucleophile X:

$$M_Y = \tfrac{1}{2} \log k_{YY}/K_{Yr}$$

$$N_X = \tfrac{1}{2} \log k_{XX}/K_{Xr}$$

Tables of M_Y and N_X values are given that permit calculation of a large number of reaction rates. The relationship of this expression to the formally similar, but empirical, Swain–Scott and Brønsted equations is of much interest.

In Chapter 4, McDonald et al. present a three-part gas-phase study (using a flowing afterglow apparatus) of nucleophilic substitution in the gas phase. In the first part, rates are measured for reaction of O_2^- with methyl bromide and chloride. These reactions are less exothermic than those studied by Brauman and according to Marcus theory must have unusually low intrinsic barriers. McDonald and co-workers consequently refer to O_2^- as a "supernucleophile". In the second section, these workers use an ingenious technique to study nucleophilic addition to carbonyl groups. Study of these reactions in the gas phase is normally prevented by the ready reversibility of the reaction. However, by use of "hypovalent" anion radicals (such as $C_6H_5N^{\cdot -}$) as nucleophiles, reversibility is prevented by subsequent β fragmentation. By use of this approach, a series of reaction rates are obtained for several different carbonyl derivatives and an intrinsic scale of carbonyl reactivity is derived. Finally, McDonald and co-workers examine reactions of $(CH_3O)_2PO^-$ with methyl halides and find the rates to be very low. Surprisingly, however, reactions of this nucleophile with CF_3I and CF_3Br are fast.

Apparently, an electron-transfer reaction is occurring. This reaction can therefore be used as a model for the first step of the $S_{RN}1$ mechanism. This competition between nucleophilic substitution and electron transfer is a subject of general interest.

The study of gas-phase reactions has obviously been of much assistance in understanding solution-phase reactions. Sometimes, however, this relationship breaks down because the reactants give a different reaction in the gas phase than in the solution phase. Caserio and Kim present such an example in Chapter 5, a thorough study of gas-phase reactions (ion cyclotron resonance) of alcohol nucleophiles with protonated carboxylates, carbonates, and phosphates. The goal of this work is to understand solvent and counterion effects on acid-catalyzed esterification. The gas-phase reaction pathways, however, turn out to be different from those in solution phase.

Much interest has been shown in studying the region between the gas and solution phases in which a few solvent molecules are bound to reactants. In Chapter 6, Henchman et al. describe gas-phase studies of methyl-transfer reactions that differ from those already described in that beam apparatus (providing translational energies) and a flow reactor (providing temperature control) are used. In addition, the *solvated* nucleophiles $(OH)^-(H_2O)_n$ and $(CH_3O)^-(CH_3OH)_n$ with $n = 1$–3 are used. These unique experiments provide information on the transition from gas phase to solution phase. One interesting point is that the double well potential surface normally seen for gas-phase reactions is observed in this transition region as well.

The study of methyl-transfer reactions has been key to application of Marcus theory to nucleophilic reactions. In Chapter 7, Ando et al. present a careful examination of transition-state (TS) variation in methyl-transfer reactions with nitrogen nucleophiles using carbon, hydrogen, and nitrogen kinetic isotope effects (KIEs) as experimental probes. Interestingly, the C KIEs did not show the bell-shaped trend with variation in bond breaking and bond making seen earlier for benzyl transfers. The authors conclude that little variation in the TS structure occurs for the methyl transfers; this conclusion is pertinent to Lewis' observation of invariant selectivity for methyl transfers. Ando and co-workers have also begun to use KIEs to investigate the effects of small amounts of water on the TS structure. In the preliminary study reported here, the effects of small amounts of water on H KIEs are measured for anion displacement on 2-octyl *p*-nitrobenzenesulfonate in chlorobenzene. Little variation of the KIE is seen under wet and dry conditions, and the conclusion is drawn that TS changes resulting from hydration are nonexistent or not of a type that can be detected by the H KIE.

One of the key goals of physical organic chemistry is to understand variation in reaction barriers resulting from reactant and solvent variation. In Chapter 8, Bernasconi uses his intriguing "principle of imperfect synchronization" to explain trends in intrinsic barriers resulting from substituent and solvent effects. According to this theory, product-stabilizing factors that develop late increase the barrier while reactant-stabilizing factors that are lost late decrease the barrier; the converse holds for early development or loss. For example, evidence exists that resonance stabilization of carbanions is late in developing, and as predicted, an increase in the intrinsic barrier to carbanion formation is observed. This approach presents the exciting possibility of using our well-developed concepts of resonance, solvation, steric, and electrostatic effects in combination with Marcus theory to predict changes in nucleophilic reactivity.

Some Comments on Theory

The symposium upon which this book is based included contributions from several leading theorists because of the obvious impact that theory (especially molecular orbital theory) has had on the study of nucleophilic reactions. Pioneering contributions in this area have come from Klopman (*10b*) and Hudson (*11*), and Hudson has reviewed his work in Chapter 13. These theoretical works are especially pertinent to the studies introduced in the present section because most of the theoretical studies have been for the gas phase and reproduce many of the key features described here. Unfortunately, several of the theorists participating in the symposium were not able to contribute to this volume, so we provide here a brief introduction to the recent literature in this area.

A number of groups have employed ab initio or semiempirical molecular orbital methods to compute energy surfaces of gas-phase S_N2 reactions; references 12–14 contain extensive references to this literature. These calculations verify the double-well energy profile observed in the gas-phase experiments described in several of the preceding chapters, and these calculations provide insight into transition-state structure and the role of solvent. For example, Chandrasekhar and Jorgensen (*14*) have performed ab initio and Monte Carlo calculations on the simple identity reaction of chloride ion with methyl chloride in the gas phase, in aqueous solution, and in dimethylformamide solution. Their results reproduce experimental data remarkably well and show that solvation can cause disappearance of the double-well potential profile. Shaik (*6*) and Pross (*15*) have published a series of papers on a novel application of valence bond theory (the configuration mixing model and the state correlation diagram model) to nucleophilic substitution reactions. This approach has permitted explanation of several poorly understood features of these reactions; in particular, the nature of the reaction barrier and the transition state has been clarified (*14*), the complex behavior of the RSP has been examined, and the concept that these polar reactions actually involve a single-electron shift has been forcefully presented (see also the discussion of the chapter by Pross).

Brønsted Equation, Hard–Soft Acid–Base (HSAB) Theory, and Factors Determining Nucleophilicity

Quantitative study of nucleophilicity can be said to have begun with the work in 1924 of Brønsted and Pederson (*16*), who noted a correlation of rates of nitramide decomposition by bases (B^-) and strength (pK_{HB}) of the basic catalysts:

$$\log k_{B^-} = pK_{HB} + C$$

This implied relationship between hydrogen basicity and hydrogen nucleophilicity (and its extended form involving other atomic centers) has been a major concern of physical organic chemists and is discussed in several of the chapters of this volume.

A single-parameter LFER for correlating nucleophilicity was presented by Swain and Scott in 1953 (*17*):

$$\log (k/k^0) = sn$$

where the nucleophilicity constants n are defined by the reaction rate of methyl bromide ($s = 1$) relative to that of water. This relationship implies that nucleophiles have the same relationship toward all reactive sites, but now ample evidence exists that this situation is not the case. The Brønsted relationship exhibited a similar problem in that nucleophiles with different attacking atoms (e.g., oxygen, carbon, and nitrogen bases) gave separate correlations. Several workers had pointed out that nucleophiles with high polarizabilities were more nucleophilic than expected on the basis of their basicities (*17, 18*). In view of these difficulties, a single-parameter equation will not provide a correlation of all nucleophiles with all centers; the Brønsted and Swain–Scott equations are, however, of much use nonetheless.

In an attempt to avoid the limitations of the one-parameter equations, Edwards (*19, 20*) in 1954 added polarizability (as measured by the oxidation potential of oxidative dimerization of nucleophiles, $E°$) as a second parameter to get a modified Brønsted equation:

$$\log (k/k_0) = \alpha E_X + \beta H_X$$

where k/k_0 is the rate constant for the reaction of a substrate with a reagent X relative to that for reaction with water, and

$$H_X = pK_{HX} + 1.74$$

$$E_X = E°(X_2) + 2.60$$

Use of this equation permitted correlation of a fair amount of data (*5*, *21*, *22*) and led Edwards and Pearson (*22*) to discover the much-discussed α effect (enhanced reactivity for nucleophiles having an unshared pair of electrons on the atom adjacent to the nucleophilic atom) as yet a third factor controlling nucleophilicity. Subsequently, Pearson et al. (*5*) compared Swain–Scott *n* values for methyl iodide and *trans*-$Pt(py)_2Cl_2$ reacting with a diverse set of nucleophiles and found no relationship between the *n* values. Attempts to use the Edwards and related equations to correlate the results also failed, and the conclusion was made that present understanding was inadequate to permit quantitative prediction of rates for a wide variety of substrates. This pessimistic conclusion still holds, but work continues in the search for the key physiochemical properties controlling nucleophilic reactivity (*23*).

Although Pearson et al. were pessimistic about quantitative prediction of nucleophilicity, "reasonable estimates" of nucleophilicity were quite possible to make. Such a qualitative approach is epitomized by Pearson's principle of hard and soft acids and bases (*24–26a*). This approach sprang from the Edwards equation in the sense that nucleophiles could be classified as polarizable, "soft", or nonpolarizable, "hard"; α is large for soft nucleophiles and β is large for hard nucleophiles. Similarly, Pearson noted that electrophiles or "acids" could be classified as responsive to soft nucleophiles (those exhibiting β values approaching unity) or responsive to hard nucleophiles (those exhibiting α values approaching unity). So the basic idea is that hard acids (electrophiles) prefer to bind to hard bases (nucleophiles) and soft acids prefer to bind to soft bases. The approach has proven to be extremely useful (*see* Chapter 15).

In Chapter 9, Bordwell et al. examine the ability of the Brønsted equation to predict S_N2 and E2 reactivity of carbon, sulfur, nitrogen, and oxygen nucleophiles in $(CH_3)_2SO$. The main point of this work is that the Brønsted relationship would not be expected to work in hydroxylic solvents because solvation of the nucleophile (showing leveling effects and varying with nucleophile size, charge, electron-pair delocalization, nature of the donor atom, etc.) is too complex to permit design of a single correlating equation. Employing $(CH_3)_2SO$ as the solvent, Bordwell and co-workers find excellent fit to the Brønsted equation for a large set of nucleophiles varying over 10 p*K* units (10 times the range in water); the different donor atom families do fit on separate correlations. Thus, under these circumstances, nucleophilicity depends only on two factors: basicity and sensitivity of the particular reaction to basicity (i.e., β). This observation constitutes one of the key advances in the study of nucleophilic reactivity and should be of immense use in determining those physiochemical factors affecting nucleophilicity.

These workers also note that carbon nucleophiles are more reactive toward carbon (S_N2) than toward hydrogen (E2). A calculation of carbon basicities places carbon on top of the thermodynamic order ($C^- > N^-$, $S^- >$

O^-), as it is with experimental hydrogen basicities; this finding suggests that the high solution reactivity of carbon nucleophiles toward carbon has a thermodynamic origin. The correlation between carbon and hydrogen basicity leads to the suggestion that this correlation is central to the operation of the Brønsted equation.

The Marcus equation is also examined in Chapter 9. As discussed previously regarding the Lewis chapter, the quadratic term of the Marcus equation leads to a dependency of rate on the square of pK_a, so that Brønsted plots would be expected to be curved. Bordwell and co-workers observe curvature in some of their Brønsted plots but conclude that the curvature is too large to be a Marcus effect and actually results from a solvation effect for some heteroatom substituents. These workers suggest that the curvature observed for Brønsted plots in water results from differential solvation.

Several workers have discussed the relationship between nucleophilicity and oxidation potential (*see* Chapters 3, 11, and 16). Bordwell and co-workers also examine this relationship and find a correlation that is described as "not precise".

The role of solvation in promoting curvature in Brønsted plots is also examined by Jencks in Chapter 10. Plots for reactions of alkoxide and phenoxide ions with esters and with C–H bonds are curved, with a decrease in β for the more basic alkoxide ions. Jencks contends that this curvature is caused by a difference in the nature of solvation for alkoxide and phenoxide (not by a difference in TSs). Alkoxide ions are said to have three waters of solvation, one of which must be removed for nucleophilic reaction. For the phenoxide, little solvation occurs initially, and nucleophilic attack can occur with little loss of solvation energy. In support of this position, Jencks looks at two more reactions. Amines reacting with phosphate esters show low or zero β despite clear-cut evidence that the displacement is a bimolecular, concerted process; quinuclidines actually give a negative β for this reaction. Jenck's explanation for this finding is that observed β is the sum for two processes, the first being nucleophilic attack with a large, positive β and the second being desolvation of the nucleophile with a negative β. The other supporting case comes from reaction of anionic nucleophiles with carbocations ($ArCH^+CH_3$). Here, Jencks finds slow rates for oxygen nucleophiles, which is consistent with the importance of desolvation for these strong bases.

One of the most intriguing reactions of nucleophiles is the simple (on paper) combination of cations and nucleophiles. In Chapter 11, Ritchie continues his study of rates and equilibria for these electrophile–nucleophile combination (ENC) reactions. Ritchie's determination of rates and equilibria for a wide range of nucleophilic reactions (particularly ENC reactions) has provided a fundamental set of data for evaluating key concepts of physical organic chemistry in general and of nucleophilicity in particular. The present work examines a correlation, observed previously (*see also* Chapters 3, 9, and 16), between one-electron oxidation potential and nucleophilicity. Here the data set is expanded to include hydrazine in water and some other nucleophiles in $(CH_3)_2SO$, and a correlation is not observed. Many points fit for water, but not all, and all $(CH_3)_2SO$ points are off the correlation line. No explanation for this result is apparent, and Ritchie concludes that as yet no firm grip exists on the physiochemical characteristics of a nucleophile that determine its reactivity.

A major contribution of Ritchie's has been his observation that a large number of nucleophiles show a constant selectivity toward a variety of electrophiles. In LFER terms, the reactivity of the nucleophile can be given by a single parameter with no selectivity coefficient (such as Brønsted β or Swain–Scott s) (*26b*):

$$\log (k/k_0) = N_+$$

In Chapter 12, Hoz examines the idea that two kinds of substrates occur in nucleophilic reactions: those that fit the Ritchie equation and those that do not but fit the Swain–Scott equation. Hoz contends that the distinction between these two types of substrates is that the Ritchie types all have a low-energy lowest occupied molecular orbital (LUMO), which is usually a π^* orbital, while the Swain–Scott types are typically S_N2 substrates where the LUMO is of high energy and is usually a σ^* orbital. Typical substrates fitting the Ritchie equation are ENC reactions of cations, carbonyls, and activated C=C, and Ritchie pointed out that the distinction between the two substrate types may simply be the presence or absence of a leaving group. Hoz, however, notes that he has found an ENC reaction that requires a selectivity parameter and a substitution reaction that gives the N_+ order of nucleophilicity. Hoz extends the low LUMO–high LUMO distinction by proposing that the low-LUMO substrates proceed through a "diradicaloid" TS (an argument similar to that of Pross, Chapter 23) while the high-LUMO substrates, being solvolyses, typically proceed through rate-determining ion-pair interconversion (in other words, these substrates are not diradicaloid).

The usefulness of the diradicaloid concept is further illustrated in Chapter 12 by application to two long-standing problems in nucleophilicity: (1) correlations of ionization potential and nucleophilicity (*see also* Chapter 11); and (2) the α effect (the familiar orbital-splitting argument is given a new twist).

Most of the chapters of the present volume consider equations designed to predict nucleophilicity. In Chapter 13, Hudson presents a derivation of a general, semiempirical equation relating nucleophilic reactivity to four parameters: solvation energy, electron affinity of the nucleophile, dissociation energy of the bond formed, and extent of bond formation in the TS. A test of validity of the equation, and illustration of its breadth, is that the equation reduces to the Brønsted and Swain–Scott equations under limiting conditions. Explanation of the α effect has long been a concern of Hudson's, and he continues this investigation in this chapter, concentrating on three points. First, he notes that a scheme for a TS effect must not produce a corresponding reactant (pK_a) effect. As illustration, Hudson notes that his orbital splitting explanation (i.e., two filled orbitals, for the nucleophilic electron pair and the adjacent electron pair, split into filled orbitals of higher and lower energy) will not explain the α effect because both the TS and the reactant are increased in energy. Second, he presents an argument for a proportionality between the magnitude of the α effect and the extent of bond formation in the TS as given by the Brønsted β, and he gives experimental evidence that the α effect is maximum at $\beta = 1$. Finally, he concludes on the basis of a molecular orbital analysis that the α effect results from abnormally high affinity of non-α nucleophiles for hydrogen and consequent relatively high affinity (based on pK_a) of α nucleophiles for carbon.

In Chapter 14, Menger explores the consequences of his proposal that nucleophile and electrophile must be within a "critical distance" for reaction to occur. The focus here is on intramolecular and enzymatic reactions, both of which constitute major areas of the unexplained; in fact, Menger claims that the large set of known intramolecular–intermolecular rate ratios constitutes "the largest and most variant body of unexplained data in physical organic chemistry." Certain intramolecular and enzymatic reactions are said to be fast because a carbon framework or a protein structure enforces residency at a critical distance. This apparently simple proposal is shown to be quite powerful by predicting reactivity of a large number of cases and explaining aspects of the role of solvent. Interestingly, the dogma that entropy can account for large intramolecular–intermolecular rate ratios is disputed.

The HSAB theory of Pearson has been one of the key organizing concepts in the study of nucleophiles. This theory is applied and examined in Chapters 15 and 16. In Chapter 15, Fuji applies the HSAB principles to design nucleophilic reagents for cleaving C–X bonds. Fuji notes that all bonds are made of a combination of Lewis acid and Lewis base and have hard–soft dissymmetry; for the typical C–X bond, the carbon is a soft acid and the X is a hard base. Thus, in accord with the HSAB principles, a soft base (the nucleophile) and a hard acid are required to cleave this bond selectively. Applying these ideas, Fuji then shows the utility of several soft base–hard acid reagents for cleaving various C–X bonds in complex molecules.

Transition metal complexes are often very good nucleophiles and qualify as being supersoft under Pearson's HSAB classification; for example, reaction with soft methyl iodide can be as much as 3×10^5 times faster than the reaction with hard methyl tosylate. Because soft nucleophiles are those with large α values in the Edwards equation, that rates for the transition metal nucleophiles are effectively correlated with oxidation potentials is not surprising. In the last chapter in this section, Chapter 16, Pearson uses recently obtained values of pK_a for transition metal complexes to test the full Edwards equation for these nucleophiles. Interestingly, the rates are also correlated by pK_a alone. Thus, the two parameters of the Edwards equation are not independent for transition metal nucleophiles. Pearson then explores this question for other nucleophiles and finds this same interdependence between oxidation potential and basicity for second-row nucleophiles (C, N, O, or F as donor atoms). However, for heavy nonmetal nucleophiles, the interdependence is broken, and at least two parameters are needed to achieve correlation. Pearson suggests that this second parameter may be the homolytic bond energy rather than the usual oxidation potential.

LFER for Solvent Nucleophilicity

Solvent effects on nucleophilic substitution reactions have long been of interest (*28b*), and LFERs have been a major tool in studying these effects. Probably the best known such LFER is that of Grunwald and Winstein (*29*):

$$\log (k/k_0) = mY \tag{2}$$

where k is the rate in some solvent, k_0 is the rate in 80% aqueous ethanol, m is the substrate response to a change in solvent "ionizing power" Y, and Y is determined from solvolysis of *tert*-butyl chloride for which m is defined as unity. *tert*-Butyl chloride was chosen as a model because it was thought to be a model S_N1 substrate, reacting without any nucleophilic solvent assistance (for a discussion of this assumption, see Chapters 17–19). S_N2 substrates, reacting with significant nucleophilic involvement, proceed through a TS in which more dispersal of charge occurs than in an S_N1 TS. Consequently, S_N2 substrates give Grunwald–Winstein plots with slopes (m values) of less than 1. Thus, the m value provides a qualitative measure of the extent of nucleophilic involvement (*30, 31*).

Swain et al. (*32*) were the first (1955) to use an LFER to quantify solvent nucleophilicity. Their approach was to extend the Grunwald–Winstein equation to include a second term for solvent nucleophilicity.

$$\log (k/k_0) = c_1 d_1 + c_2 d_2 \tag{3}$$

The c coefficients measure the substrate response to nucleophilicity, d_1, and electrophilicity, d_2. Again, 80% aqueous ethanol was chosen as the standard solvent, and a set of c values was chosen in accord with supposed mechanistic behavior of model substrates: $c_1 = 3c_2$ for methyl bromide, $c_1 = c_2$ for *tert*-butyl chloride, and $3c_1 = c_2$ for $(C_6H_5)_3CF$. This approach was largely unsuccessful, probably in large part because *tert*-butyl chloride is assumed to be equally sensitive to nucleophilicity and ionizing power (*29*). The equation predicts, for example, that tertiary, benzhydryl, and bridgehead substrates are more sensitive to solvent nucleophilicity than typical primary and secondary S_N2 substrates (*31*).

This basic approach was reexamined by Peterson et al. (*33*) and Bentley et al. (*34*), using refined mechanistic information to define more appropriate model compounds. According to the Bentley et al. equation

$$\log (k/k_0) = lN + mY \tag{4}$$

where 80% ethanol is the standard solvent and l and m measure substrate response to solvent nucleophilicity N and solvent ionizing power Y. The solvent parameters are defined by setting $m = 1$ and $l = 0$ for 2-adamantyl tosylate (or a 1-adamantyl derivative) and $l = 1$ and $m = 0.3$ for methyl tosylate. Peterson et al. (*33*) used tetramethylenechloronium or iodonium salts reacting with nucleophiles (anionic or neutral) in water or sulfur dioxide to measure nucleophilicities. Also, they altered the model assumptions for the Swain et al. equation and derived more reasonable results from this extensive data set; a comparison of Swain–Scott, Peterson et al., Bentley et al., and Swain–Mosely–Bown nucleophilicity parameters is given in this work (*33*). Work with equation 4 has been extremely valuable in analyzing important phenomena such as substrate sensitivity to nucleophilicity. The contributions in this section continue this discussion.

Of course, the study of medium effects is one of the major concerns of chemistry, and an extensive review (*35*) of the subject would be inappropriate here. However, two recent contributions are particularly pertinent to the foregoing discussion of solvent nucleophilicity; these contributions are the work of Swain et al. (*36*) and the work of Taft et al. (*37*). In these studies, the goal is to correlate *all* available medium-dependent phenomena, not just nucleophilic reactions, with a single LFER; parameters that may be related to solvent nucleophilicities are produced.

Swain et al. (*36*) have recently presented a new LFER, which superficially resembles equations 3 and 4, for evaluating medium effects:

$$P_{ij} = a_i A_j + b_i B_j + c_i \tag{5}$$

where P_{ij} is some medium-dependent phenomenon, such as log k, in reaction i and solvent j; a and b are reaction or substrate sensitivities to solvent change; A is "anion-solvating tendency" or "acity" of the solvent, and B is "cation-solvating tendency" or "basity" of the solvent. So that A and B could be evaluated, data for 77 phenomena in 61 solvents were subjected to regression analysis. These 77 phenomena include reaction rates; equilibria; and NMR, IR, ESR, and electronic spectra. Many fascinating and controversial conclusions are drawn in this work. For example, both *tert*-butyl chloride and 2-adamantyl tosylate solvolyses are said to involve significant nucleophilic solvent assistance that is in fact greater than that received by methyl bromide solvolysis! Obviously, this work has generated much heated discussion (*38–41*).

Several solvent properties could be included in an LFER for medium effects (*31*). However, the general approach, as shown previously, has been to restrict treatment to two parameters, either because of the difficulty in assessing the significance of the additional parameters (*31*) or, as in the Swain work (*36*), because it is believed that only two parameters are required. Certainly, additional parameters will give better fit to experimental data, but can this improved fit be attributed to a physical dependence on the new parameter or simply to an additional degree of adjustment? In a series of recent papers, Taft et al. (*37*) have adopted the approach of including as parameters in an LFER all those solvent characteristics for which evidence of involvement in medium effects exists; they reasoned that modern computational methods will permit statistical evaluation of parameter significance. At present, the equation in use for kinetic phenomena has four parameters:

$$\log (k/k_0) = a\alpha + b\beta + s\pi^* + d\delta_H/100 \quad (6)$$

where α is solvent hydrogen bond donor ability (or electrophilicity), β is solvent hydrogen bond acceptor ability (or nucleophilicity), π^* is solvent dipolarity–polarizability, and δ_H is the Hildebrand solubility parameter, which is thought to provide a measure of the energy required to make a cavity in a solvent for a molecule or TS. The approach is referred to as the solvatochromic method because all the parameters except δ_H were initially evaluated from solvent effects on UV–visible absorptions. The solvatochromic equation has been used to correlate a number of processes ranging from chromatographic retention times to rates of Menshutkin reactions. This method, and those discussed previously, are considered in the following chapters of this section.

In Chapter 17, Harris et al. apply the solvatochromic method of Taft et al. to solvolyses. The goal of this work is to determine if solvent electrophilicity (as expressed by α) is an important variable in solvolyses even when the leaving group is unchanged. One reason this question is important is that many methods for assessing nucleophilic solvent assistance assume that electrophilic solvent assistance does not vary with the substrate as long as the leaving group is the same. Therefore, any rate difference between some substrate and a model is assumed to be the result of variation in nucleophilic solvent assistance; obviously, if electrophilic solvent assistance is varying, these methods are flawed. In this work, a values (sensitivities to solvent electrophilicity) are measured, and electrophilic solvent assistance is determined to be substrate-dependent. Statistically significant correlations with equation 6 were found, and the conclusion was made that the Taft et al. approach is justified *if* large amounts of data are available.

In Chapter 18, Bentley reviews development of scales of solvent nucleophilicity and applies the Bentley–Schleyer equation to determine N and Y values for sulfuric acid. Also, Bentley examines here the validity of the Taft et al. assumption (Chapter 17) that solvent ionizing power (as defined by Bentley et al.) contains contributions from electrophilicity (i.e., solvent hydrogen bond donor ability α) and ionizing power (i.e., solvent dipolarity–polarizability π^*) that are not covariant; in other words, Bentley does not question the importance of these solvent properties, but rather questions whether or not an increase in one property is accompanied by an increase in the other property. An experimental test of this question is provided by plotting a series of Y values for different leaving groups against each other. Linear plots are found, even for solvents having large α values. Bentley contends that if Y included noncovariant electrophilicity and ionizing power, then nonlinear plots should have been observed. Bentley also criticizes the assumption of Harris et al. (in Chapter 17) that β values provide a measure of solvent nucleophilicity; Bentley notes that trifluoroethanol and hexafluoroisopropyl alcohol always give different N values, yet the β values for both are zero.

In Chapter 19, Kevill et al. describe their work evaluating solvent parameters by replacing the usual alkyl halides and arenesulfonates with sulfonium salts [such as 1-Ad-$S(CH_3)_2^+$]. Solvolyses with neutral leaving groups have the advantage that the effects of solvent ionizing power and solvent electrophilicity (as defined in Chapter 17) are small so that other effects, such as those from solvent nucleophilicity, are more readily revealed; for example, the rate for the 1-adamantyl sulfonium salt, for which no sensitivity to nucleophilicity occurs, varies only by a factor of 7 for a wide range of solvents that would have produced a variation of thousands for an anionic leaving group.

Kevill and co-workers first address the much-debated issue of nucleophilic involvement in solvolysis of *tert*-butyl derivatives. Interestingly, the *tert*-butyl sulfonium salt shows more rate variation with solvent changes than does the 1-adamantyl salt. In particular, the *tert*-butyl salt shows a rate increase in aqueous TFEs (where both Y and N increase) that is not found for 1-adamantyl. Because a variation in Y cannot explain the result, Kevill argues that the *tert*-butyl derivative is receiving nucleophilic solvent assistance. On the basis of the available evidence, Harris et al. (Chapter 17) propose that *tert*-butyl chloride is inaccurately indicated by some probes to receive nucleophilic solvent assistance because the model system (1-adamantyl chloride) has a different susceptibility to solvent electrophilicity. Kevill and co-workers disagree with this proposal, noting that essentially the same *tert*-butyl to 1-adamantyl rate ratio is found for the chlorides and the sulfonium salts; if solvent electrophilicity were important in one case but not the other, then the rate ratio should vary.

In the final section of Chapter 19, Kevill and co-workers use reactions of sulfonium and oxonium salts with both neutral and anionic nucleophiles to derive a scale of nucleophilicities that can be compared with other scales such as the Swain–Scott scale.

In Chapter 20, Fărcaşiu derives a novel measure of solvent nucleophilicities. First, protonation equilibria of benzene derivatives are used to measure acidity of a set of acids. Then, by measurement of the effect of weakly basic solvents (such as trifluoroacetic acid) on these equilibria, basicities of the solvents are calculated. Interestingly, these solvent basicities compare quite well with the solvent nucleophilicities calculated by Peterson (discussed previously). Fărcaşiu also considers the anion-solvating ability of hydrogen-bonding solvents and concludes, in agreement with Chapter 17 and in disagreement with Chapters 18 and 19, that solvent electrophilicity must be considered as a separate variable.

In Peterson et al.'s 1977 work (32), they showed that the early equation of Swain et al. (*31*), equation 3, could be recast in terms comparable to those of the *N*–*Y* equation, equation 4. In Chapter 21, Peterson applies similar techniques to show that the *A* and *B* values of the recent Swain LFER, equation 5, can also be converted to *N* and *Y* values. Some interesting conclusions result from this investigation. By use of the Swain data, the advantages of the statistical approach are extended to the more narrowly defined *N* and *Y* values. A major criticism of the Swain approach is that most of the *A* and *B* parameters appear, from this work, to be artifacts that arise from the extraction of two parameters from data sets that need only one solvent parameter for correlation; for most solvents, the *A* and *B* values are proportional. Only for hydroxylic and amine solvents are the *A* and *B* parameters shown to be meaningful. Peterson concludes that the *A*–*B* values represent two independent sets of data, one set that can be converted into *N* and *Y*, and another set that appears to be meaningless. For Swain's rebuttal to related arguments, *see* reference 40a.

In a study as complex as that of the origins of nucleophilic reactivity, well-understood reactions with which to test theories are critical. One of the most important and most studied of these model reactions is the solvolysis reaction of alkyl halides and esters. Although fewer researchers are continuing investigation of this reaction than in its heyday of the 1950s and 1960s, novel contributions of general significance to physical organic chemistry continue to be made; a particularly important example is the recent work of Richard et al. (*40b*) on solvolysis of 1-phenylethyl derivatives, in which improved understanding is gained of how and why reactions change mechanism. In Chapter 22, Allen et al. describe their work on kinetic and stereochemical studies of solvolysis of 1-arylethyl tosylates having electron-withdrawing substituents on the 2-carbon or the aryl ring. This system was chosen because it is in the borderline region (between S_N1 and S_N2) and thus presents the possibility of gaining insight into the manner in which the mechanistic transition is made and into the possibility of the S_N2-intermediate mechanism in which a pentavalent intermediate is formed. An impressive variation in stereochemistry is found for the trifluoromethyl derivatives, all the way from inversion in nucleophilic solvents to racemization and even retention in less nucleophilic fluorinated solvents. These workers conclude that the rate-determining step is nucleophilic attack on tight ion pair, and they find no evidence for the much-discussed pentavalent intermediate.

Complex Nucleophilic Reactions

In the discussion thus far, an attempt has been made to understand nucleophilicity by use of relatively well understood model reactions (e.g., methyl transfer) or by use of reaction conditions that reduce the number of variables affecting nucleophilicity (e.g., gas-phase reactions or reactions in nonhydroxylic solvents). In this section, we survey a group of chapters describing work with systems that could be described as being more "complex" than the preceding ones, not in the sense that the preceding studies are not at times dauntingly complex, but in the sense that the present studies deal with nucleophilic reactions that have never been observed before or that have an added element of complexity such as involvement of free radicals. Ultimately, supposedly, our theories should be powerful enough to permit prediction, before the fact, of properties of such complex systems. This group of authors is exploring new frontiers where these second-generation tests can be made.

Polar nucleophilic reactions, such as those we have been discussing so far, often compete with an alternative single-electron-transfer mechanism. This competition is discussed in Chapters 23 and 24 (*see also* Chapter 12). In Chapter 23, Pross examines the matter from a theoretical point of view. First, he argues that the polar mechanism does not involve the transfer of a pair of electrons as generally assumed, but rather involves the synchronous "shift" of a single electron from nucleophile to leaving group and a pairing of electrons between nucleophile and reactive carbon. The essential difference between the polar reaction (and its single-electron "shift") and a single-electron-transfer mechanism is said not to be the number of electrons shifted (one in both cases), but rather that the former process results in a pair of coupled electrons while the latter process gives a radical; note that the words shift and transfer are used differently. Pross then uses these ideas to examine the competition between the two reaction types by considering those factors that will favor or inhibit coupling of the two odd electrons *after* the single-electron shift. Also, the observed resistance of cation radicals to nucleophilic attack is explained.

In Chapter 24, Paradisi and Scorrano present their results on the effect of reaction conditions on the competition between nucleophilic substitution and single electron transfer for reaction of alkoxides with *p*-nitrochlorobenzene. The substitution pathway is favored by excluding oxygen and by adding a crown ether. The authors suggest that the crown enhances reactivity of the alkoxide by removing ion association and giving a free alkoxide. Interestingly, alkoxide reactivity follows the order of basicity when crown is present, but not when crown is absent. Single-electron transfer is favored by ion association, and thus the best yields are found in the absence of crown ether. Oxygen also disfavors this pathway, apparently by trapping the anion radical of the benzene reactant.

In Chapter 25, Russell and Khanna examine a novel class of nucleophilic reactions, reaction with radicals. The specific reaction is that of unsaturated anions (e.g., $R_2C{=}CO^-C_6H_5$) with radicals to give relatively stable anion radicals. Rates are obtained for the reactions, and some novel substituent effects are observed. The reactions of electron-rich $(CH_3)_3C^{\cdot}$ are especially interesting in that here the more basic nucleophiles are actually the slower to react. On the other hand, more "normal" nucleophilic behavior is seen for electron-deficient radicals (such as $C_6H_5COCH_2^{\cdot}$), where an increase in basicity of the anion leads to an increase in rate.

Phenoxide and aniline have long been included in the standard lists of nucleophiles. In Chapter 26, Buncel et al. show that these familiar reagents can behave in some unfamiliar ways when combined with highly electrophilic nitroaromatics (such as 4,6-dinitrobenzofuroxan) that yield exceptionally stable σ complexes upon nucleophile–electrophile combination. Under these conditions, the nucleophiles become subject to attack at ring carbon as well as at the more usual nitrogen or oxygen. Interestingly, adherence to the RSP is observed. This finding is somewhat surprising because these reactions appear to be much like those of Ritchie's (*26b*) in which an absence of selectivity variation with reactivity occurs.

In Chapter 27, King et al. present their work on determining reaction mechanisms for reaction of nucleophiles with sulfonyl chlorides. Many of the same phenomena observed for the more familiar nucleophilic attack at carbon are seen. Evidence is presented for pentavalent intermediates, direct displacement, elimination–addition (through sulfenes, $C{=}SO_2$), neighboring-group participation (through a β-sultone), and return processes.

The final chapter of this section is by Rappoport and is concerned with nucleophilic reactions at vinylic carbon. Two reaction types are considered, those of neutral vinyl derivatives and those of vinyl cations. Correlation of rates for these reactions with both Ritchie and Swain–Scott equations was attempted without success. Rappoport concludes that these reactions are subject to a complex blend of polar, steric, and symbiotic effects and that "a quantitative nucleophilicity scale toward vinylic carbon cannot be constructed". This conclusion is reminiscent of the earlier observation of Pearson (*see* the introduction to the section on the Brønsted equation) and the later observation of Ritchie (Chapter 11) regarding the difficulty of correlating nucleophilic reactivity with a single equation. Rappoport finds another familiar situation when he explores the relationship between reactivity and selectivity for the vinyl substrates; sometimes the RSP is obeyed and sometimes it is not.

Enhancement of Nucleophilicity

Recent years have seen impressive advances in the techniques of enhancing nucleophilic reactivity by use of host–guest complexes, phase-transfer catalysis, micelles, microemulsions, and related phenomena (*42–45*). An introduction to aspects of this work is provided in the final three chapters of this volume.

In Chapter 29, Bunton presents a brief review of micellar effects on nucleophilicity, and he describes recent work of his own in this area. A major contribution of Bunton's has been his development of a quantitative model for calculating nucleophile concentration in the pseudophase of the micelle; thus, calculation of rate constants in the micelle is possible. Using this model, Bunton finds that the reaction rates in micelles are very similar to those in water. Thus, micellar accelerations result from reactant concentration. Bunton notes that this conclusion also applies to microemulsions, vesicles, and inverse micelles. A second important contribution of this chapter is a summary of the large amount of experimental work on the contrasting effects of cationic and anionic micelles on reactions of anionic and neutral nucleophiles and on hydrolyses.

Workers in the field of phase-transfer catalysis have spent a great deal of effort elucidating the relationship between relative sizes of crown cavity and nucleophile and the strength of binding in the complex. This relationship is placed on much firmer ground by Gokel's use of stability constants to measure complex stability. In Chapter 30, Gokel et al. review this work and show how it has revealed that much of previous theory about hole-size relationships is incorrect. Also, the fascinating ability of crown "arms" (such as in "lariat ethers") to augment cation binding is reviewed.

The goal of much activity in synthesis of host–guest complexes is to mimic the selectivity and catalytic activity of enzymes. In Chapter 31, Schneider et al. examine the use of intramolecular inclusion complexes for catalysis of nucleophilic displacement reactions. The main emphasis here is on macrocyclic ammonium ions as hosts.

Literature Cited

1. Lapworth, A. *Nature* (London) **1925,** *115*, 625.
2. Ingold, C. K. *J. Chem. Soc.* **1933,** 1120.
3. Ingold, C. K. *Rec. Trav. Chim.* **1929,** *48*, 797.
4. Bunnett, J. F. *Annu. Rev. Phys. Chem.* **1963,** *14*, 271.
5. Pearson, R. G.; Sobel, H.; Songstad, J. *J. Am. Chem. Soc.* **1968,** *90*, 319.
6. Shaik, S. S. *Prog. Phy. Org. Chem.* **1985,** *15*, 197.
7. Albery, W. J.; Kreevoy, M. M. *Adv. Phys. Org. Chem.* **1978,** *16*, 87.
8. Pellerite, M.; Brauman, J. I. *J. Am. Chem. Soc.* **1980,** *102*, 5993.

9a. Lewis, E. S.; Kukes, S.; Slater, C. D. *J. Am. Chem. Soc.* **1980,** *102*, 1619.

9b. Rappoport, Z., Ed. *Isr. J. Chem.* **1985,** *26*, 303–427.

10. Klopman, G. In *Chemical Reactivity and Reaction Paths;* Klopman, G., Ed.; Wiley-Interscience: New York, 1974; Chapter 4, and references cited therein.

11. Hudson, R. F. In *Chemical Reactivity and Reaction Paths;* Klopman, G., Ed.; Wiley: New York, 1974; Chapter 5.
12. Carrion, F.; Dewar, M. J. S. *J. Am. Chem. Soc.* **1984,** *106,* 3531, and references cited therein.
13. Mitchell, D. J.; Schlegel, H. B.; Shaik, S. S.; Wolfe, S. *Can. J. Chem.* **1985,** *63,* 1642, and references cited therein.
14. Chandrasekhar, J.; Jorgensen, W. L. *J. Am. Chem. Soc.* **1985,** *107,* 2974.
15. Pross, A. *Acc. Chem. Res.* **1985,** *18,* 212.
16. Brønsted, J. N.; Pederson, K. *Z. Phys. Chim.* **1924,** *108,* 185.
17. Swain, C. G.; Scott, C. B. *J. Am. Chem. Soc.* **1953,** *75,* 141.
18. Branch, G. E. K.; Calvin, M. *The Theory of Organic Chemistry,* Prentice-Hall: New York; 1941, pp 408–423.
19. Edwards, J. O. *J. Am. Chem. Soc.* **1954,** *76,* 1540.
20. Edwards, J. O. *J. Am. Chem. Soc.* **1956,** *78,* 1819.
21. Davis, R. E.; Nehring, R.; Blume, W. J.; Chuang, C. R. *J. Am. Chem. Soc.* **1969,** *91,* 91.
22. Edwards, J. O.; Pearson, R. G. *J. Am. Chem. Soc.* **1962,** *84,* 16.
23. Ritchie, C. D. *J. Am. Chem. Soc.* **1983,** *105,* 7313.
24. Pearson, R. G. *Surv. Prog. Chem.* **1969,** *5,* 1.
25. Pearson, R. G.; Songstad, J. *J. Am. Chem. Soc.* **1967,** *89,* 1827.
26a. Pearson, R. G., Ed. *Hard and Soft Acids and Bases;* Dowden, Hutchinson, and Ross: Stroudsberg, PA; 1973.
26b. Ritchie, C. D. *Acc. Chem. Res.* **1972,** *5,* 348.
27. Ingold, C. K. *Structure and Mechanism in Organic Chemistry,* 2nd ed.; Cornell University: Ithaca, NY, 1969; p 425.
28. Grunwald, E.; Winstein, S. *J. Am. Chem. Soc.* **1948,** *70,* 846.
29. Raber, D. J.; Bingham, R. C.; Harris, J. M.; Fry, J. L.; Schleyer, P. v. R. *J. Am. Chem. Soc.* **1970,** *92,* 5977.
30. Wells, P. R. *Linear Free Energy Relationships;* Academic: London, 1968.
31. Swain, C. G.; Mosely, R. B.; Bown, D. E. *J. Am. Chem. Soc.* **1955,** *77,* 3731.
32. Peterson, P. E.; Vidrine, D. W.; Waller, F. J.; Henrichs, P. M.; Magaha, S.; Stevens, B. *J. Am. Chem. Soc.* **1977,** *99,* 7968.
33. Bentley, T. W.; Schadt, F. L.; Schleyer, P. v. R. *J. Am. Chem. Soc.* **1972,** *94,* 992.
34. Reichardt, C. *Solvent Effects in Organic Chemistry:* Verlag Chemie: Weinhein, West Germany, 1979.
35. Swain, C. G.; Swain, M. S.; Powell, A. L.; Alunni, S. *J. Am. Chem. Soc.* **1983,** *105,* 502.
36. Taft, R. W.; Abboud, J.-L.; Kamlet, M. H. *J. Solution Chem.* **1985,** *14,* 153.
37. Kevill, D. N. *J. Chem. Res.* **1984,** 86.
38. Taft, R. W.; Abboud, J.-L.; Kamlet, M. J. *J. Org. Chem.* **1984,** *49,* 2001.
39. Bentley, T. W.; Carter, G. W.; Roberts, K. *J. Org. Chem.* **1984,** *49,* 5183.
40a. Swain, C. G. *J. Org. Chem.* **1984,** *49,* 2005.
40b. Richard, J. P.; Rothenberg, M. E.; Jencks, W. P. *J. Am. Chem. Soc.* **1984,** *106,* 1361.
41. Fendler, J. H. *Membrane Mimetic Chemistry;* Wiley-Interscience: New York, 1982.
42. Starks, C. M.; Liotta, C. L. *Phase Transfer Catalysis;* Academic: New York, 1978.
43. Weber, W. P.; Gokel, G. W. *Phase Transfer Catalysis in Organic Chemistry;* Springer-Verlag: Berlin, 1977.
44. Dehmlow, E. V.; Dehmlow, S. S. *Phase Transfer Catalysis;* Verlag Chemie: Deerfield Beach, FL, 1980.

RECEIVED for review May 1, 1986. ACCEPTED August 7, 1986.

MARCUS THEORY, ALKYL TRANSFER, AND GAS-PHASE REACTIONS

2

Intrinsic Nucleophilicity

John I. Brauman, James A. Dodd, and Chau-Chung Han

Department of Chemistry, Stanford University, Stanford, CA 94305

Nucleophilic displacement reactions of anions with neutral alkyl halides were studied in the gas phase by using mass spectrometric techniques. Analysis of the rates of the reactions leads to a model of the potential surface, which has a double minimum. The energetics associated with the surface can be obtained and with the use of Marcus theory leads to an estimate of the energy of the self-exchange reaction, $Y^- + CH_3Y \rightarrow CH_3Y + Y^-$. *The energies for self-exchange are suggested as a measure of the intrinsic nucleophilicity of* Y^-. *The relationship of proton affinity and methyl cation affinity (carbon basicity) is analyzed.*

NUCLEOPHILIC DISPLACEMENT REACTIONS occupy a unique place in organic chemistry. These reactions are among the most studied and have played a critical role in numerous models of reaction chemistry such as steric effects, polar effects, solvent effects, and structure–reactivity correlations. Nucleophilic displacement reactions have been prominent in the development of paradigms for studying stereochemistry as well. In some sense, the nucleophilic displacement reaction is also among the best understood of chemical reactions, and this reaction occupies a critical place in the pedagogy of elementary organic chemistry. Nevertheless, the area still has major conceptual problems associated with it, as evidenced by the papers presented in the symposium upon which this book is based.

In fact, why some reactions are fast and others slow is often difficult to explain in simple terms, and prediction of the rates of reactions is essentially impossible without either an elaborate quantum calculation or the use of a complex set of empirical parameters. Our work in studying simple S_N2 reactions in the gas phase has been aimed at developing a view of these reactions that can be understood in relatively elementary terms. By removing the effects of solvation, we hope to discern the important factors involved in determining reaction rates. Although our efforts are far from complete, we believe that the reactions can now be understood, at least in the first approximation (*1–8*).

0065-2393/87/0215-0023$06.00/0

Analysis of Gas-Phase Dynamics

Ionic reactions in the gas phase are usually carried out in mass spectrometers, often at rather low pressures, for example, 10^{-5} torr. Under these circumstances, reactive complexes that may be formed do not undergo collisions prior to decomposition. Thus, their energy distributions are not Boltzmann, and the reactions cannot be treated in terms of the Arrhenius equation. We have discussed this problem previously (*9*, *10*), and appropriate recipes are well documented (*11*, *12*).

Our kinetic studies were carried out at low pressure in an ion cyclotron resonance spectrometer. In general, the results obtained are reproducible and in reasonably good agreement with results obtained in other laboratories using different experimental techniques (*13*, *14*). In some cases, the numbers differ modestly (the origin of these differences is not known), but the conclusions drawn from the results are not at all dependent on these differences.

A model of the potential surface is needed to understand why an S_N2 reaction is slow. The double minimum surface shown in Figure 1 can accommodate the experimental results for this system (*1–7*). Although the central barrier is at a lower energy than that of the reactants, the reaction proceeds slowly because the transition state associated with the central barrier is "tight" and the sum of states associated with it is smaller than that associated with the "loose" transition state for decomposition back to reactants. The rate constant for the reaction is given by the rate constant for formation of the complex multiplied by the fraction of complexes that go on to products. This branching fraction is the ratio of the forward step over the sum of the forward and back steps and can be related to the efficiency, which is the reaction rate divided by the collision rate.

Using Rice–Ramsperger–Kassel–Marcus (RRKM) theory (*11*, *12*), we can model the rates of these reactions as a function of the energy difference separating the two transition states. The result of the analysis is an estimate

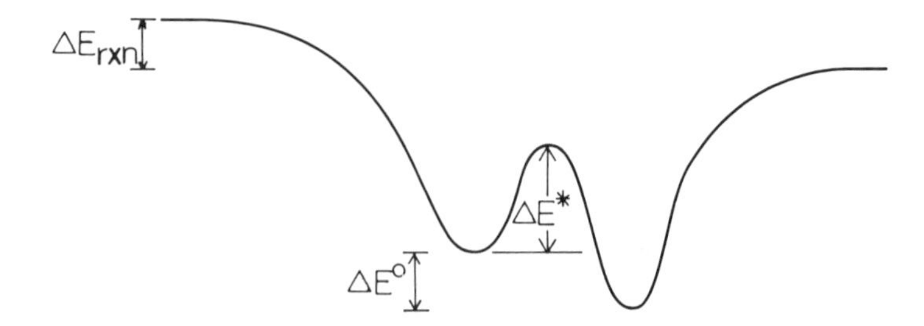

Figure 1. Double-minimum potential surface. ΔE^*, ΔE°_{rxn}, *and* ΔE° *are the energies of activation, overall reaction, and well-to-well reaction, respectively.* ΔE°_{rxn} *and* ΔE° *are often set equal to each other for purposes of simplicity.*

of the energy of the transition state. The procedure is analogous to estimating an activation energy for a high-pressure reaction by measuring the rate constant and estimating the A factor. The only difference in the low-pressure case is that the energy distribution function is that of a chemically activated complex rather than a Boltzmann distribution. Our analysis requires the frequencies and moments of inertia of the transition states; published work (*1–3*) details the choice and rationale for choosing these parameters. However, the analysis is rather insensitive to the choice of the parameters.

If the energy of the transition state is known, the activation energy from the well to the transition state can be determined if the well depth is known. In many cases, the equilibrium has been measured; in others, the well depth can be estimated fairly accurately. The rate of the reaction is not dependent on the well depth, so that the depth cannot be determined by studying the kinetics at low pressure. The well depth determines the lifetime of the complex, but this depth does not affect the reaction efficiency.

Marcus Theory

Many reactions exhibit effects of thermodynamics on reaction rates. Embodied in the Bell–Evans–Polanyi principle and extended and modified by many critical chemists in a variety of interesting ways, the idea can be expressed quantitatively in its simplest form as the Marcus theory (*15–18*). Murdoch (*19*) showed some time ago how the Marcus equation can be derived from simple concepts based on the Hammond–Leffler postulate (*20–22*). Further, in this context, the equation is expected to be applicable to a wide range of reactions rather than only the electron-transfer processes for which it was originally developed and is generally used. Other more elaborate theories may be more correct (for instance, in terms of the physical aspects of the assumptions involving continuity). For the present, our discussion is in terms of Marcus theory, in part because of its simplicity and clear presentation of concepts and in part because our data are not sufficiently reliable to choose anything else. We do have sufficient data to show that Marcus theory cannot explain all of the results, but we view these deviations as fairly minor.

The essence of the Marcus theory is that a reaction can be thought of as having an "intrinsic" activation energy, ΔE_0^*, which is the barrier that it would have in the absence of any thermodynamic driving force. The actual activation energy, ΔE^*, is given by equations 1a and 1b. The derivative with respect to $\Delta E°$, the well-to-well energy change, gives the slope of a linear free energy plot, for example, a Brønsted plot.

$$\Delta E^* = (\Delta E°)^2/16\Delta E_0^* + \Delta E_0^* + 1/2\Delta E° \quad (1a)$$

$$d\Delta E^*/d\Delta E° = \alpha = \Delta E°/8\Delta E_0^* + 1/2 \quad (1b)$$

Inspection of equations 1a and 1b shows that $\Delta E_0{}^*$ can be derived for any reaction if ΔE^* and ΔE° are known. One uniquely interesting aspect of the intrinsic activation energy is that it is equal for the forward and back directions. Thus, $\Delta E_0{}^*$ for $X^- + CH_3Y$ is identical with $\Delta E_0{}^*$ for $Y^- + CH_3X$. Immediately, the distinction between nucleophile and leaving group in nucleophilic displacement reactions is removed. This consequence of the theory is one of the most powerful and useful.

A further assumption usually made in applying Marcus theory is that the intrinsic barrier for a cross-reaction is the mean of those for the two identity reactions. $\Delta E_0{}^*$ can therefore be calculated for an identity reaction if its value is known for a particular cross-reaction and the other identity reaction.

ΔE^* was determined experimentally for a number of reactions, and when the values of ΔE° for these reactions are used, values of $\Delta E_0{}^*$ can be derived from them. With the estimate of $\Delta E^* = \Delta E_0{}^*$ for the reaction of $Cl^- + CH_3Cl$, values of $\Delta E_0{}^*$ were derived for a number of identity reactions (intrinsic nucleophilicities [Y^-, $\Delta E_0{}^*(Y^- + CH_3Y)$ (kcal/mol)]: Cl, 10; CH_3S, 24; F, 26; RO, 28; CN, 35; HCC, 41). In general, these values cannot be measured directly, because the reactions are too slow. Nevertheless, given the assumptions of the theory, these values represent the best estimate of what the activation energy would be. (The same issue arises in solution as well; a number of reactions cannot be studied simply because their activation energies are too high.)

We called $\Delta E_0{}^*$ for the identity reactions a measure of the intrinsic nucleophilicity of the reagent Y^- toward methyl centers. This definition has the advantage, as stated previously, of removing the distinction between entering group and leaving group. Moreover, only one number exists for each nucleophile (in contrast to other schemes), and within the assumptions of the Marcus theory, the set allows us to predict the intrinsic barriers for all combinations of entering and leaving groups. Finally, if the exothermicity of a given reaction is known, its overall rate constant can be predicted.

Our predictions were tested within the limited set of data available. In fact, most displacement reactions do not go at measurable rates, so we were forced to use the reactions that do go at measurable rates to derive our values of $\Delta E_0{}^*$. We can predict, however, that a number of very exothermic reactions should not occur at measurable rates, and indeed, a number of reactions do not. A list of reactions that are too slow to be measured is given in Table I.

Some quite interesting conclusions can be drawn from these values. The major conclusion is that many ions that are usually thought to be "good" nucleophiles are poor by this definition. For example, RO^- reacts readily because its reactions are usually very exothermic (especially with CH_3Br). As in solution, RO^- cannot undergo reversible displacement in the gas phase. Thus, the inertness of ethers to displacement reactions is a consequence of

Table I. Predicted Slow Reactions

Reaction	*ΔH° (kcal/mol)*	*Ref*
$D^- + CH_4 \longrightarrow H^- + CH_3D$	0	31
$CD_3O^- + CH_3OCH_3 \longrightarrow CH_3O^- + CH_3OCD_3$	0	1
$CD_3S^- + CH_3SCH_3 \longrightarrow CH_3S^- + CH_3SCD_3$	0	3
$(CH_3)_3CO^- + CH_3F \longrightarrow F^- + CH_3OC(CH_3)_3$	−7	3
$HCC^- + CH_3F \longrightarrow F^- + CH_3CCH$	−24	13
$CN^- + CH_3F \longrightarrow F^- + CH_3CN$	−5	13
$OH^- + CH_3OCH_3 \longrightarrow CH_3O^- + CH_3OH$	−6	—[a]
$NH_2^- + CH_3OCH_3 \longrightarrow CH_3O^- + CH_3NH_2$	−19	—[a]

[a]DePuy, C. H.; Bierbaum, V. M., personal communication.

the intrinsically high barrier associated with RO^- coupled with the corresponding reaction not being very exothermic. The Williamson ether synthesis works because halides have intrinsically low ΔE_0* values and reactions of alkoxides with them are very favorable thermodynamically. A corollary of this is that neither the solvent nor solvation effects control the inertness of ethers. Of course, solvent effects can be quite dramatic in determining relative reactivity, but they will not generally overwhelm the basic energy factors given here. Indeed, Albery and Kreevoy (*23, 24*) and Lewis and Hu (*25*) have shown that a similar approach to that given here works quite well for solution S_N2 reactions. (That work was done independently and at the same time as ours.) A second interesting feature is that thiolates are not especially reactive. Indeed, with the exception of Cl^- and Br^-, almost none of the nucleophiles can be said to be good. Again, reversible self-exchange does not occur in general for most compounds in solution.

Finally, theoretical studies, particularly by Wolfe et al. (*26*) and more recently by Evleth and co-workers (*27*), provided some additional justification for our analysis. Quantum calculations of barriers for cross-reactions are in agreement with values that would have been derived from a Marcus theory analysis of other calculated barriers. Jorgensen and co-workers (*28, 29*) carried this analysis further using statistical mechanics simulations that show that the gas phase potential surface indeed translates into the solution surface in the way that we predicted. (For a dicussion of an alternative way to apply Marcus theory to double-minimum surfaces, as well as a note on Cl^- + CH_3Cl, *see* reference 29a.)

Correlations

Inasmuch as we obtained activation energies for a variety of S_N2 reactions, we now try to determine whether these activation energies can be understood in terms of other thermochemical quantities. The most obvious test is to choose a system in which the intrinsic barrier remains constant. If this barrier does remain constant, then a good linear free energy plot should result. Indeed, when we measured the reaction rates for some substituted benzyl anions reacting with methyl bromide (5), all had $\Delta E_0{}^*$ values of about 23 kcal/mol, while $\Delta E°$ ranged from -48 to -60 kcal/mol. In accordance with this result, we found a good Marcus–Brønsted plot with a slope of 0.2, as predicted (Figure 2).

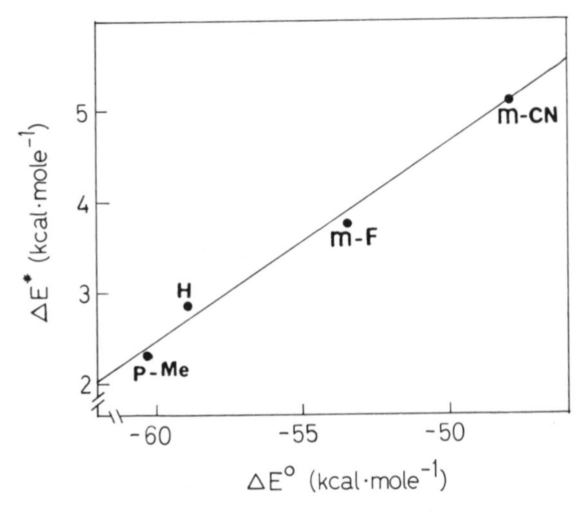

Figure 2. Marcus–Brønsted plot: ΔE^* *vs.* $\Delta E°$ *for the reaction of substituted benzyl anions with methyl bromide.*

An attempt to correlate a wider range of data in this way, however, is expected to fail. Indeed, if we plot ΔE^* versus $\Delta E°$ (or equivalently, $\Delta H°$) for a set of nucleophiles reacting with CH_3Cl, we do not expect a very good correlation, and our expectations are fulfilled (Figure 3). Clearly, a trend in the data exists, because, to the extent that Marcus theory is a correct description, ΔE^* contains a substantial contribution from $\Delta E°$. On the other hand, $\Delta E_0{}^*$ is different for each of these nucleophiles, so that not much more than a trend should be expected. The typical practice of attempting to correlate log k with $\Delta G°$ can be seen to be completely inappropriate unless $\Delta E_0{}^*$ is constant. Typically, the linear free-energy plots are "saved" by grouping the reactants into families—an exercise that is generally equivalent to choosing sets with constant intrinsic barriers.

For any sort of linear free-energy correlation, a very interesting issue to explore is that of using the pK of the nucleophile rather than the reaction exothermicity itself. Not untypically, discussions of nucleophilicity (a kinetic quantity) versus basicity or proton affinity (a thermodynamic quantity) result. In principle, no reason exists to use the pK of the nucleophile rather

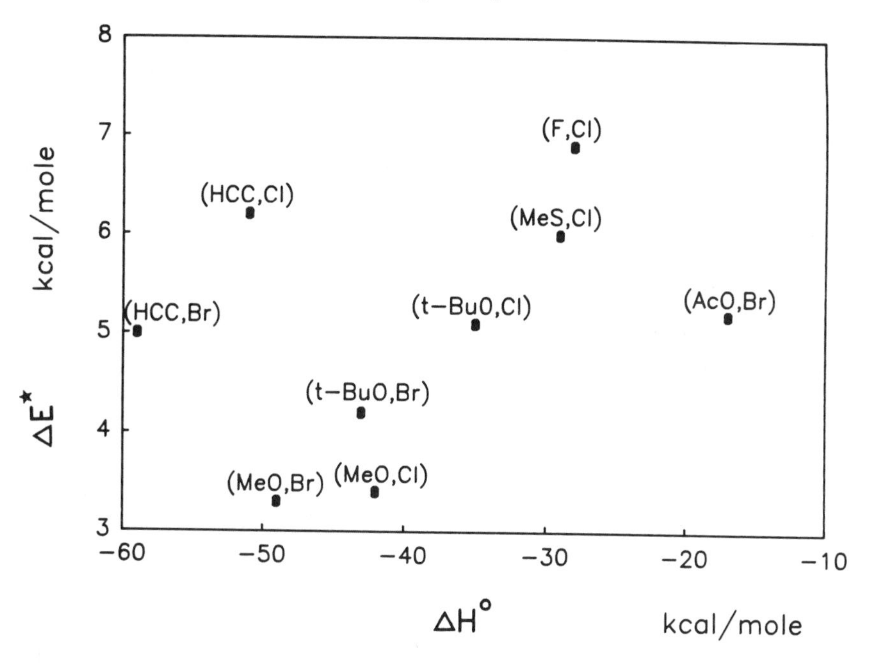

Figure 3. Activation energy vs. exothermicity (ΔE vs. ΔH°) for reactions of varying intrinsic barriers. Each point represents a (nucleophile, leaving group) combination for the S_N2 reaction at a methyl center.*

than the (correct) free-energy change for the S_N2 reaction. The pK values are often thought to be more readily available, but this is not generally true.

The relationship of proton basicity to carbon basicity was explored extensively in the past, especially in a comprehensive paper by Hine and Weimar (*30*). These authors concluded that these quantities are not well correlated and are rather structurally dependent. As a consequence, plots of S_N2 activation energy where the intrinsic barrier is constant might be expected not to correlate well with proton affinity of the nucleophile. Aficionados of these correlations recognize, however, that this situation is not the case. In fact, when correlations with $\Delta G°$ are good, plots with pK are usually also good. The reason is actually quite clear. A plot of methyl cation affinity ($\Delta H°$ for $CH_3Y \rightarrow CH_3^+ + Y^-$) against proton affinity ($\Delta H°$ for $HY \rightarrow H^+ + Y^-$) is surprisingly linear (Figure 4). Even this result is not surprising if thermochemical cycles in Scheme I are considered. If the difference between the bond dissociation energy for HY [$D°(HY)$] and the bond dissociation energy for CH_3Y [$D°(CH_3Y)$] were constant, then the methyl cation affinity would correlate with the proton affinity with unit slope. In fact, the bond energies for HY and CH_3Y are fairly well correlated but with a slope of 0.86. The slope so close to unity results in a pretty good correlation for the ionic quantities. An equivalent way of looking at the problem makes the correlation even more obvious. Consideration of the same kind of cycle shows that the only quantities that matter are the heats of formation of HY and CH_3Y. Identically with the bond energy plot, these quantities correlate remarkably well (Figure 5). Moreover, the differences in heat of formation are proportional to the values of heat of formation, so that the plot is linear although it does not have unit slope. Small differences do, of course, occur, but these differences are fairly minor on the total scale. To repeat, because the slope is 0.86, close to unity, we anticipate the correlation of proton affinity and methyl cation affinity. Hine called attention primarily to the deviations from the line of unit slope. He was aware of the correlation, but he presumably felt it to be less good than we do.

$$\begin{array}{ll}
CH_3Y \rightarrow CH_3 + Y & D°(CH_3Y) \\
CH_3 \rightarrow CH_3^+ & IP(CH_3) \\
Y + e^+ \rightarrow Y^- & -EA(Y) \\
\hline
CH_3Y \rightarrow CH_3^+ + Y^- & \text{methyl cation affinity}
\end{array}$$

$$\begin{array}{ll}
HY \rightarrow H + Y & D°(HY) \\
H \rightarrow H^+ & IP(H) \\
Y^- + e^+ \rightarrow Y^+ & -EA(Y) \\
\hline
HY \rightarrow H^- + Y^- & \text{proton affinity}
\end{array}$$

Scheme I

The effect of solvation on such plots is surprising. The same arguments made for gas-phase reactions also apply to solution. The only relevant quantities are the heats (or free energies) of formation of HY and CH_3Y. However, because the gas-phase ionization potential (*IP*) of CH_3 is lower than that of H but the solution stability of CH_3^+ is poorer than that of H^+, the intercept of the methyl cation affinity–proton affinity plot changes. In the gas phase, proton affinities are greater than methyl cation affinities; in solution, methyl cation affinities are greater. Beyond this, to the extent that HY and CH_3Y are solvated differently, we might expect our correlation to become worse. Indeed, we can predict that species that are strongly hydrogen bonded or solvated will have higher dissociation energies and thus fall off the line. For instance, good candidates for such problems are the hydrogen halides and phenols (*30*). In spite of such effects, the overall correlations do not appear to be significantly worse. Thus, although the methyl cation affinity remains the variable of choice for correlations, even in solution the proton affinity is surprisingly good.

Finally, a very interesting, nonobvious correlation discussed previously (*3*) exists. If our intrinsic nucleophilicites ($\Delta E_0{}^*$) are plotted for a series of nucleophiles, the best correlation is with the methyl cation affinity of the nucleophile. That is, those nucleophiles for which the energy for $CH_3Y \rightarrow$

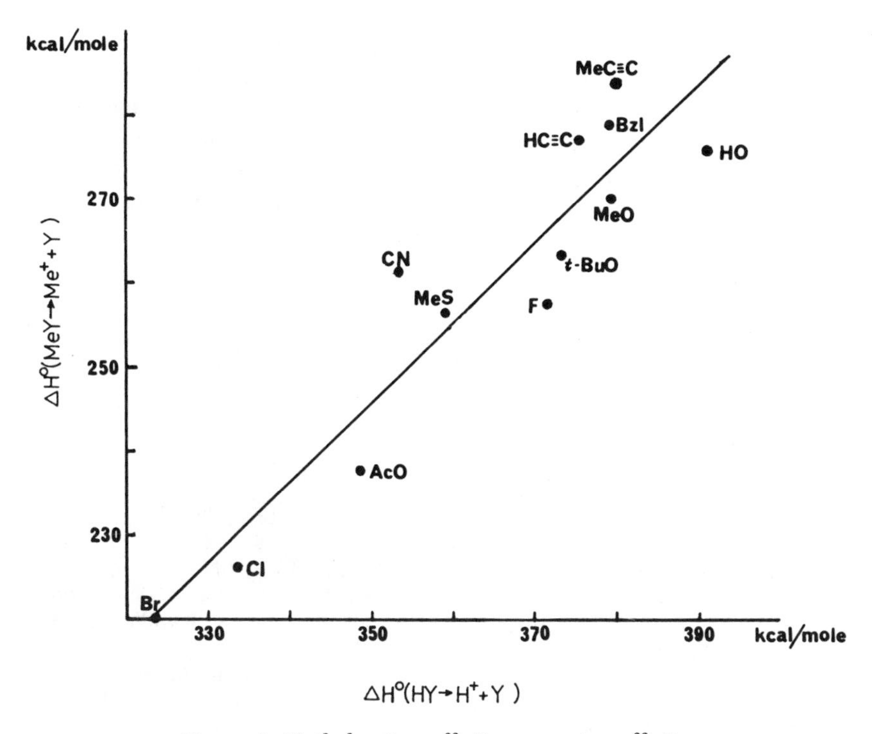

Figure 4. Methyl cation affinity vs. proton affinity.

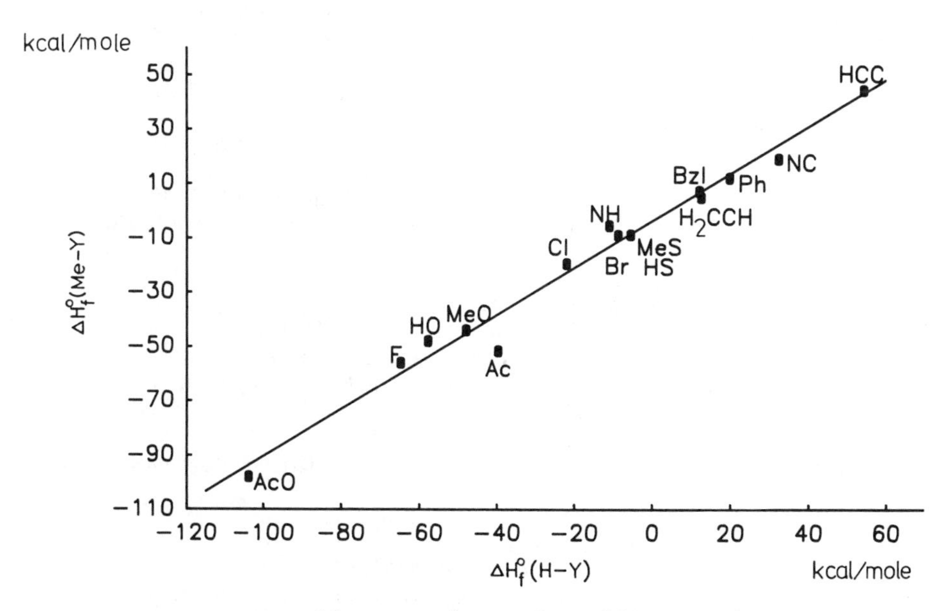

Figure 5. Heat of formation of HY vs. heat of formation of CH_3Y.

$CH_3^+ + Y^-$ is lowest have the lowest ΔE_0^* for the self-exchange reaction. No obvious theoretical reason exists to believe that such a correlation should be found, but the evidence that it exists is unequivocal. The suggestion is strong, then, that at least for these reactions the transition state may have some substantial charge-separated character. The resemblance to the corresponding proton-transfer processes becomes even more intriguing.

Acknowledgment

We are grateful to the National Science Foundation for support of this work and for a Predoctoral Fellowship to J.A.D. J.A.D. also held the F. R. Veatch Fellowship in Chemistry.

Literature Cited

1. Olmstead, W. N.; Brauman, J. I. *J. Am. Chem. Soc.* **1977,** *99*, 4219.
2. Pellerite, M. J.; Brauman, J. I. *J. Am. Chem. Soc.* **1980,** *102*, 5993.
3. Pellerite, M. J.; Brauman, J. I. *J. Am. Chem. Soc.* **1983,** *105*, 2672.
4. Pellerite, M. J.; Brauman, J. I. In *Mechanistic Aspects of Inorganic Reactions;* Rorabacher, D. B.; Endicott, J. F., Eds.; ACS Symposium Series 198; American Chemical Society: Washington, DC, 1982; Chapter 4.
5. Dodd, J. A.; Brauman, J. I. *J. Am. Chem. Soc.* **1984,** *106*, 5356.
6. Barfknecht, A. T.; Dodd, J. A.; Salomon, K. E.; Tumas, W.; Brauman, J. I. *Pure Appl. Chem.* **1984,** *56*, 1809.
7. Brauman, J. I. In *Kinetics of Ion-Molecule Reactions;* Ausloos, P., Ed.; Plenum: New York, 1979; p 153.
8. Han, C.-C.; Dodd, J. A.; Brauman, J. I. *J. Phys. Chem.* **1986,** *90*, 471.
9. Olmstead, W. N.; Lev-On, M; Golden, D. M.; Brauman, J. I. *J. Am. Chem. Soc.* **1977,** *99*, 992.
10. Jasinski, J. M.; Rosenfeld, R. N.; Golden, D. M.; Brauman, J. I. *J. Am. Chem. Soc.* **1979,** *101*, 2259.
11. Robinson, P. J.; Holbrook, K. A. *Unimolecular Reactions;* Wiley-Interscience: New York, 1972.
12. Forst, W. *Theory of Unimolecular Reactions;* Academic: New York, 1973.
13. Tanaka, K; Mackay, G. I.; Payzant, J. D.; Bohme, D. K. *Can. J. Chem.* **1976,** *54*, 1643.
14. Caldwell, G.; Magnera, T. F.; Kebarle, P. *J. Am. Chem. Soc.* **1984,** *106*, 959.
15. Marcus, R. A. *Annu. Rev. Phys. Chem.* **1965,** *15*, 155.
16. Cohen, A. O.; Marcus, R. A. *J. Phys. Chem.* **1968,** *72*, 4249.
17. Levich, V. G. *Adv. Electrochem. Electrochem. Eng.* **1966,** *4*, 249.
18. Dogonadze, R. R. In *Reactions of Molecules at Electrodes;* Hush, N. S., Ed; Wiley: London, 1971; p 135.
19. Murdoch, J. R. *J. Am. Chem. Soc.* **1972,** *94*, 4410.
20. Leffler, J. E.; Grunwald, E. *Rates and Equilibria and Organic Reactions;* Wiley: New York, 1963.
21. Leffler, J. E. *Science (Washington, D.C.)* **1953,** *117*, 340.
22. Hammond, G. S. *J. Am. Chem. Soc.* **1955,** *77*, 334.
23. Albery, W. J. *Annu. Rev. Phys. Chem.* **1980,** *31*, 227.
24. Albery, W. J.; Kreevoy, M. M. *Adv. Phys. Org. Chem.* **1978,** *16*, 87.
25. Lewis, E. S.; Hu, D. D. *J. Am. Chem. Soc.* **1984,** *106*, 3292.
26. Wolfe, S.; Mitchell, D. J.; Schlegel, H. B. *J. Am. Chem. Soc.* **1981,** *103*, 7694.

27. Cao, H. Z.; Allavena, M.; Papia, O.; Evleth, E. M. *J. Phys. Chem.* **1985,** *89,* 1581.
28. Chandrasekhar, J.; Smith, S. F.; Jorgensen, W. L. *J. Am. Chem. Soc.* **1985,** *107,* 3094.
29a. Chandrasekhar, J.; Jorgensen, W. L. *J. Am. Chem. Soc.* **1985,** *107,* 2974.
29b. Dodd, J. A.; Brauman, J. I. *J. Phys. Chem.* **1986,** *90,* 3559.
30. Hine, J.; Weimar, R. D., Jr. *J. Am. Chem. Soc.* **1965,** *87,* 3387.
31. Lieder, C. A. Ph.D. Dissertation, Stanford University. 1974.

RECEIVED for review October 21, 1985. ACCEPTED January 27, 1986.

3

A Quantitative Measure of Nucleophilic Character

Edward S. Lewis, Thomas A. Douglas, and Mark L. McLaughlin

Department of Chemistry, Rice University, Houston, TX 77251

The rate constants of the reaction $CH_3Y + X^- \rightleftarrows CH_3X + Y^-$ in sulfolane solution are described by the Marcus equation; the quadratic term contributes very little. The Marcus equation then reduces to the expression $\log k_{YX} = M_Y + N_X$, *where* M_Y *is a property of* CH_3Y *only and* N_X *is a property of* CG_3X *only. Each term includes only the identity rates and the equilibria for methylation of a reference nucleophile. The two terms are determined independently of unsymmetric rate measurements, in contrast to the Swain–Scott equation. Short tables of both terms are presented. Extension to other solvents and to other reactions including group transfers is discussed. With other alkyl groups, the simple expression may cover the continuum from elimination–addition to addition–elimination and may also cover other group transfers.*

NUCLEOPHILES CAN BE DESCRIBED qualitatively by their reactions with various electrophiles, but a single quantitative definition is not straightforward. Possible methods involve measurements of rates of reaction with electrophiles such as proton acids, Lewis acids, and alkyl halides or of the equilibria of such reactions. Unfortunately, to the extent that information is available, different methods give not only different numbers but also even different orders. In this chapter, both rate and equilibrium results are given on one series of reactions, namely, the methyl transfers or S_N2 reactions of CH_3Y, mostly restricted to the cases $m = n = 0$ or 1

$$X^{m-1} + CH_3Y^n \rightleftarrows XCH_3{}^m + Y^{n-1} \quad (1)$$

An advantage of this approach is that in such a reversible reaction the attacking nucleophile and the leaving group have exactly the same status, and the forward rate constant, the reverse rate constant, and the equlibrium

0065-2393/87/0215-0035$06.00/0

constant are not independent; any two determine the third. This approach thus treats nucleophiles and leaving groups equally.

The restriction to methyl transfers was originally chosen to reduce and make nearly constant steric effects and to avoid mechanistic complexities such as S_N1 processes, elimination reactions, and addition–elimination pathways. Limiting the scope is necessary just to make a reasonably sized experimental project. Similarly, the solvent is limited to sulfolane where possible, because sulfolane is a cheap, high dielectric, and relatively nontoxic solvent. Although sulfolane has the rather high melting point of 30 °C, lower temperatures can be attained by depressing the melting point with dimethylsulfone, which has no rate effects within our precision.

Earlier work (*1a*) on rates of S_N2 reactions in a variety of solvents in the literature abounds. In hydroxylic solvents, some classic and valuable data are available on various nucleophiles with methyl bromide, used as a basis for measuring nucleophilic character (*1b*). Another valuable and relevant study compares dipolar aprotic solvents with hydroxylic solvents (*2*). Finally, numerous systematic studies (*1a, 3a–3d*) of solvolysis reactions have been made.

The first application of the Marcus equation (*3e*) to methyl transfers was the study by Albery and Kreevoy (*4, 5*) based on data in solutions. In this study, the thermodynamic data of Abraham and McLennan (*6*) were used, and the identity barriers were more or less treated as adjustable parameters. The treatment was qualitatively plausible, but the quantitative identity barriers were not completely convincing. An important study (*7*) showed the application of the Marcus equation to methyl transfer in the gas phase.

At about the same time, we (*8*) made some equilibrium measurements in sulfolane, augmenting the earlier data of Jackman and co-workers (*9*), and new rate measurements allowed us to estimate one identity rate (that for $PhSMe_2^+$) by a linear free energy relationship (LFER) extrapolation. Assuming that the quadratic term in the Marcus equation was negligible, we estimated identity barriers for a number of other leaving groups.

Many identity rates, k_{XX}, now have been measured by more direct methods, and these rates are listed in Tables I and II, together with the equilibrium constants, K_{Xr}, for the reactions of the methylating agents with a reference benzenesulfonate ion. Table I is for neutral methylating agents; Table II is for cationic methylating agents, and the relative equilibrium values are believed to be as reliable as those in Table I, but the absolute values with respect to the benzenesulfonate nucleophile are less solid, because the measurement of these equilibria, such as reaction 2 (an example of reaction 1 with $n \neq m$), is subject to substantial salt effects. The absolute values are based on quite dilute solutions but cannot be considered extrapolated to infinitely dilute solution of zero ionic strength. These equilibria and identity rates were collected over several years by a variety of methods (*10–12*), and some further examples may be found in these references. No

Table I. Equilibrium Constants for Methylation of Benzenesulfonate Ion (K_{Xr}) and Identity Rates (k_{XX}) for Neutral Methylation Agents in Sulfolane at 35 °C

Reagent	K_{Xr}[a]	k_{XX} ($M^{-1}s^{-1}$)
$F_3CSO_3CH_3$	2.7×10^{6}	1.3×10^{-2}
FSO_3CH_3	1.7×10^{6}	(1.1×10^{-2}) [b]
$C_6F_5SO_3CH_3$	6.1×10^{5}	(7.3×10^{-3}) [b]
$CH_3OSO_3CH_3$	1.2×10^{2}	1.7×10^{-4}
p-$ClC_6H_4SO_3CH_3$	6.0	3.2×10^{-5}
$C_6H_5SO_3CH_3$	1.0	2.3×10^{-5}
$CH_3SO_3CH_3$	2.9×10^{-1}	1.2×10^{-5}
ICH_3	5.4×10^{-3}	2.6

[a] These numbers are collected from the previous papers of this series and from Jackman and co-workers (9).
[b] These identity rates are estimated from the measured methyl triflate value and the correlation $\log k_{XX} = 0.4 \log K_{Xr}$ + constant. The value 0.4 appears better for these powerful aklylating agents than the earlier arenesulfonate value (*10*) of 0.2.

Table II. Equilibrium Constants (for Methylation of Benzenesulfonate Ion) (K_{Xr}) and Identity Rates (k_{XX}) for Cationic Methylating Agents

Reagent	K_{Xr}	k_{XX}	*Solvent*
$(CH_3)_3O^+$	1.6×10^{7}	<1 [a]	sulfolane (35 °C)
p-$ClC_6H_4SPhCH_3^+$	— [b]	8.7×10^{-5}	sulfolane (100 °C)
$(C_6H_5)_2SCH_3^+$	— [b]	2.4×10^{-5}	sulfolane (100 °C)
$(C_6H_5)_2SCH_3^+$	4.5×10^{3}	2.8×10^{-8}	sulfolane (35 °C)
$(C_6H_5)_2S(CH_3)_2^+$	2.4×10^{-1}	2.5×10^{-8}	sulfolane (35 °C)
$C_5H_5NCH_3^+$ [c]	3.3×10^{-11}	—	acetonitrile[d] (35 °C)
2-$ClC_5H_4NCH_3^+$ [e]	1.1×10^{-4}	—	acetonitrile[d] (35 °C)
3-$ClC_6H_4NCH_3^{3+}$	9.2×10^{-4}	—	nitrobenzene[d] (35 °C)
3-$O_2NC_6H_4N(CH_3)_3^+$	8.5×10^{-3}	—	nitrobenzene[d] (35 °C)

[a] This value is an upper limit based on the absence of perceptible line broadening in a solution containing both trimethyl oxonium ion and dimethyl ether.
[b] Not determined.
[c] *N*-Methylpyridinium ion.
[d] Solvent differences make these values of little quantitative value; solvents are included to show that quaternary ammonium ions do have measurable equilibria. The numbers are corrected from the data of Arnett and Reich (*13*) and of Matsui and Tokura (*14*) to the reference benzenesulfonate from their iodide values by using the iodide entry in Table I.
[e] 2-Chloro-*N*-methylpyridinium ion.

identity rates in sulfolane are available from other sources, nor are there equilibrium data in this solvent, except that Jackman and co-workers' table (9) contains a number of entries not reproduced in Table I or II. However, our experience is that no gross differences exist between sulfolane and acetonitrile for rates of the charge types of reaction 1 with $n = m$, and so we have included some of the equilibrium data of Arnett and Reich (*13*) on *N*-methylpyridinium ions as well as some of that of Matsui and Tokura (*14*) on dimethylanilinium ions in nitrobenzene, both of which can be related roughly to Tables I and II by the common nucleophile I^-. However, such equilibria for $n \neq m$ can be expected to show major solvent effects, so inclusion of *N*-methylpyridinium ions and *N,N,N*-trimethylanilinium ions mostly shows that the equilibria can be measured.

$$C_6H_5S(CH_3)_2^+ + C_6H_5SO_3^- \rightleftarrows C_6H_5SMe + C_6H_5SO_3CH_3 \quad (2)$$
$$K_{Xr} = 0.24$$

The equilibrium and rate constants of Table I and a value of the Marcus work term for approach of the reagents to the reaction complex, w^R, of 1.5 kcal/mol (estimated following Albery and Kreevoy's method for water solution) can be used to calculate a number of reaction rates that can also be measured. Figure 1 shows a plot of calculated versus observed rates for cases with both equilibria and identity rates known. Figure 1 includes some experimental identity rates, which of course fit perfectly on the drawn line of slope 1, just to show how the experimental range of nonidentity reactions compares with that for the identity reactions. Not enough rate data on cationic cases are yet available to make a significant plot from the data of Table II.

Although the fit to this plot is very good, a few situations are difficult to account for. An example is the relative reactivity of dimethyl sulfate and methyl iodide. The observation (*15*) is that with first-row nucleophiles dimethyl sulfate reacts slightly more rapidly than methyl iodide, whereas with all later-row nucleophiles studied, methyl iodide is faster, usually by a small

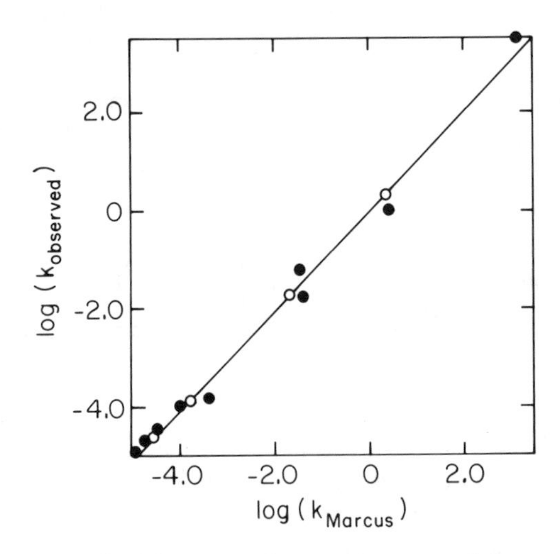

Figure 1. Comparison of measured rates of some methyl transfers compared to those calculated from the Marcus equation (closed circles). Open circles are some experimental identity rates used in the Marcus calculation that necessarily fit the line of unit slope.

factor, but in one case [the nucleophile $HFe(CO)_4^-$] by a factor of 400. Because neither equilibrium nor identity data are available on these nucleophiles, which methylating agent is out of line cannot be determined. However, the effect is consonant with Pearson's hard–soft acid–base (HSAB) principle (*16*). Examples are shown in the figure for hard–hard and hard–soft cases, with only minor deviation. This inversion is also possibly explicable in terms of Shaik's valence-bond approach (*17*), which allows some related inversions.

Inspection of Table I (and Table II) shows that no general correlation of intuitive nucleophilic character seems to occur with either k_{XX}, the identity rate, or K_{Xr}, the equilibrium methylating power of CH_3X, although methyl iodide is the major discrepancy. A correlation of k_{XX} with K_{Xr} occurs, as noted previously; again methyl iodide deviates strikingly. In other work (*12*), methylsulfonium ions and aryldimethylselenonium ions also do not fit this correlation.

In early work on methyl transfers (*18, 19*), we searched for the applicability of the reactivity–selectivity principle (RSP). Initially, small effects in this direction were found. However, with more data, counterexamples of comparable magnitude appeared (*20*). The only conspicuous loss of selectivity appeared in a reaction of $(CH_3)_3O^+$ with $C_6H_5S^-$, which was believed to be to some extent diffusionally limited (*19*). Thus, no convincing evidence was found for the RSP.

We interpreted the constancy of ρ for the reaction of substituted aromatic nucleophiles with methylating agents of varying reactivity as evidence of lack of curvature in a log k versus log K plot and hence an invariability of α (the slope of this plot) with reactivity of the methylating agent. Two entirely different factors may result in this: one factor is an inherent part of the Marcus treatment, and the other is a problem of the relation between the position of the transition state along the reaction coordinate and the slope of LFERs, such as used to define the Hammett ρ.

In the first factor, as reaction rate increases with increasing thermodynamic driving force, the Marcus equation predicts an ultimate drop in α with very large K_{YX} only if the intrinsic barrier is constant. One effect is that the range of free energy change is not very large compared with typical intrinsic barriers. Furthermore, the identity barriers for the different methylating agents are quite variable (as shown in Table I). The identity rates for a series of thiophenoxides or phenoxides are not expected to be constant, although these rates have not been measured, because the correlation of the identity rates of benzenesulfonate nucleophiles with k_{Xr} ($\rho = +0.6$) is very likely general for other nucleophiles. The consequences of an increasing barrier for better nucleophiles on the shape of a log k_{YX} versus log K_{YX} plot that follows the Marcus equation were shown (*11*). The effect is that even for quite large log K values, the ever increasing intrinsic barriers keep the slope, α, nearly constant.

The second factor is contained in the assumption of a relation between the Hammett ρ and the position of the transition state along the reaction coordinate. Conventionally (for example, in the methylation of a series of substitued thiophenoxides), the Hammett ρ is assumed to measure the extent to which the negative charge on the thiophenoxide is neutralized in the transition state, especially when normalized by dividing by $\rho_{eq,}$ which measures the substituent effect on the complete conversion of thiophenoxide to methyl phenyl sulfide. Possibly, charge change does not measure the productlike character! This possibility was suggested by Pross and Shaik (*21*), whose treatment shows that the charge distribution of reaction 1, with $n = m = 0$, is independent of X and Y. This striking conclusion may explain some of the remarkable constancies of selectivity, but this finding does not vitiate our conclusions that these transition states for methyl transfers are not very unsymmetrical anyway.

The Marcus equation, because of its quadratic term, leads to complete loss of selectivity when $\Delta G°$ (standard free energy) is four times the intrinsic barrier and is therefore compatible with the RSP. When certain plausible assumptions about the work terms (i.e., $w^R = w^P =$ constant for all X and Y, for reaction 1, $n = m$, where w^P corresponds to w^R for the reverse reaction) are made (*10*), the Marcus equation reduces to the almost experimentally verifiable form of equation 3, in which all the $\Delta G^{\ddagger}$s are from experimentally measured rates (of reaction 1 and the identity reactions), $\Delta G°_{YX}$ is from experimentally measured equilibria, and w^R is unknown, but probably not large. The absence of experimental evidence for the RSP suggested either that the Marcus equation was inapplicable or that the quadratic term was not significant, even for the most extreme cases. This latter suggestion was shown to be the case for methyl transfers, for which the identity barriers are so high in cases so far found that only with unreasonably large work terms (w^R) and very large negative values of $\Delta G°$ can significant curvature in a plot of $\Delta G^{\ddagger}_{YX}$ versus $\Delta G°_{YX}$ be expected.

$$\Delta G^{\ddagger}_{YX} = \tfrac{1}{2}(\Delta G^{\ddagger}_{XX} + \Delta G^{\ddagger}_{YY}) + \tfrac{1}{2}\Delta G°_{YX} + \Delta G°_{YX}{}^{2}/[8(\Delta G^{\ddagger}_{XX} + \Delta G^{\ddagger}_{YY} - w^R)] \quad (3)$$

Some interesting consequences of the neglect of the quadratic term result. The equation then becomes equation 4, and when equilibrium constants K_{X_r} and K_{Y_r} with a reference nucleophile (such as benzenesulfonate ion in Table I) are available, we can write

$$\log k_{YX} = \tfrac{1}{2}(\log k_{YX} + \log k_{XX}) + \tfrac{1}{2} \log K_{YX} \quad (4)$$

$$K_{YX} = K_{Y_r}/K_{X_r} \quad (5)$$

Now taking the logarithm of both sides of equation 5 and reordering the terms, we obtain

$$\log k_{YX} = \tfrac{1}{2} \log (k_{YY}K_{Yr}) + \tfrac{1}{2} \log (k_{XX}/K_{Xr}) \quad (6)$$

$$\log k_{YX} = M_Y + N_X \quad (7)$$

The first term in equation 6 is a property (M_Y) only of the methylating agent CH_3Y, and the second term is a property only of the nucleophile X^- (N_X). Although the values of M_Y and N_X depend on the choice of reference nucleophile for the equilibria, the calculated rate constant k_{YX} is independent of the reference; this rate constant only requires the same reference for CH_3X and CH_3Y.

The values of M_Y and N_X can be calculated from tables such as I and II and are shown in Tables III and IV. No fitting is required if the data are available, and the equation can be expected to apply in every case if the reaction rate in the spontaneous direction is substantially less than that of diffusion control.

Another consequence of equation 7 is that a known methylating agent with nucleophile X^- for which neither K_{Xr} nor k_{XX} is known can be used. A measurement of the rate then gives N_X, which should be the same for all methylating agents, CH_3Y, used. Values of N_X are presented in Table III; those calculated by using equation 7 are identified, and the uncertainties are based upon results with at least three methylating agents. The order is in accord with generally understood "nucleophilicities". Values of N_X calculated

Table III. Nucleophilic Power ($N_X = \tfrac{1}{2} \log k_{XX}/K_{Xr}$) in Sulfolane at 35 °C

X^-	N_X	*Source*
$CF_3SO_3^-$	−4.2	direct[a]
FSO_3^-	−4.1[b]	direct
$C_6F_5SO_3^-$	−4.1[b]	direct
$CH_3OSO_3^-$	−2.9	direct
p-$ClC_6H_4SO_3^-$	−2.6	direct
$C_6H_5SO_3^-$	−2.3	direct
$CH_3SO_3^-$	−2.2	direct
p-$O_2NC_6H_4O^-$	0.7 ± 0.6	k_{XY}[c]
I^-	1.3	direct
p-$O_2NC_6H_4S^-$	2.5 ± 0.3	k_{XY}
p-$CH_3C_6H_4S^-$	5.4 ± 0.6	k_{XY}

[a] Direct is the calculation from Table I values.
[b] Estimated.
[c] N_X was calculated by equation 7 from at least three measured rates, which agree within the standard deviations indicated.

Table IV. Kinetic Methylating Power [M_Y = ½(log $k_{YY}K_{Yr}$)] in Sulfolane at 35 °C from Values in Table I

CH_3Y	M_Y
$F_3CSO_3CH_3$	2.28
FSO_3CH_3	2.14[a]
$C_6F_5SO_3CH_3$	1.82[a]
$CH_3OSO_3CH_3$	−0.84
CH_3I	−0.92
p-$ClC_6H_4SO_3CH_3$	−1.86
$C_6H_5SO_3CH_3$	−2.32
$CH_3SO_3CH_3$	−2.73

[a] Estimated.

by equation 7 cannot be broken down into the equilibrium and identity rate parts [an attempt to separately determine k_{XX} and K_{Xr} by solving simultaneous equations including the quadratic term failed (*12*); this failure showed again that this term contributes very little]. If either component could be measured for some X^-, then the other could be calculated. Unfortunately, for most good anionic nucleophiles, the equilibrium constant K_{Xr} is apparently very small, and the identity rate is very slow. For example, in aqueous ethanol, the 35 °C identity rate for $X^- = C_6H_5S^-$ was calculated to be 10^{-12} M^{-1} s^{-1} based on very slow reactions near 200 °C (*22, 11*).

A possibility, not yet well explored, is to measure $\Delta H°$ (standard enthalpy) calorimetrically. If we can get an idea of the entropy, we can then calculate $\Delta G°$. From reference 8, both ΔG and ΔH can be found in a few cases. From these few cases, for reaction 1 with $n = m = 0$, ΔS is small, and if the data were available and were justified in generalizing this conclusion, $\Delta G° \approx \Delta H°$, and the calorimetry would allow a separation of K_{Xr} and k_{XX}. For the few cases established with $n = 0$ and $m = 1$, $\Delta S° \cong -35$ eu (presumably a consequence of solvent organization around the product ions); clearly, to try to use such data in this way would be foolhardy. We have made very few calorimetric measurements beyond those already published. A rather large body of data in other solvents on the heat of reaction of methyl fluorosulfonate with various nucleophiles was collected by R. Alder (unpublished results). Almost none of these overlap enough with our cases to allow a direct comparison, yet the principle has some potential.

Some further comments on equation 7 are appropriate. The form of the equation is reminiscent of the Swain–Scott (2) equation 8, where log k_0 can be identified with M_Y and *sn* with *N*. However, in the equation 8 applied

$$\log (k/k_0) = sn \tag{8}$$

to nucleophilic substitution, *s* is considered a property of the reaction type studied, specifically a property of CH_3Y. When the Swain–Scott equation is applied to methyl transfers in sulfolane, although their n values cannot be used because of solvent difference, the constant selectivity among different CH_3Ys implies a constant s; thus, the further identification of equations 7 and 8 is satisfactory. We differ from Swain and Scott in that the only kinetic input into equation 7 is the identity rates.

Equation 7 is also related to the Ritchie equation 9, applied to nucleophile–electrophile bond-forming reactions. The formal similarity and the apparently unusual constant selectivity common to both suggest the possibility of a closer relation. However, our N_X values are in principle related to identity rates, which as pointed out by Ritchie et al. (*23*) do not exist for these one-bond-forming reactions and cannot be a part of his N_+ values.

$$\log (k/k_0) = N_+ \tag{9}$$

As shown earlier, the systematic variation in identity rate with equilibrium methylating power could be used to estimate the charge on the methyl group in the transition state. In that case, because $\rho_{XX} > 0$, a net positive charge was assigned to the methyl group. This result was also observed in a few other cases. In both phenyldimethylsulfonium ion and diphenylmethylsulfonium ion, an increase in k_{XX} apparently occurs with chlorine substitution; a real but very small $\rho_{XX} > 0$ also is seen with aryldimethylselenonium ion. However, unfortunately at a much different temperature and solvent, the ρ_{XX} for methylation by aryl methyl sulfides is zero within experimental error (*22*). The corresponding value for aryl methyl selenides is now under investigation (T. A. Douglas, unpublished results).

Another approach to the determination of the charge on the transferred group in the transition state, not requiring measurement of k_{XX}, is the following.

A consequence of the applicability of equation 7 is that if an LFER occurs for methylation of a series of nucleophiles by different methylating agents, the slope of this LFER is given only by the substituent dependence of N_X. The value of the Hammett ρ_{YX} for reaction 1 with a variable X^- is the same for all CH_3Ys, as observed. Using the earlier treatment (*10*) and taking account of the direction of the K_{Xr} in the tables, which reverses the sign of the equilibrium ρ, we can write equation 10. In our cases, ρ_{XX} for the identity rate is >0 but smaller in absolute magnitude than the equilibrium ρ, written as ρ_{eqXr}; we therefore expect inequality 11. Naturally, if neither identity rates nor equilibria are known, the validity of this inequality is difficult to check. Nevertheless, an approximate route is open.

$$\rho_{YX} = (\rho_{XX} - \rho_{eqXr})/2 \tag{10}$$

$$0 > \rho_{YX} > -\rho_{eqXr}/2 \tag{11}$$

The effect of structural change in a nucleophile on its equilibrium properties can be estimated without knowing K_{Xr} itself by assuming that K_{Xr} varies with substituent in much the same way that $K_{HX}{}^{a}$ does. Thus, we can put ρK_{Xr} equal to ρ for the ionization of the acid, HX, which is often well-known (although less often in the solvent used for kinetics). Alternatively, the slope of a plot of log k_{YX} versus $pK_{HX}{}^{a}$, called β_{nuc}, is an approximation of the slope of the plot of log k_{YX} versus $-\log K_{Xr}$, that is, $\beta_{nuc} \cong -\rho_{XX}/\rho_{eqXr}$. Many such plots of methylation rates have been attempted, and the agreement of these rates with the inequality 11, that is, $\beta_{nuc} < 0.5$, is impressive. Thus, for thiolate ions with methyl benzenesulfonate (*24*), $\beta_{nuc} = 0.234$; for a series of oxygen nucleophiles with $CH_3O_3\ SC_6H_5$, CH_3O_3N, CH_3O_4Cl, the values of β_{nuc} are 0.236, 0.223, and 0.235, respectively (*25*). Several other β_{nuc} values, involving several other alkyl transfers as well as methyl, all between 0 and ½, are summarized by Hudson (*26a*) whose discussion of nucleophilic character is very detailed. The parallelism between $K_{HX}{}^{a}$ and K_{Xr} is considered by Hudson and by others (*26b*, *26c*). Some of the most convincing quantitative comparisons of butyl transfer rates and proton transfer equilibria are those of Bordwell (*27*), who measured both rates and basicities of a variety of sulfone anions with butyl bromide and chloride in the same solvent [$(CH_3)_2SO$]. The measured β_{nuc} was 0.40.

We previously identified ρ_{XX}/ρ_{eqXr} as the charge on the methyl group in the transition state. Using equation 10, we might then write equation 12, which could be used for cases for which the Hammett equation is inapplicable.

$$\rho_{XX}/\rho_{eqXr} = 1 - 2\beta_{nuc} \qquad (12)$$

However, because of the approximations involved, we can only say that this calculation is in agreement in both sign and magnitude with the earlier direct measurements.

The successful quantitative classifications of nucleophiles and leaving groups in methyl transfers in sulfolane is interesting, but rather limited. We shall in the following be concerned with possible extensions of both the quantitative results and the general method to a variety of other reactions, starting with systems that probably are susceptible to almost unmodified applications, proceeding to less direct applications, and then listing some nucleophilic and leaving group examples about which virtually nothing can be said, except that correlation with methyl transfers is expected to be poor.

The smallest change that appears to be of some interest is isotopic substitution in the methyl group. The interest is in the isotope effect (for example, for CD_3 transfer) to mesh our identity rate work with Schowen and co-workers' (*28*) isotope effect criterion of transition-state looseness. Clearly, the error of the fit to the Marcus treatment (as shown in Figure 1) is well outside any conceivable isotope effect, but isotope effects should nevertheless be measurable on the identity reactions, with only minor effects on the equilibria.

The change from methyl transfer to the transfer of other alkyl groups could be of great utility. The identity rates are quite sensitive to the nature of the alkyl group; for example, in acetone, the identity rate for methyl iodide is about 10^4 greater than that for 2-iodooctane (*29, 30*). Nevertheless, we can expect (except for steric effects) to find a very similar sequence of identity rates for primary and secondary alkyl groups. Thus, Streitwieser (*1a*) suggested long ago that in S_N2 reactions, different alkyl groups might have a constant rate ratio, with (averaged over several temperatures, solvents, nucleophiles, and leaving groups) ethyl slower than methyl by a factor of about 30. To the extent that this finding is general, at least when limited to a single solvent, charge type, and temperature, this result would apply to identity reactions. Thus, identity reactions for any alkyl group can be roughly estimated, given the value for methyl identity rate with the same leaving group. However, possibly the Marcus equation will not work as well with the larger steric interactions of some alkyl groups.

The equilibria listed in Tables I and II can be expected to hold fairly well for other alkyl groups. The independence of alkyl group R of these equilibrium constants is equivalent to assuming $K = 1$ for equation 13. This assumption is plausible if X is attached to an sp^3-hybridized carbon in R, so that no major resonance interaction occurs. Deviations from unity may be induced by electronegativity, steric (especially F-strain effects), and other differences, but perhaps not very serious ones.

$$RX + CH_3O_3SC_6H_5 \rightleftarrows RO_3SC_6H_5 + CH_3X \qquad (13)$$

Thus, possibly identity rates and equilibrium constants can be estimated well enough to permit the estimation of rates for any S_N2 reaction in sulfolane at 35 °C. Enough temperature dependencies are available so that correction to some other nearby temperature should not be too difficult.

Solvent changes also introduce important effects. As long as only reactions involving no charge change (equation 1, $m = n$) are used, solvent effects can be small. Thus, rates in acetonitrile and sulfolane are similar. Hydrogen-bonding effects in protic solvents may produce major effects, although the ions in Table I do not include those ions (like ^-OR, Cl^-, F^-) that are most notoriously influenced by the change to protic solvents (*3*).

We expect that equations analogous to equation 7 will be useful for S_N2 reactions in most solvents and for most R groups if new tables of equilibria and identity rates are determined. Even now such tables can be roughly estimated.

Similarly, we can expect the Marcus equation with similar simplifications to apply to other group transfers occurring in a single step, for in these identity reactions group transfers exist. The S_N1 reaction, elimination reactions, and nucleophilic attack on carbonyls, which are alike in that no identity reaction occurs, are discussed next.

The S_N1 reaction of RY is believed to have a rate-determining ionization of RY, equation 14, and the transition state is believed to be very productlike. Thus, a close parallelism occurs between rate and stability of the product ions, although this statement is generally made with respect to variation in R rather than in Y.

$$RY \underset{k_{-14}}{\overset{k_{14}}{\rightleftharpoons}} R^+ + Y^- \quad (14)$$

Product stability can refer to the equilibrium constant K_{14}, and for variation in Y as discussed, K_{14} differs from K_{Yr} by a constant factor and the variation in rate can be expected to follow K_{Yr} from Table I.

This k_{14}–K_{Yr} parellelism implies a constant k_{-14} (perhaps diffusion-controlled). However, Ritchie (*31a*) found many much slower reactions between various R^+ with various Y^-, and the rates of these correlate very poorly with the equilibrium constants. The present approach can therefore say nothing about these fascinating and puzzling reactions, and we cannot expect wide applicability to the rate equilibrium reactions for the forward process. We also neglect the intermediacy in the S_N1 of ion pairs and the fact that the S_N1 does not really take place at all with sulfolane or other aprotic solvents of low nucleophilie character.

The literature (*31b, 31c*) on aliphatic nucleophilic substitution has drifted from pure S_N1 and S_N2 to a consideration of borderline mechanisms. Interestingly equation 7 contains within itself the elements needed to cover this borderline region.

When an alkyl group is able to stabilize a positive charge much better than a methyl group, this ability should show up in the dependence of the identity rates on the structure of the leaving group or the attacking group. Thus, if log k_{YX} correlates with log K_{Yr}, with a slope greater than ½

$$\log k_{YY} = a \log K_{Yr} + b \quad (15)$$

where $0 < a < 1$, and we can substitute in equation 6 to give

$$\log k_{YX} = \tfrac{1}{2}(b + a \log K_{Yr}) + \tfrac{1}{2}K_{Yr} + \tfrac{1}{2}(b + a \log K_{Xr}) - \tfrac{1}{2}K_{Xr} \quad (16)$$

$$= b/2 + (\log K_{Yr})(1 + a)/2 + b/2 + (a - 1)(\log K_{Xr})/2$$

$$= b + (a + 1)/2 \log K_{Yr} + (a - 1)/2 \log K_{Xr} \quad (17)$$

In the limit of unit charge on the alkyl group, $a = 1$, equation 17 reduces to equation 18, in which the rate is independent of X.

$$\log k_{YX} = b + \log K_{Y_r} \tag{18}$$

In the limit of $a = -1$, equation 17 now reduces to an as yet unobserved case, in which the rate reduces to equation 19, in which the rate is independent of Y

$$\log k_{YX} = b - \log K_{X_r} \tag{19}$$

an effect reminiscent of the "element effect" seen in activated nucleophilic aromatic substitution. The parallel is imperfect, because the aromatic substitution is considered a two-step process via the Meisenheimer intermediate.

Thus, the borderline cases with a substantial but not quite $+1$ value of a are clearly contained within the treatment, and the intuitive position of the borderline is contained within the a parameter.

The leaving group sequence in the E2 reaction would appear to be complex. The sequence is not expected to correlate with M_Y, or the equilibrium tables, because the coupling with a proton loss presumably makes the rate dependent on the acid-strenthening effect of the β-Y. Furthermore, although a sort of intrinsic barrier to the departure of Y may exist, this barrier cannot be rigorously associated with the methyl-transfer identity rates, nor indeed to any other identity rate. The only present data that seem to have a probable relation to E2 reactions are the equilibrium constants of Tables I and II; these constants become more and more relevant as the mechanism approaches the E1 limit.

The well-known failure of eliminations rates to correlate with S_N2 leaving-group properties is especially conspicuous in the 1,2 eliminations studied by Stirling and co-workers (*32*), as well as 1,3 eliminations (*33*), which mostly follow the E1cB mechanism. Apparently, the loss of the leaving groups from the conjugate base is fast and very exothermic so that the constant selectivity that is associated with reactions well below the diffusion controlled rate is no longer always applicable. Thus, groups can be lost that are virtually never lost in methyl transfers. Except for the special charge type of E1cB mechanisms, cases are imaginable in which the equilibrium order of leaving groups would be followed, as in the S_N1 mechanism mentioned, but they appear to be well outside the range of current experience.

The rates of addition of nucleophiles to carbonyl groups and the rates of elimination from the tetrahedral intermediates constitute another class, probably similar to the activated aromatic nucleophilic substitution. The carbonyl group is an electrophile, and no obvious source of any barrier exists, outside of desolvation. Therefore, a resemblance to Ritchie's systems is found. No obvious relation between our kinetic nucleophilic characters (N_X) and the additions occurs, but a possible parallel to the equilibrium methylating powers, K_{YX} (in Tables I and II), of the conjugate methylating agent of the nucleophile exists, in that the transition state may strongly resemble the product and the product at least has carbon bonded to the nucleophile. The rates of loss of leaving groups from the tetrahedral intermediate do not appear to be simply correlated to any methyl-transfer factors; indeed, the loss of groups like ^{-}OR, $^{-}CCl_3$, and enolate ions (as in the reverse Claisen) has no S_N2 precedent.

Little can be said about nucleophilic substitution, attack, or leaving groups on other elements except the one-step substitution reactions, in which cases identity reactions can exist. Some analogies to the carbon cases may occur, but a parallel cannot be expected.

Acknowledgment

We thank the National Science Foundation and The Robert A. Welch Foundation for support of this work.

Literature Cited

1a. Streitwieser, A., Jr. *Chem. Rev.*, **1956,** *56*, 571.
1b. Swain, C. G.; Scott, C. B. *J. Am. Chem. Soc.* **1953,** *75*, 141.
2. Parker, A. J. *Chem. Rev.* **1969,** *69*, 1.
3a. Robertson, R. E. *Prog. Phys. Org. Chem.*, **1967,** *4*, 213.
3b. Bensley, B.; Kohnstam, G. *J. Chem. Soc.* **1956,** 289.
3c. Murr, B.; Shiner, V. J. Jr. *J. Am. Chem. Soc.* **1962,** *84*, 4672.
3d. Thornton, E. R. *Solvolysis Mechanisms*, Ronald: New York, 1964.
3e. Marcus, R. A. *J. Phys. Chem.* **1956,** *24*, 966.
4. Albery, W. J.; Kreevoy, M. M. *Adv. Phys. Org. Chem.* **1978,** *16*, 87.
5. Albery, W. J. *Annu. Rev. Phys. Chem.* **1980,** *31*, 227.
6. Abraham, M. H.; McLennan, D. J. *J. Chem. Soc., Perkin Trans. 2* **1977,** 873.
7. Pellerite, M.; Brauman, J. I. *J. Am. Chem. Soc.* **1980,** *102*, 5993.
8. Lewis, E. S.; Kukes, S.; Slater, C. D. *J. Am. Chem. Soc.* **1980,** *102*, 1619.
9. Wong, C. R.; Jackman, L. M.; Portman, R. G. *Tetrahedron Lett.* **1974,** 921.
10. Lewis, E. S.; Hu, D. D. *J. Am. Chem. Soc.* **1984,** *106*, 3292.
11. Lewis, E. S.; McLaughlin, M. L.; Douglas, T. A. *Isr. J. Chem.* **1985,** *26*, 331.
12. Lewis, E. S.; McLaughlin, M. L.; Douglas, T. A. *J. Am. Chem. Soc.* **1985,** *107*, 6668.
13. Arnett, E. M.; Reich, R. *J. Am. Chem. Soc.* **1980,** *102*, 5892.
14. Matsui, T.; Tokura, N. *Bull. Chem. Soc. Jpn.* **1970,** *43*, 1751.
15. Lewis, E. S; McLaughlin, M. L. Presented at Southwestern Regional Meeting of the American Chemical Society, Lubbock, TX, 1984.
16. Pearson, R. G.; Songstad, J. *J. Am. Chem. Soc.* **1967,** *89*, 1827.
17. Shaik, S. S. *Nouv. J. Chim.* **1983,** *7*, 201.
18. Lewis, E. S.; Vanderpool, S. H. *J. Am. Chem. Soc.* **1977,** *99*, 1946.
19. Lewis, E. S.; Vanderpool, S. H. *J. Am. Chem. Soc.* **1978,** *100, 6421*.
20. Lewis, E. S.; Kukes, S.; Slater, C. D. *J. Am. Chem. Soc.* **1980,** *102*, 303.
21. Pross, A.; Shaik, S. S. *Tetrahedron Lett.* **1982,** *23*, 5467.
22. Lewis, E. S; Kukes, S. *J. Am. Chem. Soc.* **1979,** *101*, 417.
23. Ritchie, C. D.; Kubistz, C.; Ting, G.-Y. *J. Am. Chem. Soc.* **1983,** *105*, 279.
24. Kyllönen, A. Koskikallio, J. *Suom. Kemistil. B* **1972,** *45*, 212.

25. Koskikallio, J. *Acta Chem. Scand.* **1971,** *26,* 1201.
26a. Hudson, R. F. In *Chemical Reactivity and Reaction Paths;* Klopman, G., Ed.; Wiley: New York, 1974; Chapter 5.
26b. Jencks, W. P. *Chem. Rev.* **1985,** *85,* 511.
26c. Lewis, E. S., In *Rates and Mechanisms of Reaction;* Bernasconi, C., Ed. Wiley-Interscience: New York, 1985; pp 888, 889.
27. Bordwell, F. G. *Isr. J. Chem.*, in press (I thank Professor Bordwell for an early copy of this work).
28. Gray, C. H.; Coward, S. K; Schowen, K. B.; Schowen, R. L. *J. Chem. Soc.* **1979,** *101,* 4351.
29. Swart, E. R.; LeRoue, L. J. *J. Chem. Soc.* **1957,** 406.
30. Hughes, E.D.; Juliusberger, F.; Scott, A. D.; Topley, B.; Weiss, J. *J. Chem. Soc.* **1936,** 1173.
31a. Ritchie, C. D. *Acc. Chem. Res.* **1972,** *5,* 348.
31b. Ingold, C. K. *Structure and Mechanism in Organic Chemistry;* Cornell University: Ithaca, 1953; pp 310–313.
31c. Sneen, R. A. *Accts. Chem. Res.*, **1973,** *6,* 46.
32. Marshall, D. R.; Thomas, P. J.; Stirling, C. J. M. *J. Chem. Soc., Perkin Trans. 2* **1977,** 1898.
33. Issari, B.; Stirling, C. J. M. *J. Chem. Soc., Perkin Trans. 2* **1984,** 1043.

RECEIVED for review October 21, 1985. ACCEPTED January 28, 1986.

4

Nucleophilic Reactivity in Gas-Phase Anion–Molecule Reactions

Richard N. McDonald, A. K. Chowdhury, W. Y. Gung, and K. D. DeWitt

Department of Chemistry, Kansas State University, Manhattan, KS 66506

Three different topics of nucleophilic anions reacting with neutral molecules are discussed. (1) $O_2^{\cdot -}$ is established intrinsically and kinetically as a super nucleophile in S_N2 reactions with CH_3X molecules. (2) The intrinsic reactivity scale of nucleophilic additions to carbonyl centers is developed by using $C_6H_5N^{\cdot -}$ as the nucleophile and can be extended with $(C_6H_5)_2C^{\cdot -}$. (3) The phosphoryl anion $(CH_3O)_2PO^-$ is shown to be a poor nucleophile in S_N2 reactions with CH_3X molecules but reacts rapidly with ICF_3 and $BrCF_3$ by initial electron transfer; mainly $(CH_3O)(X)P_2^-$ negative ions result.

TOPICS OF THE GENERAL AREA OF NUCLEOPHILIC REACTIONS OF ANIONS with neutral substrates in the gas phase included in this chapter are (1) the nucleophilicity of $O_2^{\cdot -}$ in S_N2 reactions, (2) development of an intrinsic reactivity scale for nucleophilic reactions with organic carbonyl-containing molecules, and (3) investigations of $(CH_3O)_2PO^-$ in S_N2 reactions with CH_3X reactants and electron-transfer processes with XCF_3 molecules.

Experimental Section

Experiments were carried out in a previously described (*1*, *2*) flowing afterglow (FA) apparatus at 298 K (*see* Figure 1). Briefly, the ion of interest is produced continuously in a fast flow of helium buffer gas in the upstream end of the flow tube by electron impact on small concentrations of added reagents via inlets 1–5. The fast flow (v = 80 m/s, P_{He} = 0.5 torr) is maintained by a large, fast pumping system. Following thermalization of the ion of interest by collisions with the buffer gas in the next 20–45 cm of the flow tube, neutral reactant molecules are added via the inlet located about halfway down the flow tube, and the ion–molecule reaction occurs in the final 65 cm of the flow tube. The flow is sampled into a differentially pumped compartment (10^{-7} torr) containing the quadrupole mass filter and electron multiplier, which continuously monitor the ion composition of the flow. Kinetics of these bimolecular ion–molecule reactions are determined under pseudo-first-order conditions with the concentration of the added neutral reactant in large excess compared to the ion concentration by methods already given (*1*).

0065-2393/87/0215-0051$06.00/0

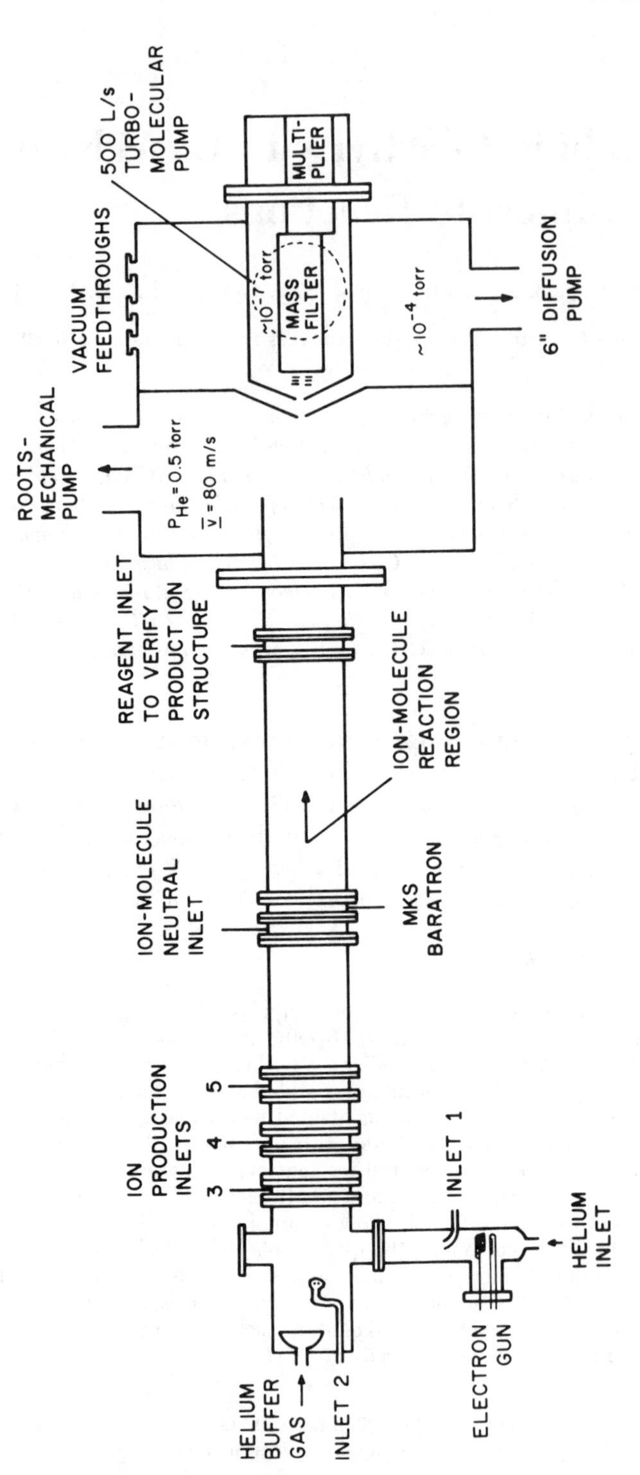

Figure 1. Diagram of the flowing afterglow apparatus.

$O_2^{\cdot -}$ as a Super S_N2 Nucleophile

Ion Generation. The generation of $O_2^{\cdot -}$ in the FA involved the sequence of reactions (*3, 4*)

$$NH_3 + e^- \rightarrow H_2N^- + \cdot H \tag{1}$$

$$H_2N^- + CH_3CH{=}CH_2 \rightarrow C_3H_5^- + NH_3 \tag{2}$$

where $k = 2.5 \times 10^{-10}$ cm^3 molecule^{-1} s^{-1}, reaction efficiency equals 0.43, and $\Delta H° = -14 \pm 5$ kcal/mol, and

$$C_3H_5^- + O_2 \rightarrow O_2^{\cdot -} + C_3H_5^{\cdot -} \tag{3}$$

where $k = 3.0 \times 10^{-10}$ cm^3 molecule^{-1} s^{-1}, reaction efficiency equals 0.43, and $\Delta H° = -1.4 \pm 0.7$ kcal/mol (*5, 6*). Rate constants, reaction efficiencies (defined as k_{obsd}/k_{col} where k_{col} is the calculated collision limited rate constant), and reaction exothermicities are given (*3–6*). Ammonia was added via inlet 1, propene via inlet 2, and dioxygen via inlet 4 in Figure 1.

S_N2 Reactions of $O_2^{\cdot -}$ with CH_3X Molecules. Bohme and co-workers (*7*) established a kinetic nucleophilicity scale for gas-phase anions in S_N2 reactions based on their rates of reaction with CH_3X (X = Br, Cl, or F). For this comparison, the reactions of $O_2^{\cdot -}$ with CH_3Br and CH_3Cl (Table I) were used because the displacement of F^- from CH_3F by $O_2^{\cdot -}$ is strongly endothermic. Both of these reactions occurred at close to the collision limit; thus, $O_2^{\cdot -}$ is placed in the category of gas-phase anions of high nucleophilicity. Other members are H^-, F^-, CH_3O^-, HO^-, and H_2N^-. However, the considerably lower exothermicities for these two $O_2^{\cdot -}$ reactions distinguish it from the other high nucleophilic anions in that all of them have much larger reaction exothermicities with these two CH_3X molecules.

From the Pellerite and Brauman (*8*) application of Marcus theory to S_N2 methyl-transfer reactions, the kinetic barrier is made up of an intrinsic barrier for the reaction that is decreased by the magnitude of the reaction exothermicity. Because the exothermicities of the reactions with $O_2^{\cdot -}$ are the lowest (by upward of 20 kcal/mol) compared to the other anions of high kinetic nucleophilicity, the intrinsic barriers for these S_N2 reactions with $O_2^{\cdot -}$ are the smallest. Thus, $O_2^{\cdot -}$ can be called a *super* S_N2 nucleophile.

The remaining reactions of $O_2^{\cdot -}$ with CH_3X substrates in Table I occur primarily or exclusively by S_N2 displacement. Their rates given as reaction efficiencies vary from reaction occurring on essentially every collision with $CF_3CO_2CH_3$ to $CH_3CO_2CH_3$ where 2 out of every 1000 collisions, on the average, yield $CH_3CO_2^-$. In accordance with the S_N2 mechanism for their reactions, the reaction efficiencies are correlated with the reaction exothermicities and by the anionic leaving group ability as modeled by the proton affinity of the departing anion.

Table I. Kinetic and Thermochemical Data for Reactions of $O_2^{\cdot -}$ with CH_3X Molecules

Reaction	S_N2 *Channel*	$\Delta H°$ *(kcal/mol)*	E_{rxn}[a]	$\Delta H°_{acid}$ *(HA)* $=A_p(A^-)$ *(kcal/mol)*[b]
$O_2^{\cdot -} + CF_3CO_2CH_3 \rightarrow CF_3CO_2^- + CH_3O_2^{\cdot -}$	0.97	−32.5	0.92	323
$O_2^{\cdot -} + CH_3Br \rightarrow Br^- + CH_3O_2^{\cdot -}$	1.00	−24.9	0.76	324
$O_2^{\cdot -} + CH_3Cl \rightarrow Cl^- + CH_3O_2^{\cdot -}$	1.00	−17.0	0.39	333
$O_2^{\cdot -} + HCO_2CH_3 \rightarrow HCO_2^- + CH_3O_2^{\cdot -}$	0.90	−12.1	0.05	345
$O_2^{\cdot -} + H_2C{=}CHCO_2CH_3 \rightarrow H_2C{=}CHCO_2^- + CH_3O_2^{\cdot -}$	1.00	—[c]	0.03	—[c]
$O_2^{\cdot -} + CH_3CO_2CH_3 \rightarrow CH_3CO_2^- + CH_3O_2^{\cdot -}$	0.79	− 7.7	0.002	349

[a] k_{obsd}/k_{col} equals reaction efficiency (E_{rxn}).
[b] Reference 12; errors are ±2 kcal/mol.
[c] A_p ($H_2C{=}CHCO_2^-$) is unknown.

Intrinsic Reactivity Scale for Nucleophilic Addition Reactions at Carbonyl Centers

The intrinsic reactivity scale is a topic that led us to begin studies of gas-phase reactions in 1978. However, the problem of reversibility of the nucleophilic addition from the tetrahedral intermediate is even more severe in the gas-phase pressure regimes than it is in solution.

Our approach to this problem involved generation of a new class of reactive intermediates called hypovalent anion radicals. Hypovalent anion radicals contain *less* than the normal number of substituents attached to the central atom found in the corresponding neutral free radical, and these anion radicals have both the electron pair of the anion and the spin unpaired electron of the radical formally located on the central atom. Phenylnitrene anion radical ($C_6H_5N^{\cdot -}$) is a member of this class of nitrogen-centered species and is readily formed from $C_6H_5N_3$ added at inlet 1 (Figure 1) by dissociative electron attachment (equation 4) (9). The idea was to shut down the reverse of the nucleophilic addition to the carbonyl group from the tetrahedral intermediate by allowing the faster follow-up chemical reaction of radical β fragmentation to occur; the acylanilide anions plus the radicals R_1 or R_2 are obtained:

$$C_6H_5N_3 + e^- \rightarrow \underset{m/z\ 91}{C_6H_5N^{\cdot -}} + N_2 \qquad (4)$$

$$PhN^{\cdot -} + R_1{-}\overset{\overset{O}{\|}}{C}{-}R_2 \rightarrow \left[Ph\dot{N}{-}\overset{\overset{O^-}{|}}{\underset{\underset{R_2}{|}}{C}}{-}R_1 \right] \rightarrow \begin{cases} PhN{=}C(O^-)R_2 + \cdot R_1 \\ PhN{=}C(O^-)R_1 + \cdot R_2 \end{cases} \qquad (5)$$

We examined the reactions of C_6H_5 N·$^-$ with the series of carbonyl-containing molecules listed in Table II (*10*). Because H^+ transfer is a competing reaction channel in some of these reactions, we factored out of the total rate constants that part due to carbonyl addition and radical fragmentation, and these rate constants were made relative to that for acetone; these $k_{rel}^{C=0}$ values are given in the middle column of Table II.

Table II. Relative Rate Constants and Reaction Exothermicities of Carbonyl Addition–Radical Fragmentation of Carbonyl-Containing Molecules with $C_6H_5N\cdot^-$

Substrate	$k_{rel}^{C=O}$ [a]	ΔH° (*kcal/mol*)
CH_3COCH_3	1	−19
$CH_3COC_2H_5$	4	−22
Cyclobutanone	11	—[b]
CH_3COCF_3	80	−22
CF_3COCF_3	83	−31
CH_3CHO	12	−16
C_2H_5CHO	31	−18
$(CH_3)_3CCHO$	67	−24
HCO_2CH_3	0.2	−6
$CH_3CO_2CH_3$	0.02	−4
$CF_3CO_2CH_3$	157	−15
$CH_3COCOCH_3$	108	−32
$CH_3COCO_2CH_3$	113	—[b]

[a] These relative rate constants are k_{total} (sum of the fractions of those channels yielding addition adducts or acylanilide anions) with each substrate relative to $k^{C=O}$ for acetone.
[b] Not determined.

Three observations from these data are noteworthy. (1) The range of $k_{rel}^{C=O}$ is about 8,000, which is nearly the full range of kinetic reactivity available in our FA experiments, from reactions occurring on every collision ($k_{obsd} = k_{col}$) to those occurring in one out of 10^4 collisions. (2) We observe that for "normal" substituents on the carbonyl center, the order of reactivity is aldehydes (CH_3CHO) > ketones [$(CH_3)_2C{=}O$] > esters ($Ch_3CO_2CH_3$). This reactivity difference holds for the intramolecular comparison with methyl pyruvate where the total reaction occurred on every other collision, but 7 times more addition–fragmentation occurred at the keto CO than at the ester CO (*10*). (3) No correlation was observed between the $k_{rel}^{C=O}$ values and the overall reaction exothermicity in forming the acylanilide anion and the radical (*see* equation 5). This latter point is most clearly seen by comparing the slow reaction of $C_6H_5N\cdot^-$ with acetone, which is 19 kcal/mol exothermic, with the very fast reaction of $C_6H_5N\cdot^-$ while $CF_3CO_2CH_3$ is only 15 kcal/mol exothermic.

Point 3 and certain other results suggest that the potential energy versus reaction coordinate diagram for these carbonyl addition–radical fragmentation reactions with $C_6H_5N\cdot^-$ occur by the triple minimum surface shown in Figure 2. The interaction of $C_6H_5N\cdot^-$ with a dipolar carbonyl-containing substrate molecule is attractive even at rather long distances and leads to formation of loose collision complexes given as a in Figure 2. Such complexes are held together by ion–dipole and ion-induced dipole forces with well depths of 10–20 kcal/mol. Closer approach of the complex components is repulsive until net bonding takes over with formation of the tetrahedral intermediate (b). Radical β-fragmentation is considered to have a lower

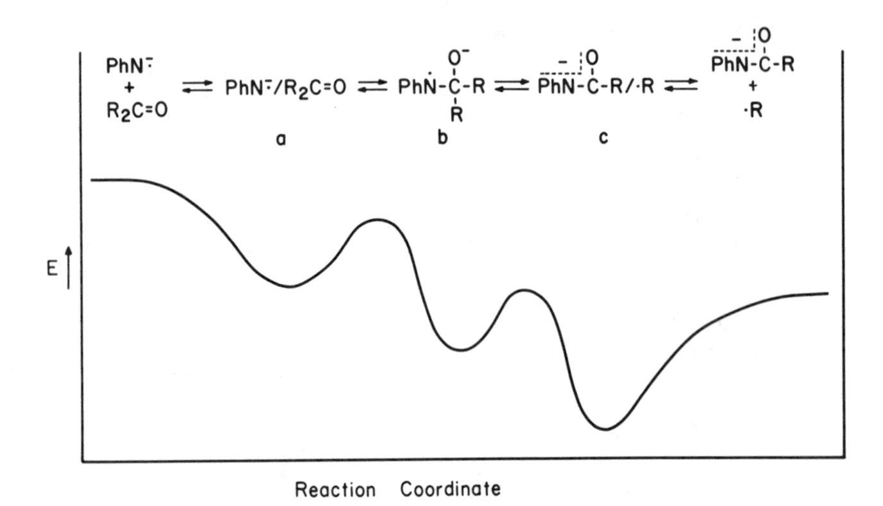

Figure 2. Proposed triple-minimum potential energy vs. reaction coordinate diagram for the reactions of PhN·⁻ with carbonyl-containing molecules. Reproduced from reference 10. Copyright 1983 American Chemical Society.

barrier than that of nucleophilic addition and leads to the loose complex (c), which separates, primarily due to entropic forces, to yield the observed reaction products. This model can be viewed simply in terms of the conversion of reaction a → reaction b being rate-limiting. Because this barrier and the rate of nucleophilic addition have little to do with the overall reaction exothermicity, no correlation between $k_{rel}^{C=O}$ values and $-\Delta H°$ is observed (or expected) (*10*).

One problem with the reactions of $C_6H_5N\cdot^-$ with organic carbonyl containing molecules was that the carbonyl group reactivity scale could not be extended past the simple esters HCO_2CH_3 and $CH_3CO_2CH_3$ because their rates were already at the lower limit of determination in the FA. To get around this limitation, we examined other hypovalent species such as

$(C_6H_5)_2C\cdot^-$, which is cleanly generated from $(C_6H_5)_2C{=}N_2$ added at inlet 1 (Figure 1) by dissociative electron attachment:

$$(C_6H_5)_2C{=}N_2 + e^- \rightarrow \underset{m/z\ 166}{(C_6H_5)_2C\cdot^-} + N_2 \qquad (6)$$

Our *first* efforts were directed toward determination of the proton affinity (A_p) of $(C_6H_5)_2C\cdot^-$ (*11*). This determination is accomplished by using the bracketing method. Potential proton donors of known gas-phase acidity (Table III) are added to the flow containing $(C_6H_5)_2C\cdot^-$ until H^+ transfer is no longer observed with the weaker acids; H^+ transfer is judged to occur by attenuation of the $(C_6H_5)_2C\cdot^-$ signal *and* observation of the signal for the conjugate base of the acid. This transition occurs between $CH_3C{\equiv}CH$ and *p*-xylene; assignment of A_p $[(C_6H_5)_2C\cdot^-]$ = 382 ± 2 kcal/mol results (for A_p values of other organic anions, *see* reference 12). Because $(C_6H_5)_2CH\cdot$ is the product of protonation and $\Delta H_f^\circ[(C_6H_5)_2CH\cdot]$ = 69 ± 2 kcal/mol (*13*), $\Delta H_f^\circ[(C_6H_5)_2C\cdot^-]$ can be calculated to be 358 ± 2 kcal/mol. The A_p of $(C_6H_5)_2C\cdot^-$ is similar to that of CH_3O^- (A_p = 379 ± 2 kcal/mol) (*12*), and $(C_6H_5)_2C\cdot^-$ is a stronger base than the corresponding carbanion $(C_6H_5)_2CH^-$ (A_p = 364.5 ± 2 kcal/mol) (*12*) by 18 kcal/mol.[1]

Table III. Data for Bracketing $A_p[(C_6H_5)_2C\cdot^-]$ in H^+-Transfer Reactions with HA Molecules

HA	*H^+ Transfer*	ΔH°_{acid} *(HA)*[a] *(kcal/mol)*
$C_6H_5CH_3$	yes	381.2
CH_3OH	yes	381.4
$CH_3C{\equiv}CH$	yes	381.8
p-$CH_3C_6H_4CH_3$	no	382.7
$CH_3CH{=}CH_2$	no	390.0
H_2O	no	390.8

[a]Reference 12; errors are ±2 kcal/mol.

The reactions of $(C_6H_5)_2C\cdot^-$ with CH_3Br and CH_3Cl were studied to determine the kinetic nucleophilicity of $(C_6H_5)_2C\cdot^-$ in S_N2 displacements:

$$(C_6H_5)_2C\cdot^- + CH_3Br \rightarrow Br^- + (C_6H_5)_2\dot{C}CH_3 \qquad (7)$$

where $k = 3.9 \times 10^{-10}$ cm^3 molecule^{-1} s^{-1}, reaction efficiency equals 0.35, and $\Delta H^\circ = -65.5$ kcal/mol, and

[1] For a similar relationship in the A_p values of c-$C_5H_4\cdot^-$ and c-$C_5H_5^-$, *see* reference 1.

$$(C_6H_5)_2C\cdot^- + CH_3Cl \rightarrow Cl^- + (C_6H_5)_2\dot{C}CH_3 \qquad (8)$$

where $k = 3.2 \times 10^{-11}$ cm^3 molecule^{-1} s^{-1}, reaction efficiency equals 0.03, and $\Delta H° = -60.0$ kcal/mol. These results show that $(C_6H_5)_2C\cdot^-$ is of medium kinetic S_N2 nucleophilicity (*7*).

The reaction of $(C_6H_5)_2C\cdot^-$ with $CF_3CO_2CH_3$ was examined to determine the kinetic nucleophilicity of $PH_2C\cdot^-$ toward carbonyl addition in competition with the highly exothermic S_N2 methyl transfer:

$$Ph_2C\cdot^- + CF_3CO_2CH_3 \rightarrow \left[Ph_2\dot{C}\text{-}C(O^-)(CF_3)\text{-}OCH_3 \right] \xrightarrow{0.81} Ph_2C{=}C(O^-)CF_3 + \cdot OCH_3 \qquad (9)$$

$$\left[Ph_2\dot{C}\text{-}C(O^-)(CF_3)\text{-}OCH_3 \right] \xrightarrow{0.14} Ph_2C{=}C(O^-)OCH_3 + \cdot CF_3 \qquad (10)$$

$$Ph_2C\cdot^- + CF_3CO_2CH_3 \xrightarrow{0.05} CF_3CO_2^- + Ph_2\dot{C}CH_3 \qquad (11)$$

where $k = 1.0 \times 10^{-9}$ cm^3 molecule^{-1} s^{-1}, reaction efficiency equals 0.71, $\Delta H°(9) = -25$ kcal/mol, $\Delta H°(10) = -23$ kcal/mol, and $\Delta H°(11) = -72.5$ kcal/mol. Carbonyl addition–radical fragmentation wins out over S_N2 displacement by a factor of 19 to 1 in this very fast reaction. The favored radical fragmentation pathway from the tetrahedral intermediate formed by nucleophilic addition to the carbonyl center is that of loss of the more weakly bound $CH_3O\cdot$ compared to the $F_3C\cdot$.

The reaction of $(C_6H_5)_2C\cdot^-$ with HCO_2CH_3 occurred exclusively by the addition–fragmentation pathway:

$$(C_6H_5)_2C\cdot^- + HCO_2CH_3 \rightarrow (C_6H_5)_2C{=}C(O^-)H \qquad (12)$$

where $k = 1.6 \times 10^{-10}$ cm^3 molecule^{-1} s^{-1}, reaction efficiency equals 0.11, and $\Delta H°(12) = -12.0$ kcal/mol. The intriguing result of this reaction is not only the fact of the exclusivity of the addition–fragmentation channel but also that the rate constant is $>10^2$ faster than that of the reaction of $C_6H_5N\cdot^-$ with HCO_2CH_3 (*10*). Thus, the kinetic reactivity scale of carbonyl centers can be readily extended beyond that of the simple esters by using $(C_6H_5)_2C\cdot^-$ and other hypovalent anion radicals presently under investigation.

Proton Affinity, $\Delta H_f°$, and Reactions of $(CH_3O)_2PO^-$

Substitution reactions by anions at carbon are also known to occur by initial electron transfer. The mechanism of such transformations was first characterized by Russell and Danen (*14*) and Kornblum et al. (*15*), and Bunnett (*16*) significantly developed its applications and named it the $S_{RN}1$ reaction; an example of aromatic substitution is given in equation 13 (*17*). Following a discussion with Jim Swartz of Grinnell College concerning these aromatic substitution reactions involving the phosphoryl anion (e.g., equation 13), we determined the thermochemical properties and gas-phase reactions of the phosphoryl anion.

$$ArI + (EtO)_2PO^- K^+ \xrightarrow[NH_3]{h\nu} ArP(=O)(OEt)_2 + KI \quad (13)$$

Generation and Thermochemical Properties of $(CH_3O)_2PO^-$ *(18a).* For our studies, dimethyl phosphonate $[(CH_3O)_2P(=O)H]$ was selected as our reagent because it is more volatile than the diethyl ester. The phosphonate structure was shown to be 6.5 kcal/mol more stable than that of its trivalent tautomer, $(CH_3O)_2P(OH)$ (*18b*). A variety of gas-phase anionic bases can be used to remove a proton from the phosphonate ester forming the phosphoryl anion (*m/z* 109), but the use of $C_6H_5N\cdot^-$ as the base produces no other primary product negative ions:

$$(CH_3O)_2P(=O)H + C_6H_5N\cdot^- \rightarrow \underset{m/z\ 109}{(CH_3O)_2PO^-} + C_6H_5NH\cdot \quad (14)$$

Our initial studies were directed toward determination of the A_p of *m/z* 109. The bracketing method previously described was employed for determination of $A_p[(C_6H_5)_2C\cdot^-]$ (Table IV). The transition from yes to no for observed proton transfer to *m/z* 109 occurred with C_2H_5SH and CH_3NO_2. Although proton transfer did not occur from CH_3NO_2, slow formation of a cluster ion $(CH_3O)_2PO^-$–CH_3NO_2 was observed, which is probably the hydrogen-bonded complex negative ion; similar cluster ions are formed when CH_3SH and CF_3CH_2OH are added to the flow containing *m/z* 109. Thus, $A_p[(CH_3O)_2PO^-] = 358 \pm 2$ kcal/mol; if proton transfer is assumed to occur at phosphorus, $\Delta H^\circ_{acid}[(CH_3O)_2P(=O)H] = A_p[(CH_3O)_2PO^-]$.

Using Benson's tables (*19*), we calculated $\Delta H_f^\circ[(CH_3O)_2P(=O)H] = -198.1$ kcal/mol. From the equation for ionization of $(CH_3O)_2PO^-$ in equa-

Table IV. Data for Bracketing $A_p[(CH_3O)_2PO^-]$ in H^+-Transfer Reactions with HA Molecules

HA	*H^+ Transfer*	*$\Delta H^\circ_{acid}(HA)$[a] (kcal/mol)*
c-C_5H_6	yes	356.1
C_2H_5SH	yes	357.4
CH_3NO_2	no	358.7
CH_3SH	no	359.0
CF_3CH_2OH	no	364.4

[a]Reference 12; errors are ±2 kcal/mol.

tion 15, we calculated ΔH_f° [$(CH_3O)_2PO^-$] = -207.3 ± 2 kcal/mol. This value can be used to calculate ΔH° for various reactions of the phosphoryl anion.

$$(CH_3O)_2P(=O)H \xrightarrow{\Delta H^\circ_{acid}} (CH_3O)_2PO^- + H^+ \quad (15)$$

S_N2 Nucleophilicity of $(CH_3O)_2PO^-$. To determine the intrinsic S_N2 kinetic nucleophilicity of $(CH_3O)_2PO^-$, we investigated the reactions of the anion with the series of CH_3X molecules listed in Table V. The rates of these reactions vary from modest with the most reactive CH_3I molecules to slow with CH_3Br to no observed reaction with CH_3Cl. From the results, we conclude that the phosphoryl anion is kinetically a poor nucleophile in S_N2 displacement reactions.

Table V. Kinetic and Thermochemical Data for S_N2 Displacement Reactions of $(CH_3O)_2PO^-$ with CH_3X Molecules

Reaction	ΔH° (*kcal/mol*)	k_{obsd} (*cm*^{3} *molecule*$^{-1}$ *s*$^{-1}$)	k_{obsd}/k_{col}
$(CH_3O)_2PO^- + CH_3I \rightarrow I^- + (CH_3O)_2P(=O)CH_3$	−52.2	6.7×10^{-11}	— [a]
$(CH_3O)_2PO^- + CH_3I \rightarrow I^- + (CH_3O)_3P$	−9.3	— [a]	0.06
$(CH_3O)_2PO^- + CH_3Br \rightarrow Br^- + (CH_3O)_2P(=O)CH_3$	−45.6	1.8×10^{-12}	— [a]
$(CH_3O)_2PO^- + CH_3Br \rightarrow Br^- + (Ch_3O)_3P$	−2.7	— [a]	0.002
$(CH_3O)_2PO^- + CH_3Cl \rightarrow Cl^- + (CH_3O)_2P(=O)CH_3$	−37.7	$<10^{-13}$	— [a]
$(CH_3O)_2PO^- + CH_3Cl \rightarrow Cl^- + (CH_3O)_3P$	5.2	— [a]	— [a]

[a]Not determined.

The phosphoryl anion is an ambident species and the calculated ΔH_f° values (*19*) of the tautomers $(CH_3O)_2P(=O)CH_3$ (−210.2 kcal/mol) and $(CH_3O)_3P$ (−167.6 kcal/mol) differ by 43 kcal/mol. Whether methyl transfer occurs to phosphorus of the anion (thermodynamic control) where a large intrinsic barrier exists or at oxygen under kinetic control with a smaller intrinsic barrier is not clear at this time; a much lower reaction exothermicity exists in this latter mode.

Reactions Involving Electron Transfer. To model the first step in the $S_{RN}1$ mechanism of electron transfer from the anion to the neutral substrate, we examined the reactions of the phosphoryl anion with the XCF_3 molecules where X = I, Br, or Cl:

$$(CH_3O)_2PO^- + ICF_3 \;(m/z\ 109) \longrightarrow \begin{cases} \xrightarrow{0.85} (CH_3O)(I)PO_2^- + CF_3CH_3 \\ \xrightarrow{0.09} I^- + (CH_3O)_2PO\cdot + \cdot CF_3 \\ \xrightarrow{0.06} \text{adduct}^- \end{cases} \quad (16)$$

where $k = 6.4 \times 10^{-10}$ cm^3 molecule^{-1} s^{-1} and reaction efficiency equals 0.70. The fast reaction with ICF_3 (equation 16) occurring in 7 out of every 10 collisions yields mainly the anion *m/z* 221. This ion is the product of iodine transfer to and loss of CH_3 from *m/z* 109. The minor negative ion products are 9% I^- and 6% of an adduct *m/z* 305; one possible structure for the adduct may be the four-coordinate anion, $(CH_3O)_2P(I)OCF_3^-$.

The reaction of *m/z* 109 with $BrCF_3$ was almost as fast with >99% of the product channels giving the isotopic doublet at *m/z* 173 and 175 for the product of bromine transfer and dimethylation (equation 17) along with a trace of Br^-. No reaction was observed between $(CH_3O)_2PO^-$ and $ClCF_3$; this finding indicates that the rate constant is less than our lower limit, $\geq 10^{-13}$ cm^3 molecule^{-1} s^{-1}.

The dramatic change in the rate constants in this series of reactions between $BrCF_3$ and $ClCF_3$ is not in keeping with the simple halogen atom transfer because the C–X and P–X bond energies should exhibit a closer parallel relationship. However, this sudden break would be expected if the reaction mechanism involves initial electron transfer and the electron affinity (A_e) of $ClCF_3$ is too small to allow this step to occur. We suggest that this is the correct explanation and that the reactions of the phosphoryl anion with ICF_3 and $BrCF_3$ yield either $[(CH_3O)_2P(=O)\cdot X^-/\cdot CF_3]$ or $[(CH_3O)_2P(=O)\cdot/\cdot -XCF_3]$ as collision complexes following electron transfer (/ is the termolecular, and /– the bimolecular ion-dipole complex in the gas phase). The dimethyl phosphoryl radical could undergo radical β-fragmentation with loss of a methyl radical; methyl metaphosphate (CH_3OPO_2) is produced to which I^- or Br^- add to yield the major product ions.

$$(CH_3O)_2PO^- + BrCF_3 \longrightarrow \begin{cases} (CH_3O)(Br)PO_2^- + CF_3CH_3 \quad (m/z\ 173,\ 175) \\ Br^- + (CH_3O)_2PO\cdot + \cdot CF_3 \end{cases} \tag{17}$$

The results and this interpretation suggest that electron transfer to ICF_3 ($A_e = 1.57 \pm 0.2$ eV) (5) from $(CH_3O)_2PO^-$ is thermoneutral or slightly exothermic and thus may occur at a relatively longer distance of approach than is present in the collision complex. Such would account for the larger amount of I^- observed in this reaction. This result would require that electron transfer from $(CH_3O)_2PO^-$ to $BrCF_3$ ($A_e = 0.91 \pm 0.2$ eV) (5) is endothermic and would be allowed only in the attractive anion–neutral collision complex, which would have a well depth of 10–20 kcal/mol. The follow-up reactions of Br^- transfer and dimethylation must then be sufficiently exothermic to allow for the observed products in an overall fast reaction. That these processes, including the electron transfer, may only occur in the collision complex would account for the fact that only a trace of Br^- is observed. Although the A_e of $ClCF_3$ has not been reported, $A_e(ClCF_3)$ must be below the lower limit for endothermic electron transfer to occur within its collision complex with $(CH_3O)_2PO^-$.

Conclusions

The results of this research produce the following conclusions.

1. $O_2^{\cdot -}$ is kinetically and intrinsically a powerful S_N2 nucleophile in its reactions with CH_3X molecules. In most respects, this conclusion agrees with studies of S_N2 reactions with $O_2^{\cdot -}$ in the condensed phase (*20*).
2. An intrinsic scale of the reactions of carbonyl centers of organic molecules with nucleophiles has been established through the use of $C_6H_5N^{\cdot -}$. Although the lower limit with $C_6H_5N^{\cdot -}$ is the simple esters, HCO_2CH_3 and $CH_3CO_2CH_3$, this finding can now be extended to less reactive carbonyl centers by using other hypovalent anion radicals including $(C_6H_5)_2C^{\cdot -}$.
3. The dimethyl phosphoryl anion $[(CH_3O)_2PO^-]$ was readily prepared by H^+ transfer from $(CH_3O)_2P({=}O)H$ to a number of anionic bases. $A_p[(CH_3O)_2PO^-] = 358 \pm 2$ kcal/mol was determined. If protonation at phosphorus of the anion by H^+ donors was assumed, $\Delta H_f^\circ[(CH_3O)_2PO^-] = -207.3 \pm 2$ kcal/mol was calculated. Although $(CH_3O)_2PO^-$ is kinetically a poor S_N2 nucleophile with CH_3X molecules, $(CH_3O)_2PO^-$ reacts rapidly with XCF_3 where X = I and Br to form principally $(CH_3O)(X)PO_2^-$ negative ions. No reaction was observed with $ClCF_3$. The reactions with ICF_3 and $BrCF_3$ are considered to involve initial electron transfer between the phosphoryl anion and the neutral substrate. The A_e $(ClCF_3)$ must be too low to allow electron transfer to occur.

Acknowledgments

We are grateful to the National Science Foundation for support of this research and to Sam McManus and Milton Harris for the invitation to participate in the symposium.

Literature Cited

1. McDonald, R. N.; Chowdhury, A. K.; Setser, D. W. *J. Am. Chem. Soc.* **1980,** *102,* 6491–6498.
2. McDonald, R. N.; Chowdhury, A. K. *J. Am. Chem. Soc.*, **1983,** *105,* 7267–7271.
3. McDonald, R. N.; Chowdhury, A. K. *J. Am. Chem. Soc.* **1985,** *107,* 4123–4128.
4. Bohme, D. K.; Young, L. B. *J. Am. Chem. Soc.* **1970,** *92,* 3301–3309.

5. Drazaic, P. S.; Jeffrey, M.; Brauman, J. I. In *Gas Phase Ion Chemistry;* Bowers, M. T., Ed.; Academic Press: New York, 1984; Vol. 3, Chapter 21.
6. Oakes, J. M.; Ellison, G. B. *J. Am. Chem. Soc.* **1984,** *106,* 7734–7741.
7. Tanaka, K.; Mackay, G. I.; Payzant, J. D.; Bohme, D. K. *Can J. Chem.* **1976,** *54,* 1643–1659.
8. Pellerite, M. J.; Brauman, J. I. *J. Am. Chem. Soc.* **1983,** *105,* 2672–2680.
9. McDonald, R. N.; Chowdhury, A. K.; Setser, D. W. *J. Am. Chem. Soc.* **1981,** *103,* 6599–6603.
10. McDonald, R. N.; Chowdhury, A. K. *J. Am. Chem. Soc.* **1983,** *105,* 198–207.
11. W. Y. Gung, M.S. Thesis, Kansas State University, 1984.
12. Bartmess, J. E.; McIver, R. T. In *Gas Phase Ion Chemistry;* Bowers, M. T., Ed.; Academic Press: New York, 1979; Vol. 2, Chapter 11.
13. Rossi, M. J.; McMillen, D. F.; Golden, D. M. *J. Phys. Chem.* **1984,** *88,* 5031–5039.
14. Russell, G. A.; Danen, W. C. *J. Am. Chem. Soc.* **1966,** *88,* 5663–5665.
15. Kornblum, N.; Michel, R. E.; Kerber, R. C. *J. Am. Chem. Soc.* **1966,** *88,* 5662–5663.
16. Bunnett, J. F. *Acc. Chem. Res.* **1978,** *11,* 413–420, and references cited therein.
17. Bunnett, J. F.; Creary, X. *J. Org. Chem.* **1974,** *39,* 3612–3614.
18a. K. D. DeWitt, M.S. Thesis, Kansas State University, 1986.
18b. Pietro, W. J.; Hehre, W. J. *J. Am. Chem. Soc.* **1982,** *104,* 3594–3595.
19. Benson, S. W. *Thermochemical Kinetics,* 2nd ed.; Wiley: New York, 1976.
20. Sawyer, D. T.; Valentin, J. S. *Acc. Chem. Res.* **1981,** *14,* 393–400, and references cited therein.

RECEIVED for review October 21, 1985. ACCEPTED January 27, 1986.

5

Comparative Behavior of Nucleophiles in Gas and Solution Phase: Acylation, Alkylation, and Phosphorylation

Marjorie C. Caserio and Jhong K. Kim

Department of Chemistry, University of California, Irvine, CA 92717

The present study describes the reactions of neutral alcohols with protonated carboxylic, carbonate, and organophosphorus acids or derivatives generated as gaseous ions under ion cyclotron resonance conditions. Evidence is presented on the mechanisms of these gas-phase ion–molecule reactions, which formally resemble acid-catalyzed acylation and phosphorylation commonly observed in solution. Gas-phase acylation appears to be a displacement process involving acylium ion transfer. Phosphorylation was not observed as such, but ions corresponding to dimethyl metaphosphate reacted rapidly with methanol in an exchange process.

ACYLATION AND PHOSPHORYLATION OF NUCLEOPHILES are among the most important and widely studied reactions in chemistry. Formation and hydrolysis of phosphate esters and carboxylic acid derivatives are of great importance in both organic and bioorganic processes and, in fact, are vital to the functioning of living systems. The focus of this paper is on the behavior of *neutral* nucleophiles, mostly alcohols, with *gaseous cations* derived from oxyacids of carbon and phosphorus in reactions that resemble, at least superficially, esterification and hydrolysis commonly observed in condensed phase. The objective of studying such reactions in the gas phase is to compare them wherever possible to the corresponding processes in solution; thereby, something is learned about the role of solvent and counterions in moderating the reactions of ions with neutral molecules.

Generation of Ions

In this study, gaseous organic cations were generated by the technique of ion cyclotron resonance (ICR) spectroscopy, which is a form of mass spectroscopy

0065-2393/87/0215-0065$06.00/0

(*1–5*). Ions are produced by electron impact (from a heated filament) on a neutral sample at low pressure (10^{-7} torr). The ions enter the ICR cell under the influence of external electric and magnetic fields, which constrain them to move in circular orbits with a cyclotron frequency ω that is dependent on the mass (m) and charge (z) of the ion and on the strength of the applied magnetic field H:

$$\omega = zH/mc \tag{1}$$

The ions are trapped within the cell by applying appropriate voltages to the cell plates: thereby, reactive encounters can occur between ions and neutral molecules:

$$X^+ + M \rightarrow Y^+ + N \tag{2}$$

Both reactant and product ions can be detected in a manner resembling the detection of magnetic nuclei in an NMR experiment. The ions are exposed to a radiofrequency field ω_{rf} from an external oscillator, and while sweeping the magnetic field H, the ions absorb energy from the applied field when the resonance condition of equation 1 is satisfied (i.e., $\omega = \omega_{rf}$).

The progress of reaction can be monitored as a function of time and ion intensity (Figures 1–3) or as a mass spectrum at a particular time interval (Figure 4–6). The technique is limited to the detection of ions only, and because of the low sample pressures, reactions that are endothermic or have high activation energies are not generally observed.

Acylation

Proton transfers are among the most rapid of ion–molecule reactions in the gas phase. Depending on the relative proton affinities of the reactant and product neutrals (M and A in equation 3), an acidic fragment cation commonly transfers a proton to the neutral parent to form MH^+.

$$\left.\begin{array}{ll} M \xrightarrow{eV} AH^+ & \text{primary fragment ions} \\ AH^+ + M \rightarrow MH^+ + A & \text{secondary ions} \end{array}\right\} \tag{3}$$

As an example, acetyl derivatives usually fragment under electron impact to produce acetylium ions, CH_3CO^+, which, in turn, react with the neutral parent:

$$CH_3COX \xrightarrow{eV} CH_3CO^+ \xrightarrow{CH_3COX} [CH_3COX]H^+ + CH_2{=}C{=}O \tag{4}$$

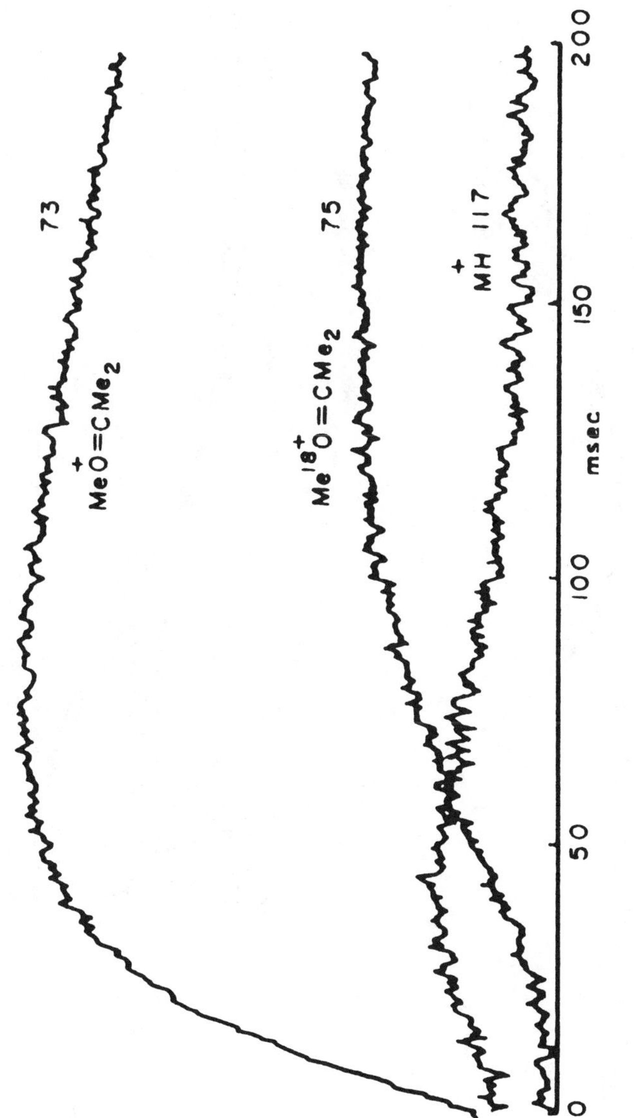

Figure 1. Time plot of ion intensity in the 19-eV ICR spectrum of isopropenyl methyl carbonate at 8×10^{-7} torr with $Me^{18}OH$ at 1×10^{-6} torr. The numbers are m/z *values. The horizontal scale represents 200 ms. Only the major ions present are shown.*

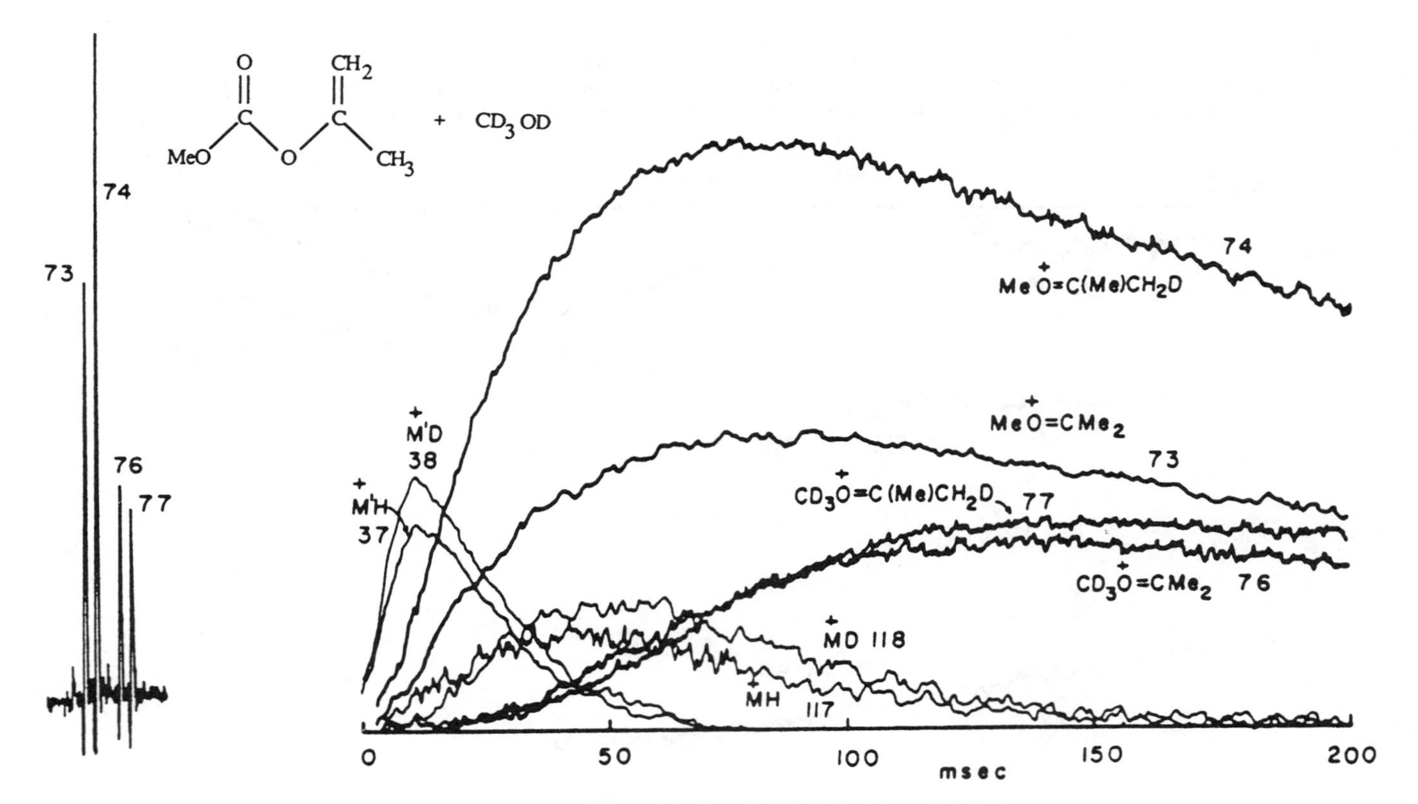

Figure 2. As in Figure 1 but with CD_3OD at 1×10^{-6} torr.

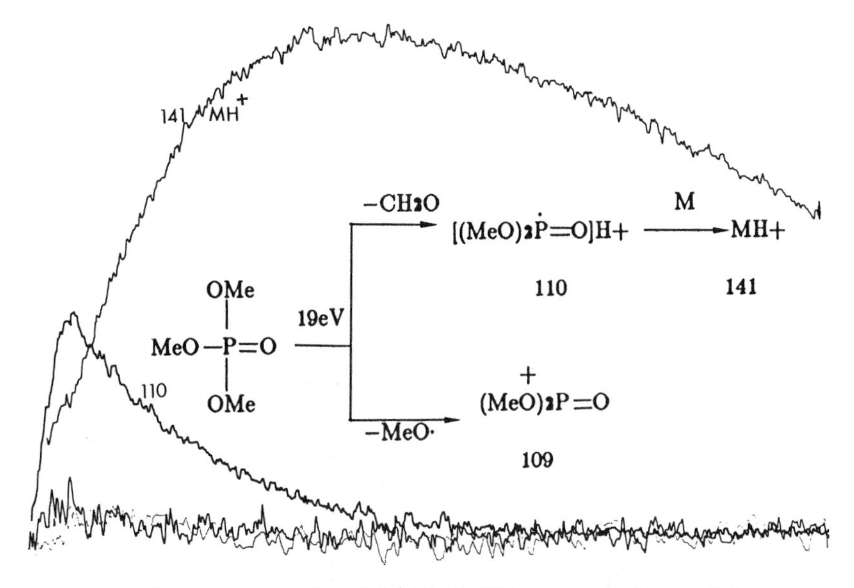

Figure 3. Time plot of the 19-eV ICR spectrum of trimethyl phosphate. The horizontal scale is 200 ms. Only the major ions are shown.

The question of interest is whether the ions produced in this manner behave like their counterparts in solution, which are the conjugate acids of the parent carboxyl derivative. In particular, do they react with added nucleophiles to transfer the acyl group in a manner related to acid-catalyzed acyl transfer commonly observed in condensed phase?

The most prevalent pathway for acyl transfer catalyzed by acids, bases, or enzymes in condensed phase is an addition–elimination sequence involving the formation of tetrahedral intermediates, **I** (Scheme I) (*6*). A similar

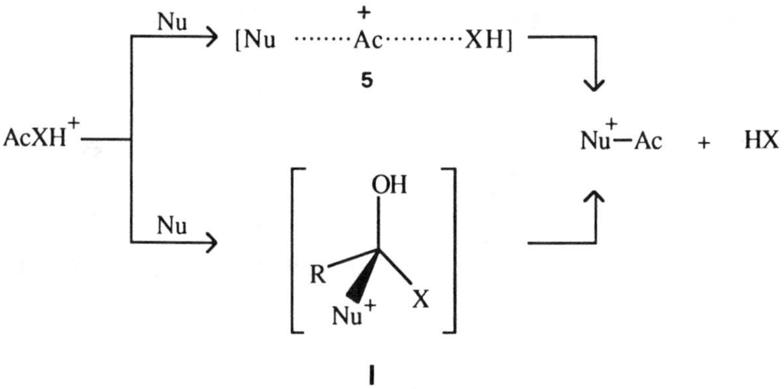

$$\mathrm{AcXH^+} \xrightarrow{\mathrm{Nu}} [\mathrm{Nu} \cdots \overset{+}{\mathrm{Ac}} \cdots \mathrm{XH}] \;(\mathbf{5}) \longrightarrow \mathrm{Nu^+{-}Ac} + \mathrm{HX}$$

$$\mathrm{AcXH^+} \xrightarrow{\mathrm{Nu}} \left[\mathrm{R(OH)(X)C{-}Nu^+}\right] \;(\mathbf{I}) \longrightarrow \mathrm{Nu^+{-}Ac} + \mathrm{HX}$$

Scheme I. Acyl transfer.

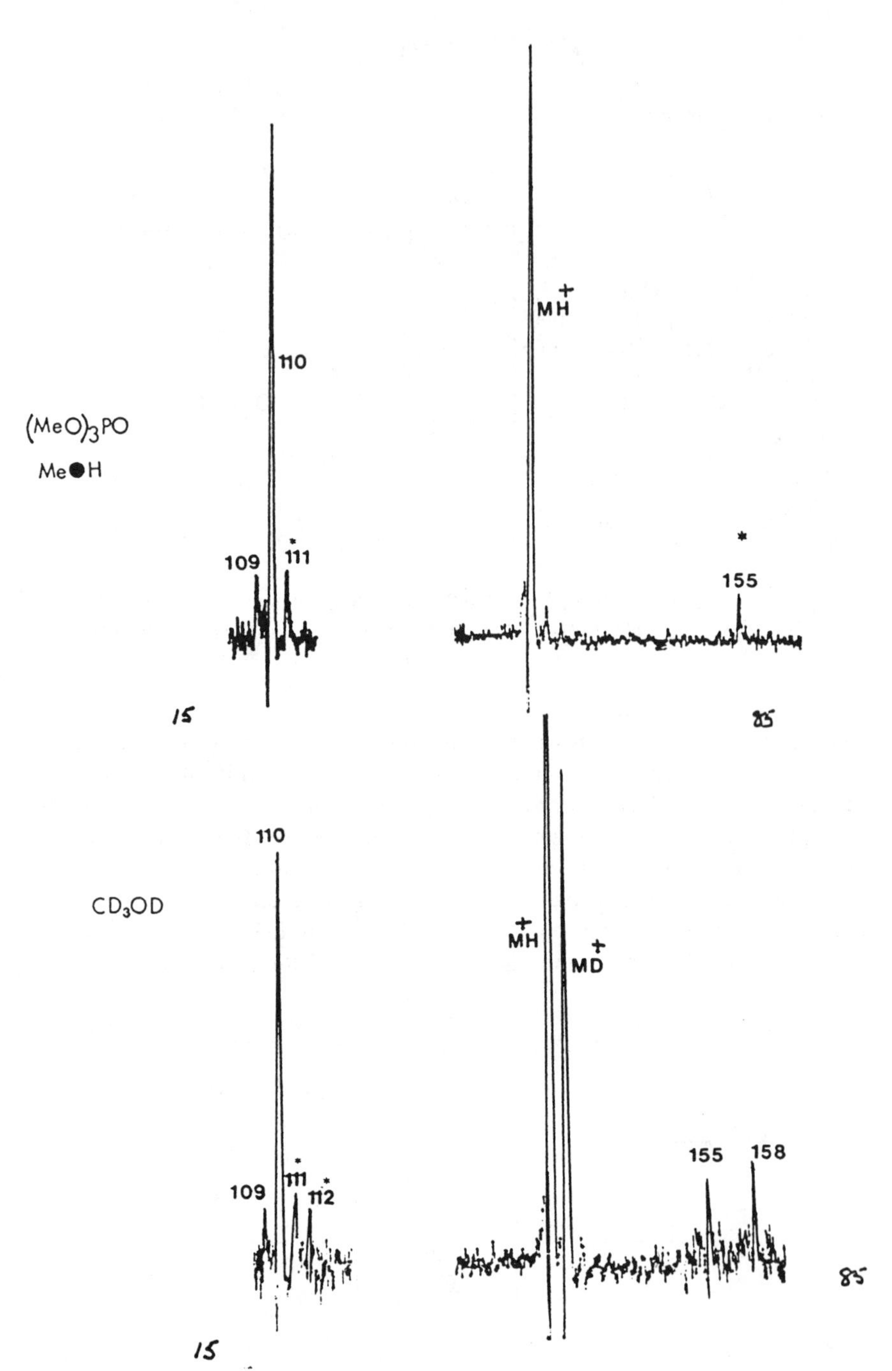

Figure 4. ICR mass spectrum of trimethyl phosphate in the presence of $Me^{18}OH$ (top) and CD_3OD (bottom).

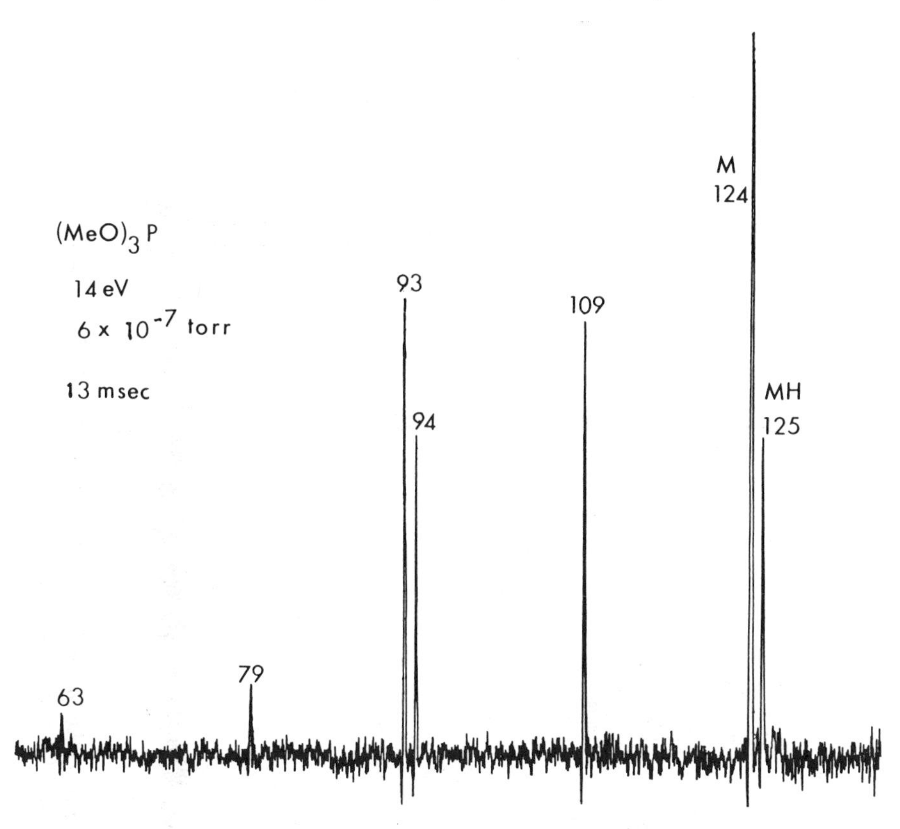

Figure 5. ICR mass spectrum of trimethyl phosphite.

process has been found in the gas-phase acylation of anionic nucleophiles (*7–9*), but the acylation of neutral neucleophiles by gaseous cations appears to take a different pathway (*10–14*).

Thus, in the course of studying the reactions of acetyl derivatives with various nucleophiles under ICR conditions, in no instance have we observed cleavage of the carbonyl C–O bond (*10, 11*). This finding is surprising because, if a tetrahedral intermediate **I** is formed, it would be expected to partition among possible routes corresponding to C–O, C–X, and C–Nu cleavage in proportion to the relative energies of the products, yet only C–X or C–Nu cleavage occurs even when C–O cleavage is energetically favored.

Take, for example, the reaction of methanol or methanethiol with protonated thiolacetic acid (Scheme II). The observed product of acyl transfer is **II** corresponding to loss of H_2S. If **II** is formed by way of an addition intermediate, then formation of the intermediate by an independent route should give the same product **II**. However, reaction of the thionic ester **III** with water failed to give **II** or any product that could be ascribed to the intervention of a tetrahedral addition intermediate **I** (*15*).

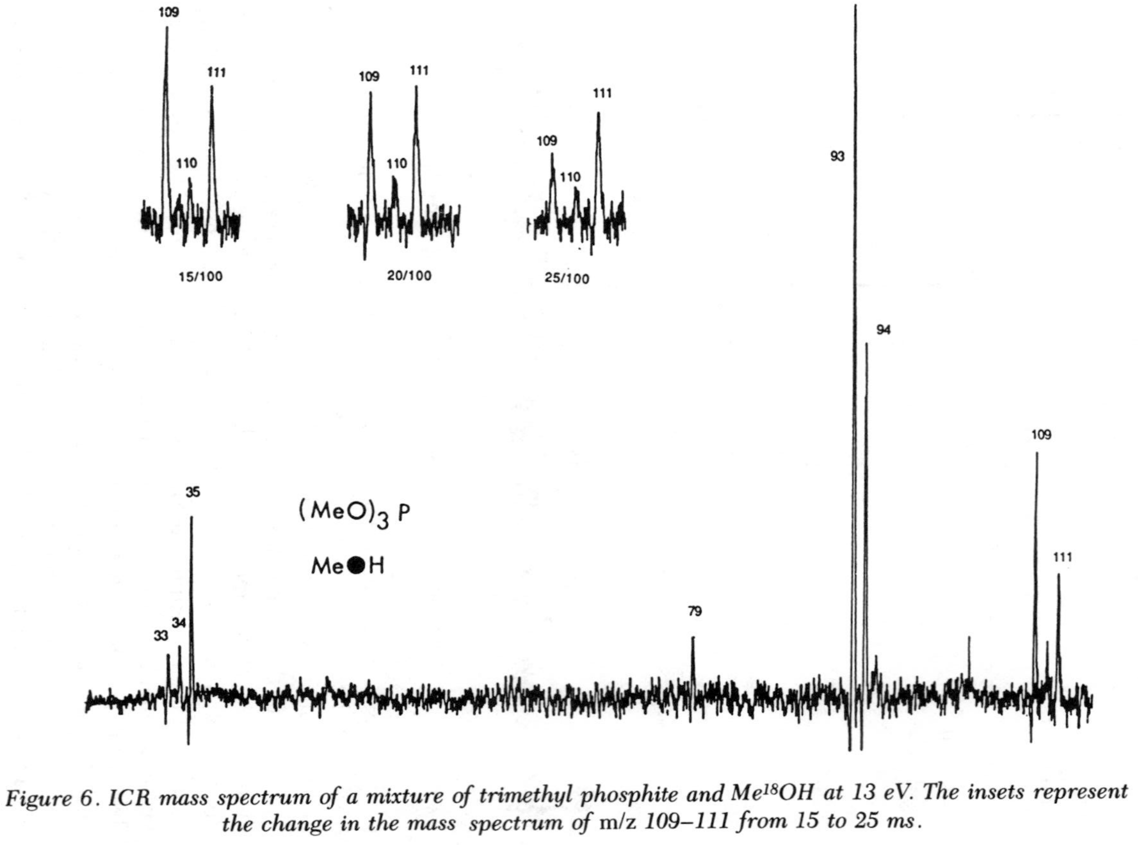

Figure 6. ICR mass spectrum of a mixture of trimethyl phosphite and Me^{18}*OH at 13 eV. The insets represent the change in the mass spectrum of* m/z *109–111 from 15 to 25 ms.*

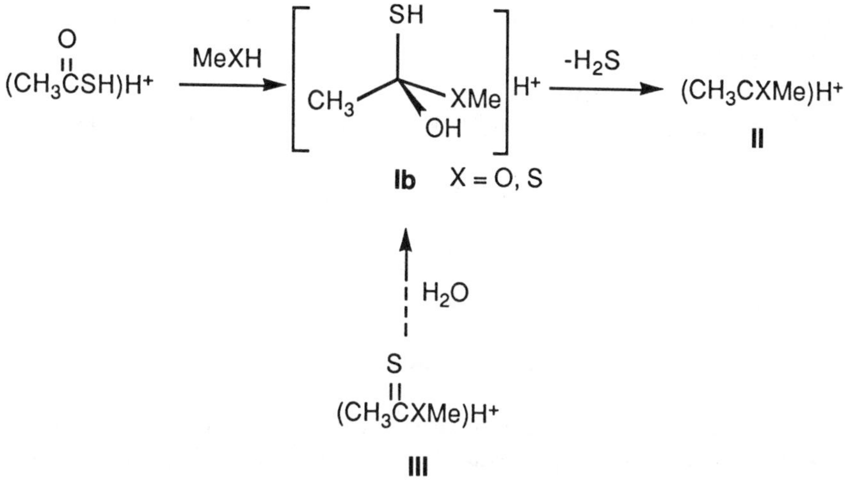

$$(CH_3\overset{O}{\overset{\|}{C}}SH)H^+ \xrightarrow{MeXH} \left[CH_3C(SH)(OH)XMe \right]H^+ \xrightarrow{-H_2S} (CH_3CXMe)H^+$$

Ib X = O, S; II

$$(CH_3\overset{S}{\overset{\|}{C}}XMe)H^+ \xrightarrow{H_2O} \text{Ib}$$

III

Scheme II

Acylation of sulfide and ether nucleophiles by protonated acyl derivatives is uncommon in condensed phase; but, in the gas phase, ions of composition Ac^+XMe_2 are readily formed. For example, vinyl and isopropenyl acetates acylate ethers under ICR conditions (*11*):

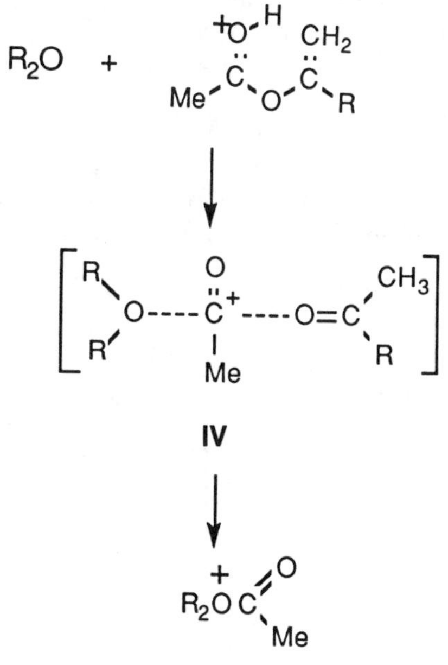

$$R_2O + Me\overset{\overset{+}{O}-H}{\overset{\vdots}{C}}-O-\overset{CH_2}{\overset{\vdots}{C}}-R \longrightarrow \left[R_2O\text{----}\overset{O}{\overset{\|}{C}}{}^{+}(Me)\text{----}O{=}C(CH_3)R \right] \longrightarrow R_2\overset{+}{O}C({=}O)Me \qquad (5)$$

IV; V

The product ions have the oxonium ion structure **IV** based on the observation that they acylate the parent neutral and because the alternative dialkoxy structure **VI**, when generated independently, as in the EI cleavage of orthoesters or protonation of ketene acetals, does not acylate added nucleophiles:

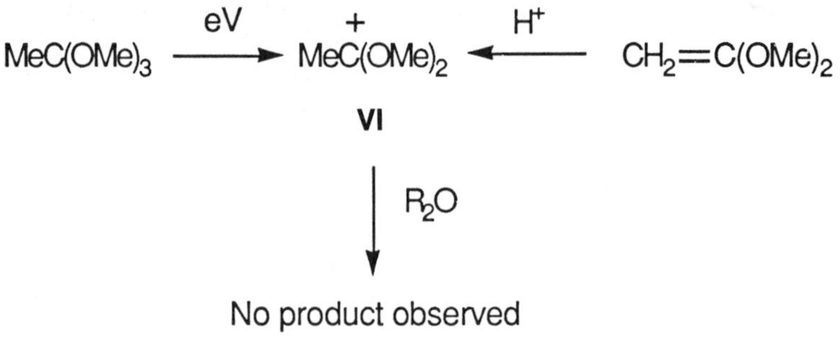

$$MeC(OMe)_3 \xrightarrow{eV} \overset{+}{MeC}(OMe)_2 \ (\mathbf{VI}) \xleftarrow{H^+} CH_2{=}C(OMe)_2$$

$$\mathbf{VI} \xrightarrow{R_2O} \text{No product observed} \quad (6)$$

On thermodynamic grounds, acylation by way of **I** is considered unlikely because the addition step is estimated to be *endothermic* by about 20 kcal/mol. However, acylation by a displacement pathway involving an acylium ion complex **V** appears energetically favorable because of energy gained by association of the acylium ion with two nucleophilic "solvent" molecules. The association energy may be as high as 20–40 kcal/mol. This combined with the experimental fact that gaseous acylation preserves the integrity of the acyl C–O bond leads us to conclude that acylation does not occur by an addition–elimination sequence but rather by a displacement route involving acylium ion complexes of the type $R_2Y \cdots Ac^+ \cdots XR_2$.

Carbonate Esters

The chemistry of neutral nucleophiles with protonated carbonate esters differs in interesting ways from that of related carboxylate esters. In the first place, acyl transfer of methoxycarbonyl, $MeOCO^+$, to added nucleophiles does *not* occur (*16*):

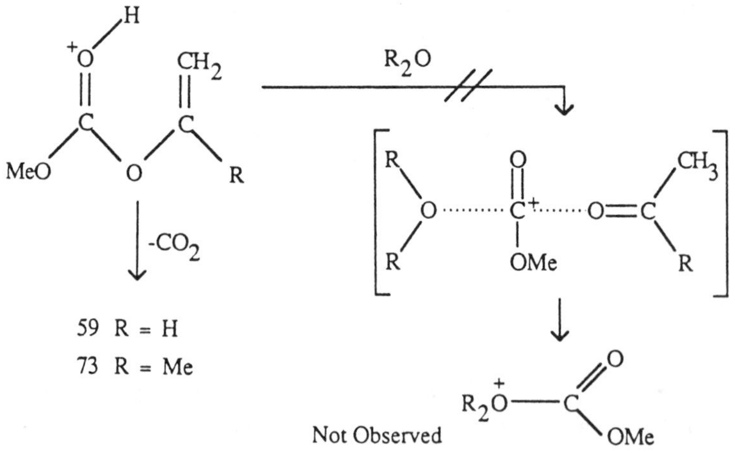

$MeO{-}C({=}\overset{+}{O}H){-}O{-}C({=}CH_2){-}R \xrightarrow{R_2O}$ (not formed) $[R_2O \cdots \overset{+}{C}({=}O)(OMe) \cdots O{=}C(CH_3)R] \rightarrow R_2\overset{+}{O}{-}C({=}O)OMe$ Not Observed (7)

$\downarrow -CO_2$

59 R = H

73 R = Me

The dominant product ions from mixtures of methanol with methyl vinyl carbonate, methyl isopropenyl carbonate, and methyl *tert*-butyl carbonate are *m/z* 59, 73, and 73, respectively. In the case of the vinylic esters, ions *m/z* 59 and *m/z* 73 are each formed by two independent routes, decarboxylation and condensation of the protonated ester with methanol. Using deuterium and O-18-labeled methanol, we were able to unravel the reaction pathways involved.

To illustrate, Figure 1 shows the change in ion intensity for the reaction of 98% O-18-labeled methanol with isopropenyl acetate. The appearance of *m/z* 75 means that the oxygen from the alcohol is incorporated in the product ion—the precursor ion being the protonated ester *m/z* 117. Yet, an abundance of the *unlabeled* ion *m/z* 73 exists that is clearly formed at a faster rate by a pathway that is different from that producing the O-18-labeled ion, *m/z* 75. The key to both processes is a step involving proton transfer to the vinylic carbon of the ester; *m/z* 73 arises from dissociation of the C-protonated ester (Scheme III), and *m/z* 78 arises from the condensation of the C-protonated ester with methanol (Scheme IV).

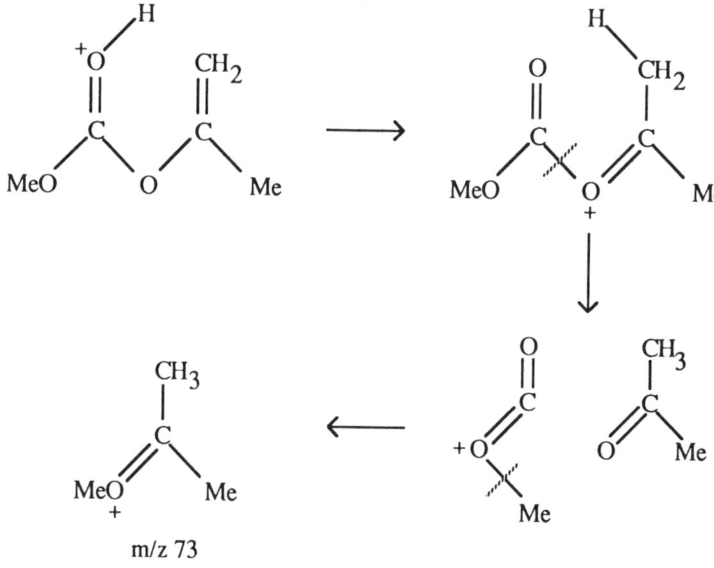

m/z 73

Scheme III. Isopropenyl methyl carbonate: pathways to m/z *73.*

Support for Schemes III and IV comes from the reaction of isopropenyl acetate with CD_3OD (Figure 2). The major product ions, besides MH^+ and MD^+, are *m/z* 73, 74, 76, and 77, corresponding respectively to decarboxylation of MH^+ (*m/z* 117) and MD^+ (*m/z* 118) and condensation of CD_3OD with MH^+ and MD^+.

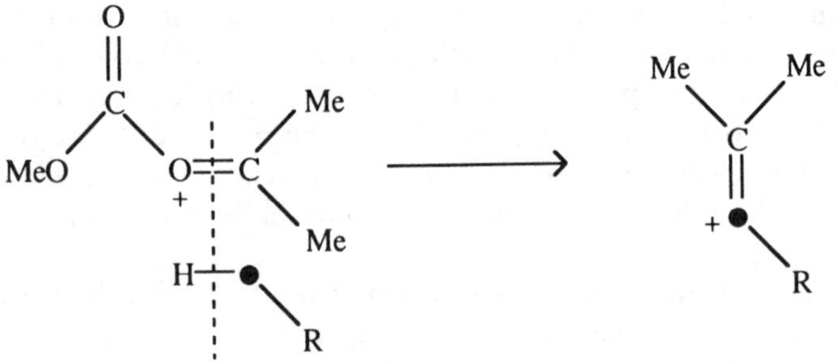

O C MeO O=C Me Me + H R → Me Me C + R

R^{18}●H 75

CD_3OD 76, 77

Scheme IV. Pathways to m/z *73, 75, 76, and 77.*

Moreover, when the C-protonated ester *m/z* 117 is generated independently by EI cleavage of *tert*-butyl methyl carbonate in the presence of CD_3OD, *m/z* 73 and 76 are formed, as expected, from independent decarboxylation and condensation reactions (Scheme V).

As mentioned, carbonate esters, unlike acetate esters, do not acylate neutral nucleophiles to form ions of the type $AcNu^+$. Some additional differences between the ion chemistry of methyl acetate and dimethyl carbonate are noteworthy. Acyl cations RCO^+ are major fragment ions from both

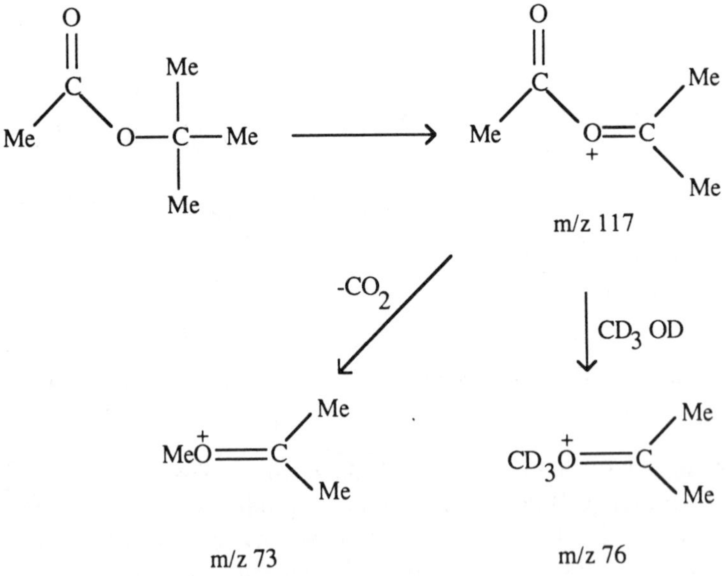

O C Me O—C—Me Me Me → O C Me O=C Me Me +

m/z 117

$-CO_2$

CD_3OD

$Me\overset{+}{O}=C(Me)_2$

$CD_3\overset{+}{O}=C(Me)_2$

m/z 73

m/z 76

Scheme V. tert-*Butyl methylcarbonate cleavage.*

esters; but, the acetylium ion (R = Me) from acetates is acidic and rapidly protonates the parent ester whereas the methoxycarbonyl ion (R = MeO) from dimethyl carbonate is a methylating agent that reacts with the parent ester to form MCH_3^+ (*m/z* 105):

$$(MeO)_2C{=}O \xrightarrow{eV} MeOC{=}\overset{+}{O} \quad m/z\ 59 \tag{8}$$

$$Me\overset{+}{OC}{=}O + (MeO)_2C{=}O \rightarrow (MeO)_3^+ + CO_2 \quad m/z\ 105$$

Another methylation product is Me_3O^+, *m/z* 61. Formation of *m/z* 61 is best explained as the result of methyl transfer to the *ether* oxygen of the parent ester, rather than to the *carbonyl* oxygen. The ion–molecule complex thus formed may be expected to decarboxylate to give *m/z* 61 (*see also* Scheme III):

$$\underset{\mathbf{X}}{MeO{-}C({=}O){-}OMe} \longrightarrow \underset{m/z\ 59}{Me\overset{+}{OC}{=}O} \xrightarrow{(MeO)_2C=O} \underset{m/z\ 105}{\left[Me_2\overset{+}{O}{\cdot}\overset{O}{\overset{\|}{C}}{\cdot}OMe\right]} \tag{9}$$

$$\underset{m/z\ 61}{Me_3O^+} \xleftarrow{-CO_2} \underset{\mathbf{XI}}{\left[Me_2O{-}{-}{-}Me\overset{+}{OC}{=}O\right]}$$

To our knowledge, comparable methylation reactions of methyl carbonate esters in condensed phase have not been reported (*17, 18*).

Phosphorylation

The two major pathways for phosphorylation reactions in condensed phase are strikingly similar to those for the corresponding acylation reactions (*19*). One pathway is an addition–elimination process whereby a nucleophile *adds* to the phosphoryl group to give a pentacoordinate intermediate that collapses to product by *elimination* of a nucleophile (mechanism A, Scheme VI). The other pathway is a displacement process in which the phosphoryl group is transferred as tricoordinate *metaphosphate* to the attacking nucleophile (mechanism B, Scheme VI). Numerous studies have documented the intervention of metaphosphate in phosphorylation reactions in condensed phase, although metaphosphate has eluded direct detection (*19–24*). The purpose of the ICR study reported here was to see if gas-phase phosphorylation can be achieved and whether metaphosphate intermediates are involved.

MECHANISM A

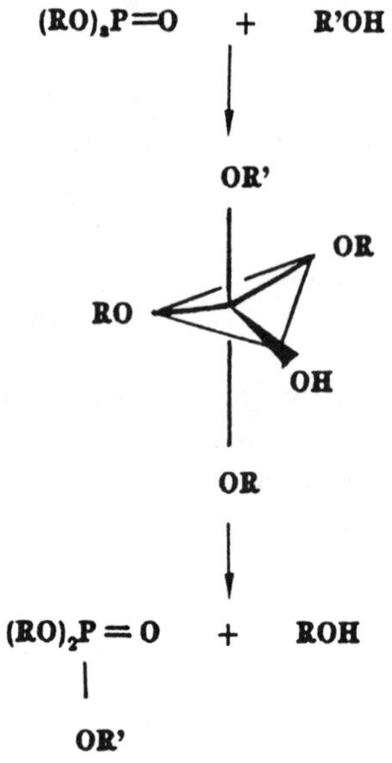

MECHANISM B

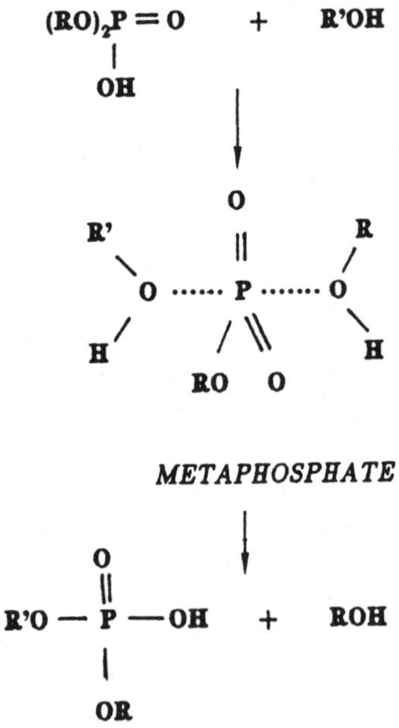

Scheme VI. Phosphoryl transfer.

ICR studies of phosphorus esters have been reported (25–27). However, phosphorylation of neutral nucleophiles is not well documented. Accordingly, we examined the positive-ion chemistry of alcohols with phosphorus esters.

The dominant chemistry of trimethyl phosphate under ICR conditions is the formation of MH^+ (*m/z* 141) by proton transfer to the parent ester from the fragment ion *m/z* 110 (Figure 3). In the presence of 98% O-18-labeled methanol, *no* products were formed that could be ascribed to methanol-exchange phosphorylation of the type

$$(MeO)_3P{=}OH^+ + MeO^*H \rightarrow (MeO)_2P(O^*Me){=}OH^+ + MeOH$$

However, a minor fragment ion *m/z* 109 underwent rapid methoxyl exchange as evidenced by the appearance of *m/z* 111 (from O-18-labeled methanol), and *m/z* 112 (from CD_3OD) after only 15 ms. At longer reaction times, methyl transfer to M gave *m/z* 155 (from *m/z* 109 and 111), and *m/z* 158 (from *m/z* 112) (Figure 4). These results suggest that *m/z* 109 is the tricoordinate dimethyl metaphosphate cation VII that reacts according to Schemes VII and VIII.

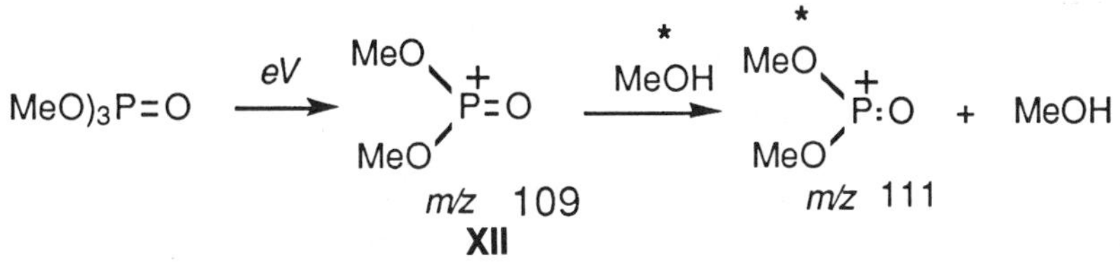

$$MeO)_3P{=}O \xrightarrow{eV} (MeO)_2\overset{+}{P}{=}O \; (m/z\ 109,\ \mathbf{XII}) \xrightarrow{MeO^*H} (MeO^*)(MeO)\overset{+}{P}{:}O \; (m/z\ 111) + MeOH$$

Scheme VII

$$(MeO)_2\overset{+}{P}{=}O \; (\mathbf{7},\ 109) \xrightarrow{CD_3OD} (MeO)(CD_3O)\overset{+}{P}{=}O \; (112) + MeOD$$

$$109 \xrightarrow{M} (MeO)_4\overset{+}{P} \; (155)$$

$$112 \xrightarrow{M} (MeO)_3\overset{+}{P}{-}OCD_3 \; (158)$$

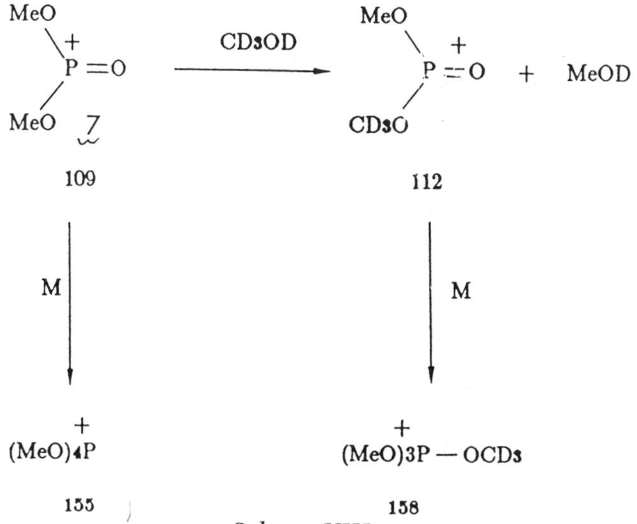

Scheme VIII

On the basis of thermochemical calculations (28, 29) addition of methanol to dimethyl metaphosphate is likely to be exothermic by at least 26 kcal/mol, but addition is unlikely to lead to a stable adduct under the low-pressure conditions of the ICR experiment. Indeed, no adduct was observed, and its formation can only be inferred from the appearance of dissociation products of OMe exchange (Scheme VI).

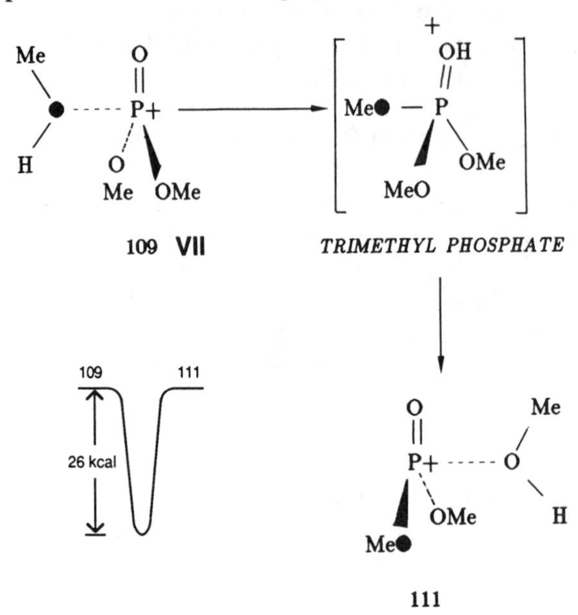

Scheme VI. Addition of methanol to dimethyl metaphosphate.

So that the methanol exchange reaction of Schemes VII and VIII could be verified, a source of m/z 109 in greater abundance was needed. Attempts to generate m/z 109 by EI cleavage of dimethyl chlorophosphonate, $(MeO)_2P(O)Cl$, and dimethyl hydrogen phosphonate, $(MeO)_2P(O)H$, were unsuccessful. However, m/z 109 is formed in significant abundance from fragmentation of trimethyl phosphite (equation 10) (Figure 5) (27).

$$(MeO)_2P\text{–}OMe \xrightarrow{eV} \underset{m/z\ 109}{(MeO)_2P^+{=}0} \qquad (10)$$

Reassuringly, m/z 109 from the *phosphite* ester behaved in a similar manner to m/z 109 from the *phosphate* ester. In the presence of O–18–labeled methanol, the fragment ion m/z 109 rapidly produced m/z 111 (Figure 6), and both ions were consumed within 85 ms by subsequent reactions (methyl transfer).

That m/z 109 reacts rapidly with methanol whereas the structurally related ion m/z 110 is unreactive (except in proton transfer) may appear to be surprising. However, m/z 110 is a radical cation, and after analysis of the exchange process in terms of single electron shifts according to the method of Pross (*30*), group coupling to form a bond to phosphorus is favorable only with m/z 109.

$$\text{Nu:} + \overset{+}{\text{P}}{=}\text{O} \rightarrow \overset{\uparrow}{\text{Nu}}+ \; +\overset{\downarrow}{\text{P}}{=}\text{O} \rightarrow \text{Nu}{-}\overset{+}{\text{P}}{=}\text{O}$$

$$\text{Nu:} + \overset{\uparrow}{\text{P}}{=}\text{OH}+ \rightarrow \overset{\uparrow}{\text{Nu}}+ \; +\overset{\downarrow\uparrow}{\text{P}}{-}\text{OH} \not\rightarrow \text{Nu}{-}\overset{+\;\uparrow}{\text{P}}{-}\text{OH}$$

If our interpretation of the results is correct, it is amusing to reflect on the contrast between gas-phase and solution-phase phosphorylation. In solution, the observables are tetracoordinate orthophosphates with metaphosphate as an inferred intermediate. In the gas phase under ICR conditions, the reverse is true: metaphosphates are the observables, and orthophosphate is the intermediate.

Alkylation

Certain alcohols, notably tertiary alcohols, are not readily acylated or phosphorylated in the condensed phase under conditions that normally succeed for primary alcohols. Likewise, tertiary alcohols or thiols are not acylated or phosphorylated in the gas phase. Alkylation is the more general reaction. For example, *tert*-butyl alcohol under ICR conditions condenses with protonated carbonyl and phosphoryl compounds to produce ions of the type $X{=}O^{+}Bu_{t}$, where X = C or P. The process has been described previously as a displacement reaction of the type shown in equation 11 (*10, 16*).

$$X{=}OH + t\text{-}BuOH \rightarrow [X{=}O \cdots \overset{+}{Bu} \cdots OH_2] \rightarrow X{=}\overset{+}{O}Bu_t + H_2O \qquad (11)$$

However, tert-butylation of phosphoryl oxygen is notably slower than butylation of carbonyl oxygen.

Literature Cited

1. Beauchamp, J. L. *Annu. Rev. Phys. Chem.* **1971,** *22,* 527.
2. Baldeschwieler, J. D. *Science (Washington, D.C.)* **1968,** *159,* 263.
3. McIver, R. T., Jr. *Rev. Sci. Instrum.* **1977,** *49,* 111.

4. McIver, R. T., Jr. *Rev. Sci. Instrum.* **1970,** *41,* 555.
5. Lehman, T. A.; Bursey, M. M. *Ion Cyclotron Resonance Spectrometry;* Wiley: New York, 1976.
6. Euranto, E. K. In *The Chemistry of Carboxylic Acids;* Patai, S., Ed.; Interscience; New York, 1969; Chapter 11.
7. Nibbering, N. M. M. *NATO Adv. Study Inst. Ser., Ser. B* **1979,** *40,* 165.
8. McDonald, R. N.; Chowdhury, A. K.; Gung, W. Y.; DeWitt, K. D. *Abstracts of Papers,* 190th National Meeting of the American Chemical Society; Sept. 8–13, Chicago; American Chemical Society: Washington, DC, 1985; ORGN 133 (see also accompanying paper in this volume).
9. Takashima, K.; Riveros, J. M. *J. Am. Chem. Soc.* **1978,** *100,* 6128.
10. Kim, J. K.; Caserio, M. C. *J. Am. Chem. Soc.* **1982,** *104,* 4624.
11. Kim, J. K.; Caserio, M. C. *J. Am. Chem. Soc.* **1981,** *103,* 2124.
12. Tiedemann, P. W.; Riveros, J. M. *J. Am. Chem. Soc.* **1974,** *96,* 185.
13. McMahon, T. B. *Can. J. Chem.* **1978,** *56,* 670.
14. Beauchamp. J. L. *NATO Adv. Study Inst. Ser., Ser. B.* **1975,** *8,* 418.
15. Caserio, M. C.; Kim, J. K. *J. Am. Chem. Soc.* **1983,** *105,* 6896.
16. Caserio, M. C.; Kim, J. K. *Spectrosc. Int. J.* **1983,** *2,* 207.
17. Shah, A. A.; Connors, K. A. *J. Pharm. Sci.* **1968,** *57,* 283.
18. Tillett, J. G.; Wiggins, D. E. *Tetrahedron Lett.* **1971,** 911.
19. Westheimer, F. H. *Chem. Rev.* **1981,** *81,* 313.
20. Quin, L. D.; Marsi, B. G. *J. Am. Chem. Soc.* **1985,** *107,* 3389.
21. Henchman, M.; Viggiano, A. A.; Paulson, J. F. *J. Am. Chem. Soc.* **1985,** *107,* 1453.
22. Haake, P.; Allen, G. W. *Bioorg. Chem.* **1980,** *9,* 325.
23. Skoog, M. T.; Jencks, W. P. *J. Am. Chem. Soc.* **1983,** *105,* 3356.
24. Rebek, J., Jr.; Gavina, F.; Navarro, C. *J. Am. Chem. Soc.* **1978,** *100,* 8113.
25. Ausbiojo, O. I.; Braumann, J. I. *J. Am. Chem. Soc.* **1977,** *99,* 7707.
26. Hodges, R. V.; Sullivan, S. A.; Beauchamp, J. L. *J. Am. Chem. Soc.* **1980,** *102,* 935.
27. Hodges, R. V.; McDonnell, T. J.; Beauchamp, J. L.. *J. Am. Chem. Soc.* **1980,** *102,* 1327.
28. Guthrie, J. P. *J. Am. Chem. Soc.* **1978,** *100,* 5892.
29. Guthrie, J. P. *J. Am. Chem. Soc.* **1977,** *99,* 3991.
30. Pross, A. *Acc. Chem. Res.* **1985,** *18,* 212.

RECEIVED for review November 6, 1985. ACCEPTED February 28, 1986.

6

Nucleophilic Displacement in the Gas Phase as a Function of Temperature, Translational Energy, and Solvation Number

Michael Henchman[1], Peter M. Hierl[2], and John F. Paulson

Air Force Geophysics Laboratory, Hanscom Air Force Base, MA 01731

Nucleophilic displacement reactions $X^- + CH_3Y \rightarrow CH_3X + Y^-$ were studied in the gas phase [$X^- = OH^-(H_2O)_n$ or $CH_3O^-(CH_3OH)_n$; Y = Cl or Br] under conditions of variable temperature (200–500 K) in a flow reactor and of variable translational energy in a beam apparatus. In both cases, the solvation number of the ionic reactant was varied: $0 \leq n \leq 3$. Topics treated include the competition between nucleophilic displacement and proton transfer, the use of solvate as a stereochemical marker to probe mechanism, and the comparison between the gas phase and the solution of the reaction thermodynamics and kinetics.

NUCLEOPHILIC DISPLACEMENT OFFERS EXCITING OPPORTUNITIES to relate reactivity in the gas phase to reactivity in solution—perhaps to a greater extent than any other reaction type (*1–4*). The nature of that relationship, and the reasons for it, are developed in this and several other chapters in this volume, for example, by Brauman, Caserio, and McDonald. Exploring the relationship is not just an academic exercise: this relationship is clearly fundamental to any general treatment of reactivity. Reaction in the gas phase reveals intrinsic or solvent-free reactivity (*5*). In different solutions, this intrinsic reactivity is variously transformed, through the differing action of the solvents. Viewed thus, the intrinsic reactivity is the element common to each, and as such, it plays the central role.

Study of organic reactions in the gas phase should therefore reveal much of interest to the organic chemist working in solution. This aim has guided our own studies. Unfortunately, relating the gas-phase chemistry to solutions

[1] AFSC–URRP Visiting Professor, 1984–1986. Current address: Department of Chemistry, Brandeis University, Waltham MA 02254.
[2] SCEEE Fellow/AFOSR Summer Faculty Research Associate (1984). Current address: Department of Chemistry, University of Kansas, Lawrence KS 66045.

0065-2393/87/0215-0083$06.00/0

is not always straightforward. One deterrent for the solution chemist is the complexity and unfamiliarity of the techniques used in the gas phase—and that is most certainly true of the beam and flow-reactor techniques used in our laboratory. Actually, the techniques do not have to be understood, and even though they may require a decade to develop, the details are really only of interest to the users. Solution chemists, interested solely in results, simply need to appreciate *what* it is that is measured and *why* it is measured. In general, new techniques are developed to achieve better control over an increasing number of important variables. In our beam studies, we vary translational energy and we therefore study the effect of translational energy on reactivity. In the flow reactor, we vary temperature and study its effect on reactivity. For both the beam and flow reactor techniques, we vary the solvation number of the reactant to explore its influence on reactivity. The apparatus appears complicated because of the need to control simultaneously translational energy, temperature, and solvation number. What is gained is the ability to control variables that cannot be varied systematically in solution—solvation number and translational energy.

What are the principal features and advantages of these two techniques? In the flow reactor (6), the solvent of a solution is replaced by a bath of inert gas (typically helium) (Figure 1). The bath gas cannot of course solvate the reactants: what the bath gas does is to undergo many millions of collisions with the reactants; thus, the reactants are at the bath temperature. This bath temperature can be varied (typically from 100 to 500 K) to measure *rate constants as a function of temperature*. Because the flow reactor succeeds in defining the temperature unambiguously, it offers advantages over ion cyclotron resonance techniques.

In the beam technique (7, 8), one reactant is accelerated to a particular translational energy and fired at a second reactant (Figure 2). Only one collision is allowed to occur. What is then measured is the *reaction probability per collision as a function of translational energy.* (The reaction probability, or cross section, may also be expressed in terms of an effective rate constant.) The translational energy range for these experiments spans the range of chemical bond energies, extending to ~200 kcal/mol (corresponding to ~10 eV or translational temperatures of ~100,000 K). From measurement of the lowest translational energy at which chemical reactions occur (the threshold energy), the energy barriers of chemical reactions can be determined.

What distinguishes the gas phase from solution is the absence of solvent and solvation. Study of selectively solvated reactants in the gas phase offers the opportunity of relating the reactivity in the two phases (9). For both of our gas-phase techniques, we are, however, able to solvate one reactant with up to three solvate molecules. We can therefore begin to simulate in small part—with solvate but without bulk solvent—some of the conditions prevailing in solution. Thus, in the beam experiments, we can simultaneously vary

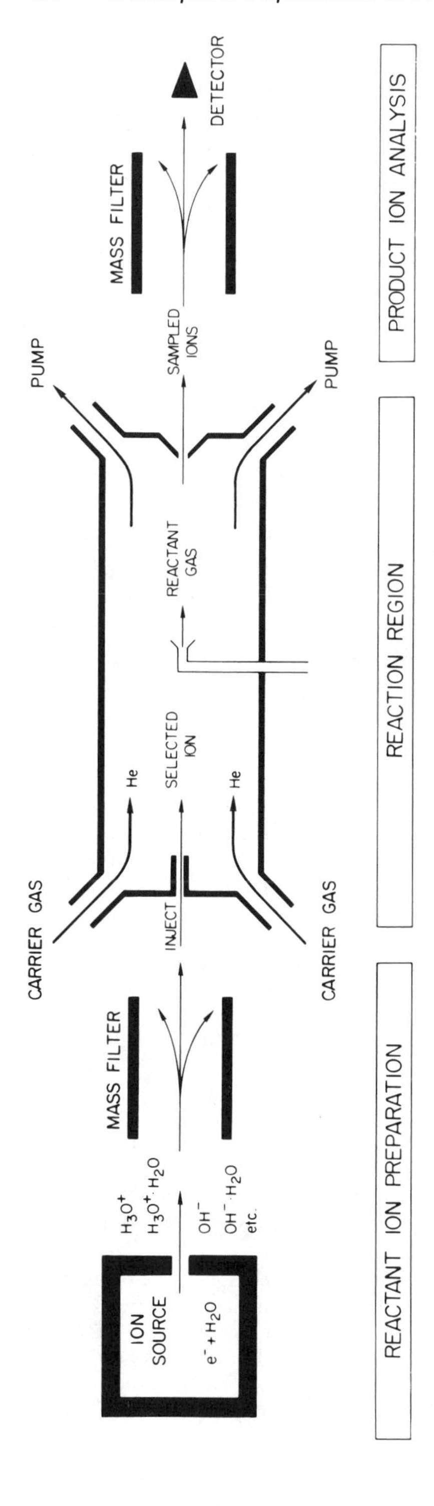

Figure 1. A schematic diagram of the flow reactor.

translational energy and solvation number, whereas in the flow reactor, we can vary temperature and solvation number.

In the experiments to be described here, we have been interested in a number of basic questions: (1) When is a nucleophile nucleophilic? Under what conditions does a Lewis base act as a Brønsted base and under what conditions does it act as a nucleophile? (2) If a nucleophile is "tagged" with one or two solvate molecules, can these "tags" be used as stereochemical markers to probe the mechanism? (3) What relationship does reaction in the gas phase, conducted with solvate but no bulk solvent, bear to reaction, conducted with both, in solution?

The choice of system deserves a few comments. Hydroxide ion and methyl halides were chosen for the present study. Nucleophilic displacement reactions have already been studied extensively in the gas phase (for reviews, *see* references 1 and 9–13); and current techniques are limited to reactions where only one of the two reactants is charged—substrate or nucleophile. Reactions with a negatively charged nucleophile offer an attractive choice because they have been investigated so extensively in solution. Methyl halides are the substrates of choice because elimination is not a possible pathway (*4*). Hydroxyl is a convenient nucleophile because its large hydration energies (*14*) minimize decomposition of the hydrated ions during their preparation and reaction.

Experimental Section

Because the techniques used are unfamiliar, a brief general description is given, indicating what properties are measured and how these may be interpreted.

Both techniques use the same ion source to prepare the negatively charged nucleophile: OH^-, $OH^-(H_2O)$, and $OH^-(H_2O)_2$ (Figures 1 and 2). The ion source is a metal box in which water vapor, at a pressure of ~1 torr, is ionized by electrons. Negative ions are extracted through a slit in the box by an electric field. While the ions are in the box, they undergo sufficient collisions to form hydrated ions, for example

$$OH^- + H_2O = OH^-(H_2O)$$

$$OH^-(H_2O) + H_2O = OH^-(H_2O)_2$$

Once out of the box, the negative ions are focused, as a beam, through a mass filter, which will only transmit ions of a particular mass-to-charge ratio. In the experiments described here, the mass filter selects either OH^- or $OH^-(H_2O)$ or $OH^-(H_2O)_2$ to be the reactant—for either the beam technique or the flow reactor.

Beam Technique. In the beam technique, the beam of sorted reactant ions (e.g., OH^-) is injected into a tiny collision chamber, containing the neutral reactant (e.g., methyl bromide) at a pressure of ~1 mtorr (Figure 2). The collision chamber is so thin (~2 mm) that no ion undergoes more than one collision and both reactant and product ions escape through an exit slit—to be analyzed, again by mass in another mass filter, and counted in a particle counter. In the illustrative example, unreacted

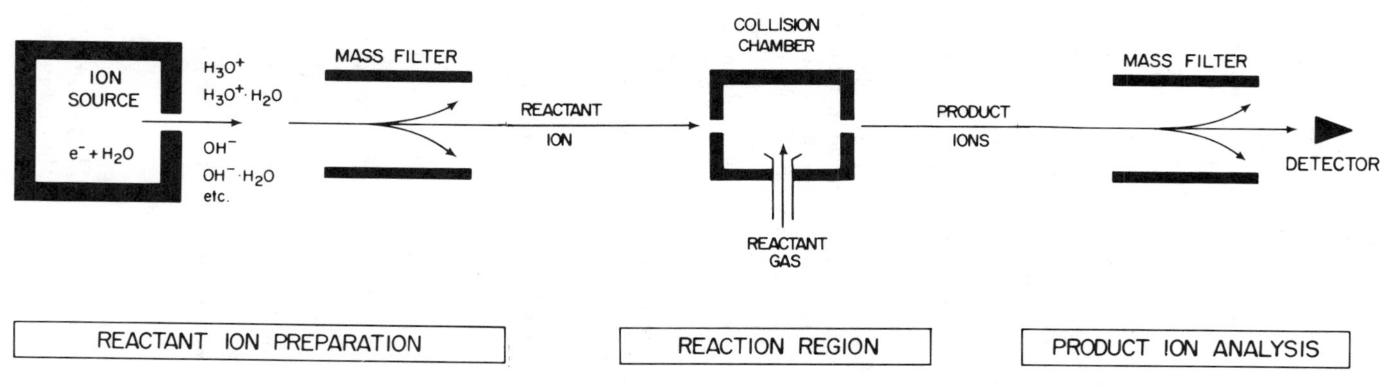

Figure 2. A schematic diagram of the ion beam apparatus. The translational energy of the ions is controlled by the voltage maintained between the ion source and the collision chamber.

reactant OH^- and the product Br^- formed via $OH^- + CH_3Br \rightarrow CH_3OH + Br^-$ would be measured (*see* Figure 3 and the following discussion). The attenuation of the OH^- beam by reaction can be visualized as the methyl bromide molecules presenting an effective "hitting area" to the OH^- projectiles. This effective "hitting area" is the reaction cross section, σ. σ provides a measure of the reaction probability, and in these experiments, it reaches a maximum value of ~100 Å^2. (As a consequence of the powerful electrostatic forces drawing together the reactants—ion and polarizable molecule—these cross sections are 10 times larger than geometric cross sections.) The cross section σ can also be expressed as an effective rate constant k by the expression $k = \sigma v_r$, where v_r is the relative velocity.

In the beam experiments, the reaction cross sections are measured as a function of E_T, the relative translational energy, which is that part of the translational energy that is available to drive chemical reactions. (Some of the translational energy is "unavailable" because of the kinematic requirement to conserve momentum.) The parameter that controls the translational energy of the ions, E, is the potential difference applied between the ion source and the collision chamber. The relative energy, E_T, is then calculated from the translational energy, E, by using conservation of energy and momentum: $E_T/E = \text{mass}_{target}/(\text{mass}_{target} + \text{mass}_{ion})$.

Flow Reactor. In the flow reactor (Figure 1), the selected reactant ion beam is injected into a gas stream that is flowing rapidly down a tube (*15*). (This is another application of the Venturi effect, familiar to bench chemists in the action of the water pump or aspirator.) Speeds of ~10^4 cm/s along a tube of 1-m length give residence times in the flow tube of ~10 ms. The substrate is introduced into the tube downstream, and an aliquot of the flowing gas is sampled through a pinhole at the far end of the flow tube. Again, this sample is analyzed in the same way as in the beam experiment. From the known flow rate of the substrate and from the reaction time derived from the flow characteristics, a rate constant can be derived appropriate to the temperature of the reactor. As mentioned previously, the temperature is variable (100–500 K).

Brønsted Base or Nucleophile? Ambivalence in the Lewis Base

The system

$$OH^- + CH_3Br \begin{cases} \rightarrow H_2O + CH_2Br^- & \text{proton transfer} \quad (1) \\ \rightarrow CH_3OH + Br^- & \text{nucleophilic displacement} \quad (2) \end{cases}$$

is a representative example of a Lewis base in competing roles of Brønsted base and nucleophile. We consider first the thermodynamic properties of the two channels, in the gas phase and in solution. Our interest then is in the competition between the two channels—as a kinetic probe of basicity versus nucleophilicity.

Thermodynamics. No mystery surrounds the reaction between OH^- and CH_3Br in solution. The pK_a is so much larger for CH_3Br than for water that hydroxide acts not as a Brønsted base but as a nucleophile. Displace-

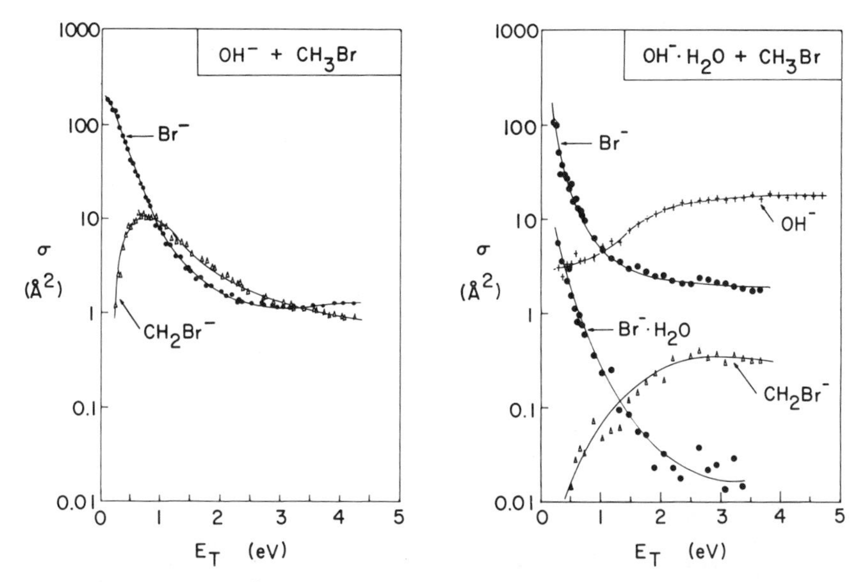

Figure 3. Dependence of the reaction cross section (reaction probability) upon relative translational energy, E_T, *for the products indicated (20). The reactants are as indicated:* $OH^- + CH_3Br$ *and* $OH^- \cdot H_2O + CH_3Br$.

ment prevails in the competition between proton transfer and nucleophilic displacement. In thermodynamic terms, the barrier to reaction is lower for nucleophilic displacement than for proton transfer. Quantitative data are given in Figure 4.

The barrier height (Arrhenius activation energy) for displacement is 23 kcal/mol (*16*), and a lower bound on the barrier height for proton transfer is set by its free energy change, $\Delta G° \approx +40$ kcal/mol [derived by estimating $pK_a[CH_3Br] \approx 45$ (*17*)]. However, even though displacement occurs with a large free-energy decrease, $\Delta G° = -23$ kcal/mol (*18*), the large activation energy reduces the rate constant at 300 K by a factor of 10^{-17}. Proton transfer does not compete because the activation energy of the displacement reaction is, in equivalent terms, much smaller than the ΔpK_a for the proton-transfer reaction.

The thermodynamics are very different in the gas phase (Figure 4). Proton transfer is much less endothermic ($\Delta H° = +6$ kcal/mol, *see* under Kinetics); nucleophilic displacement is much more exothermic ($\Delta H° = -56$ kcal/mol) (*1*), and the measured rate constant demonstrates that the barrier must be low.

Why do the thermodynamics differ so markedly for the two phases? Consider first the proton-transfer reaction. In the gas phase, the heats of

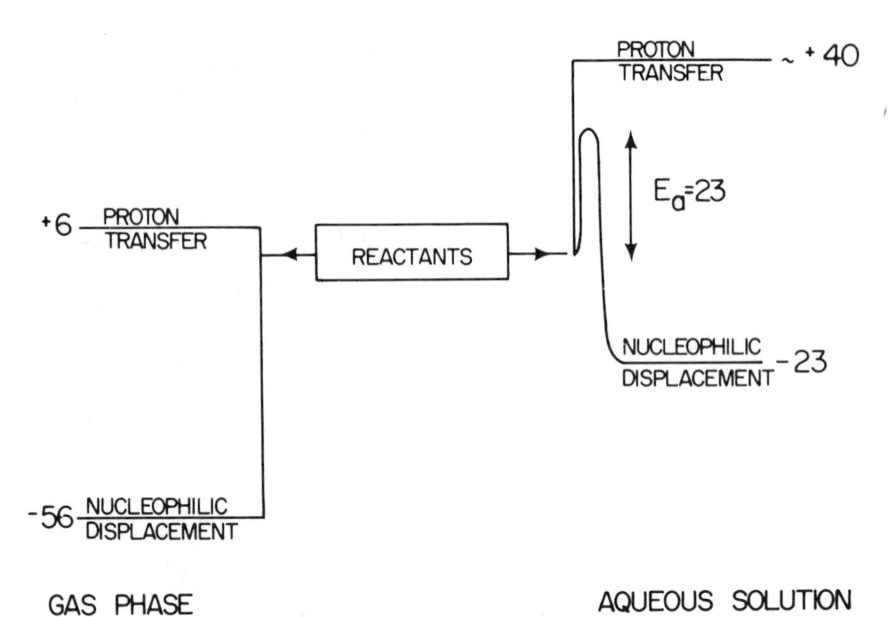

Figure 4. A comparison between the gas phase and solution for the reactants OH^- + CH_3Br undergoing proton transfer and nucleophilic displacement. The vertical scale (not shown) is in kcal/mol, relative to the reactants. The gas-phase values are enthalpies, ΔH°, and the solution values free energies, ΔG°. Because the entropy changes are small, the enthalpies and the free energies may be compared.

deprotonation of H_2O [$\Delta H°$ = 391 kcal/mol (*19*)] and of CH_3Br [$\Delta H°$ = 397 kcal/mol (*20*)] are rather similar; in solution, the corresponding pK_a values (16 and ~45, respectively) reflect the low hydration energy of CH_2Br^- compared to that of OH^-. Now consider the nucleophilic displacement reaction—most conveniently, consider it as a member of the homologous series $OH^- + CH_3X \rightarrow CH_3OH + X^-$, where X is a halogen. The reaction may be viewed as a sequence of steps—breaking the C–X bond, transferring the electron from OH to X, and forming the C–OH bond—so that the reaction enthalpy is the sum of the differences of the bond energies and the electron affinities. Table I shows why, in the gas phase, the displacement reactions become increasingly exothermic as X changes from fluorine to iodine. The thermodynamics differ in solution, the exothermicity being less and effectively constant throughout the series. *Significant differences in the intrinsic thermodynamic driving force (gas phase) are thus effectively quenched in solution*. In the series X = F → I, the increasing exothermicity in the gas phase appears to be offset in solution by the decreasing solvation energy of X^-.

Table I. Thermodynamic Properties of the Reaction $OH^- + MeX \rightarrow MeOH + X^-$

X	*D(Me–X)*[a]	*(X)*[b]	ΔBDE[c]	*Δ(EA)*[d]	ΔH °[e] *(gas phase)*	ΔG °[f] *(solution)*
F	109	78	18	−36	−18	−22
Cl	84	83	−7	−41	−48	−22
Br	70	77	−21	−35	−55	−23
I	56	71	−35	−29	−64	−21

NOTE: All values in kcal/mol. D(Me–X) is the bond energy; EA(X) is the electron affinity.
[a] Reference 21.
[b] Reference 22.
[c] $\Delta BDE = D(\text{Me–X})\text{-}D(\text{Me–OH})$ where $D(\text{Me–OH}) = 91$ (*21*).
[d] Δ(EA) = EA(OH)—EA(X) where EA(OH) = 43 (*23*).
[e] $\Delta H° = \Delta(\text{EA})—\Delta(BDE)$. The numbers here differ from those in the text by 1–2 kcal/mol, being based on reference 21. This is a measure of the uncertainty.
[f] Aqueous solution (*18*). The entropy changes are small, so $\Delta H° \cong \Delta G°$.

Kinetics. We now explore the kinetics of the competition. In general, in solution only one channel can be seen—the channel with the lowest barrier. All the other channels, with significantly higher barriers, are excluded by the Maxwell–Boltzmann energy distribution. In the gas phase, the translational energy can be varied (Figure 2) to examine the reactivity of *all* the channels throughout the *entire* energy range.

The competition between reactions 1 and 2 in the gas phase is shown in Figure 3a, as a function of relative energy. At the lowest energy, only displacement is observed. (This result is what is expected from experiments at thermal energy using the flow reactor.) When the energy is increased to ~0.3 eV (~7 kcal/mol or a translational temperature of ~3000 K), proton transfer begins to compete (Figure 4). (From this measured energy threshold, the enthalpy change $\Delta H° = +6$ kcal/mol is derived for reaction 1.) The fractional yields of the proton-transfer channel are shown in Figure 5, as a function of relative energy. The remarkable kinetic result is that the endothermic channel (reaction 1) prevails over the highly exothermic channel (reaction 2) even though reaction 2 occurs on nearly every collision at thermal energies.

When the hydroxide is hydrated with a water molecule, the displacement channel is unchanged but the proton-transfer channel is reduced by one order of magnitude (Figure 3b compared with Figure 3a). Hydration suppresses the competition of endothermic proton transfer with exothermic nucleophilic displacement (Figure 5). These two results—a competition that is contrary to the thermodynamics and the suppression of this competition by one water of hydration—are summarized in Figure 5 and have been explained in terms of the relative energies of the reaction intermediates (*24*).

Considerable evidence now exists that reactions such as 1 or 2 proceed via two intermediates in the gas phase. [The reaction energy profile shows

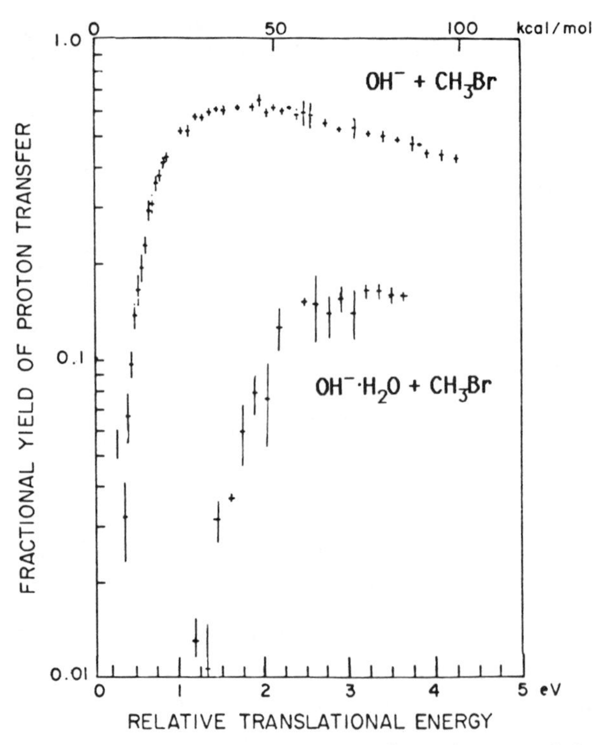

Figure 5. Fraction of proton transfer measured as a function of the relative translational energy, E_T, *for the reactants* $OH^- + CH_3Br$ *and* $OH^- \cdot H_2O + CH_3Br$.

two minima separated by a low barrier (*see,* for example, reference 1, Figure 1).] For the proton-transfer reaction (1), the two intermediates, II and III, can be written as

$$\underset{\textbf{I}}{HO^- + CH_3Br} \rightleftharpoons \underset{\textbf{II}}{HO^- \cdot CH_3Br} \rightleftharpoons \underset{\textbf{III}}{HOH \cdot CH_2Br^-} \rightleftharpoons \underset{\textbf{IV}}{H_2O + CH_2Br^-} \quad (3)$$

Analysis of the reaction energetics shows that, even though the overall reaction (**I** ⟶ **IV**) is endothermic, transfer of the proton *within the reaction intermediate* (**I** ⟶ **III**) is exothermic. The overall reaction is only endothermic because of the energy needed to separate the products (**III** ⟶ **IV**). In marked contrast, when the OH^- is hydrated, *both* processes (**I** ⟶ **IV**) and (**I** ⟶ **III**) are endothermic. Translational energy is then needed both to drive the proton-transfer reaction and to separate the products: when unhydrated, translational energy is needed simply to separate products that are already formed. The energetics can thus provide a possible explanation for the unusual competition between reactions 1 and 2, shown in Figure 5.

These results extend our knowledge of the competition between basicity and nucleophilicity. In solution, proton transfer preempts nucleophilic displacement wherever it is possible thermodynamically. This situation is the simple consequence of low barriers for the former and high barriers for the latter. In the gas phase, proton transfer also preempts nucleophilic displacement. Here, there are no barriers for proton transfer—reaction occurs on every collision—and low barriers for nucleophilic displacement. [The relationship between the barriers in the two phases has been explored in a series of important papers by Brauman and co-workers and reviewed (*25, 26*).] What has therefore been established in the gas phase is that proton transfer always wins when it is exothermic and that nucleophilic displacement is only efficient when proton transfer is endothermic (*10*). What is added here is that *endothermic* proton transfer can still win, when energy is available. If the explanation advanced is correct, this situation is only possible where the actual proton transfer remains exothermic.

Chemistry as a Function of Solvation Number

Beam techniques may be used to probe energy barriers and chemistry driven by translational energy: the flow reactor is used to study rates and mechanisms at thermal energies. Rates studied as a function of temperature reveal barriers to reaction. Rates studied as a function of solvation number reveal the kinetic role of solvate, in the absence of bulk solvent. We illustrate this behavior for the nucleophilic displacement reaction

$$OH^- + CH_3Cl \rightarrow CH_3OH + Cl^- \qquad \Delta H^\circ = -50 \text{ kcal/mol} \qquad (4)$$

The substrate CH_3Cl differs from CH_3Br in a slightly different exothermicity (Table I). The data presented here for CH_3Cl extend our published data for the CH_3Br system (*27*).

Rates. Rate data for reaction 4 are shown in Figure 6. What is plotted is not a rate constant for the reaction but a reaction efficiency per collision. Analysis shows that the rate constant for an ion and a polar molecule simply to collide is itself temperature-dependent (*28*). What is of chemical interest is the property plotted—the fraction of the collisions that actually result in reaction. This reaction efficiency per collision is the ratio of the experimental rate constant to the calculated collision rate constant (*29*).

Although reaction 4 is exceedingly exothermic, it does not occur on every collision, and the efficiency *decreases* with increasing temperature (Figure 6). This finding is characteristic of a reaction where the potential surface shows two wells (i.e., two intermediates, as in reaction 3) (*4*). With modeling, these data may be used to derive intrinsic nucleophilicities (*5*) although we have yet to do this for reaction 4.

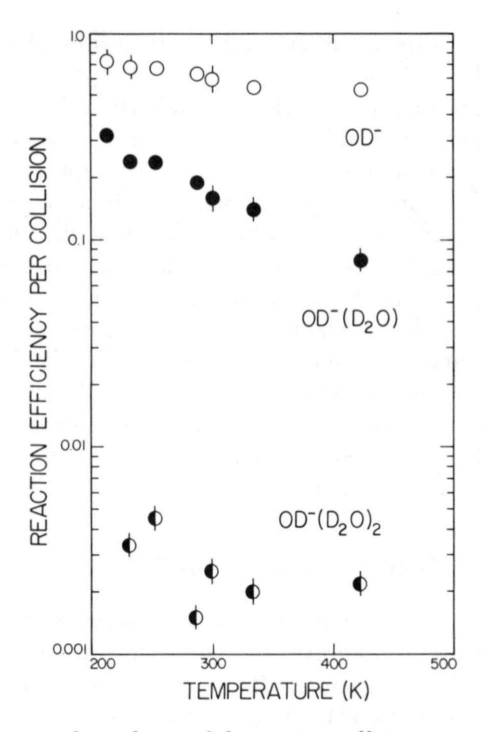

Figure 6. Temperature dependence of the reaction efficiency per collision for the reactions of OD^- + CH_3Cl (open circles), OD^-D_2O + CH_3Cl (filled circles), and $OD^-(D_2O)_2$ + CH_3Cl (half-filled circles). The reaction efficiency per collision is the experimental rate constant divided by the calculated collision rate constant, calculated by Clary using the adiabatic capture centrifugal sudden approximation (ACCSA) (28). For experimental reasons (29), the measurements were made with completely deuterated anions.

Adding one water of hydration to the OH^- reactant has two effects: the reaction efficiency is decreased by half an order of magnitude and the temperature dependence is increased (Figure 6). Adding two waters decreases the efficiency by *two* orders of magnitude, and it may, within the limits of scattered data, increase the temperature dependence further. These findings agree with the results previously obtained by Bohme at 300 K (2). The parallelism between the decrease in the reaction efficiency, on the one hand, and the magnitude of the temperature dependence, on the other, is again qualitatively consistent with the two-well potential model (29).

Product Distributions. When the reactant ion is hydrated, different reaction channels are possible because different numbers of waters may hydrate the product ion. For example, the reactant $OH^-\cdot H_2O$ in reaction 4 can form either Cl^- or $Cl^-\cdot H_2O$. One channel is available for the unhydrated reactant, two for the singly hydrated, three for the doubly hydrated, and so on. The possible channels are shown in Figure 7, together with their enthalpy changes.

Figure 7 also shows the experimental product distributions that were observed in the flow reactor. A consistent pattern emerges. However hydrated the reactant, the predominant product ion is Cl^- *without* hydration. Forming $Cl^- \cdot H_2O$ is apparently a very unlikely process (collision efficiencies of <1%), and $Cl^- \cdot (H_2O)_2$ is not observed at all.

Why no reaction is found for the triply hydrated reactant is now apparent. If the product ion has to be Cl^-, this channel is now endothermic by 6 kcal/mol. The more hydrated the reactant ion, the more endothermic will be the reaction channel leading to the Cl^- product. Hydration quenches the reaction if the reactant has more than two waters of hydration. When the reactant has only one or two waters, Figure 7 suggests why the reactivity decreases as shown in Figure 6. Addition of the two waters successively decreases the exothermicity: this result, in turn, raises the barrier for nucleophilic displacement—through application of Marcus theory (*24*, *25*).

Formation of Hydrated Products. According to the enthalpies shown in Figure 7, the most exothermic channel would transfer all the hydration from reactant to product. That channel also has the greatest free energy decrease. Paradoxically, experiment shows the dominant channel to be the one *least* favored thermodynamically. Is the experimental result an artifact of

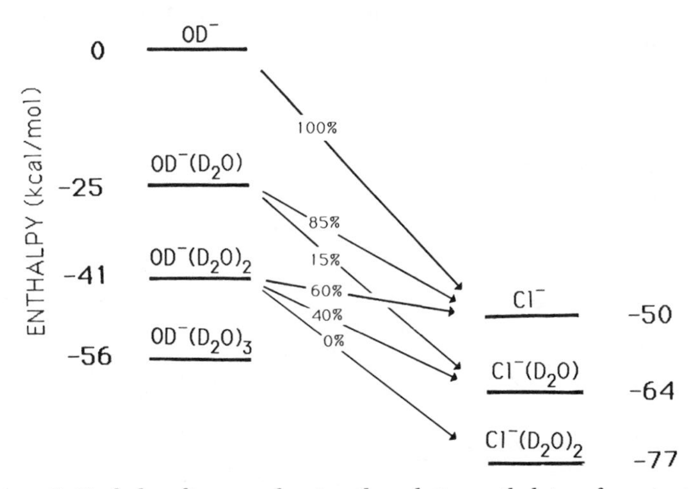

Figure 7. Enthalpy diagram showing the relative enthalpies of reactants and products, relative to the enthalpy of the dehydrated reactants OD^- + CH_3Cl. In each case the label indicates only the ionic reactant or product. Thermochemical values are for undeuterated species from references 2 and 14.

the technique? Could it be that the waters do transfer efficiently but then "boil off" (*30*), because the reaction is so exothermic? For example, for the singly hydrated reacant, could reactions 5 and 6 occur sequentially, in 85% of the collisions that result in reaction?

$$OH^-\cdot H_2O + CH_3Cl \rightarrow CH_3OH + Cl^-\cdot H_2O \qquad \Delta H^\circ = -39 \text{ kcal/mol} \qquad (5)$$

$$Cl^-\cdot H_2O \rightarrow Cl^- + H_2O \qquad \Delta H^\circ = +14 \text{ kcal/mol} \qquad (6)$$

No definitive answer can be given to these questions at the present time (2). Circumstantial evidence, however, suggests strongly that solvate does not transfer from reactant to product, that reaction 5 does not occur. First, the experimental data are reliable: product distributions, measured with the beam technique, extrapolate to those, measured in the flow reactor, in the limit of thermal energies (Figure 8). [The same agreement is found with CH_3Br as the substrate (27), and the raw beam data are shown in Figure 3.] Second, if the mechanism indeed were reactions 5 and 6, the product distribution would be expected to show a greater dependence on translational energy than is shown in Figure 8 (*see also* reference 27). Third, if the sequential mechanism 5 and 6 were operating, why is no doubly hydrated product seen for the doubly *and triply* hydrated reactants? Finally, when methoxide is used as the nucleophile and is solvated with methanol, similar product distributions are found. The reaction enthalpies of the various channels for $OH^-\cdot H_2O$ as nucleophile differ from those for $CH_3O^-\cdot CH_3OH$ as the nucleophile. With these constraints, it is hard to see how the "boil-off" mechanism, 5 and 6, could yield similar product distributions for the two different nucleophiles.

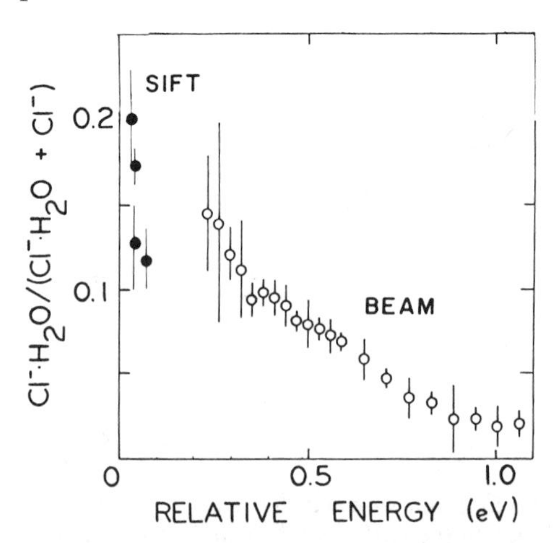

Figure 8. Fraction of Cl^--hydrated product, as $Cl^-\cdot H_2O$, as a function of E_T, *relative energy, for the reactants $OH^-\cdot H_2O + CH_3Cl$. Open circles: beam measurements. Filled circles: flow reactor measurements.*

Reaction Mechanism: Gas Phase and Solution. What mechanistic information is provided by the solvated-ion experiments in the gas phase? If thermodynamic control should yield solvated products, what is the nature of the kinetic control that apparently yields unsolvated ones?

We have argued (*27*) that a nucleophilic displacement between $OH^{-}\cdot H_2O$ and CH_3Cl could involve two steps—Walden inversion of the methyl group as the OH^{-} nucleophile displaces the Cl^{-} leaving group and hydrate transfer from the nucleophile to the leaving group. Three mechanisms are then possible, depending on the sequence of the two steps. These mechanisms are shown in Figure 9, which is drawn in the configuration of a More O'Ferrall plot, as used for S_N2 reactions in solution (*31*). One mechanism consists of hydrate transfer followed by inversion (**V** ⟶ **VI** ⟶ **VII**), another of the reverse sequence, inversion and hydrate transfer (**V** ⟶ **VIII** ⟶ **VII**), and the third of both concerted (**V** ⟶ **VII**). By considering the heights of the barriers for these mechanisms, we have eliminated the first and we

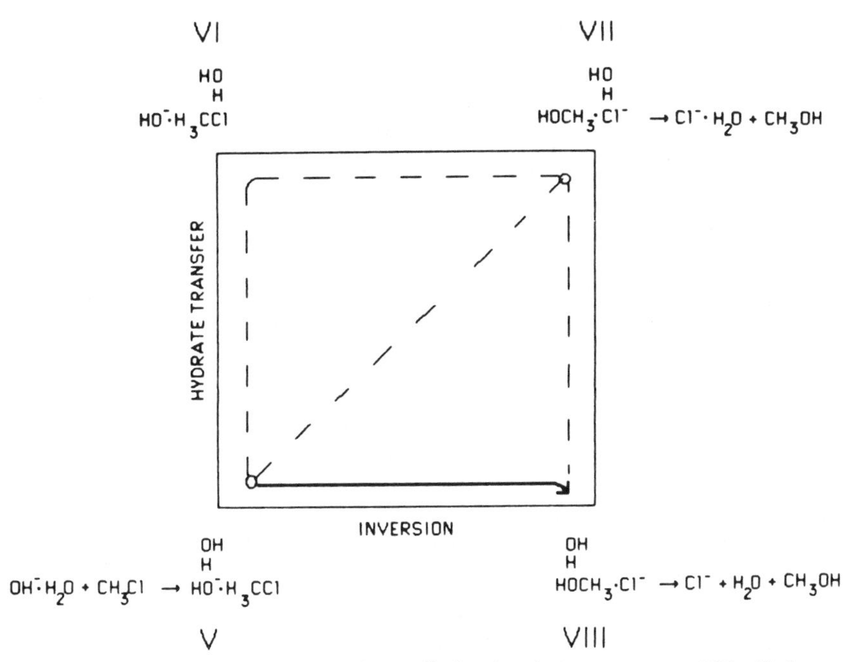

Figure 9. A schematic More O'Ferrall plot for the S_N2 reaction $OH^{-}\cdot H_2O + CH_3Cl \longrightarrow CH_3OH + Cl^{-}\cdot H_2O$, with reactants at the lower left and products at the upper right. The mechanism is resolved into two steps: inversion (horizontal axis) and hydrate transfer from nucleophile to leaving group (vertical axis). The horizontal and vertical dashed lines show the two sequential mechanisms; the diagonal line shows the concerted mechanism. The mechanism observed—inversion + desolvation—is shown by the solid line.

have argued that the second would be favored over the third (*27*). Our conclusion, therefore, is that the first step would be inversion (**V** → **VI**), which would liberate ~50 kcal/mol from the reaction exothermicity. With this much excitation energy, the intermediate **VI** would simply desolvate and yield Cl^-.

Central to this description is the idea that hydrate does not transfer from nucleophile to leaving group because the transition state **IX** is unfavorable—unfavorable energetically and unfavorable entropically.

$$HO^{\delta-} \cdots C(H)(H)(H) \cdots Cl^{\delta-} \quad (H \; H \; O \; H \; H \text{ bridge})$$

IX

$$CH_3O^{\delta-} \cdots C(H)(H)(H) \cdots Cl^{\delta-} \quad (H \; H \; O \; H \; CH_3 \text{ bridge})$$

X

Nevertheless, for this reaction of $OH^- \cdot H_2O + CH_3Cl$, some 15% of the hydrated product $Cl^- \cdot H_2O$ is formed. We consider the hydrated product unlikely to be formed via the transition state **IX,** because in the analogous reaction $CH_3O^- \cdot CH_3OH + CH_3Cl$, the yield of solvated product $Cl^- \cdot CH_3OH$ is remarkably similar (~20%). In the methoxide reaction, the corresponding transition state **X** cannot provide a route for solvate transfer. The binding energies of methyl groups to anions are too low (*32*) to form a solvate bridge. Water can form two hydrogen bonds to bridge; methanol can only form one.

How the minor yield of solvated product is formed remains an interesting question. If mechanisms involving inversion and solvate transfer are excluded, attention turns to direct mechanisms involving frontside attack (*33*). Ample precedent for such processes can be found in neutral chemistry, but at higher energies and not for solvated reactants (*34*).

Finally, we consider the relevance of these solvated-ion studies in the gas phase to the corresponding reactions in solution. In the gas phase, the products are predominantly unsolvated: in solution, they are completely solvated. In the gas phase, reaction 4 is apparently quenched by solvating the reactant with more than two solvate molecules: in solution, the reaction proceeds when the reactants are infinitely solvated. This highlights the importance of the bulk solvent in solution. All pervasive, the bulk solvent can always enable the *concerted* desolvation of the nucleophile and solvation of the leaving group. In contrast, in the gas phase, the same solvate molecules that are released in the desolvation must be used in the solvation; and if the solvate does not transfer, solvation must stop the reaction. In the gas phase, solvate molecules, unconstrained by the presence of bulk solvent, have much greater maneuverability and hence a greater opportunity to participate in chemical reaction. In the reactions discussed here, ironically, this enhanced capability is not useful. Instead, it serves to emphasize the key kinetic role played in solution by the bulk solvent.

Conclusions

Beam techniques, which have not been applied extensively to physical organic chemistry, are used here (1) to explore higher-energy pathways, (2) to measure energy barriers, and (3) to probe mechanisms through varying translational energy. Specifically, (1) the role of translational energy has been explored in influencing the competition between nucleophilicity and basicity; (2) some basic thermochemistry has been determined; and (3) certain products were identified as being primary rather than secondary in origin. It should be clear that beam experiments of this type have a significant role to play in physical organic chemistry, for example, in the *direct* measurement of intrinsic nucleophilicity and for testing *directly* the validity of the Marcus postulates (*5*). As shown in Figures 3 and 7, it is important in these beam experiments to be able to work at the lowest possible energies (~0.2 eV). Unfortunately, such measurements at low energy present the greatest experimental difficulties, and the thermal-energy measurements in the flow reactor provide an important check (Figure 8).

The rate measurements in the flow reactor, as a function of temperature and solvation number, provide new data to support the current picture of nucleophilic displacement in the gas phase, characterized by a potential surface with two wells, and the system undergoing Walden inversion in passing from the first intermediate to the second. The extent of solvate transfer reveals new information about the geometry of the transition state. If solvate is to pass from reactant ion to product ion, it must be able to track the moving charge. Pathways where the solvate and charge are separated must have a higher energy and be less favored; this is true for nucleophilic displacement. As the charge moves from nucleophile to leaving group, down the backbone of the intermediate, the solvate can only follow by circumnavigating the perimeter, and it doesn't. [A striking contrast is provided in proton transfer reactions where the different geometry of the transition state allows efficient solvate transfer (*34*)].

Both techniques provide insights into the relationship between reactivity in the gas phase and reactivity in solution (*3*). Beam techniques in the gas phase explore the reactions that are not seen at thermal energies and provide a basis of understanding (Figure 4). In the flow reactor at thermal energies, chemistry studied as a function of solvation number allows some distinction to be made between the roles of solvate and bulk solvent in solution.

Acknowledgments

We are most grateful to D. C. Clary for calculating the collision rate constants used in Figure 6, to W. P. Jencks for discussion, and to our colleagues, Fred Dale, Carol Deakyne, and Al Viggiano, for invaluable assistance.

Literature Cited

1. Olmstead, W. N.; Brauman, J. I. *J. Am. Chem. Soc.* **1977,** *77*, 4219.
2. Bohme, D. K.; Raksit, A. B. *J. Am. Chem. Soc.* **1984,** *106*, 3417.
3. Chandrasekhar, J.; Smith, S. F.; Jorgensen, W. L. *J. Am. Chem. Soc.* **1984,** *107*, 154.
4. Caldwell, G.; Magnera, T. F.; Kebarle, P. *J. Am. Chem. Soc.* **1984,** *107*, 959.
5. Pellerite, M. J.; Brauman, J. I. *J. Am. Chem. Soc.* **1983,** *105*, 2672.
6. DePuy, C. H.; Bierbaum, V. M. *Acc. Chem. Res.* **1981,** *14*, 146.
7. Paulson, J. F.; Dale, F.; Studniarz, S. A. *Int. J. Mass Spectrom. Ion Phys.* **1970,** 5, 113.
8. Futrell, J. H.; Tiernan, T.O. In *Ion-Molecule Reactions;* Franklin, J. L., Ed.; Plenum: New York, 1972; Vol. II, p 485.
9. Bohme, D. In *Ionic Processes in the Gas Phase;* Almoster Ferreira, M. A., Ed.; Reidel: Dordrecht, 1984; p 111.
10. Beauchamp, J. L. In *Interactions Between Ions and Molecules;* Ausloos, P., Ed.; Plenum: New York, 1975, p 413.
11. Nibbering, N. M. M. In *Kinetics of Ion-Molecule Reactions;* Ausloos, P., Ed.; Plenum: New York, 1979; p 165.
12. McIver, R. T., Jr. *Sci. Am.* **1980,** *243(5)*, 186.
13. DePuy, C. H.; Grabowski, J. J.; Bierbaum, V. M. *Science (Washington, D.C.)* **1982,** *218*, 955.
14. Kebarle, P. *Annu. Rev. Phys. Chem.* **1977,** *28*, 445.
15. Smith, D.; Adams, N. G. In *Gas Phase Ion Chemistry;* Bowers, M. T., Ed.; Academic: New York, 1979; Vol. 1, p 1.
16. Bathgate, R. H.; Moelwyn Hughes, E. A. *J. Chem. Soc.* **1959,** 2642.
17. Streitwieser, A., Jr.; Heathcock, C. H. *Introduction to Organic Chemistry*, 2nd ed.; MacMillan: New York, 1981; p 1198.
18. Albery, W. J.; Kreevoy, M. M. *Adv. Phys. Org. Chem.* **1978,** *16*, 87.
19. Bartmess, J. E.; McIver, J. T., Jr. In *Gas Phase Ion Chemistry;* Bowers, M. T., Ed.; Academic: New York, 1979; Vol. II, p 88.
20. Hierl, P. M.; Henchman, M.; Paulson, J. F. in preparation.
21. Streitwieser, A., Jr.; Heathcock, C. H. *Introduction to Organic Chemistry*, 2nd ed.; MacMillan: New York, 1981, p 1194.
22. Mead, R. D.; Stevens, A.; Lineberger, W. C. In *Gas Phase Ion Chemistry;* Bowers, M. T., Ed.; Academic: New York, 1984; Vol. III, p 214.
23. Drzaic, P. S.; Marks, J.; Brauman, J. I. In *Gas Phase Ion Chemistry;* Bowers, M. T., Ed.; Academic: New York, 1984; Vol. III, p 168.
24. Henchman, M.; Hierl, P. M.; Paulson, J. F. *J. Am. Chem. Soc.* **1985,** *107*, 2812.
25. Brauman, J. I. In *Kinetics of Ion-Molecule Reactions;* Ausloos, P., Ed.; Plenum: New York, 1979; p 153.
26. Barfknecht, A. T.; Dodd, J. A.; Salomon, K. E.; Tumas, W.; Brauman, J. I. *Pure Appl. Chem.* **1984,** *56*, 1809.
27. Henchman, M.; Hierl, P. M.; Paulson, J. F. *J. Am. Chem. Soc.* **1983,** *105*, 5509.
28. Clary, D. C. *Mol. Phys.* **1985,** *54*, 605.

29. Hierl, P. M.; Ahrens, A. F.; Henchman, M. J.; Viggiano, A. A.; Paulson, J. F.; Clary, D. C. *J. Am. Chem. Soc.* **1986,** *108,* 3142.
30. Kebarle, P. *Mass Spectrometry in Inorganic Chemistry;* Margrave, J. L., Ed.; Advances in Chemistry 72; American Chemical Society: Washington, DC, 1968; p 24.
31. Harris, J. M.; Shafer, S. G.; Moffatt, J. R.; Becker, A. R. *J. Am. Chem. Soc.* **1979,** *101,* 3295.
32. Yamabe, S.; Hirao, K. *Chem. Phys. Lett.* **1981,** *84,* 598.
33. Schlegel H. B.; Mislow, K.; Bernardi, F.; Bottoni, A. *Theor. Chim. Acta.* **1977,** *44,* 245.
34. Henchman, M.; Wolfgang, R. *J. Am. Chem. Soc.* **1961,** *83,* 2991.
35. Hierl, P. M.; Ahrens, A. F.; Henchman, M. J.; Viggiano, A. A.; Paulson, J. F.; Clary, D. C. *J. Am. Chem. Soc.* **1986,** *108,* 3140.

RECEIVED for review October 21, 1985. ACCEPTED March 24, 1986.

7

Variation in the Transition-State Structure of Aliphatic Nucleophilic Substitution

Takashi Ando[1], Takahide Kimura[2], and Hiroshi Yamataka[2]

[1]Department of Chemistry, Shiga University of Medical Science, Seta Tsukinowa, Otsu, Shiga 520–21, Japan
[2]The Institute of Scientific and Industrial Research, Osaka University, Mihogaoka, Ibaraki, Osaka 567, Japan

Carbon kinetic isotope effects in two series of methyl transfers (methyl brosylate with substituted N,N-*dimethylanilines and methyl iodide with substituted pyridines) were all large and did not show the bell-shaped behavior previously observed in benzyl transfers. This finding indicates the importance of the Walden inversion motion in methyl transfers in which transition states hold total bonding constant. Nitrogen kinetic isotope effects in several reactions of benzyl benzenesulfonates with* N,N-*dimethylanilines were all small and almost constant. Thus, the early–late characters of the transition states were difficult to distinguish. Hydration of nucleophilic anions (Cl^-, Br^-, I^-, SCN^-, and N_3^-) in the reaction with* n-*octyl* p-*nitrobenzenesulfonate altered the kinetic parameters but not the α-deuterium isotope effects. Dehydration during activation without significant change in the transition-state structure was proposed.*

Aliphatic nucleophilic substitution played a central role in the recent debate on variable transition states (TS) (*1–20*). In previous papers (*1–4*), we successfully interpreted the variation in the TS structure of the Menschutkin-type reaction of benzyl benzenesulfonates with *N*,*N*-dimethylanilines (equation 1), by the extensive use of carbon-14 and tritium isotope effects.

0065-2393/87/0215-0103$06.00/0

$$\text{Y-C}_6\text{H}_4\text{N(CH}_3)_2 + \text{Z-C}_6\text{H}_4\text{CH}_2\text{OSO}_2\text{C}_6\text{H}_4\text{-X} \xrightarrow[\text{Acetone}]{35\,^\circ\text{C}} [\text{Y-C}_6\text{H}_4\text{N(CH}_3)_2\text{CH}_2\text{C}_6\text{H}_4\text{-Z}]^{+}\ {}^{-}\text{OSO}_2\text{C}_6\text{H}_4\text{-X} \quad (1)$$

Qualitative rules used to link the variation in the TS and that in the experimentally observed kinetic isotope effect (KIE) are as follows: (1) The carbon-14 KIE becomes maximum when the TS is symmetrical in terms of the force constants, $f_{C\text{–}N}$ and $f_{C\text{–}O}$, of the two reacting bonds and decreases when the TS becomes reactant-like or product-like upon changing X, Y, and Z. (2) As a general trend, the α-tritium KIE increases when the TS becomes loose and decreases when it becomes tight (Figure 1). The bell-shaped plots of the carbon-14 KIE, observed for several series with varied X or Y, have been attributed to TS variations crossing the diagonal connecting the C^+ and C^- corners in Figure 1; this diagonal corresponds to symmetrical TS structure. Concurrent observation of a small and monotonic change in the α-tritium KIE has been regarded as an indication of a variation along the tight–loose direction. All the variations have been discussed in terms of Thornton's rules

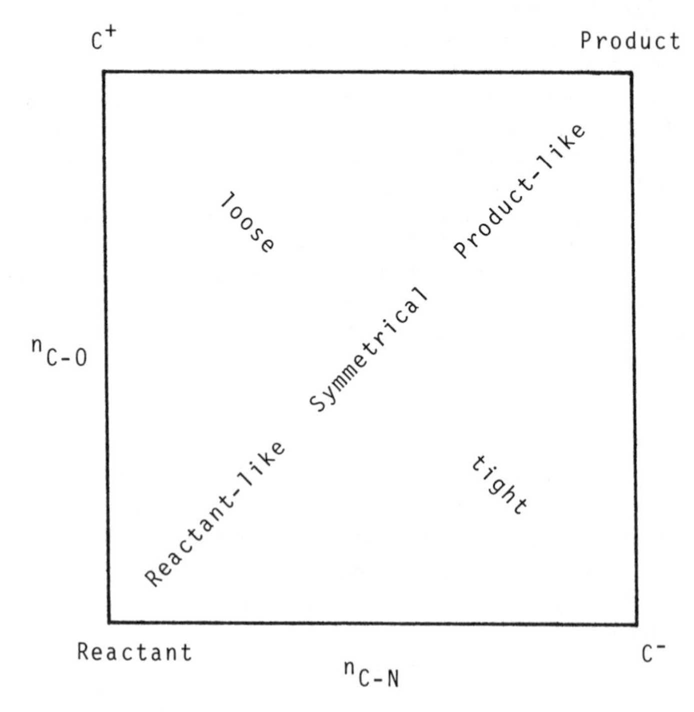

Figure 1. More O'Ferrall diagram of the Menschutkin-type reaction.

(*21*) and More O'Ferrall-type analysis. Our previous work (*1–4*), however, raised the following questions: (1) Is the bell-shaped behavior of the carbon KIE limited only to the benzyl system? (2) Does any experimental evidence distinguish between reactant-like and product-like TSs, both of which show smaller than maximum carbon KIE? (3) What is the role of solvent molecules and how do they behave in the TS of an S_N2 reaction? This chapter describes our recent studies aimed at answering these questions.

Kinetic Isotope Effects in Methyl-Transfer Reactions

The simplest S_N2 reaction involves methyl transfer between heteroatoms; thus, the possibility of bell-shaped behavior of the carbon KIE for such a substitution reaction was explored. No systematic measurements of KIEs for a series of multilabeled compounds were previously reported for methyl transfer. Carbon-14 and α-deuterium KIEs for reactions of methyl-^{14}C brosylate and methyl-d_3 brosylate with substituted *N,N*-dimethylanilines (equation 2; Y = *p*-CH_3O, *p*-CH_3, H, *p*-Br, and *m*-Br) were measured. Reactions were carried out in acetonitrile at 55 °C with 0.05 mol L^{-1} of methyl brosylate and 0.10 mol L^{-1} of nucleophiles. Reactions showed good second-order rate plots ($r > 0.999$) through at least 70% reaction. Carbon-14 KIEs

$$Y\text{-}C_6H_4N(CH_3)_2 + \overset{**}{C}H_3OSO_2C_6H_4Br \xrightarrow[CH_3CN]{55\,°C} [Y\text{-}C_6H_4N(CH_3)_2\overset{**}{C}H_3]^{+}\ {}^{-}OSO_2C_6H_4Br \quad (2)$$

Y = *p*-CH_3O, *p*-CH_3, H, *p*-Br, *m*-Br

Table I. Rate Constants and KIEs in the Reaction of Methyl Brosylate with Y-Substituted *N,N*-Dimethylanilines

Y	10^4k_2 ($L\ mol^{-1}\ s^{-1}$)	$^Hk/^Dk$ [a]	$^{12}k/^{14}k$
p-CH_3O	37.40 ± 0.5	0.983 ± 0.010	1.149 ± 0.001[b]
p-CH_3	18.80 ± 0.2	0.971 ± 0.009	1.152 ± 0.002[b]
H	6.91 ± 0.02	0.971 ± 0.010	1.162 ± 0.003[b]
p-Br	1.71 ± 0.02	0.953 ± 0.002	1.162 ± 0.009[c]
m-Br	0.965 ± 0.028	0.958 ± 0.011	1.163 ± 0.001[c]

NOTE: 0.05 mol L^{-1} in methyl brosylate and 0.10 mol L^{-1} in nucleophile in acetonitrile at 55.00 ± 0.02 °C.

[a] Average of two or three runs.
[b] For uncertainties, see reference 22.
[c] Average of two runs.

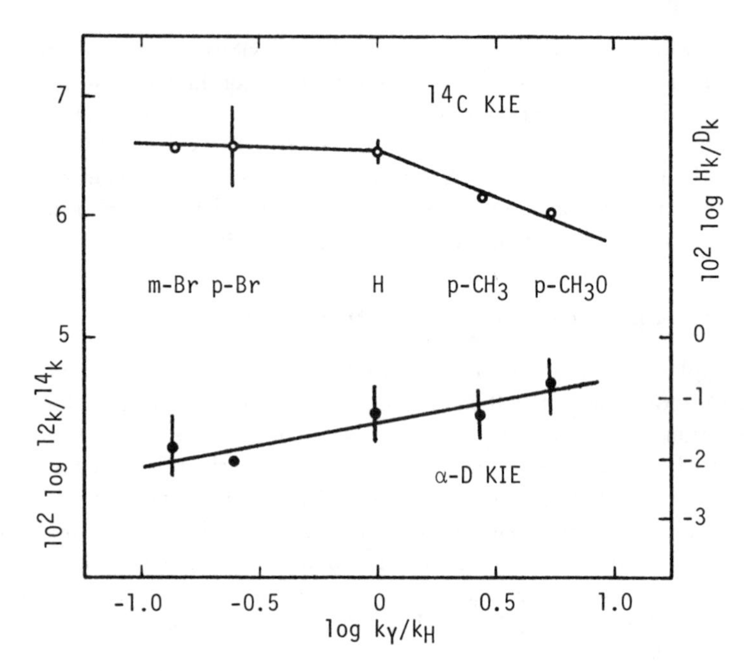

Figure 2. Carbon-14 and α-deuterium KIEs for the reaction of methyl brosylate with substituted N,N-*dimethylanilines plotted against relative rates.*

were determined by the standard method (*22*). Accurate radioactivities were measured by liquid scintillation counting of methyl brosylate recovered at various stages of reaction. Results are shown in Table I, and KIEs are plotted against relative rates in Figure 2. Deuterium KIEs are all smaller than unity, showing the typical S_N2 character of the methyl system, and decrease as the reaction rates decrease. Carbon-14 KIEs are all large, and the value of 1.163 at 55 °C is the largest ever reported. Apparently, a bell-shaped variation of the carbon KIEs is not observed; instead, the value increases and then levels at about 1.16 as the nucleophilicity of the aniline decreases. This observation may indicate that bell-shaped behavior is only inherent in the benzyl systems. However, the measurement of carbon KIEs for the methyl system has not been as extensive as for the benzyl one. Second-order rate constants varied by about 190 times for the benzyl derivatives but only 40 times for methyl brosylate.

Harris et al. reported (*11*) a large variation of the deuterium KIE in the Menschutkin reaction of methyl iodide with substituted pyridines (equation 3; X and Y = CH_3, H, and Cl). Because this large variation seems to be good evidence for a change of TS in this series of reactions, the carbon KIE was measured. Reactions were carried out under the same conditions as reported

$$\text{X,Y-C}_5\text{H}_3\text{N} + {}^{**}\text{CH}_3\text{I} \xrightarrow[2\text{-PrNO}_2]{25\,°\text{C}} \left[\text{X,Y-C}_5\text{H}_3\text{N-}{}^{**}\text{CH}_3\right]^+ \text{I}^- \quad (3)$$

X,Y = CH_3, H, Cl

previously (*11, 13*) by using both methyl iodide enriched with 99% ^{13}C and normal methyl iodide. Conductometric determination of rate constants gave the results shown in Table II and plotted in Figure 3 together with Harris et al.'s results (*11*).

Figure 3 shows that variation in the carbon KIE observed for reaction 3 is quite similar to that for reaction 2. Carbon-13 KIEs are all large, and the value slightly increases as the nucleophilicity of the pyridine decreases. Again, no maximum is observed in spite of the large variation in the rate constants (340-fold) and the concomitant large variation in the α-deuterium KIE. Although the similarity between the behavior of the carbon KIEs in the two series might be coincidental, possibly large and rather flat carbon KIEs are general for methyl transfer. The question then remains as to how the difference between the benzyl and methyl derivatives can be explained.

Model Calculation of Kinetic Isotope Effects

We quantitatively demonstrated previously (*23*) that, in the model calculations of KIEs for the benzyl derivatives, large carbon-14 KIEs (up to 1.16 at 35 °C) cannot be obtained with a simple symmetrical model (*24*) but can only be obtained with a modified model in which the Walden inversion motion was explicitly included as part of the reaction coordinate. Heavy-atom KIEs are approximated as a product of two factors: the temperature-independent imaginary frequency factor and the temperature-dependent zero-point energy factor. The previous calculations showed that the ratio of the imaginary frequencies ($\nu_{1L}^{\ddagger}/\nu_{2L}^{\ddagger}$) describing transmission over the TS barrier must be large. Furthermore, so that the large variation in the carbon KIE of the benzyl derivatives could be simulated, the contribution of the Walden inversion was varied with the relative strength of the two reacting bonds. The reaction-coordinate frequency should be largest when the TS is symmetrical and rapidly decrease when the TS becomes unsymmetrical. In contrast to the benzyl system, the large and rather flat carbon KIEs of the methyl derivatives suggest that the contribution of the Walden inversion is important throughout the series. As the Walden inversion motion is an essential part of the S_N2 reaction, methyl transfer, as a representative of S_N2, is probably always accompanied by the large Walden inversion contribution.

Model calculations of KIE for reactions 2 and 3 were carried out by use of the program BEBOVIB-IV (*25*) in the same manner as for reaction 1 (*23*). TS structures able to simulate the experimentally observed carbon and

Table II. Rate Constants and KIEs in the Reaction of Methyl Iodide with 3-X-5-Y-Substituted Pyridines

X	Y	$(10^4)^{12}k_2$ ($L\ mol^{-1}\ s^{-1}$)[a]	$(10^4)^{13}k_2$ ($L\ mol^{-1}\ s^{-1}$)[a]	$^{12}k/^{13}k$[b]	$^{H}k/^{D}k$[c]
CH_3	CH_3	4.247 ± 0.009	3.994 ± 0.013	1.063 ± 0.004	0.908
CH_3	H	2.394 ± 0.001	2.254 ± 0.005	1.062 ± 0.002	0.851
H	H	1.620 ± 0.003	1.520 ± 0.006	1.066 ± 0.005	0.850
Cl	H	0.1174 ± 0.0001	0.1093 ± 0.0002	1.074 ± 0.002	0.835
Cl	Cl	0.01249 ± 0.00012	0.01161 ± 0.00013	1.076 ± 0.016	0.810

NOTE: In 2-nitropropane at 25.00 ± 0.01 °C; 0.09–0.22 mol L^{-1} in pyridines, 0.0771 mol L^{-1} in methyl iodide, and 0.0766 mol L^{-1} in methyl iodide (99%, KOR and MSD).

[a] Average of at least three runs.

[b] Calculated by taking account of the ^{13}C content (99%).

[c] Reference 11.

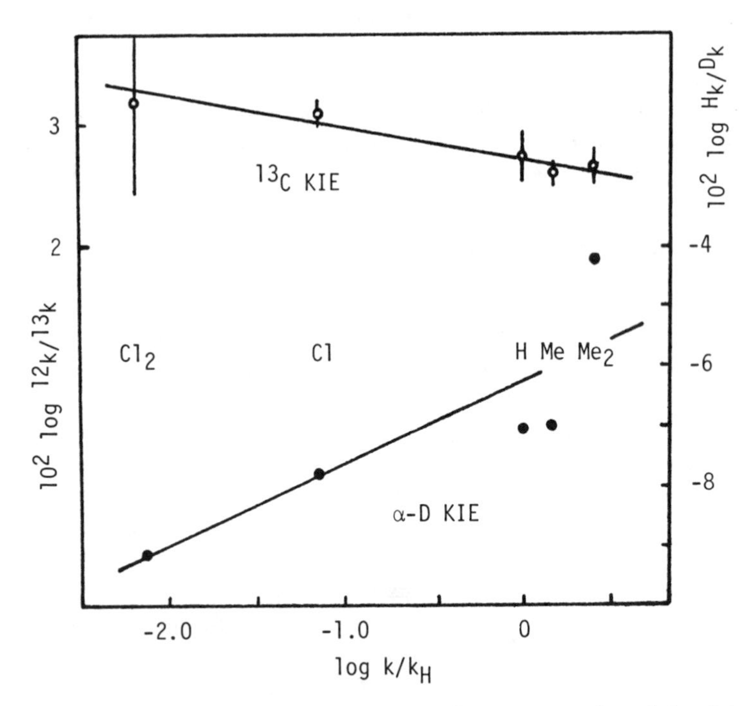

Figure 3. Carbon-13 and α-deuterium KIEs for reactions of methyl iodide with substituted pyridines plotted against relative rates.

hydrogen KIEs for the three reactions are mapped on the More O'Ferrall diagram in Figure 4. The conclusions obtained from this type of calculation are barely quantitative, but their semiquantitative conclusions demonstrated visually are quite informative. Figure 4 shows that (1) TSs for the benzyl derivatives are in the cationic region and looser than those for the methyl

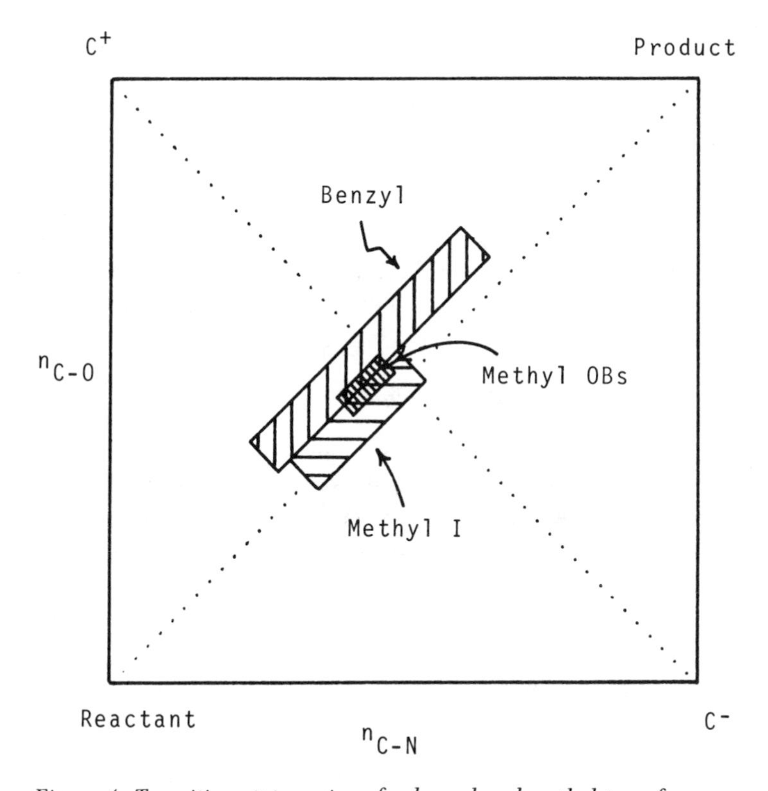

Figure 4. Transition-state regions for benzyl and methyl transfers.

derivatives, (2) TSs for the methyl derivatives are along the diagonal with constant total bonding and not in the anionic region as proposed by Harris et al. (*11*), and (3) variation in the TS structures shown by the shaded areas is greater for the benzyl derivatives than for the methyl ones. This variation indicates that the TS for benzyl transfer is more flexible than that for methyl transfer, presumably because of the greater polarizability of the phenyl group compared with that of hydrogen.

TS areas for the two methyl series overlap with each other in spite of the apparent difference in the deuterium KIEs; the KIEs are much smaller and more variable for reaction 3 than for reaction 2. These results are rationalized with the idea that secondary α-deuterium KIEs in S_N2 are primarily determined by the H/D fractionation factors of the reactant and the product (*26–28*). The much larger increase in the H/D fractionation factor for the I → N methyl transfer than for the O → N methyl transfer inevitably results in the observed behavior for similar TS structures.

These calculations were carried out with the TS model in which the contribution of the Walden inversion is varied with the symmetry of the TS.

For methyl-transfer reactions, the calculations could be performed with another TS model incorporating the constant contribution of the inversion motion. These latter calculations for reactions 2 and 3 produced somewhat longer TS regions extended into the reactant corner. Which model is most appropriate is difficult to determine, but the qualitative conclusion is the same: that is, a large inversion motion always accompanies methyl transfer but is important only for strictly symmetrical TSs in benzyl transfer. In other words, marked bell-shaped behavior of carbon KIEs may be characteristic of reactions in which TSs are shifted away from the diagonal of constant total bonding.

Nitrogen-15 Kinetic Isotope Effects

As an experimental method to answer question 2 (does any experimental evidence distinguish between reactant-like and product-like TSs), measurements of KIEs for atoms in the leaving group or the nucleophile may be promising. Thus, nitrogen-15 KIEs were measured for several substrates with varying X, Y, and Z in reaction 1 (*29*) (Table III).

Table III. Nitrogen-15 KIEs in the Reaction of Z-Benzyl X-Benzenesulfonates with *N,N*-Dimethyl-Y-anilines

X	Y	Z	$^{14}k/^{15}k$	10^4k_2 ($L\ mol^{-1}\ s^{-1}$)
p-CH_3O	*p*-CH_3	H	1.0028 ± 0.0004	2.36
H	*p*-CH_3	H	1.0028 ± 0.0004	7.82
p-Cl	*p*-CH_3	H	1.0027 ± 0.0002	27.2
p-Cl	*p*-Br	H	1.0038 ± 0.0003	3.88
p-Cl	*p*-CH_3O	H	1.0019 ± 0.0003	62.6
p-Cl	*p*-CH_3	*m*-Br	1.0020 ± 0.0004	16.3

NOTE: In acetone at 35.00 ± 0.02 °C; 0.008–0.015 mol L^{-1} in ester and 0.016–0.060 mol L^{-1} in nucleophiles.

These results were not previously discussed in detail because the KIEs are all close to unity and the variation in them is small. The variations in the KIE for nucleophiles and leaving groups, in several series of S_N2 reactions, are all much smaller than expected (*6, 15–20*) and a large change in the nitrogen nucleophile KIE once reported (*16*) was later admitted to be in error (J.L. Kurz, personal communication). In our previous paper (*29*), the imaginary frequency factor (1.01) times the zero-point energy factor (0.99) produced a KIE close to unity. The only detectable difference outside the error limit was observed for varied Y with fixed X (*p*-Cl) and Z (H); a smaller isotope effect was obtained for an electron-donating substituent (*p*-CH_3O $<$ *p*-CH_3 $<$ *p*-Br). As the TS becomes product-like, the zero-point energy factor should decrease because of an increasing bonding and the imaginary frequency factor is claimed to decrease to unity (*30, 31*). Thus, this trend in the nitrogen-15 KIE may be an indication of a later TS for a more electron-donating substituent, which is the so-called "anti-Hammond" behavior.

If this conclusion is correct, all the TS maps that we proposed so far should be reversed in the early–late character. However, possibly the observed, small variation in the nitrogen KIE could be consistent with the Hammond behavior.

In benzyl-transfer reactions, the TS is located in the cationic region because of the stability of the hypothetical intermediate, benzyl cation. In these reactions, the C–O bond breaking and N–C bond forming are not synchronous; the reaction coordinate for a reactant-like TS would be largely a change in the C–O bonding while that for a product-like TS would be largely a change in the N–C bonding (Figure 5) (*32*). If this idea is correct, then the imaginary frequency factor of the nitrogen KIE might be small for a reactant-like TS and large for a product-like TS, which is opposite to the previous estimation. Compensation by change in the zero-point energy factor might produce a difference of only 0.002. This difference is observed only on varying Y because substituent Y on the nucleophile has the strongest influence in determining the TS, as was seen by the sharp variation in the carbon KIE on varying Y (*1, 3*). Thus, we should wait, to answer question 2, until accurate KIEs for nucleophiles and leaving groups in various alkyl transfers are accumulated.

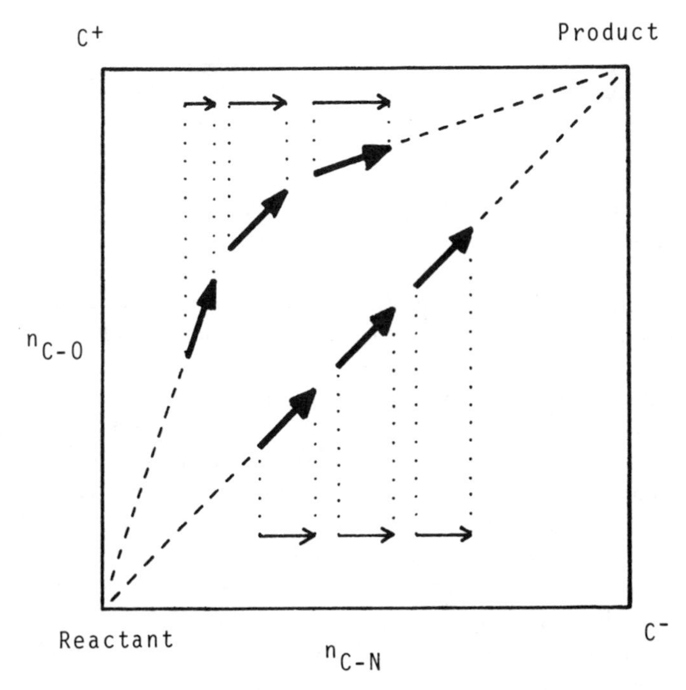

Figure 5. Schematic reaction coordinate for benzyl transfer.

Solvation Effects on Transition States

Because aliphatic nucleophilic substitution is ionic by nature, solvent molecules must play an important role as reactants proceed to TSs. In KIE studies on nucleophilic substitutions, however, solvent molecules have only been taken into consideration in a few cases. Solvation to the leaving chlorine atom in the TS of the solvolysis of *tert*-butyl chloride (*33*) and solvation to the β-hydrogens in the TS of the borderline solvolysis of an isopropyl ester (*34*) are typical examples. Kurz et al. (*35*) described methyl-transfer reactions by a two-dimensional diagram (one axis corresponds to changes in internal structure and the other to changes in solvent polarization) to interpret the deuterium KIE of deuterated hydroxylic solvents. Italian researchers (*36*) clearly demonstrated that hydration of nucleophilic anions by a small number of water molecules, under phase-transfer catalysis conditions in nonpolar solvent, causes a decrease of their reactivity in aliphatic nucleophilic substitution. To obtain some information on the possibility that hydration to nucleophiles may cause changes in the internal structure of the TS and that possible reorganization of hydrating water molecules during activation may correlate with the decreased reactivity, we measured α-deuterium KIEs for reaction 4 under hydrated (wet) and unhydrated (dry) conditions.

$$\underline{n}\text{-}C_7H_{15}\overset{**}{C}H_2OSO_2C_6H_4NO_2 + Q^+Y^- \longrightarrow \underline{n}\text{-}C_7H_{15}\overset{**}{C}H_2Y + Q^{+-}OSO_2C_6H_4NO_2 \quad (4)$$

where $Q^+ = n\text{-}C_{16}H_{33}P^+(C_4H_9)_3$ and $Y^- = SCN^-, I^-, Br^-, Cl^-$, or N_3^-. Reactions were carried out in chlorobenzene at 50 °C, with 3×10^{-1} mol L^{-1} of ester and $(6\text{–}10) \times 10^{-3}$ mol L^{-1} of nucleophiles, and followed by spectrophotometry. For wet conditions, the chlorobenzene solutions of Q^+Y^- were equalibrated with 4 mol L^{-1} of aqueous solution of KY before use (Table IV).

The data in Table IV show that α-deuterium KIEs under dry and wet conditions are almost the same (*37*). Although the values for SCN^- show some difference, the result from wet conditions contains a large uncertainty. These constant KIEs show that any change of the TS by hydration, if it occurs, is not of a type that can be detected by α-deuterium KIEs. Because the TS of this reaction has a delocalized negative charge, nucleophilic hydration to α-hydrogens may be unimportant. The tight–loose character of the TS is not affected by hydration.

Activation parameters under dry and wet conditions in cyclohexane were also measured for reaction 4 (Table V). Reactions were carried out at

20–40 °C with 2 × 10^{-4} mol L^{-1} of ester and 2 × 10^{-3} mol L^{-1} of nucleophiles. Wet conditions were prepared by adding (6–8) × 10^{-3} mol L^{-1} of water to the cyclohexane solution. The results indicate that the decreased reactivities under wet conditions are attributed to the increase in activation enthalpies, even though activation entropies increase by hydration. This tendency is more noticeable as the nucleophile becomes harder, that is, I < Br < Cl. These observations are consistent with the idea that the activation process under wet conditions includes desolvation or reorganization of hydrated water molecules.

Our study on the solvation to the TS of aliphatic nucleophilic substitution is now in its infancy, and the results are only preliminary. We hope that measurements of KIEs will also be effective in this study, as they were in our previous studies.

Table IV. Rate Constants and α-Deuterium KIEs in the Reaction of *n*-Octyl *p*-Nitrobenzenesulfonate with Quaternary Phosphonium Salts in Dry and Wet Chlorobenzene at 50 °C

X^-	k_2 *(dry)* ($L\ mol^{-1}\ s^{-1}$)	k_2 *(wet)* ($L\ mol^{-1}\ s^{-1}$)	${}^Hk/{}^Dk$ *(dry)*	${}^Hk/{}^Dk$ *(wet)*
SCN	0.02062 ± 0.00013	0.01477 ± 0.00014	1.090 ± 0.007	1.066 ± 0.031
I	0.1009 ± 0.0010	0.0840 ± 0.0002	1.141 ± 0.018	1.154 ± 0.018
Br	0.2407 ± 0.0024	0.1425 ± 0.0002	1.106 ± 0.012	1.095 ± 0.002
Cl	0.5494 ± 0.0001	0.1387 ± 0.0009	1.087 ± 0.010	1.082 ± 0.012
N_3	1.549 ± 0.018	0.805 ± 0.007	1.070 ± 0.019	1.059 ± 0.012

Table V. Activation Parameters in the Reaction of *n*-Octyl *p*-Nitrobenzene sulfonate with Quaternary Phosphonium Halides in Dry and Wet Cyclohexane

Parameter	*dry*			*wet*		
	Cl^-	Br^-	I^-	Cl^-	Br^-	I^-
$k_2^\ddagger$(at 30 °C; L mol^{-1} s^{-1}	1.0607	0.6123	0.3073	0.549	0.344	0.249
$\Delta H^\ddagger$ (kcal mol^-)	14.3	15.1	16.0	16.5	16.3	16.5
$\Delta S^\ddagger$ (cal deg^{-1} mol^{-1})	−10.5	−9.8	−8.1	−5.0	−6.8	−7.0

Literature Cited

1. Ando, T.; Tanabe, H.; Yamataka, H. *J. Am. Chem. Soc.* **1984,** *106*, 2084–2088.
2. Yamataka, H.; Ando, T. *Tetrahedron Lett.* **1975,** *16*, 1059–1062.
3. Yamataka, H.; Ando, T. *J. Am. Chem. Soc.* **1979,** *101*, 266–267.
4. Yamataka, H.; Ando, T. *Tetrahedron Lett.* **1982,** *23*, 4805–4808.
5. Albery, W. J.; Kreevoy, M. M. *Prog. Phys. Org. Chem.* **1978,** *13*, 87–157.
6. Westaway, K. C.; Ali, S. F. *Can. J. Chem.* **1979,** *57*, 1354–1367.
7. Young, P. R.; Jencks, W. P. *J. Am. Chem. Soc.* **1979,** *101*, 3288–3294.

8. Harris, J. M.; Shafer, S. G.; Moffatt, J. R.; Becker, A. R. *J. Am. Chem. Soc.* **1979,** *101,* 3295–3300.
9. Ahsan, M.; Robertson, R. E.; Blandamer, M. J.; Scott, J. M. W. *Can J. Chem.* **1980,** *58,* 2142–2145.
10. Vitullo, V. P.; Grabowski, J.; Sridharan, S. *J. Am. Chem. Soc.* **1980,** *102,* 6463–6465.
11. Harris, J. M.; Paley, M. S.; Prasthofer, T. W. *J. Am. Chem. Soc.* **1981,** *103,* 5915–5916.
12. Waszczylo, Z.; Westaway, K. C. *Tetrahedron Lett.* **1982,** *23,* 143–146.
13. Arnett, E. M.; Reich, R. *J. Am. Chem. Soc.* **1980,** *102,* 5892–5902.
14. Johnson, C. D. *Tetrahedron Lett.* **1982,** *23,* 2217–2218.
15. Westaway, K. C.; Waszczylo, Z. *Can. J. Chem.* **1982,** *60,* 2500–2520.
16. Kurz, J. L.; Seif El-Nasr, M. M. S. *J. Am. Chem. Soc.* **1982,** *104,* 5823–5824.
17. Swain, C. G.; Thornton, E. R. *J. Org. Chem.* **1961,** *26,* 4808–4809.
18. Grimsrud, E. P.; Taylor, J. W. *J. Am. Chem. Soc.* **1970,** *92,* 739–741.
19. Hargreaves, R. T.; Katz, A. M.; Saunders, W. H., Jr. *J. Am. Chem. Soc.* **1976,** *98,* 2614–2617.
20. Friedberger, M. P.; Thornton, E. R. *J. Am. Chem. Soc.* **1976,** *98,* 2861–2865.
21. Thornton, E. R. *J. Am. Chem. Soc.* **1967,** *89,* 2915–2927.
22. Ando, T.; Kim, S. G.; Matsuda, K.; Yamataka, H.; Yukawa, Y.; Fry, A.; Lewis, D. E.; Sims, L. B.; Wilson, J. C. *J. Am. Chem. Soc.* **1981,** *103,* 3505–3516.
23. Yamataka, H.; Ando, T. *J. Phys. Chem.* **1981,** *85,* 2281–2286.
24. Sims, L. B.; Fry, A.; Netherton, L. T.; Wilson, J. C.; Reppond, K. D.; Crook, S. W. *J. Am. Chem. Soc.* **1972,** *94,* 1364–1365.
25. Sims, L. B.; Burton, G. W.; Lewis, D. E. *Quantum Chem. Program Exch.* **1977,** *No. 337,* BEBOVIB-IV.
26. Hartshorn, S. R.; Shiner, V. J., Jr. *J. Am. Chem. Soc.* **1972,** *94,* 9002–9012.
27. Buddenbaum, W. E.; Shiner, V. J., Jr. *Isotope Effects on Enzyme-Catalyzed Reactions;* Cleland, W. W.; O'Leary, M. H.; Northrop, D. B., Eds.; University Park: Baltimore, MD, 1977; pp 1–36.
28. Gray, C. H.; Coward, J. K.; Schowen, K. B.; Schowen, R. L. *J. Am. Chem. Soc.* **1979,** *101,* 4351–4358.
29. Ando, T.; Yamataka, H.; Wada, E. *Isr. J. Chem.* **1985,** *26,* 354–356.
30. Bigeleisen, J.; Wolfsberg, M. *J. Chem. Phys.* **1954,** *22,* 1264.
31. Melander, L.; Saunders, W. H., Jr. *Reaction Rates of Isotopic Molecules;* Wiley-Interscience: New York, 1980, pp 315–318.
32. Kurz, J. L. *J. Org. Chem.* **1983,** *48,* 5117–5120.
33. Burton, G. W.; Sims, L. B.; Wilson, J. C.; Fry, A. *J. Am. Chem. Soc.* **1977,** *99,* 3371–3379.
34. Yamataka, H.; Tamura, S.; Hanafusa, T.; Ando, T. *J. Am. Chem. Soc.* **1985,** *107,* 5429–5434.
35. Kurz, J. L.; Lee, J.; Rhodes, S. *J. Am. Chem. Soc.* **1981,** *103,* 7651–7653.
36. Landini, D.; Maia, A.; Montanari, F. *J. Am. Chem. Soc.* **1978,** *100,* 2796–2801.
37. Landini, D.; Maia, A.; Montanari, F.; Rolla, F. *J. Org. Chem.* **1983,** *48,* 3774–3777.

RECEIVED for review October 21, 1985. ACCEPTED February 10, 1986.

8

Nonperfect Synchronization in Nucleophilic Additions to Carbon–Carbon Double Bonds

Claude F. Bernasconi

Thimann Laboratories of the University of California, Santa Cruz, CA 95064

The principle of nonperfect synchronization states that any product-stabilizing factor that develops late along the reaction coordinate, or any reactant-stabilizing factor that is lost early along the reaction coordinate, has the effect of increasing the intrinsic *barrier of a reaction. Much of the structure–reactivity behavior in nucleophilic additions to activated C=C double bonds can be understood as the manifestation of this principle. This behavior includes the effect of resonance by substituents adjacent to the site of negative charge development (or its disappearance) and the effect of remote substituents, the solvent, steric crowding, intramolecular hydrogen bonding, and intramolecular electrostatic stabilization.*

THE NATURE OF THE INTERACTION MECHANISMS that stabilize the carbanion profoundly influences the rates of reactions that lead to the generation of carbanions. For the most thoroughly studied carbanion-forming reactions, the ionization of carbon acids, this fact has been recognized for a long time and discussed in a number of reviews (*1–7*).

$$\mathrm{RCHXY} + \mathrm{B}^{v} \underset{k_{-1}}{\overset{k_1}{\rightleftharpoons}} \mathrm{RC}\langle^{X}_{Y}\rangle^{-} + \mathrm{BH}^{v+1} \qquad (1)$$

One characteristic feature of these reactions is that k_0, the intrinsic rate constant,[1] tends to become smaller [or the intrinsic barrier[1] increases] as X

[1] The intrinsic rate constant, k_0, is usually defined as $k_1 = k_{-1}$ when $pK_a^{BH} = pK_a^{CH}$, although sometimes statistical factors are included, $k_0 = k_1/q = k_{-1}/p$ when $pK_a^{BH} = pK_a^{CH} + \log(p/q)$. This intrinsic barrier, $\Delta G_0^‡$, is $\Delta G_1^‡ = \Delta G_{-1}^‡$ for $\Delta G° = 0$ or its statistically corrected counterpart.

0065-2393/87/0215-0115$06.50/0

Table I. log k_0 for Deprotonation of CH_2XY by Amines and for Amine Addition to $H_5C_6CH{=}CXY$ in 50% $(CH_3)_2SO$–50% Water at 20 °C

$<^{X}_{Y}$	$\log k_0(C\text{–}H)$	$\log k_0(C{=}C)$
$<^{CN}_{CN}$	~7.00[a]	4.94[b]
$<^{CN}_{C_6H_4\text{-}4\text{-}NO_2}$	3.95[c]	3.35[d]
$<^{CN}_{C_6H_3\text{-}2,4\text{-}(NO_2)_2}$	2.90[c]	2.65[d]
$<^{COCH_3}_{COCH_3}$	2.75[e]	0.30[f]
$<^{H}_{NO_2}$	0.73[g]	2.55[h]
$<^{C_6H_5}_{NO_2}$	−0.25[g]	1.42[i]

NOTE: Amines are piperidine and morpholine.
[a] Estimated from Hibbert's (5) data.
[b] Reference 28.
[c] Reference 32.
[d] Reference 29.
[e] Reference 15.
[f] C. F. Bernasconi and A. Kanavarioti, unpublished results.
[g] C. F. Bernasconi, A. Mullin, and D. Kliner, unpublished results.
[h] Reference 30.
[i] C. F. Bernasconi and R. A. Renfrow, unpublished results.

and Y become better π acceptors (Table I). This trend in k_0 has generally been attributed to a greater need for structural and solvational reorganization that accompanies the formation of a resonance-stabilized ion (*1–7*), factors that presumably enhance the intrinsic barrier. The poorer hydrogen-bonding capabilities of carbon acids and of the carbanionic carbon when X and Y become better π acceptors have also frequently been suggested as additional important factors (*1–7*). However, this feature has more often been invoked to explain why carbon acids as a class are ionized more slowly than normal acids rather than to rationalize differences among different types of carbon acids. We shall return to this point.

Another striking feature of these reactions actually helps us visualize how this need for reorganization can contribute to the intrinsic barrier. This feature is the disparity or imbalance in measured structure–reactivity coefficients such as Brønsted α and β values. Indeed, the Brønsted α values obtained by varying a remote substituent in the carbon acid, for example, Z in **I** (equation 2), are usually larger than the Brønsted β values obtained by varying the base, that is, $I = \alpha - \beta > 0$ (Table II).

Table II. α and β Values in Proton Transfers

C–H Acid	*Base*	α	β	$I = \alpha - \beta$
$ArCH_2(CN)_2$[a]	$RCOO^-$	0.98	1.00	~0.00
$ArCH_2CH(COCH_3)_2$[a]	$RCOO^-$	0.58	0.44	0.14
$ArCH_2CH(COCH_3)COOCH_2CH_3$[a]	$RCOO^-$	0.76	0.44	0.32
$ArCH_2C_6H_3$-2,4-$(NO_2)_2$[b]	R_2NH	0.87	0.45	0.42
$ArCH(CH_3)NO_2$[c]	R_2NH	1.29	0.55	0.74

NOTE: In water at 25 °C unless stated otherwise.
[a] Reference 31.
[b] In 50% $(CH_3)_2SO$–50% water at 20 °C; reference 33.
[c] Reference 41.

These imbalances can be interpreted in terms of a transition state (**II**) in which delocalization of the negative charge into X and Y and concomitant solvation (equation 2) have made little progress compared to the degree of proton transfer. The low intrinsic rate constants for the formation of strongly resonance-stabilized carbanions thus seem to be a consequence of a transition state that is of high energy compared to the product (**III**), because the transition state is not able to benefit from the resonance stabilization and solvation of **III**.

$$\underset{\textbf{I}}{Z\text{-}C_6H_4\text{-}CHXY} + B^{\nu} \rightleftharpoons \underset{\textbf{II}}{\left[B^{\nu+\delta}\cdots H\cdots \overset{-\delta}{C}XY\text{-}C_6H_4\text{-}Z\right]^{\ddagger}} \rightleftharpoons \underset{\textbf{III}}{(HO\text{-}H\cdots)X\overset{-}{\cdots}C\cdots Y(\cdots H\text{-}OH)\text{-}C_6H_4\text{-}Z} \qquad (2)$$

In searching for a (semi)quantitative formalism to describe these observations, we expressed (8) the decrease in k_0 that arises from the late development of resonance by

$$\delta \log k_0^{res}(C^-) = (\alpha_{res}^{C^-} - \beta)\delta \log K_1^{res}(C^-) \tag{3}$$

where $\delta \log k_0^{res}(C^-)$ is the change in $\log k_0$ relative to a (hypothetical) reaction in which no resonance stabilization of the carbanion by X and Y occurs. $\delta \log K_1^{res}(C^-)$ is the increase in the equilibrium constant of reaction 1 induced by the resonance effect, again relative to the case where X and Y have no π acceptor capabilities. β is the experimental Brønsted coefficient obtained by varying the base, while $\alpha_{res}^{C^-}$, not to be confused with the Brønsted α, is a parameter that measures the progress of resonance stabilization in the transition state. $\alpha_{res}^{C^-}$, which is not experimentally accessible, is defined (8) by

$$\alpha_{res}^{C^-} = \delta \log k_1^{res}(C^-)/\delta \log K_1^{res}(C^-) \tag{4}$$

where $\delta \log k_1^{res}(C^-)$ is the increase in the rate constants, k_1, of reaction 1 induced by the resonance effect.

If resonance development is retarded relative to what is being measured by β (presumably the charge change $+\delta$ on the base), that is, $\alpha_{res}^{C^-} < \beta$, the intrinsic rate consant is reduced because $\delta \log K_1^{res}(C^-) > 0$ (equation 3). On the other hand, if resonance development were synchronous with charge change ($\alpha_{res}^{C^-} = \beta$), no effect on k_0 would occur; if resonance development were ahead ($\alpha_{res}^{C^-} > \beta$), k_0 would be higher than that in the reference reaction.

If $\alpha_{res}^{C^-} - \beta$ does not vary greatly from one system to another, then, according to equation 3, the lower intrinsic rate constants for carbon acids with better π acceptors are a consequence of a larger $\delta \log K_1^{res}(C^-)$.

Equation 3 can be generalized to apply to any factor (f) that affects the free energy of reactants or products. $\delta \log k_0^f$ and $\delta \log K_1^f$ describe the change induced by f compared to a reference reaction where f is absent. If f is in the product, α_f measures the progress in the development of the factor in the transition state; if f is in the reactant, α_f is a measure of how much of it has been lost in the transition state.

$$\delta \log k_0^f = (\alpha_f - \beta)\delta \log K_1^f \tag{5}$$

On the basis of equation 5, we (8) formulated the principle of nonperfect synchronization (PNS)[2] as follows: A product-stabilizing factor ($\delta \log K_1^f > 0$) always lowers k_0 if it develops late ($\alpha_f < \beta$) but increases k_0 if it develops early

[2] In reference 9, we called it the principle of imperfect synchronization.

($\alpha_f > \beta$). A reactant-stabilizing factor ($\delta \log K_1^f < 0$) always lowers k_0 if it is lost early ($\alpha_f > \beta$) but increases k_0 if it is lost late ($\alpha_f < \beta$). For product- and reactant-destabilizing factors, the opposite relations hold.

Besides resonance stabilization of the carbanion, solvation has a major effect on k_0 of proton-transfer reactions. In a first approximation, we only consider the solvation of ionic species. For the solvation of the carbanion, equation 5 takes on the form

$$\delta \log k_0^{sol}(C^-) = (\alpha_{sol}^{C^-} - \beta)\delta \log K_1^{sol}(C^-) \quad (6)$$

For the solvation of BH^{v+1} with $v = 0$

$$\delta \log k_0^{sol}(BH^+) = (\alpha_{sol}^{BH^+} - \beta)\delta \log K_1^{sol}(BH^+) \quad (7)$$

while for the desolvation of B^v with $v = -1$, equation 5 becomes

$$\delta \log k_0^{des}(B^-) = (\alpha_{des}^{B^-} - \beta)\delta \log K_1^{des}(B^-) \quad (8)$$

A growing body of evidence indicates that solvation lags behind (desolvation is ahead of) bond changes (*7–16*). Hence, each of these solvational factors depresses k_0 [$\alpha_{sol}^{C^-} < \beta$, $\delta \log K_1^{sol}(C^-) > 0$; $\alpha_{sol}^{BH^+} < \beta$, $\delta \log K_1^{sol}(BH^+) > 0$; and $\alpha_{des}^{B^-} > \beta$, $\delta \log K_1^{des}(B^-) < 0$].

Nucleophilic Addition to C=C Double Bonds

Another important carbanion forming process is the addition of a nucleophile to an activated olefin.

$$\mathrm{>C{=}C(X)(Y)} + Nu^v \underset{k_{-1}}{\overset{k_1}{\rightleftharpoons}} \mathrm{-\overset{|}{\underset{Nu^{v+1}}{C}}-\overset{(-)}{C}(X)(Y)} \quad (9)$$

In view of the similarity between the carbanions formed in reactions 9 and 1, reaction 9 might be expected to show similar reactivity patterns as reaction 1. However, an important difference between the two reactions is that in the nucleophilic addition the procarbanionic carbon is sp^2 hybridized while in reaction 1 it is sp^3 hybridized. This difference is likely to bring about some modifications in the reactivity pattern of the nucleophilic additions. Also, as will become apparent, several factors not observed in proton transfers play an important role in nucleophilic addition reactions. These factors can also be understood in the context of the PNS.

Intrinsic Rate Constants as a Function of X and Y. The largest set of data obtained under comparable conditions refers to the addition of piperidine and morpholine to benzylidene-type substrates in 50% $(CH_3)_2SO$–50% water at 20 °C.

$$\text{Z-C}_6\text{H}_4\text{-CH=CXY} + \text{R}_2\text{NH} \underset{k_{-1}}{\overset{k_1}{\rightleftharpoons}} \text{Z-C}_6\text{H}_4\text{-CH(R}_2\overset{+}{\text{N}}\text{H)-}\bar{\text{C}}\text{XY} \qquad (10)$$

IV V

k_0 for these reactions was obtained as $k_0 = k_1 = k_{-1}$ by interpolation or extrapolation of plots of log k_1 (or log k_{-1}) versus log K_1 to log $K_1 = 0$[3]. Table I lists log k_0 values along with data for the corresponding proton transfers. Figure 1 shows a plot of log k_0 for reaction 10 versus log k_0 for reaction 1 in which $R' = H$.

Except for the strongly deviating point for benzylideneacetylacetone to be discussed, a remarkably good correlation between log $k_0(C{=}C)$ and log $k_0(C{-}H)$ exists: X and Y indeed affect the intrinsic rate constants for both reactions in a qualitatively similar way.

Two different lines are present in Figure 1. One line is the best least-squares line through all points except that for benzylideneacetylacetone. This line has a slope of 0.45 ± 0.06 and is based on the notion that the deviations from the correlation are random or caused by poorly understood factors.

The second line is through the points for benzylidenemalononitrile and β-nitrostyrene only. This line has a slope of 0.38 and is based on the hypothesis that all other points deviate negatively because of a steric effect that lowers log $k_0(C{=}C)$.

Regardless of which line is preferred, the small slopes show that the effect of X and Y on reaction 10 is strongly attenuated compared to that on proton transfers.

Similar conclusions are reached when comparing the reactivity of benzylidene-type substrates, **IV**, with respect to attack by hydroxide ion. If adjustments for the dependence of the equilibrium constants on X and Y are made (*18*, *19*), the relative intrinsic rate constants can be estimated (Table III). As with the amine reactions, the k_0 values follow the same order as for

[3] This definition of k_0 creates a problem of units because k_1 is in units of $M^{-1}\,s^{-1}$ and k_{-1} in s^{-1}. A possible solution to the problem was suggested by Hine (*17*) but this solution suffers from the disadvantage of having to assume the same equilibrium constant for encounter complex formation between nucleophile and electrophile in all reactions. In terms of *relative* k_0 values, little difference exists between our and Hine's definitions.

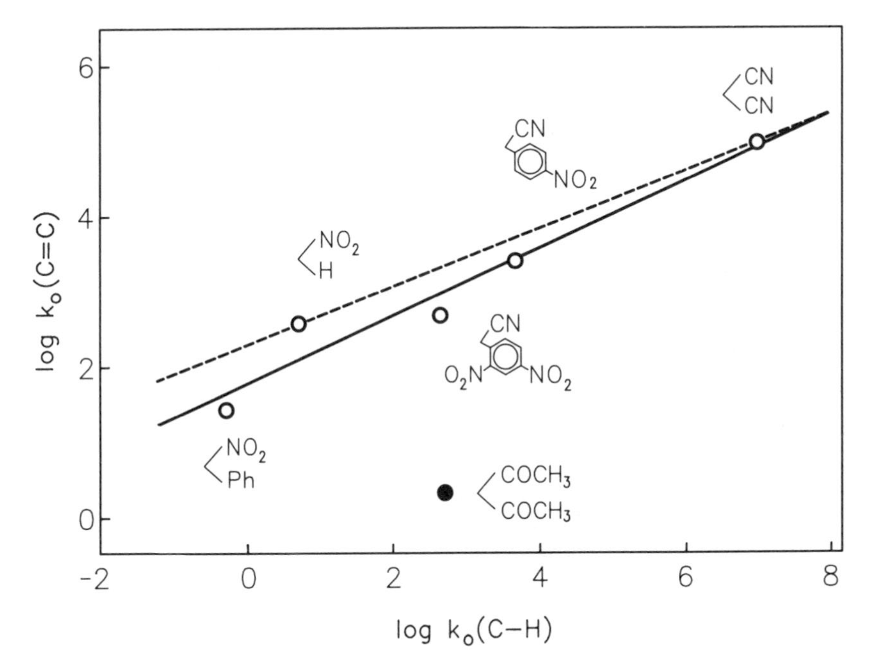

Figure 1. Plot of intrinsic rate constants for piperidine and morpholine addition to $C_6H_5CH{=}CXY$ versus intrinsic rate constants for deprotonation of CH_2XY by piperidine and morpholine. The data are from Table I. The solid circle corresponds to $XY = (COCH_3)_2$.

proton transfers, and a correlation of log $[k_0^{(CN)_2}/k_0X^{XY}]_{C=C}$ versus log $[k_0^{(CN)_2}/k_0^{XY}]_{C-H}$ (not shown) yields a slope of 0.57 ± 0.06; this slope shows again the attenuation of the effect of X and Y. However, the attenuation appears to be significantly less than in the amine reactions (slope of 0.45 ± 0.06 or 0.38). In a later section, we shall argue that the smaller slope for the amine reactions can be attributed to internal stabilization of **V** by an electrostatic effect.

Further examples demonstrating the reactivity pattern $k_0^{(CN)_2} \gg k_0^{(NO_2)H}$ were recently provided by Hoz (*20*, *21*) in the reactions of various nucleophiles with activated 9-methylenefluorenes.

Imbalance as a Function of X and Y. Nucleophilic additions to olefins also show imbalances in the structure–reactivity coefficients. In analogy to the proton transfers, $I = \alpha_{nuc}^{\,n} - \beta_{nuc}^{\,n}$ is defined as a measure of the imbalance: $\beta_{nuc}^{\,n}$ is the normalized β_{nuc} value;[4] for $\alpha_{nuc}^{\,n}$ (measure of negative

[4] $\beta_{nuc}^{\,n} = \beta_{nuc}(k_1)/\beta_{eq}(K_1)$; $\rho_n = \rho(k_1)/\rho(K_1)$.

Table III. Relative Intrinsic Rate Constants for OH^- Addition to $H_5C_6CH{=}CXY$ and for Deprotonation of CH_2XY Expressed as log $[k_0(CN)_2/k_0XY]$

X ⟨ Y	*C–H*	*C=C*
CN ⟨ CN	0	0
COO CH$_3$[a] ⟨ ✕ COO CH$_3$	~3.4	~2.2
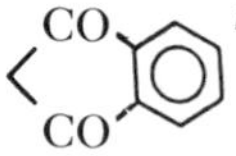 CO (benzene ring) [b] ⟨ CO	~3.9	~1.9
H[a] ⟨ NO_2	~7.3	~4.2

NOTE: Reproduced from reference 19. Copyright 1985 American Chemical Society.
[a] In water at 25 °C.
[b] In 50% $(CH_3)_2SO$–50% water at 20 °C.

charge development), ρ_n, which is the normalized ρ value obtained by varying Z in substrates like **IV** or **VI**, can be used in some cases.

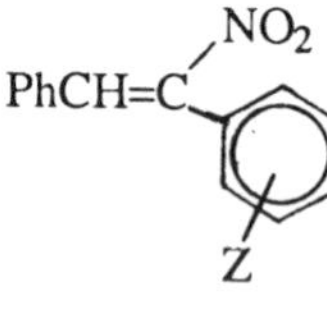

$PhCH{=}C(NO_2)(C_6H_4Z)$

VI

However, in reactions where the nucleophile is an amine, the developing positive charge on the nitrogen leads to an exalted ρ_n value, which overestimates $\alpha_{nuc}{}^n$ (22). In such cases, $\alpha_{nuc}{}^n$ is equated with $\rho_n(C^-)$, which is the component of ρ_n that can be attributed to the *negative* charge (22).

Examples of $\alpha_{nuc}{}^n$ and $\beta_{nuc}{}^n$ values are summarized in Table IV. Just as $I = \alpha - \beta > 0$ in proton transfers, $I = \alpha_{nuc}{}^n - \beta_{nuc}{}^n > 0$ in nucleophilic additions except that the imbalances in the latter reactions are generally significantly smaller. The only exception is with XY = $(CN)_2$; here I for the proton transfer seems abnormally small, probably because this reaction behaves almost like a diffusion-controlled proton transfer (5).

Table IV. α_{nuc}^{n} and β_{nuc}^{n} Values in Nucleophilic Additions to Olefins

Olefin	*Nucleophile*	α_{nuc}^{n}	β_{nuc}^{n}	$I = \alpha_{nuc}^{n} - \beta_{nuc}^{n}$
$ArCH{=}C(CN)_2$ [a,b]	R_2NH	0.43[i]	0.30	0.13
$ArCH{=}C(COO)_2C(CH_3)_2$ [d,e]	R_2NH	0.19[i]	0.07	0.12
$ArCH{=}C(COO)_2(CH_3)_2$ [d,f]	ArO^-	0.59[h]	0.39	0.20
$ArCH{=}CHNO_2$ [d,h]	R_2NH	0.51[b]	0.25	0.26
$C_6H_5CH{=}C(Ar)NO_2$ [a,i]	R_2NH	0.65[i]	0.34	0.31

[a] In 50% $(CH_3)_2SO$–50% water at 20 °C.
[b] $\alpha_{nuc}^{n} = \rho_n(C^-)$, see the text.
[c] C. F. Bernasconi and R. A. Renfrow, unpublished results.
[d] C. F. Bernasconi and R. B. Killon, unpublished results.
[e] C. F. Bernasconi and M. Panda, unpublished results.
[f] In water at 25 °C.
[g] Reference 22.
[h] Reference 42.
[i] $\alpha_{nuc}^{n} = \rho_n$, see the text.

The most obvious explanation for the smaller imbalances in the olefin reactions is that the lag in resonance development and solvation of the carbanion is less extreme than in proton transfers because the sp^2 hybridization makes it difficult to localize the negative charge on carbon, unless the carbon assumes substantial sp^3 character in the transition state. Possibly, the negative charge does not accumulate at all on carbon and the imbalances may be entirely due to late solvation.

In the context of this interpretation, the smaller sensitivity of log k_0(C=C) to X and Y compared with that of log k_0(C–H) can be understood as a direct consequence of the smaller imbalances. This correlation is easily seen if equations 3 and 6 (after replacing β with β_{nuc}^{n}) are applied to the present situation. A small imbalance implies that the $\alpha_{res}^{C^-} - \beta_{nuc}^{n}$ and $\alpha_{sol}^{C^-} - \beta_{nuc}^{n}$ terms in equations 3 and 6, respectively, are not strongly negative. Because, on the other hand, $\delta \log K_1^{res}(C^-)$ and $\delta \log K_1^{sol}(C^-)$ for the nucleophilic additions are expected to be quite comparable to the same quantities in the proton transfers, the smaller sensitivity of log k_0(C=C) to X and Y must be mainly the consequence of $\alpha_{res}^{C^-} - \beta_{nuc}^{n}$ and $\alpha_{sol}^{C^-} - \beta_{nuc}^{n}$ being smaller than the corresponding terms for the proton transfers.

This interpretation is not the whole story, though. In reactions of the type

$$\underset{\displaystyle \mathrm{H}\;\;\mathrm{O^-}}{\overset{\displaystyle \mathrm{H_5C_6}}{\mathrm{CCHXY}}} \rightleftharpoons \mathrm{C_6H_5CH{=}O} + \mathrm{HC}\!\left(\begin{matrix}\mathrm{X}\\ -\\ \mathrm{Y}\end{matrix}\right. \tag{11}$$

the sensitivity of log k_0 to X and Y is comparable to that of log k_0(C=C) (*18*, *19*). A large uncertainty in the relative k_0 values of these reactions exists, which renders this conclusion rather tentative, but even when these uncertainties are taken into consideration, the sensitivity of log k_0 to X and Y in these reactions does not approach that of log k_0(C–H). This fact is surprising because in reaction 11 the procarbanionic carbon is sp^3-hybridized just as in a proton transfer, and therefore, reaction 11 would be expected to behave more like a proton transfer than a nucleophilic addition.

These results suggest that in proton transfers an additional factor, not present in the other carbanion-forming processes, determines relative intrinsic rate constants. We tentatively suggest that this factor is the difference in the hydrogen-bonding capabilities of the carbon acids and the carbanionic carbon mentioned in the introduction. In other words, the hydrogen-bonding factor may be important not only in distinguishing most carbon acids as a class from the normal acids but also in discriminating among various types of carbon acids according to the π-acceptor ability of X and Y. In fact, our results, if they can be corroborated, may eventually constitute the most compelling evidence for the importance of the hydrogen-bonding factor.

Electrostatic Effects on k_0. As mentioned earlier, for the reactions with amine nucleophiles, the sensitivity of log k_0(C=C) to X and Y is smaller (slope of 0.38 or 0.45) than that in the reactions with hydroxide ion (slope of 0.57). This smaller sensitivity can be understood in terms of an electrostatic effect. The internal electrostatic stabilization of the zwitterionic adduct should enhance the equilibrium constant of the reaction. If this effect were poorly developed in the transition state ($\alpha_{el} < \beta_{nuc}{}^n$), it would depress k_0 according to

$$\delta \log k_0{}^{el} = (\alpha_{el} - \beta_{nuc}{}^n)\delta \log K_1{}^{el} \tag{12}$$

Equation 12 raises two questions: (1) Is it reasonable to assume that $\alpha_{el} < \beta_{nuc}{}^n$? (2) Does $\delta \log k_0{}^{el}$ depend on X and Y in such a way as to decrease $\partial \log k_0(\mathrm{C{=}C})/\partial \log k_0(\mathrm{C{-}H})$ for amine addition compared with the addition of an anionic nucleophile?

As to the first question, because the stabilization energy is proportional to the product of the charges, this stabilization cannot be very extensive before the charges are fully developed, and thus the energy of the transition state should not be affected very much. This problem of small charges is somewhat counteracted by having the negative charge closer to the positive charge in the transition state than in the adduct, although this advantage in turn is reduced by the longer C–N bond in the transition state. On balance, apparently $\alpha_{el} < \beta_{nuc}{}^{n}$.

Regarding the second question, the electrostatic stabilization energy of the zwitterionic adduct should be reduced when X and Y become better π acceptors because the center of gravity of the negative charge is farther away from the center of the positive charge. Hence, $\delta \log K_1{}^{el}$ is reduced in equation 12, which renders $\delta \log k_0{}^{el}$ less negative for better π acceptors. As a consequence, a plot of $\log k_0$(C=C) for amine addition versus $\log k_0$(C–H) should have a smaller slope than a similar plot for the addition of an anion. This fact is illustrated schematically in Figure 2.

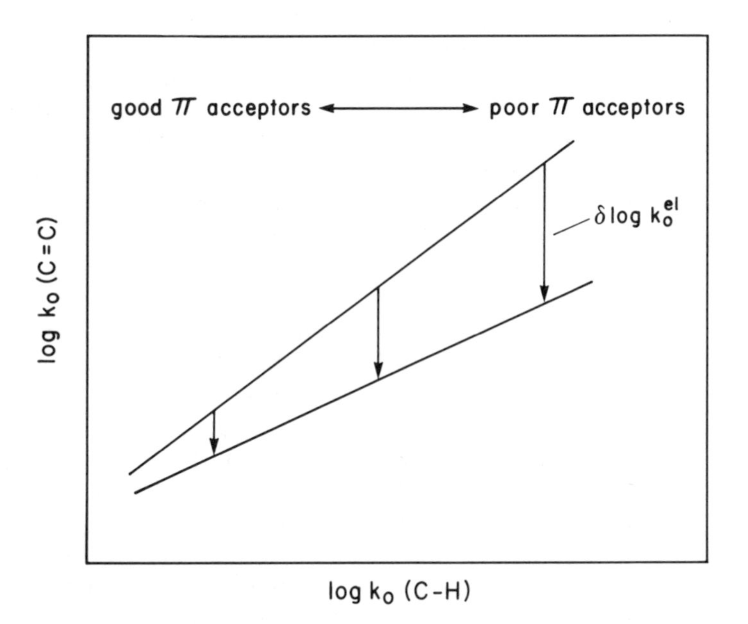

Figure 2. Schematic representation of the electrostatic effect on a plot of log k_0(C=C) versus log k_0(C–H). The upper line refers to addition of an anionic nucleophile where electrostatic stabilization does not occur; the lower line refers to amine addition. $\delta \log k_0{}^{el}$ becomes less negative for better π acceptors.

Polar Effects of Remote Substituents. Even though the identity of X and Y is the major factor in determining k_0, intrinsic rate constants show a small dependence on the remote substituent Z in **IV** or **VI.** In analogy to the situation in proton transfers (8), the effect of Z on $\log k_0$ can be expressed by

$$\delta \log k_0{}^{pol}(Z) = (\alpha_{nuc}{}^{n} - \beta_{nuc}{}^{n})\delta \log K_1{}^{pol}(Z) \qquad (13)$$

In cases where $\alpha_{nuc}{}^n = \rho_n$, $\delta \log K_1{}^{pol}(Z)$ is simply the experimentally measurable change in $\log K_1$ induced by a change in Z. When $\alpha_{nuc}{}^n \neq \rho_n$, experimental $\delta \log K_1{}^{pol}(Z)$ in equation 13 can still be used, but $\alpha_{nuc}{}^n$ must be replaced by ρ_n. Alternatively, if $\alpha_{nuc}{}^n$ is used, $\delta \log K_1{}^{pol}(Z)$ refers to the contribution that comes only from the interaction of Z with the negative charge of the adduct (*22*).

Equation 13 indicates that electron-withdrawing substituents increase k_0 because $\delta \log K_1{}^{pol}(Z) > 0$ while electron-donating substituents decrease it. However, just as for proton transfers, the effects of Z on k_0 are quite small compared to the effects of X and Y.

Solvent Effects on k_0. Only a few studies so far allow an assessment of solvent effects of k_0. The change from water to 50% $(CH_3)_2SO$–50% water enhances $\log k_0$ from 2.10 to 2.55 in the reaction of piperidine and morpholine with β-nitrostyrene (*22*) and from 4.55 to 4.94 with benzylidenemalononitrile (C. F. Bernasconi and R. B. Killion, unpublished results). On the other hand, the change from water to acetonitrile either leaves k_0 unaffected or even decreases it slightly in the reaction of benzylidene-substituted Meldrum's acid with the same amines (*23, 24*); the uncertainty is due to a lengthy extrapolation in water that renders k_0 uncertain in this solvent.

The observed increases in k_0 upon addition of $(CH_3)_2SO$ are consistent with similar increases in proton-transfers (*13, 15, 16, 25–27*) and are easily understood in terms of a reduced $\delta \log K_1{}^{sol}(C^-)$ in the $(CH_3)_2SO$-containing medium (equation 6, $\beta_{nuc}{}^n$ instead of β). The magnitude of the effects is smaller than in the proton-transfer reactions, though. For example, $\delta \log k_0 = 0.45$ for nucleophilic addition to β-nitrostyrene contrasts with $\delta \log k_0 = 1.35$ for the same solvent change in the deprotonation of nitromethane anion by the same amines (C. F. Bernasconi, A. Mullin, and D. Kliner, unpublished results). This attenuation of the solvent effect is another manifestation of the smaller imbalances in the olefin reactions.

An additional factor that may reduce the solvent effect on k_0 with *amine* nucleophiles is the internal electrostatic stabilization of the zwitterion. This electrostatic effect decreases k_0 according to equation 12. This reduction should be larger in a less polar solvent because electrostatic stabilization in the zwitterion is stronger (larger $\delta \log K_1{}^{el}$), and hence, the solvent effect on k_0 should be correspondingly reduced. This factor may be a contributing reason why k_0 for the reaction of amines with β-nitrostyrene increases so little upon addition of $(CH_3)_2SO$.

The very small or nonexistent solvent effect on k_0 for the benzylidene-substituted Meldrum's acid reaction upon transfer from water to acetonitrile is surprising because acetonitrile is a poorer solvator than $(CH_3)_2SO$–water mixtures. In principle, the smallness of this effect could be attributed to an unusually large electrostatic effect that completely compensates for the normal solvent effect on k_0. More likely, though, intramolecular hydrogen bonding enters as an additional factor in this reaction, as discussed in the next section.

Effect of Intramolecular Hydrogen Bonding on k_0. In the reaction of amines with olefins, the possibility exists of intramolecular hydrogen bonding between the NH proton and one of the electronegative atoms in X or Y of the zwitterionic adducts. With the adducts of benzylidene Meldrum's acid (**VII**) and benzylideneacetylacetone (**VIII**), this intramolecular hydrogen bonding appears quite strong.

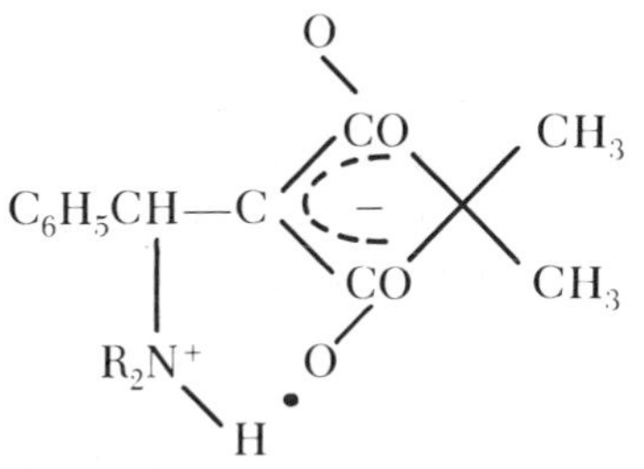

O
CO CH$_3$
C_6H_5CH—C ⁻
CO CH$_3$
R_2N^+ O
H

VII

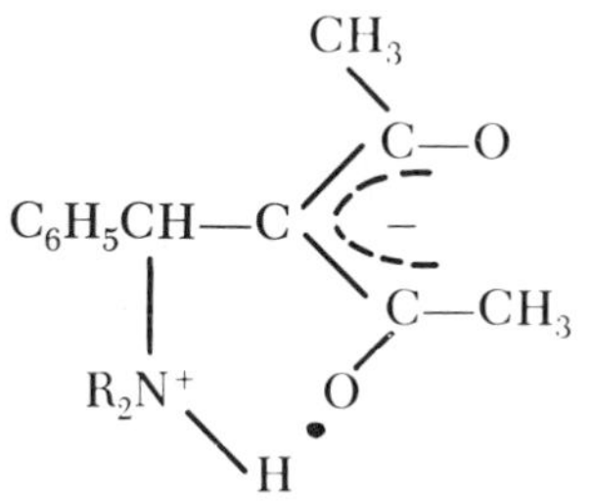

CH$_3$
C—O
C_6H_5CH—C ⁻
C—CH$_3$
R_2N^+ O
H

VIII

The main evidence for its existence is the high pK_a value of the NH proton (p$K_a^{\pm}$) in **VII** and **VIII.** For most adducts of type **V** (equation 10), p$K_a^{\pm}$ is lower than the p$K_a^{R_2NH_2^+}$ of the parent amine. For example, p$K_a^{\pm}$ − p$K_a^{R_2NH_2^+}$ = −0.72 for XY = $(CN)_2$ (*28*), −2.33 for XY = $(CN)C_6H_4$-4-NO_2 (*29*), and −2.70 for XY = $H(NO_2)$ (*30*). These values contrast with p$K_a^{\pm}$ − p$K_a^{R_2NH_2^+}$ = + 0.24 for **VII** (*24*) and +2.50 for **VIII** (C. F. Bernasconi and A. Kanavarioti, unpublished results).

The intramolecular hydrogen bond adds stability to the adduct and hence a late development would decrease, and an early development increase, k_0 according to

$$\delta \log k_0^{HB} = (\alpha_{Hb} - \beta_{nuc}{}^{n})\delta \log K_1^{Hb} \quad (14)$$

Because the stability of the hydrogen bond depends on the nearly full development of both the acidic properties of the NH proton and the basic properties of the acceptor oxygen, little stabilization in the transition state is expected, that is, $\alpha_{Hb} < \beta_{nuc}{}^{n}$.

The two sets of experimental data that allow us to test this prediction are consistent with it. The first data set is the observation that in the reaction of amines with benzylidene-substituted Meldrum's acid, k_0 is essentially unchanged or even slightly reduced upon transfer from water to acetonitrile

(*23, 24*). As discussed in the previous section, the normal solvent effect would be to increase k_0. The anomalous solvent effect may be a consequence of the intramolecular hydrogen bond being stronger in acetonitrile than in water for which independent evidence exists (*23*). The stronger hydrogen bond is tantamount to a larger $\delta \log K_1^{Hb}$, which leads to a correspondingly more negative $\delta \log k_0^{Hb}$ in acetonitrile.

The second example is the reaction of benzylideneacetylacetone with piperidine and morpholine in 50% $(CH_3)_2SO$–50% water (C. F. Bernasconi and A. Kanavarioti, unpublished results). As noted earlier, k_0 for this reaction shows a strong negative deviation from the correlation in Figure 1. Probably, part of this deviation is caused by intramolecular hydrogen bonding, although a steric contribution also occurs, as discussed next.

Steric Effects on k_0. At least two types of steric effects occur: direct repulsion between the attacking nucleophile and the olefin and steric interference with the optimal π overlap among X, Y, and the carbanionic carbon in the adduct. Both effects have their counterpart in proton transfers, but as long as R′ = H in R′CHXY, these effects are small unless very bulky bases are used (*32, 33*).

In proton transfers and transfer reactions in general (e.g., S_N2), direct repulsion between reactants can occur only in the transition state, and hence the result is always to lower k_0. In an addition reaction, repulsion occurs in the transition state as well, but in the adduct, repulsion is even more severe. Hence, according to equation 15 ($\delta \log K_1^{st} < 0$), the effect on k_0 depends on whether

$$\delta \log k_0^{st} = (\alpha_{st} - \beta_{nuc}^{\,n})\delta \log K_1^{st} \qquad (15)$$

the steric interaction develops ahead of C–Nu bond formation ($\alpha_{st} > \beta_{nuc}^{\,n}$, $\delta \log k_0^{st} < 0$) or lags behind it ($\alpha_{st} < \beta_{nuc}^{\,n}$, $\delta\ log\ k_0^{st} > 0$). No conclusive data exist yet that bear on the timing of this direct steric effect.

With respect to hindrance of the π overlap, we have encountered two systems where this effect is clearly affecting the stability of the adduct. These systems are the reactions of α-cyano-2,4-dinitrostilbene (*29*) and benzylideneacetylacetone (C. F. Bernasconi and A. Kanavarioti, unpublished results) with piperidine and morpholine.

In theses reactions, the equilibrium constant for the formation of the anionic adduct (**IX**), which is the product of K_1 for zwitterion formation (equation 10) and $K_a^{\pm}$, the acid dissociation constant of the zwitterion,[5] is abnormally low, particularly so for the benzylideneacetylacetone reaction (C.

[5] $K_1K_a^{\pm}$ is a better gauge of the steric effect than K_1 because K_1 may include a contribution from intramolecular hydrogen bonding, as is the case for the benzylideneacetylacetone reaction.

F. Bernasconi and A. Kanavarioti, unpublished results). This reaction is also the one for which log k_0(C=C) shows a dramatic negative deviation in Figure 1. This deviation appears too large to be accounted for by intramolecular hydrogen bonding only, and hence we conclude that part of it is caused by steric hindrance of the π overlap.

$$C_6H_5CH{=}CXY + R_2NH \underset{}{\overset{K_1Ka^{+-}}{\rightleftharpoons}} C_6H_5\underset{R_2N}{C}HC(X)(Y)^- + H^+ \quad (16)$$

IX

The decrease in k_0 can be expressed by the same equation used for the direct steric effect (equation 15 with $\alpha_{st} > \beta_{nuc}{}^{n}$). However, the negative δ log $k_0{}^{st}$ will be counteracted by a less negative δ log $k_0{}^{res}$ because of the diminished resonance in the adduct. The results indicate that the former effect dominates, at least in the case of benzylideneacetylacetone.

Further evidence for steric hindrance to π overlap comes from the crystallographic structure of (4-methoxybenzylidene)acetylacetone, which indicates that one of the acetyl groups is strongly turned out of the plane defined by the C=C double bond and the second acetyl group. If no good π overlap occurs in either the reactant or the product, probably π overlap does not occur in the transition state, and thus α_{st} may be ≈ 1.0.

We suspect that in all olefinic systems listed in Table I except for β-nitrostyrene and benzylidenemalononitrile a slight to moderate steric hindrance to optimal π overlap in the adduct occurs that is not present in the corresponding proton transfer. This could be the reason all the points in Figure 1 deviate negatively from the line defined by β-nitrostyrene and benzylidenemalononitrile. On the other hand, possibly this steric effect is of minor importance in all except for the α-cyano-2,4-dinitrostilbene and of course the benzylideneacetylacetone systems, as reflected in the negative deviations for these two compounds from the least-squares (dashed) line in Figure 1.

Generalizations. Are There Exceptions?

In the quest for generalizations or a search for exceptions, generalizations that simply spring from the definitions that underlie the PNS must first be distinguished from those that bear on the question of what happens during a chemical reaction.

In the first category, we have the fact that as long as the PNS is defined on the basis of equation 6 (or its equivalent for nucleophilic additions) the PNS is universal, that is, it can have no exceptions. This universality, however, does not mean that every change in k_0 can be attributed to a PNS effect. Factors that are present in the transition state but not in the reactants or products do influence k_0, but these factors cannot be treated in the framework of the PNS. The factors mainly occur in proton transfers and include direct steric repulsion, hydrogen bonding to the carbanion carbon, and electrostatic effects (*8*).

However, the validity of equations such as equation 6 does not bring us any closer to an understanding of what transition-state property is being measured by β or β_{nuc}^{n}. The usual working hypothesis is that those parameters are an approximate measure of the charge change on the base or nucleophile or of the degree of bond formation between the base and the proton (or the nucleophile and carbon). Even if this working hypothesis were proven wrong in the future (*34*),[6] the validity of equation 5 would remain intact.

The second category of generalizations is the one of real interest because it, along with possible exceptions from typical behavior, provides insights into how chemical reactions occur. One of the safest such generalizations is that the development of resonance and the concomitant solvation of the carbanion invariably appear to lag behind other bond changes. These PNS effects thus typically lead to a lowering of k_0. The possible reasons why reactions proceed in this fashion are discussed elsewhere (*8, 37–40*). These reasons include quantum mechanical (resonance) and entropy effects (solvation).

Are there exceptions? We have uncovered one exception thus far. This exception refers to water (rather than OH^-) addition to benzylidene-type substrates.

$$C_6H_5CH{=}CXY + H_2O \rightleftharpoons C_6H_5CH(OH)\text{–}C^{-}XY + H^+ \qquad (17)$$

In this reaction, k_0 for benzylidene-substituted Meldrum's acid is higher than for benzylidenemalononitrile $\{\log [k_0^{(CN)_2}/k_0^{XY}] \approx -0.7\}$, while for benzylidene-1,3-indandione, k_0 is about the same as for the malononitrile derivative (*19*). This finding contrasts with the "normal" behavior where k_0 for benzylidene-substituted Meldrum's acid and benzylidene-1,3-indandione should be much lower than that for benzylidenemalononitrile (Table III).

The likely reason for the unusually high reactivity of the Meldrum's acid and 1,3-indandione derivatives is that the transition state is subject to extra stabilization by intramolecular hydrogen bonding solvation as shown in **X** for

[6] Increasing evidence indicates that β and β_{nuc}^{n} may contain contributions that are unrelated to charge change. These contributions include solvation effects (*11, 12, 14–16, 34, 35*) as well as others (*36*).

the benzylidene-1,3-indandione case. This intramolecular hydrogen bonding essentially provides a way to avoid the late solvation of one of the oxygens of the 1,3-indandione moiety and with it its k_0-lowering effect.

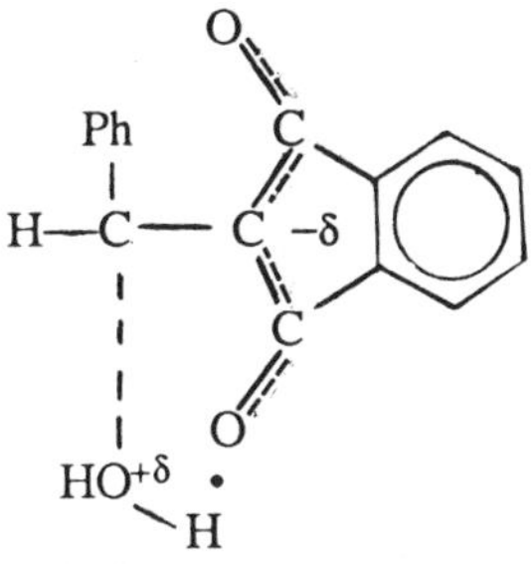

X

Another generalization that appears safe, even though it is based on only two examples so far, is that intramolecular hydrogen bonding in the formation of zwitterionic amine adducts such as **V** should always be late and hence lower k_0. We believe this generalization is safe because hydrogen-bonding stabilization cannot be extensive before the acidic properties of the NH proton and the basic properties of the acceptor atom are nearly fully developed. For similar reasons, electrostatic stabilization of zwitterionic adducts probably always develops late.

The intramolecular hydrogen bonding in water additions (**X**) represents quite a different situation from that encountered in amine additions. In the water addition, the hydrogen-bonding proton is eventually lost to the solvent. This result means we are not dealing with a PNS effect because hydrogen bonding is present only in the transition state and thus can only lead to an increase in k_0.

Whether steric hindrance always develops early and thus always lowers k_0 cannot be determined on the basis of the limited data obtained so far. For steric hindrance of π overlap, the problem is complicated by the fact that the reduced resonance in the adduct leads to a less negative $\delta \log k_0^{res}$, which tends to counteract the decrease in k_0 according to equation 15. Thus, whether k_0 increases or decreases depends on a delicate balance between these two effects.

For the direct steric interaction between the approaching nucleophile and electrophile, no data are available.

Acknowledgment

This research was funded by Grant CHE–8315374 from the National Science Foundation.

Literature Cited

1. Eigen, M. *Angew. Chem., Int. Ed. Engl.* **1964,** *3,* 1.
2. Ritchie, C. D. In *Solute-Solvent Interactions;* Coetzee, J. F.; Ritchie, C. D., Eds.; Dekker: New York, 1969; p 219.
3. Bell, R. P. *The Proton in Chemistry;* Cornell University: Ithaca, NY, 1973; Chapter 10.
4. Kresge, A. J. *Acc. Chem. Res.* **1975,** *8,* 354.
5. Hibbert, F. *Compr. Chem. Kinet.* **1977,** *8,* 97.
6. Hine, J. *Adv. Phys. Org. Chem.* **1977,** *15,* 1.
7. Bernasconi, C. F. *Pure Appl. Chem.* **1982,** *54,* 2335.
8. Bernasconi, C. F. *Tetrahedron* **1985,** *41,* 3219.
9. Jencks, W. P. *Catalysis in Chemistry and Enzymology;* McGraw-Hill: New York, 1969; p 178.
10. Bordwell, F. G.; Bartmess, J. E.; Hautala, J. A. *J. Org. Chem.* **1978,** *43,* 3107.
11. Hupe D. J.; Wu, D. *J. Am. Chem. Soc.* **1977,** *99,* 7653.
12. Hupe, D. J.; Wu, D.; Shepperd, P. *J. Am. Chem. Soc.* **1977,** *99,* 7659.
13. Keeffe, J. R.; Morey, J.; Palmer, C. A.; Lee, J. C. *J. Am. Chem. Soc.* **1979,** *101,* 1295.
14. Jencks, W. P.; Brant, S. R.; Gandler, J. R.; Fendrich, A.; Nakamura, C. *J. Am. Chem. Soc.* **1982,** *104,* 7045.
15. Bernasconi, C. F.; Bunnell, R. D. *Isr. J. Chem.* **1985,** *26,* 420.
16. Bernasconi, C. F.; Paschalis, P. *J. Am. Chem Soc.* **1986,** *108,* 2969.
17. Hine, J. *J. Am. Chem. Soc.* **1971,** *93,* 3701.
18. Bernasconi, C. F.; Howard, K. A.; Kanavarioti, A. *J. Am. Chem. Soc.* **1984,** *106,* 6827.
19. Bernasconi, C. F.; Laibelman, A.; Zitomer, J. L. *J. Am. Chem. Soc.* **1985,** *107,* 6563.
20. Hoz, S.; Speizman, D. *J. Org. Chem.* **1983,** *48,* 2904.
21. Hoz, S.; Gross, Z.; Cohen, D. *J. Org. Chem.* **1985,** *50,* 832.
22. Bernasconi, C. F.; Renfrow, R. A.; Tia, P. R. *J. Am. Chem. Soc.* **1986,** *108,* 4541.
23. Schreiber, B.; Martinek, H.; Wolschann, P.; Schuster, P. *J. Am Chem. Soc.* **1979,** *101,* 4708.
24. Bernasconi, C. F.; Fornarini, S. *J. Am. Chem. Soc.* **1980,** *102,* 5329.
25. Cox, B. G.; Gibson, A. *J. Chem. Soc., Chem. Commun.* **1974,** 638.
26. Cox, B. G.; Gibson, A. *Symp. Faraday Soc.* **1975,** *No. 10,* 107.
27. Bernasconi, C. F.; Kanavarioti, A. *J. Org. Chem.* **1979,** *44,* 4829.
28. Bernasconi, C. F.; Fox, J. P.; Fornarini, S. *J. Am. Chem. Soc.* **1980,** *102,* 2810.
29. Bernasconi, C. F.; Murray, C. J.; Fox, J. P.; Carré, D. J. *J. Am. Chem. Soc.* **1983,** *105,* 4349.
30. Bernasconi, C. F.; Carré, D. J.; Fox, J. P. In *Techniques and Applications of Fast Reactions in Solution,* Gettins, W. J.; Wyn-Jones, E., Eds.; Reidel: Dordrecht, Holland, 1979; p 453.
31. Bell, R. P.; Grainger, S. *J. Chem. Soc., Perkin Trans. 2* **1976,** 1367.
32. Bernasconi, C. F.; Hibdon, S. A. *J. Am. Chem. Soc.* **1983,** *105,* 4343.
33. Terrier, F.; Lelievre, J.; Chatrousse, A. P.; Farrell, P. G. *J. Chem. Soc., Perkin Trans. 2* **1985,** 1479.
34. Jencks, W. P.; Haber, M. T.; Herschlag, D.; Nazaretian, K. L. *J. Am. Chem. Soc.* **1986,** *108,* 479.
35. Bordwell, F. G.; Branca, J. C.; Cripe, T. A. *Isr. J. Chem.* **1985,** *26,* 357.
36. Pross, A. *J. Org. Chem.* **1984,** *49,* 1811.
37. Kresge, A. J. *Can. J. Chem.* **1975,** *52,* 1897.

38. Pross, A.; Shaik, S. S. *J. Am. Chem. Soc.* **1982,** *104,* 1129.
39. Pross, A. *Adv. Phys. Org. Chem.* **1985,** *21,* 99.
40. Dewar, M. J. S. *J. Am. Chem. Soc.* **1984,** *106,* 209.
41. Bordwell, F. G.; Boyle, W. J., Jr. *J. Am. Chem. Soc.* **1972,** *94,* 3907.
42. Bernasconi, C. F.; Leonarduzzi, G. D. *J. Am. Chem. Soc.* **1982,** *104,* 5133.

RECEIVED for review October 21, 1985. ACCEPTED January 27, 1986.

BRØNSTED EQUATION, HARD-SOFT ACID-BASE THEORY, AND FACTORS IN NUCLEOPHILICITY

9

Nucleophilicity, Basicity, and the Brønsted Equation

Frederick G. Bordwell, Thomas A. Cripe, and David L. Hughes

Department of Chemistry, Northwestern University, Evanston, IL 60201

Application of the Brønsted relationship reveals that, when measurements of rate and equilibrium constants are made in dimethyl sulfoxide solution using families of bases wherein donor atom and steric effects are kept constant, nucleophilicities depend on only two factors, (a) the relative basicities of Nu^- *and (b) the sensitivities of the rates to changes in these basicities (the Brønsted* β*). All combinations of nucleophiles and electrophiles appear to fit this pattern. Points for* para π*-acceptor substituents deviate from the Brønsted lines because of enhanced solvation effects that introduce a kinetic barrier that is not modeled properly by the equilibrium acidities. The carbon basicities of carbanion, nitranion, oxanion, and thianion families were calculated in the gas phase and shown to correlate linearly with their experimental gas-phase hydrogen basicities. Evidence is presented for a rough, general linear correlation between log* k *and the oxidation potential of anions in nucleophile–electrophile combinations.*

THE TERM NUCLEOPHILICITY refers to the relative rate of reaction of an electron donor with a given electrophile, as distinct from basicity, which refers to its relative affinity for a proton in an acid–base equilibrium. A quantitative relationship between rate and equilibrium constants was discovered by Brønsted and Pedersen (*1*) in 1924. These authors found that the rate constants for the catalytic decomposition of nitramide by a family of bases, such as carboxylate ions ($GCH_2CO_2^-$), could be linearly correlated with the acidities of their conjugate acids, pK_{HB}. This observation led to the discovery of general base catalysis and the first linear free-energy relationship, which later became known as the Brønsted equation:

0065-2393/87/0215-0137$06.00/0

$$\log k_{B^-} = \beta pK_{HB} + C \qquad (1)$$

Hammett (2) recognized the general nature of the Brønsted relationship in 1935 and showed that this relationship could be applied to several other kinds of reactions, including a methyl-transfer ($CH_{3,T}^+$) reaction between $ArN(CH_3)_2$ and CH_3I, as well as to proton-transfer (H_T^+) reactions. (He pointed out that the $CH_{3,T}^+$ equilibrium constant would no doubt provide a better model than the H_T^+ equilibrium for the rate constants for $CH_{3,T}^+$, but that such equilibrium constants were not available.) Until recently, however, the Brønsted equation has been used primarily to correlate rate-equilibrium data for H_T^+ reactions in aqueous media.[1] Now that the mechanism of the decomposition of nitramide is understood to involve a base-promoted deprotonation of a tautomer of nitramide, accomplished by elimination of hydroxide ion (equation 2) (*3, 4*), it is clear that Brønsted and Pedersen were measuring the relative nucleophilicities of various families of bases toward hydrogen with reference to the relative acidities of their conjugate acids. The Brønsted relationship can be cast in the form of the Hammett equation to bring out this feature (equation 3). Therefore, the Hammett equation is really a special case of the Brønsted relationship.

$$B^- + HN{=}\overset{+}{N}(O^-)(OH) \longrightarrow B\text{-}H + N{\equiv}\overset{+}{N}{-}O^- + HO^- \qquad (2)$$

$$\log (k/k_0) = \beta \log (K/K_0) \qquad (3)$$

In 1953 Swain and Scott (5) assumed that nucleophilicity (n) in S_N2 and related reactions was an inherent property, which could be defined by equation 4, where s is the sensitivity of the rate constants to variations in (or of) the electrophile. S_N2 reactions of nucleophiles with CH_3Br in water were used as a standard ($s = 1.0$).

$$\log (k/k_0) = sn \qquad (4)$$

It was soon realized, however, that this simple definition of nucleophilicity would not suffice. Shortly afterward, Edwards (6) attempted to define relative nucleophilicities in terms of two parameters, H (basicity) and E_{ox} (oxidation potential), using the (variable) coefficients α and β to relate these properties to changes in the electrophile (equation 5). Later, Edwards and Pearson substituted a polarizibility parameter, P, for E_{ox}. In essence, equation 5 is a Brønsted equation with a second parameter added.

[1] For other applications of the Brønsted equation to S_N2 reactions, see Smith, G. F. *J. Chem. Soc.* **1943,** 521–523. Hudson, R. F.; Klopman, G. *J. Chem. Soc.* **1962,** 1062–1067. Hudson, R. F. *Chemical Reactivity and Reaction Paths;* Klopman, G., Ed.; Wiley-Interscience: New York, 1974; Chapter 5. For applications of the Brønsted equation to acyl transfer and other reactions, see Hammett (2) and Jencks, W. P. *Catalysis in Chemistry and Enzymology;* McGraw-Hill: New York, 1969.

$$\log (k/k_0) = \alpha H + \beta E_{ox} \quad (5)$$

The demonstration by Parker (*8*) that nucleophilicities in S_N2 reactions could be enhanced by as much as 10^8 by changing from a hydroxylic solvent [H_2O, CH_3OH, $HOC(O)CH_3$, etc.] to a non-H-bond donor "aprotic" solvent [$(CH_3)_2SO$, CH_3CN, $HCON(CH_3)_2$, etc.][2] called attention once again to the dominant role that solvation plays in rates of nucleophile–electrophile combinations in solution (*9, 10*). Because stabilization by H-bond-donor solvents varies greatly with nucleophile size, charge, extent of electron pair delocalization, and the nature of the donor atom, the solvation parameter alone makes it difficult, if not impossible, to design an equation capable of quantitative correlation of rate-equilibrium data in such media. In contrast, solvation, as well as other factors controlling nucleophilicity, can be held relatively constant in dimethyl sulfoxide and like non-H-bond donor solvents. This result can lead to a simple relationship between nucleophilicity and basicity.

Precision and Scope of the Brønsted Equation. The key to eliminating factors, other than basicity, that dictate nucleophilicity lies in the use of families of nucleophiles in dipolar non-H-bond donor solvents. By a family of nucleophiles, we generally mean a family of anionic bases, A^-, wherein the basicity can be changed by remote substitution. For example, the basicity of fluorenide carbanions can be changed by 10 or more pK units in $(CH_3)_2SO$ solution by introducing remote substitutents (*11*). [By contrast, with H-bond-donor solvents, such as water, the solvent leveling effects usually restrict basicity changes in a family to a practical limit (1–2 pK units)[3].] At the same time, the nucleophile donor atom (carbon) is kept constant, and steric and solvation effects are kept nearly constant. Brønsted plots of the rate constants (log k) for fluorenide ions (A^-) reacting with electrophiles plotted against equilibrium acidities of their conjugate acids (pK_{HA}), both in $(CH_3)_2SO$ solution, exhibit excellent linearity for all electrophiles studied to date. Table I summarizes the results of Brønsted correlations with families of fluorenide ions and related carbanions (C^-), as well as other families of anions (nitranions, N^-, oxanions, O^-, and thianions, S^-). The reaction types include S_N2 (*12–14*), S_N2' (A. H. Clemens and J.-P. Cheng, unpublished results), E2 (*15*), H_T^+ (*16*), S_NAr (*12*), and e_T^- (*17, 18*).

For all 20 of the combinations of nucleophiles and electrophiles shown in Table I, and others that we have studied, the relative rate constants depend on only two factors: (a) the basicity of the anion as defined by the acidity of its conjugate acid, pK_{HA}, and (b) the sensitivity of the rate constant

[2] These solvents all contain protons that react with strong bases. The term "aprotic" is a misnomer that should be abandoned.

[3] For reasons elaborated in the section on solvation effects on nucleophilicity, use of substituents of the type p-NO_2, p-CN, p-SO_2R, and p-COR to extend the lines in most Brønsted plots is impractical.

Table I. Brønsted Correlations for Reactions of Anion Families with Electrophiles in $(CH_3)_2SO$ Solution

Reaction	*Electrophile*[a]	β[b]	*Nu⁻ Families*[c]
S_N2	RX[d]	0.2–0.5	C^-, N^-, O^-, S^-
S_N2'	$CH_2{=}CHCRR'Cl$[e]	0.2–0.4	C^-, N^-
E2	c-$C_6H_{11}Br$	0.4–0.5	N^-, O^-, S^-
H_T^+	$CH_2{=}CHCH_2CN$	0.7–1.0	C^-, N^-, O^-, S^-
S_NAr	p-$NO_2C_6H_4X$	0.5–0.7	C^-, N^-, O^-, S^-
e_T^-	$R_2C(NO_2)Z$[f]	~1.0	C^-, N^-
	$C_6H_5SO_2CH_2X$	~1.0	C^-, N^-

[a]X = F, Cl, Br, or I.
[b]The slope of the Bronsted plot.
[c]9-Substituted fluorenide ions, substituted cyclopentadienide ions, $ArCHCN^-$, $ArCHSO_2C_6H_5^-$, $ArAr'CH^-$, $ArC(CN)_2^-$, $ArCHCOCH_3^-$, and $ArCOCH_2^-$; $ArNH^-$, $ArNCH_2CH_3^-$, $ArAr'N^-$, $ArNCOCH_3^-$, $Ar'NCOAr^-$, carbazolide ions, and phenothiazide ions; and ArO^-, 2-NpO^-, and ArS^-.
[d]R = $C_6H_5CH_2$, $CH_3(CH_2)_3$ $(CH_3)_2CH(CH_2)_2$, C_6H_{11}, $(CH)_3C_3$, $ArCOCH_2$, $CNCH_2$, or $CH_2(CO)_2(CH_2)_3$.
[e]R = CH_3; R′ = H or CH_3.
[f]Z = NO_2 or SO_2Ar.

to changes in basicity as measured by the slope of the Brønsted coefficient, β. As mentioned earlier, this simplicity is achieved by keeping the nucleophile donor atom constant and steric and solvation effects nearly constant. The electrophile may have a dramatic effect on the rate of reaction and often has a marked effect on the sensitivity (β) of the reaction to changes in basicity.

From the results summarized in Table I, apparently the Brønsted relationship will hold for all combinations of nucleophiles and electrophiles. Because, as pointed out previously, the Hammett equation is really a special case of the Brønsted relationship, all the legion of nucleophile–electrophile, rate-equilibrium Hammett correlations that have been studied also fall under the scope of the Brønsted relationship. For example, nucleophilicities of ArO^-, ArS^-, $ArC(CN)_2^-$, and the other families listed in footnote *c* of Table I have generally been correlated by the Hammett equation, where the acidities of benzoic acids in water are used as a model for substituent interactions with the reaction site (σ), and the variable parameter ρ is used to define the sensitivity of the rate constants to these substituent effects. The Brønsted equation (equation 3) offers a much more precise relationship of the same kind, because this equation does not depend on an arbitrary model and allows rate and equilibrium constants to be measured in the same solvent. Furthermore, the Brønsted relationship is also applicable to families of aliphatic bases such as carboxylate ions ($GCH_2CO_2^-$), alkoxide ions (GCH_2O^-), and amines (GCH_2NH_2). In addition, other correlations of a kinetic parameter (log k, $\Delta G‡$, E_a, etc.) can be included with various thermodynamic parameters (pK_a, $\Delta G°$, E_{ox}, etc.) under the Brønsted label.

Viewed in this way, the Brønsted relationship becomes the most general of all the linear free-energy relationships. This relationship is also the most precise. This statement will come as a surprise to some readers because the Brønsted relationship, as practiced for H_T^+ reactions in aqueous solution, has become notorious for scatter. This scatter is caused, however, by the necessity of mixing families of bases with different donor atoms together with orphan nucleophiles in constructing extended plots. Linearity can be achieved in $(CH_3)_2SO$ solution over extended ranges, however, when donor atom and solvation effects are kept constant. For example, the Brønsted plots of S_N2 reactions for 9-substituted fluorenide ions reacting with $C_6H_5CH_2Cl$ in $(CH_3)_2SO$ solution cover a range of over 25 kcal/mol in pK_{HA} (Figure 1).

Nucleophile Steric Effects in S_N2 Reactions. Examination of Figure 1 shows a series of parallel lines, each of which represents a family of 9-substituted fluorenide ions, 9-G-Fl^-, in which the nature of G has been altered and the basicity within the family has been changed by placing substituents in the fluorene ring. The displacements of the lines, one from the other, represent small changes in the steric or electronic environment at the reaction site caused by the presence of the substitutent, G. Most of the displacements are not large, amounting to only two- to threefold changes in rate constants. Remarkably enough, these large, flat, highly delocalized ions have only small steric requirements. The 9-G-fluorenide ion lines can be extended by adding two 9-phenylxanthenide ion points. Also (*see* Figure 2), in a Brønsted plot for reactions with $C_6H_5CH_2Cl$, α-cyano carbanions of all shapes and sizes fit near to the extension of the 9-$CH_3CO_2Fl^-$ ion family line (*13*). Groups such as $(CH_3)_2CH$, $(CH_3)_3C$, *o*-$CH_3C_6H_4$, and 2,4,6-$(CH_3)_3C_6H_2$ (Mes), which protrude on both sides of the fluorenide ion ring, cause sizable steric effects, however. An equation has been derived that provides a quantitative expression of these steric effects (*13*).

Solvation Effects on Nucleophilicities. The strict linearity of plots such as those shown in Figure 1 is testimony to the constancy of solvation for the individual members of these nucleophile families. [Our contention is that the curvature claimed for Brønsted plots in aqueous solution over comparable $\Delta G°$ ranges is usually an artifact caused by the differential solvation of the nucleophiles (*14*).] The constancy of solvation in $(CH_3)_2SO$ for highly delocalized carbanions of this type has been demonstrated by the linearity of plots of relative acidities in the gas phase and in $(CH_3)_2SO$ solution (*19, 20*), for example, Figure 3.

The acidities of the hydrocarbons shown in Figure 3 are increased by over 300 kcal/mol by going from the gas phase to $(CH_3)_2SO$ solution because of the ability of the solvent to solvate the proton and the anion. The slope of unity for the plot tells us, however, that solvation effects are held essentially constant for these large delocalized hydrocarbon anions over the entire range

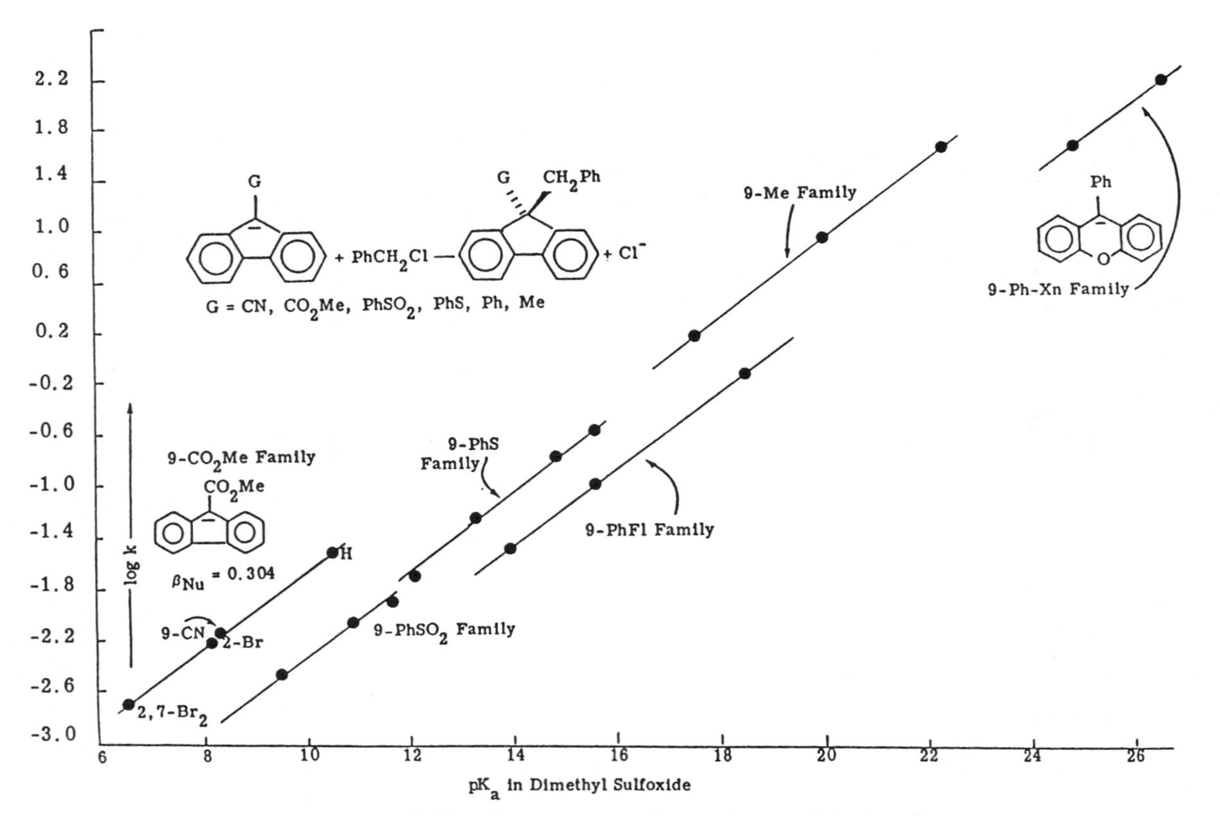

Figure 1. Brønsted plot of S_N2 reactions for 9-substituted fluorenide ion families with benzyl chloride in $(CH_3)_2SO$ at 25 °C.

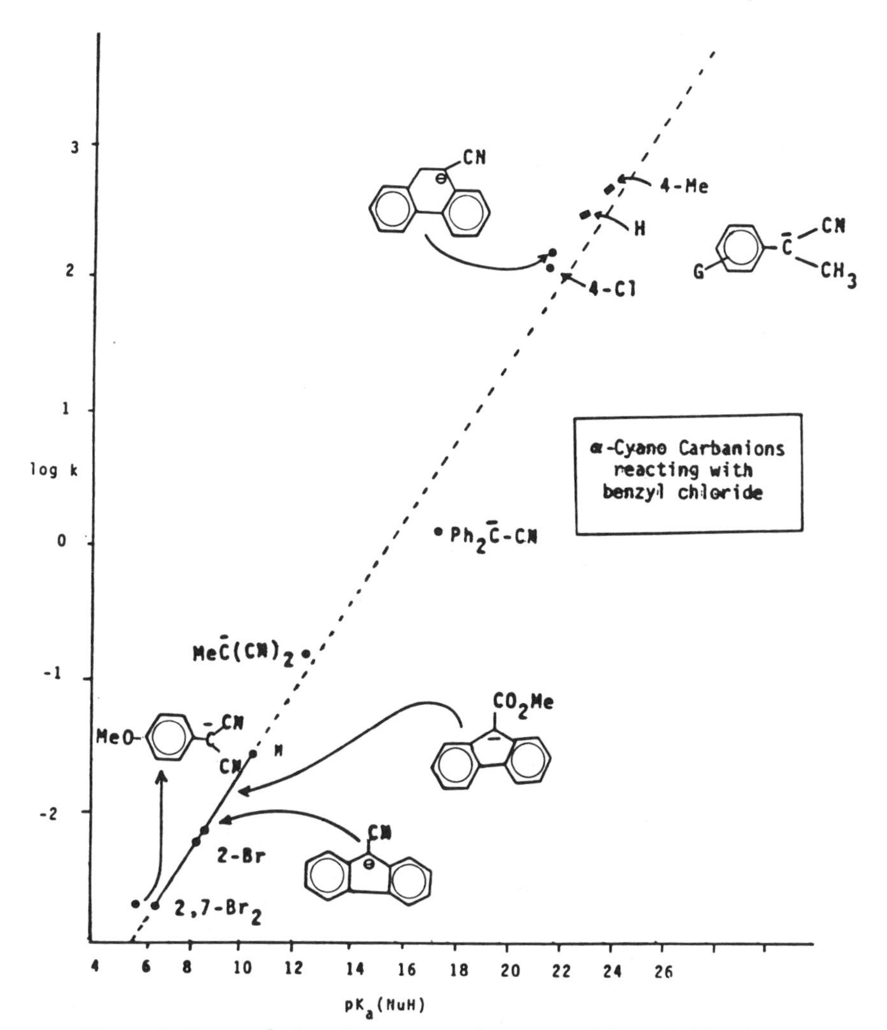

Figure 2. Brønsted plot of α-cyano carbanions with benzyl chloride in $(CH_3)_2SO$ at 25 °C. (Reproduced from reference 13. Copyright 1983 American Chemical Society.)

of about 40 kcal/mol. Points for carbanions derived from the carbon acid families $ArCH_2CN$, $ArCH_2SO_2C_6H_5$, $ArCH_2COC_6H_5$, and $ArCH_2NO_2$ are displaced from the line to the extent to which the negative charge is delocalized to the heteroatoms present in the functional group (*20*). Relatively little displacement occurs for α-CN and α-$SO_2C_6H_5$ carbanions, more for enolate ions, and most for nitronate ions (*20*). An extended Brønsted plot for the S_N2 reaction of a family of $ArCHSO_2C_6H_5^-$ carbanions with *n*-BuCl in $(CH_3)_2SO$ is shown in Figure 4 (*21*).

Close inspection of Figure 4 shows that inclusion of all the points would give a curved plot. Curvature is expected on the basis of the Hammond –

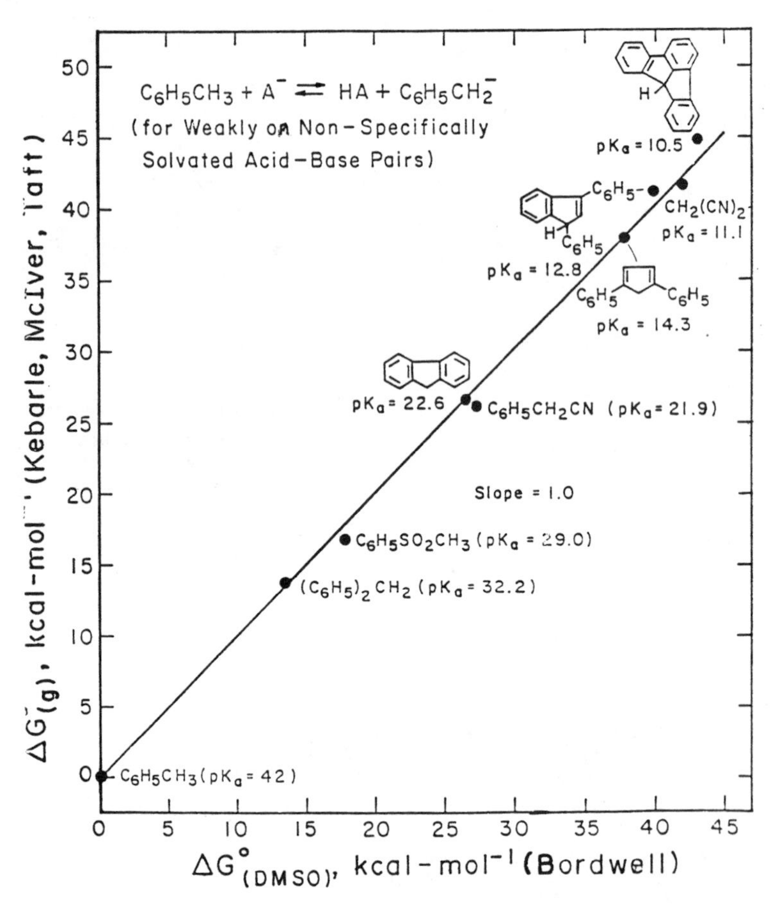

Figure 3. Plot of acidities of hydrocarbons in the gas phase vs. $(CH_3)_2SO$ solution phase. (Reproduced with permission from reference 21. Copyright 1986 Weizmann Science Press.)

Leffler and reactivity–selectivity postulates, which predict that the selectivity should decrease and the Brønsted β approach unity as the reactions become more endergonic. The curvature in this plot is much greater than predicted by the Marcus equation (equation 6) (22), however, and is believed to be an artifact caused by enhanced solvation of π-acceptor para substituents such as CN, COC_6H_5, and NO_2 (*21*). [The Marcus equation, which has gained wide acceptance in the interpretation of electron-transfer reactions, is represented in equation 6 as a Brønsted relationship with an exponential term added to take into account curvature (23)].

$$\log (k/k_0) = a(pK_a)^2 + \beta(pK_a) + c(pK_a)^0 \tag{6}$$

Further examination of Figure 4 shows that the para points deviating from the line are those for substituents bearing heteroatoms to which the negative charge from the oxide ion may be delocalized. Similar deviations have been observed in Brønsted plots for S_N2 reactions for (a) the p-NO_2 point for ArS^- ions reacting with n-BuCl (*14*), (b) p-CN and p-CF_3 points for $ArC(G)CN^-$ ions (G = H, c-NC_4H_8O) reacting with n-BuCl (*24*), and p-$COCH_3$, p-COC_6H_5, p-CO_2CH_3, p-CH_3SO_2, p-C_6H_5, and p-NO_2 points for ArO^- ions reacting with $C_6H_5CH_2Cl$ (*24*). Similar deviations observed for such para substituents in a plot of gas-phase versus $(CH_3)_2SO$ solution-phase acidities for phenol equilibrium acidities were attributed to enhanced solvation of these substituents (*25*). Their presence in Brønsted plots is caused by the failure of pK_{HA} values to provide a proper model for the kinetic phenom-

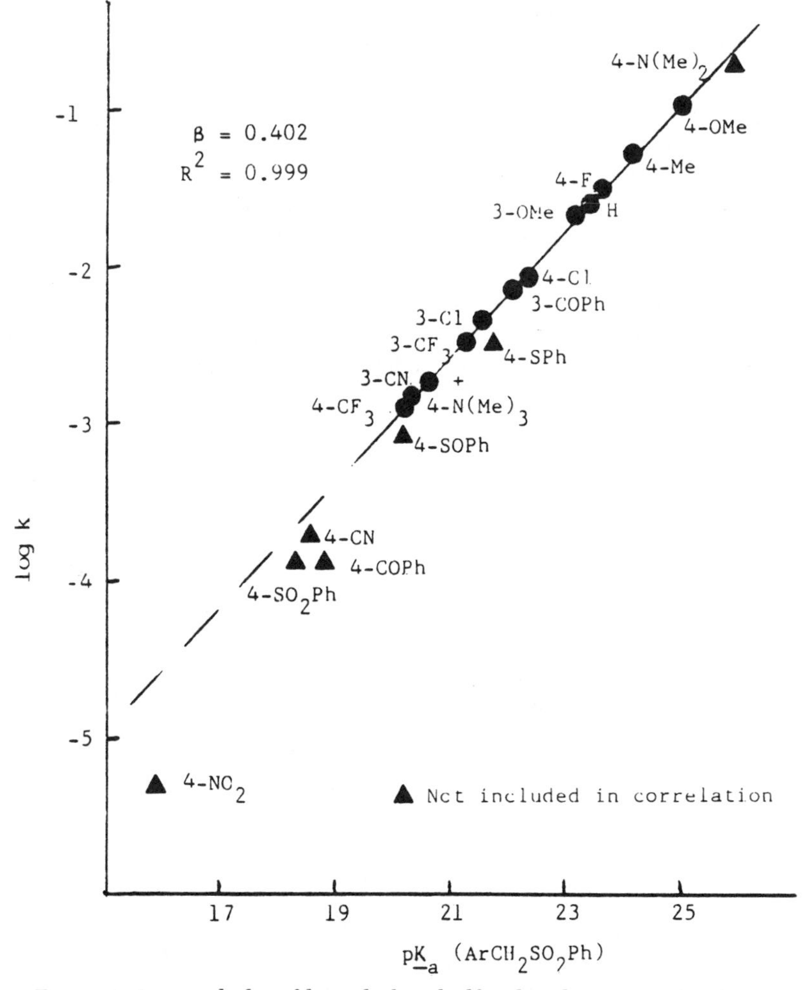

Figure 4. Brønsted plot of benzyl phenylsulfonyl carbanions with n-*butyl chloride in* $(CH_3)_2SO$ *at 25 °C. (Reproduced with permission from reference 21. Copyright 1986 Weizmann Science Press.)*

ena, involving desolvation, that occur when anions bearing these substituents interact with the electrophile. We anticipate finding similar deviations in most, if not all, of the reactions listed in Table I. Examination of the literature further confirmed the generality of this phenomenon (*24*).

Donor Atom Effects. When families of nucleophiles with different kinds of donor atoms react with a given electrophile, they often give parallel, or nearly parallel, Brønsted lines. The displacements of the lines, one from the other, provide a measure of the relative reactivity of the nucleophiles at the same basicity. For example, in S_N2 reactions with $C_6H_5CH_2Cl$ in $(CH_3)_2SO$ solution, the order of reactivities at the same basicity was observed to be thiophenoxides (ArS^-) ≫ 9-methylfluorenide ions (9-CH_3Fl^-) > 2-naphthoxide ions (2-NpO^-) > carbazolide (Cb^-) or phenothiazinide (Pz^-) ions (*12*, *26*). The ArS^- and 9-CH_3Fl^- ion families differ in reactivity by a factor of about 10^3, the 9-CH_3Fl^- ion family is about 10 times more reactive than the 2-NpO^- ion family, and the 2-NpO^- family is about 3–4 times more reactive than the Cb^- or Pz^- ion families. A similar comparison of the ArS^-, ArO^-, $ArCHCN^-$, and $ArNR^-$ ion families reacting with *n*-$CH_3(CH_2)_3Cl$ gives a slightly different order, ArS^- ≫ ArO^- > $ArCHCN^-$ > $ArNR^-$ (Figure 5) (*24*). The amount of scatter would be enormous if points from the lines in Figure 5 were combined into a single Brønsted line.)

The order of donor atom effects, $S^- > O^-$, $C^- > N^-$, appears to be general for S_N2 reactions with alkyl halides inasmuch as this order holds for

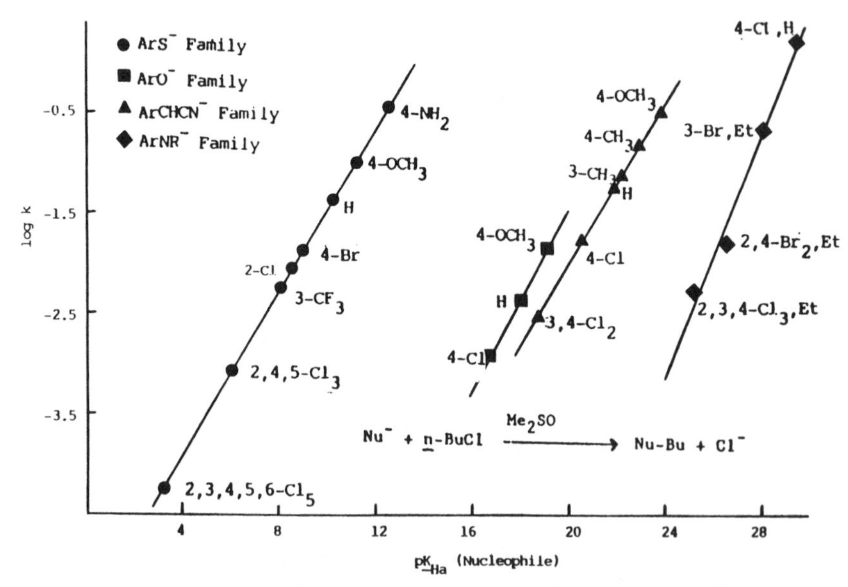

Figure 5. Donor atom effects in reactions of anions with n-*butyl chloride in* $(CH_3)_2SO$ *at 25 °C (24).*

n-BuCl, *n*-BuBr, *n*-BuI, $C_6H_5CH_2Cl$, and $ClCH_2CN$ (*12–14*). The reactivity order between O^- and C^- can change for 9-G-Fl^- ions with the nature of G because rate constants for 9-G-Fl^- ions vary over a range of about 10-fold and decrease with changes in G in the order CN, CO_2CH_3 > CH_3 > C_6H_5 > $H_5C_6SO_2$ > C_6H_5 (Figure 1) (*11*). The 9-CH_3Fl^- ion family, which is at about the midpoint of reactivity, has been chosen as a standard. A composite α-cyano carbanion family including $ArC(CN)_2^-$, 9-CN-Fl^-, and $ArCHCN^-$ ions is slightly more reactive than the 9-CH_3Fl^- ion family (Figure 2) (*13*). Sterically hindered carbanions, such as 9-*t*-$BuFl^-$ and $C_6H_5C(R)SO_2C_6H_5^-$ ion families, react much more slowly, however (*11*).

The relative reactivities of anion nucleophiles depend somewhat on the nature of the electrophile. For example, for S_N2 reactions with *n*-$CH_3(CH_2)_2OTs$, the Cb^- nitranion, as well as $C_6H_5O^-$ and 2-NpO^- ions, are more reactive than the 9-CH_3Fl^- ion family (Figure 6) (*12*).

Examination of Figure 6 shows that the orphan nucleophile ions Br^- and Cl^- are "supernucleophiles", like thianions, when comparisons are made at the same basicity. On the other hand, CH_3O^- and HO^- ions are unusually poor nucleophiles, no doubt because these highly basic ions are paired to the

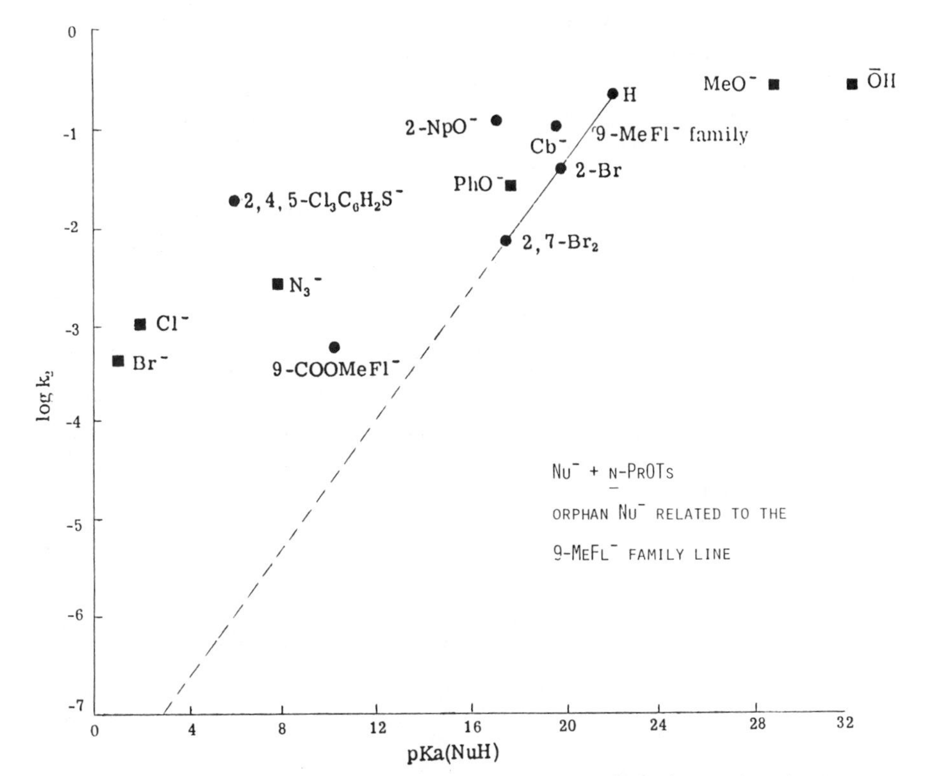

Figure 6. Brønsted plot of anions reacting to n-*propyl tosylate in* $(CH_3)_2SO$ *at 25 °C (12).*

counterion and homo-H-bonded to their conjugate acids (*12*). The high reactivity of Cl^- and Br^- was confirmed by the results of a study with $n\text{-}CH_3(CH_2)_3I$ electrophile (*12*).

The reactivity order at the same basicity for families of nucleophiles with different donor atoms changes to $S^- > O^- > N^- > C^-$ when attack is on hydrogen in an E2 reaction (*15*), and a similar order, S^-, $O^- > N^- > C^-$, was observed for attack on hydrogen in the base-catalyzed isomerization of butenenitrile (*16*).

Why does the Brønsted relationship hold for these diverse reactions and give promise of holding for all combinations of nucleophiles and electrophiles (Table I)? In other words, why does the thermodynamics of proton transfer serve as a model, at least as a first approximation, not only for rates of reactions involving proton transfers (H_T^+) but also for rates of reactions involving alkyl transfers (R_T^+), bromine atom transfers (Br_T^+), e_T^-, and other transfers? A partial answer to this question comes from recent results of Arnett and co-workers (*28–30*), who found in carbanion–carbocation combination equilibria that the hydrogen basicities for the carbanions correlated linearly with their carbon basicities. Also, in our laboratory, Cripe (*24*) observed that the intrinsic, gas-phase, hydrogen basicities for several anion families correlate linearly with their carbon basicities. Furthermore, there is good reason to believe that these relative intrinsic basicities for anion families observed in the gas phase carry over to the solution phase (*24*). These conclusions were arrived at by using a modification of a method developed by Hine and Weimar (*31*) to calculate gas-phase (intrinsic) carbon basicities.[4]

Basicities of Anions toward Hydrogen and toward Carbon. Hine and Weimar used the position of the equilibrium in equation 7 to define carbon basicity, relative to hydrogen basicity.

$$CH_3OH + HA \xrightleftharpoons{K_{HA}^{CH_3A}} CH_3A + H_2O \quad (7)$$

By multiplying $K_{HA}^{CH_3A}$ by the relative hydrogen basicity, $K_{A^-}^{H}$ (equation 8), the basicity of various anions, A^-, toward carbon, $K_{A^-}^{CH_3}$, could be determined, relative to that of hydroxide ion:

$$H_2O + A^- \xrightleftharpoons{K_{A^-}^{H}} HA + HO^- \quad (8)$$

$$K_{A^-}^{CH_3} = K_{HA}^{CH_3A} K_{A^-}^{H} \quad (9)$$

$$CH_3OH + A^- \xrightleftharpoons{K_{A^-}^{CH_3}} CH_3A + HO^- \quad (10)$$

[4] The Brønsted equation, which is empirical, can be viewed as being a special case of the Marcus equation, which was derived from first principles. The latter theory has the important added feature of defining intrinsic activation barriers.

The values of $K_{A^-}^{CH_3}$ provide a measure of the ability of A^- to effect a methyl cation transfer $[(CH_3)_T^+]$, just as $K_{A^-}^{H}$ values provide a measure of the ability of A^- to effect a proton transfer (H_T^+). Using this method, Hine and Weimar found some remarkable differences between carbon and hydrogen basicities. For example, although CN^- is about 10 times less basic than $C_6H_5O^-$ toward H_T^+, CN^- is about 10^{14} times *more* basic toward $CH_{3,T}^+$ (in aqueous solution). The CH_3O^-:CH_3S^- basicity ratio toward H_T^+ is 10^5, but toward $CH_{3,T}^+$ this ratio is 10^{-8}. Although these reversals in anion affinity for incipient cationic species (H_T^+ and $CH_{3,T}^+$) are remarkable, relatively little follow up has occurred on Hine's method. A likely reason is that the values for $K_{HA}^{CH_3A}$ (equation 7) are easily available only in the gas phase while values for $K_{A^-}^{H}$ (equation 8) have been available, until recently, only in aqueous solution. Application of the method has therefore required conversion of gas-phase free energies of formation into aqueous solution. Fortunately, now that a gas-phase acidity scale has been established (*32*), the comparisons of hydrogen and carbon basicity can be made in the gas phase without recourse to aqueous solution. Table II summarizes the H_T^+ and $CH_{3,T}^+$ data for 30 anions calculated by the Hine method (*23*).

In Figure 7, the $\Delta G°(CH_{3,T}^+)$ values are plotted against $\Delta G°(H_T^+)$ values for the 30 anions in Table II. Least-squares lines with slopes near unity were drawn through the points for the carbanion, oxanion, and thianion families, and a line of similar slope was drawn through the closely spaced points for the nitranion family. When the orphan anions are included, the intrinsic carbon basicity at the same hydrogen basicity is observed to decrease in the order $H^- > C^- > S^- >$, N^-, I^-, and $Br^- > Cl^- > O^-$ and F^- with the total range being almost 20 orders of magnitude. This represents a difference of 27 kcal/mol in the free energies of reactions of these anions with a common electrophile (CH_3OH), even after the differences associated with pK_{HA} are factored out.

The difference in oxanion and carbanion basicities toward carbon at the same hydrogen basicity is about 16 kcal/mol (for example, compare $C_6H_5O^-$ and CN^- in Table II). We might wonder whether or not solvation effects are likely to mask this intrinsically greater carbon basicity of carbanions than oxanions. Comparison of equations 8 and 10 shows that the effect on the equilibria of going from the gas phase into the solution phase will depend primarily on the relative solvation energies of the two different anions, because (a) HO^- is common to the two equations and (b) solvation effects on the neutrals are expected to be small and to differ but little. Fortuitously, the free energies of aqueous solvation of $C_6H_5O^-$ and CN^- ions are almost identical (-74 and -73 kcal/mol, respectively) (*33*), which means that the ~16 kcal/mol greater basicity of CN^- ion toward carbon will be retained in solution. Other solvation differences will be larger, but the relative intrinsic basicities of anions toward hydrogen and carbon revealed by the gas-phase data will be retained in solution because solvation stabilization of the anions will affect equations 8 and 10 to exactly the same degree.

Table II. Experimental Free-Energies of Reactions $\Delta G°(H^+_T)$ and Equilibrium Constants ($K^H_{A^-}$) for the Reaction in Equation 8, and Calculated Equilibrium Constants, K^{Me}_A, for Reaction in Equation 10

No.	A^-	$\Delta G°(H^+_T)$[a]	$K^H_{A^-}$ [b]	$K_{A^-}^{Me}$ [b]	$\Delta G°(Me^+_T)$
1	Me^-	−25.0	1.77×10^{18}	4.43×10^{26}	−36.5
2	$PhCH_2^-$	11.7	2.88×10^{-9}	3.20×10^{1}	−2.1
3	$CH_2CHCH_2^-$	0.2	7.15×10^{-1}	8.37×10^{9}	−13.6
4	$C_6H_5^-$	−12.0	5.74×10^{8}	4.35×10^{21}	−29.6
5	$CH_2{=}CH^-$	−20.0	3.97×10^{14}	1.21×10^{27}	−37.1
6	$HC{\equiv}C^-$	16.8	5.46×10^{13}	7.21×10^{1}	−2.5
7	$MeC{\equiv}C^-$	12.1	1.47×10^{-9}	1.91×10^{4}	−5.9
8	$O_2NCH_2^-$	32.0	4.39×10^{-24}	6.72×10^{-13}	16.7
9	$O_2NCH(Me)^-$	32.3	2.65×10^{-24}	2.22×10^{-11}	14.6
10	CN^-	38.2	1.31×10^{-28}	9.75×10^{-14}	17.8
11	$CNCH_2^-$	19.6	4.94×10^{-15}	7.71×10^{-2}	1.5
12	$MeC(O)CH_2^-$	22.4	4.46×10^{-17}	9.14×10^{-7}	8.3
13	$CHOCH_2^-$	24.4	1.55×10^{-18}	1.72×10^{-7}	9.3
14	F_3C^-	16.1	1.77×10^{-12}	1.48×10^{3}	−4.3
15	HO^-	(0)	(1)	(1)	(0)
16	MeO^-	11.4	4.77×10^{-9}	3.86×10^{-6}	7.4
17	EtO^-	14.5	2.61×10^{-11}	1.56×10^{-26}	10.7
18	PhO^-	39.5	1.47×10^{-29}	2.73×10^{-26}	35.0
19	HCO_2^-	43.4	2.09×10^{-32}	2.99×10^{-30}	40.4
20	H_2N^-	−12.1	6.79×10^{8}	6.93×10^{11}	−16.2
21	$MeNH^-$	−11.7	3.47×10^{8}	7.08×10^{13}	−19.0
22	Me_2N^-	−5.2	6.25×10^{3}	8.81×10^{9}	−13.6
23	HS^-	36.9	1.16×10^{-27}	3.72×10^{-20}	26.6
24	MeS^-	31.3	1.42×10^{-23}	6.72×10^{-15}	19.4
25	PhS^-	56.0	1.33×10^{-41}	4.50×10^{-33}	44.3
26	H^-	~−10	$\sim 2 \times 10^{7}$	$\sim 5.2 \times 10^{27}$	−38.0
27	F^-	18.3	4.39×10^{-14}	6.50×10^{-13}	16.7
28	Cl^-	56.9	2.93×10^{-42}	3.78×10^{-37}	49.9
29	Br^-	68.5	1.00×10^{-50}	1.36×10^{-43}	58.7
30	I^-	77.8	1.63×10^{-57}	2.17×10^{-48}	65.3

[a] Values taken from the following two references: Bartmess, J. E.; McIver, R. T., Jr. "Gas Phase Ion Chemistry Volume 2"; Bowers, M. T., Ed.; Academic Press: New York, 1979, Chapter 11. Taft, R. W. *Prog. Phys. Org. Chem.* **1983,** *14,* 305–346.
[b] $K_A^{Me} = K_A^H \times K_{HA}^{MeA}$.

In the previous section, anions of the same basicity in $(CH_3)_2SO$ solution showed a different order of reactivities when the anions attacked hydrogen than when they attacked carbon. The major difference was an enhanced nucleophilicity for carbanions when forming a bond to carbon. This difference may have a thermodynamic origin because, in the gas phase, carbanions have an enhanced carbon basicity, relative to nitranions and oxanions, at the same hydrogen basicity. Although the rates (nucleophilicities) of the S_N2 reactions for different donor-atom anions in $(CH_3)_2SO$ were compared at the same hydrogen basicity (pK_{HA}), the order is not intrinsic, that is, a nucleophilic order where the rates have been adjusted for $\Delta G°$ differences. Indeed, when the n-$CH_3(CH_2)_3Cl$ S_N2 reactivities are adjusted for estimated

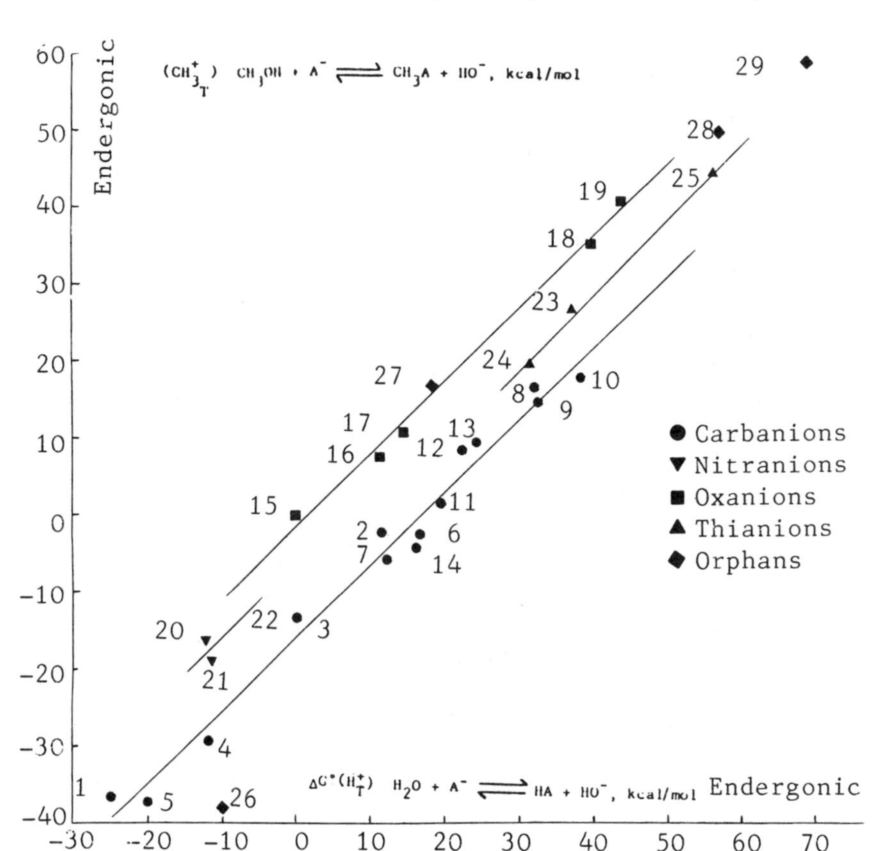

Figure 7. Plot of $\Delta G°(CH_{3,T}{}^+)$ values against $\Delta G°(H_T{}^+)$ values for the 30 anions in Table II (24).

differences in gas-phase basicities toward carbon at the same hydrogen basicity, by using the families of lines in Figure 7 as a model, the intrinsic nucleophilicity order toward carbon in $(CH_3)_2SO$ at constant hydrogen basicity and with $\Delta G° = 0$ (*22*, *23*), will probably be $S^- > O^- > N^- > C^-$.[5] This order is the same as the intrinsic nucleophilicity order of these donor-atom anions toward hydrogen in $(CH_3)_2SO$ (*16*). Other manifestations of the high carbon basicity of carbanions, relative to oxygen, include the well-known tendency for enolate ions to alkylate on carbon and the tendency of carbanions to effect S_N2 substitution on cyclohexyl substrates under conditions where oxanions of the same hydrogen basicity effect E2 elimination.

[5] Because S^- and N^- family lines appear to be nearly colinear and about midway between the O^- and C^- family lines in Figure 7, their relative reactivities in Brønsted plots will not change on converting the *x* axis from hydrogen to carbon basicity. But the O^- family is less basic toward carbon and the C^- family is more basic toward carbon, at the same hydrogen basicity, than the S^- and N^- families. The values obtained by taking the difference in carbon basicities of the O^- and C^- lines, relative to the N^- and S^- lines, and multiplying by the β value indicate that the apparent C^- rates are about 10^2 too fast and the apparent O^- rates are about 10^2 too slow when corrected for the thermodynamic driving force.

Nucleophilicity, Basicity, and Redox Potentials. Several authors have shown that nucleophilicities in various reactions can be correlated with redox potentials of nucleophiles. Edwards (*6*) used E_{ox} values in water in his two-parameter equation (equation 5). Dessy et al. (*34*) observed a linear correlation between oxidation potentials of various transition metal nucleophiles, ML_n^-, and the rate constants for reactions with CH_3I. Recently, Ritchie (*33*) found a correlation between oxidation potentials of nucleophiles and their rates of combination with pyronin cation. Linear correlations between oxidation potentials of anions and the pK_{HA} values of their conjugate acids were also observed by a number of investigators (*35–37*). In our laboratory, a good linear correlation between oxidation potentials of 2-fluorenide ions and the pK_{HA} values of their conjugate acids was found and a similar, but poorer, correlation for meta-substituted phenylcyanomethide ions was observed. Points for para donor substituents deviate from the lines in these plots because the pK_{HA} values fail to take into account the radical-stabilizing abilities of these substituents (*18*). The radical-stabilizing effects of remote substituents on radicals are relatively small, however (~0.5–3.0 kcal/mol) (*18*). As a consequence, in view of the general correlations observed between log k and pK_{HA} for nucleophile–electrophile combinations (Table I), a general, but not precise, relationship between log k and E_{ox} is expected.

Acknowledgment

We are grateful to the National Science Foundation for support of this research.

Literature Cited

1. Brønsted, J. N.; Pedersen, K. J. *Z. Phys. Chem., Stoechiom. Verwandschaftsl.* **1924,** *108,* 185–235.
2. Hammett, L. P. *Chem. Rev.* **1935,** *17,* 125–136.
3. Pedersen, K. J. *J. Phys. Chem.* **1934,** *38,* 581–600.
4. Bell, R. P. *The Proton in Chemistry,* 2nd ed.; Cornell University: Ithaca, NY, 1973; p 161.
5. Swain, C. G.; Scott, C. B. *J. Am. Chem. Soc.* **1953,** *75,* 141–147.
6. Edwards, J. O. *J. Am. Chem. Soc.* **1954,** *76,* 1540–1547.
7. Edwards, J. O.; Pearson, R. G. *J. Am. Chem. Soc.* **1962,** *84,* 16–24.
8. Parker, A. J. *Chem. Rev.* **1969,** *69,* 1–32.
9. Ogg, R. A.; Polanyi, M. *Trans. Faraday Soc.* **1935,** *31,* 604–620.
10. Glew, D. N.; Moelwyn-Hughes, E. A. *Proc. R. Soc. London, Ser. A* **1952,** *211,* 254–265.
11. Bordwell, F. G.; Hughes, D. L. *J. Org. Chem.* **1980,** *45,* 3314–3320, 3320–3325.
12. Hughes, D. L. Ph.D. Dissertation, Northwestern University, 1981.
13. Bordwell, F. G.; Hughes, D. L. *J. Org. Chem.* **1983,** *48,* 2206–2215.
14. Bordwell, F. G.; Hughes, D. L. *J. Org. Chem.* **1982,** *47,* 3224–3232.
15. Bordwell, F. G.; Mrozack, S. B. *J. Org. Chem.* **1982,** *47,* 4813–4815.

16. Bordwell, F. G.; Hughes, D. L. *J. Am. Chem. Soc.* **1985,** *107,* 4737–4744.
17. Bordwell, F. G.; Clemens, A. H. *J. Org. Chem.* **1985,** *50,* 1151–1156.
18. Bordwell, F. G.; Bausch, M. J. *J. Am. Chem. Soc.,* **1986,** *108,* 1979–1988.
19. Bordwell, F. G.; Bartmess, J. E.; Drucker, G. E.; Margolin, Z.; Matthews, W. S. *J. Am. Chem. Soc.* **1975,** *97,* 3226–3227.
20. Taft, R. W. *Prog. Phys. Org. Chem.* **1983,** *14,* 246–350.
21. Bordwell, F. G.; Branca, J. C.; Cripe, T. A. *Isr. J. Chem.* **1985,** *26,* 357–366.
22. Marcus, R. A. *J. Phys. Chem.* **1968,** *72,* 891–899.
23. Hassid, A. I.; Kreevoy, M. M.; Liang, T.-M. *Symp. Faraday Soc.* **1975,** *10,* 69–77.
24. Cripe, T. A. Ph.D. Dissertation, Northwestern University, Sept 1985.
25. Mishima, M.; McIver, R. T., Jr.; Taft, R. W.; Bordwell, F. G.; Olmstead, W. N. *J. Am. Chem. Soc.* **1984,** *106,* 2717–2718.
26. Bordwell, F. G.; Hughes, D. L. *J. Org. Chem.* **1982,** *47,* 169–170.
27. Bordwell, F. G.; Hughes, D. L. *J. Am. Chem. Soc.* **1984,** *107,* 3234–3239.
28. Arnett, E. M.; Troughton, E. B. *Tetrahedron Lett.* **1983,** *24,* 3299–3302.
29. Arnett, E. M.; Troughton, E. B.; McPhail, A. T.; Molter, K. E. *J. Am. Chem. Soc.* **1983,** *105,* 6172–6173.
30. Troughton, E. B.; Molter, K. E.; Arnett, E. M. *J. Am. Chem. Soc.* **1984,** *106,* 6726–6735.
31. Hine, J.; Weimar, R. D., Jr. *J. Am. Chem. Soc.* **1965,** *87,* 3387–3396.
32. Bartmess, J. E.; McIver, R. F., Jr. *Gas Phase Ion Chemistry;* Bowers, M. T., Ed.; Academic: New York, 1979; Vol. 2, Chapter 11.
33. Ritchie, C. D. *J. Am. Chem. Soc.* **1983,** *105,* 7313–7318.
34. Dessy, R. E.; Pohl, R. L.; King, R. B. *J. Am. Chem. Soc.* **1966,** *88,* 5121–5129.
35. Breyer, B. *Ber. Dtsch. Chem. Ges.* **1938,** *71,* 163–171.
36. Kern, J. M.; Sauer, J. D.; Federlin, P. *Tetrahedron* **1982,** *38,* 3032–3033.
37. Bank, S.; Schepartz, A.; Giammateo, P.; Zubieta, J. A. *J. Org. Chem.* **1983,** *48,* 3458–3464.

RECEIVED for review October 21, 1985. ACCEPTED February 10, 1986.

10

Effects of Solvation on Nucleophilic Reactivity in Hydroxylic Solvents

Decreasing Reactivity with Increasing Basicity

William P. Jencks

Graduate Department of Biochemistry, Brandeis University, Waltham, MA 02254

Three situations are described in which the removal of a solvating water molecule from an oxygen anion or an amine can have significant effects on the observed kinetics and structure–reactivity correlations of nucleophilic reactions. A decrease in the rate constants with increasing basicity of the nucleophile in reactions of substituted quinuclidines with phosphate monoester dianions and phosphorylated pyridine provides evidence for an initial desolvation step of the hydrated amine that becomes more difficult with increasing amine basicity. Limiting rate constants for the reactions of trifluoroethoxide and acetate anions with unstable 1-phenylethyl carbocations are 1 order of magnitude slower than the diffusion-controlled reactions of azide and thiol anions and free thiols. This finding is consistent with a requirement for a kinetically significant desolvation step in the solvent-separated ion pair before O–C bond formation. Brønsted plots for the reactions of alkoxide and phenoxide ions with esters and carbon acids are generally curved, with a significant decrease in β for basic alkoxide anions. The decrease in slope can be explained by a requirement for the removal of water from basic RO^- anions before reaction. Less basic anions are solvated by fewer water molecules and can react with less or no desolvation.

SOLVATION OF NUCLEOPHILIC REAGENTS AND LEAVING GROUPS causes enormous differences in the behavior of reactions in the gas phase and in solution (*1*). However, it is not so clear how solvation effects are manifested in the experimental data for reactions in a given solvent. A tendency for

0065-2393/87/0215-0155$06.00/0

solvation effects on reactants and transition states to cancel exists (2), and direct evaluation of the effects of changing solvation on nucleophilic reactions in water is difficult. This chapter describes three relatively small effects that influence observed nucleophilic reactivity in water and may be attributed directly to a requirement for desolvation.

Monosubstituted Phosphate Derivatives

Reactions of phosphate monoester dianions and of phosphorylated pyridines with water and other nucleophiles proceed through a mechanism that shows many of the characteristics expected for reaction through an intermediate metaphosphate monoanion, and several investigators (3–5) have suggested that these and related reactions proceed through such an intermediate. However, recent work has shown that metaphosphate, if it is formed, has too short a lifetime in several hydroxylic solvents to diffuse through the solvent before reaction, so that the reaction must occur by a stepwise or concerted preassociation mechanism. Methanolysis of the 2,4-dinitrophenyl phosphate dianion occurs with inversion, for example (6).

Phosphoryl transfer between pyridines has been examined with attacking pyridines of both larger and smaller basicity compared with the leaving pyridine. If the reaction occurred through an intermediate, as shown in the lower pathway of equation 1, a change in the rate-limiting step must occur at the point at which the attacking pyridine changes from being less nucleophilic to being more nucleophilic than the leaving group. This change in the rate-limiting step should produce a change in the sensitivity of the rate constant to the basicity of the nucleophile and a break in a Brønsted-type correlation of rate constants against basicity. However, no such break is observed. This finding provides evidence for a concerted reaction mechanism that occurs in one step through a single transition state (7–9). Examination of a larger series of nucleophiles including primary amines has provided additional evidence that the reaction does not proceed through two steps, with nucleophilic assistance to metaphosphate formation (*10*).

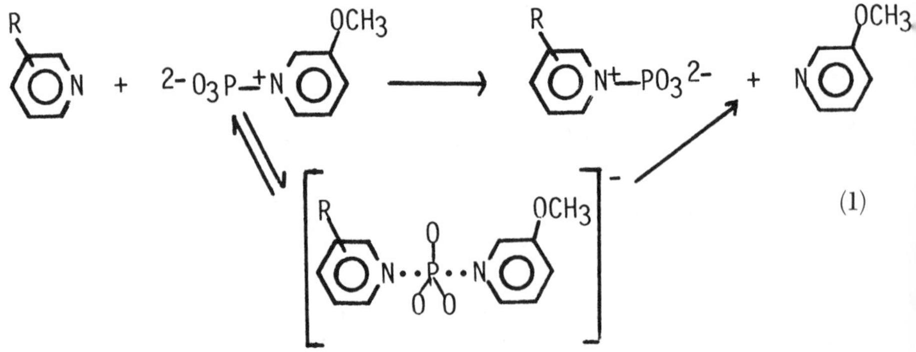

(1)

Nevertheless, the transition states of these reactions resemble the metaphosphate monoanion, with a relatively small amount of bond formation and a large amount of bond breaking in the transition state. Brønsted-type correlations give values of $\beta_{nuc} = 0.17–0.28$ (*10*). The value of β_{nuc} for the reaction of pyridines with *p*-nitrophenyl phosphate is small (*5*), and the value of β_{nuc} is 0 for the reaction of 2,4-dinitrophenyl phosphate dianion with pyridines, over a range of 9 p*K* units (*11*). This finding might seem to suggest that no bond formation to the nucleophile occurs in the transition state for these reactions. However, these are clear-cut second-order reactions that are much faster than the reaction with water; a significant amount of bond formation to the attacking pyridine in the transition state must occur. This bond formation should cause charge development on the pyridine, which should be reflected in an increase in reactivity with increasing basicity.

This surprising result suggested that the positive charge development in the transition state, which should be manifested as a positive value of β_{nuc}, might be offset by a requirement for desolvation of the attacking nucleophile, which would have a negative β value:

$$\rangle N\cdot HOH \overset{K_d}{\rightleftarrows} \rangle N: \overset{k}{\longrightarrow} \rangle N^{+}—PO_3^{2-} \qquad (2)$$

$$\beta_{obsd} = -0.2 + 0.1 = -0.1$$

Hydrogen bonding to a hydroxylic solvent is expected to become stronger with increasing basicity of a pyridine, so that the desolvation step should become more difficult with increasing basicity. This result could offset the rate increase in the nucleophilic step with increasing basicity in such a way as to give no net change in observed reactivity with basicity. To test this hypothesis, we examined the reactions of quinuclidines, for which some evidence of decreasing reactivity with increasing basicity already exists (*10, 11*) with monosubstituted phosphate derivatives. Amines are less strongly solvated than alkoxide ions, but amines have only a single electron pair so that complete desolvation is required before nucleophilic attack can occur (equation 2).

Figure 1 shows that indeed a decreased reactivity with increasing basicity of a series of substituted quinuclidines is observed for the reactions with 2,4-dinitrophenyl phosphate and *p*-nitrophenyl phosphate. The observed rate constants follow Brønsted correlations with slopes of approximately −0.1 and −0.05, respectively (*12*). The upper line for the reaction of the dinitrophenyl phosphate–calcium complex shows that the two positive charges of calcium do not change the result. Therefore, the negative slopes do not arise from an electrostatic interaction between polar substituents on the nucleophile and the two negative charges of the phosphate ester. The upper line in Figure 1 also shows that catalysis by calcium ion does not bring

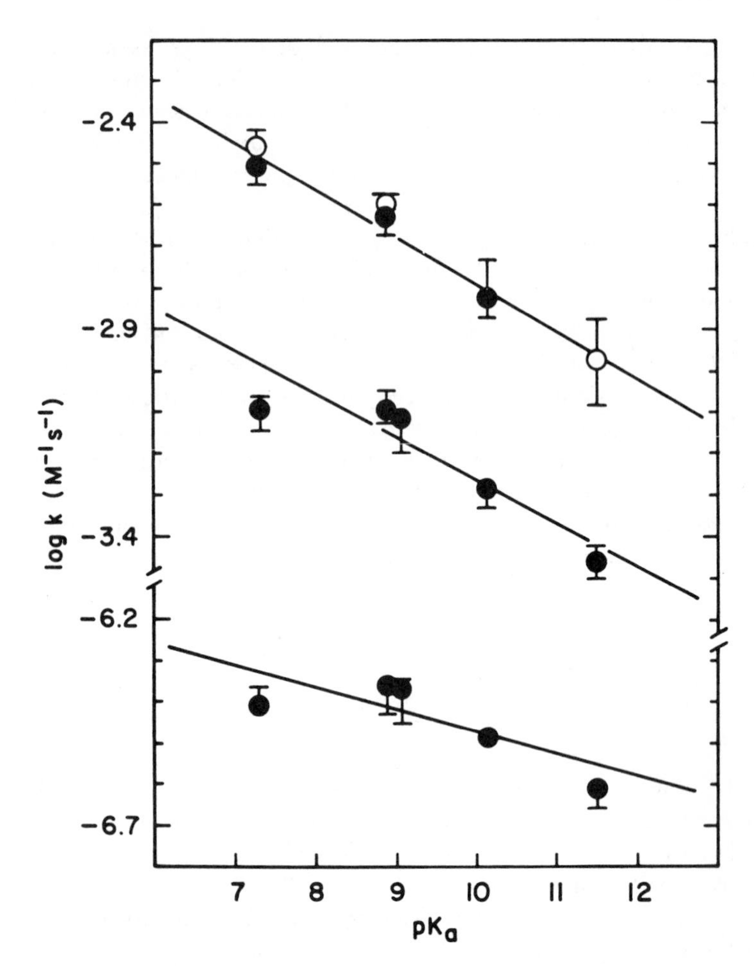

Figure 1. Brønsted-type plots for the reactions of substituted quinuclidines with the dianions of 2,4-dinitrophenyl phosphate complexed with calcium (upper line), 2,4-dinitrophenyl phosphate (middle line), and p-*nitrophenyl phosphate (lower line) at 39 °C, ionic strength 1.0 (KCl) (●) and ionic strength 1.5 (KCl) (○) (Reproduced from reference 12. Copyright 1986 American Chemical Society.)*

about a significant change in the nature of the transition state for nucleophilic substitution on monosubstituted phosphates.

Very similar results are found for reactions of quinuclidines with phosphorylated pyridine, which follow a Brønsted slope of $\beta_{nuc} = -0.1$. The less reactive phosphorylated 4-morpholinopyridine follows a slope of $\beta_{nuc} = -0.01$ (*12*). The changes in β_{nuc} for these phosphorylated pyridines and phosphate esters with a poorer leaving group (lg) represent Hammond effects that are described by the cross-interaction coefficient, p_{xy}, of equation 3. This cross coefficient is larger than the direct coefficient that describes curvature in Brønsted-type plots, $p_x = \partial\beta_{nuc}/\partial pK_{nuc}$, as predicted by energy contour diagrams for these reactions (*10*).

$$p_{xy} = \partial\beta_{nuc}/\partial pK_{lg} = \partial\beta_{lg}/\partial pk_{nuc} \quad (3)$$

These results are consistent with the observation that increasing basicity leads to a more favorable solvation energy for transfer of pyridines from the gas phase to water. The available data are consistent with a value of β_d for desolvation of approximately -0.2 (*10, 13–16*). The observed β value of -0.1 for the reactions of quinuclidines with 2,4-dinitrophenyl phosphate and phosphorylated pyridine would then reflect the difference between the value of $\beta_d = -0.2$ and a value of $\beta_{nuc} = 0.1$ for the bond-forming step (equation 2).

The more negative values of β_{nuc} for quinuclidines, compared with those for pyridines and amines, are not a consequence of the greater basicity of the quinuclidines because different slopes are observed for the different compounds over the same range of basicity and no significant curvature of the Brønsted plots occurs. Possibly, the smaller β values for the quinuclidines reflect a relatively early transition state because of steric hindrance for the reactions of tertiary amines.

This requirement for desolvation means that observed values of β_{nuc} underestimate the amount of bond formation in the transition state for reactions in which relatively little bond formation occurs in the transition state. The correction is small for large values of β_{nuc} because the protonated amine with a charge of $+1.0$ is the common reference point for the pK and for complete bond formation. Intermediate values of β can be corrected according to

$$\beta_{corr} = 0.2 + 0.8\beta_{obsd} \quad (4)$$

Fast Reactions of Unstable Carbocations

The solvolysis of 1-phenylethyl derivatives substituted with electron-donating substituents proceeds through an intermediate carbocation that has a significant lifetime in 50% trifluoroethanol–water. The carbocation intermediate can be trapped by azide ion or thiol anions at the same diffusion-controlled rate (equation 5). The same limiting rate constant, estimated to be 5×10^9 M^{-1} s^{-1}, is observed for propanethiol for less stable carbocations with fewer electron-donating substituents (Figure 2). However, trifluoroethoxide anion and acetate anion approach limiting rate constants that are about 10-fold smaller as the cation becomes less stable (Figure 2) (*17*).

$$Nu^- + (H)(CH_3)C^+\text{—}Ar \xrightarrow{k} Nu\text{—}C(H)(CH_3)\text{—}Ar \quad (5)$$

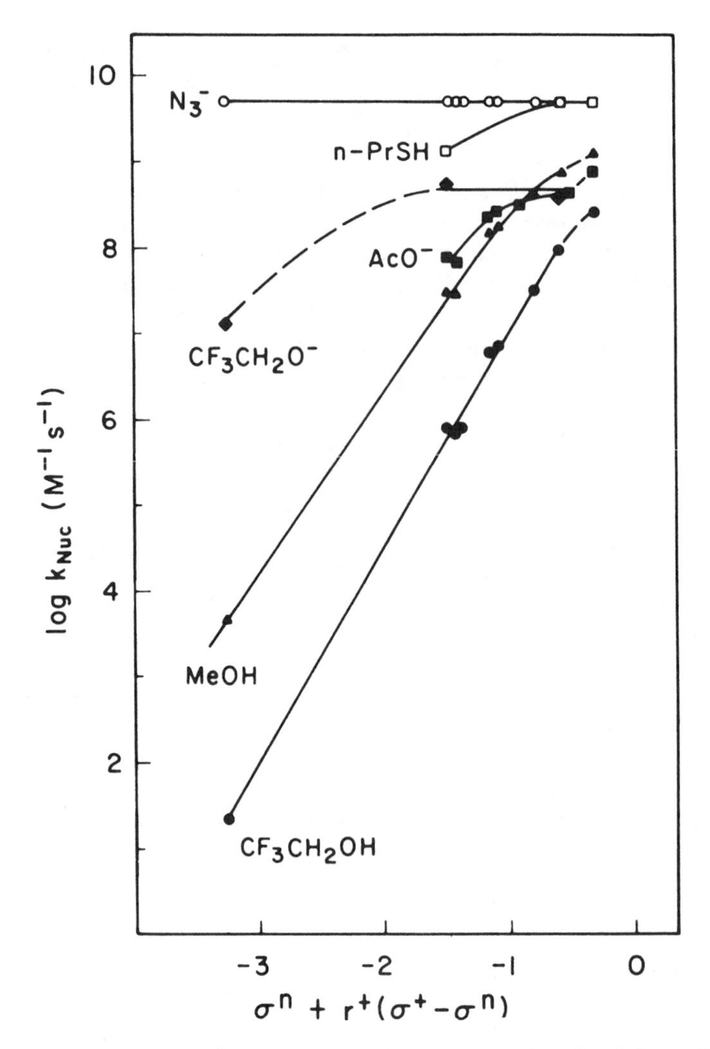

Figure 2. Estimated rate constants for reactions of nucleophiles with substituted 1-phenylethyl carbocations, plotted against the effective Hammett constant of the ring substituent with $r^+ = 2.1$*: (●) trifluoroethanol; (▲) methanol; (■) acetate anion; (◆) trifluoroethoxide anion; (□) propanethiol; (○) azide. (Reproduced from reference 17. Copyright 1984 American Chemical Society.)*

This 10-fold rate decrease means that these anionic oxygen nucleophiles react only approximately one time in 10 when they diffuse up to the carbocation. If diffusion of the solvated nucleophile toward the carbocation gives the solvent-separated ion pair **I** in equation 6 as the initial product, the ion pair diffuses apart about 10 times for every time that it loses water to form the intimate ion pair, which collapses rapidly to product. This finding suggests that the requirement for desolvation of these oxygen anions causes a 5–10-fold decrease in the limiting rate constants for their nucleophilic reactions; a decrease of approximately twofold could be an orientation or steric effect (*17*).

$$\mathrm{RO^{-}\cdot HO} + \overset{\diagdown + \diagup}{\underset{|}{\mathrm{C}}} \underset{k_{-1}}{\overset{k_1}{\rightleftharpoons}} \mathrm{RO^{+}\cdot H\underset{\mathrm{H}}{O}\cdot}\overset{\diagdown + \diagup}{\underset{|}{\mathrm{C}}} \xrightarrow{k_2} \mathrm{RO^{+}\cdot}\overset{\diagdown + \diagup}{\underset{|}{\mathrm{C}}} \xrightarrow{\text{fast}} \mathrm{RO{-}}\overset{|}{\underset{|}{\mathrm{C}}}{-} \quad (6)$$

I

This system provides information about the behavior of solvent-separated ion pairs in a largely aqueous solvent and provides one reason that limiting "diffusion-controlled" rate constants for reactions of different bases are not always the same; the final step of diffusion represents the difference between microscopic and macroscopic diffusion. A similar requirement for desolvation can account for the decrease of approximately 10-fold in the limiting rate constants for proton abstraction from hydrogen cyanide by basic amines in water (*18*). In contrast to many acids in which the proton is bound to an electronegative atom, proton transfer from hydrogen cyanide occurs only directly, not through an intermediate water molecule, so that proton transfer requires loss of the solvating water from the base before proton removal (*19*). Grunwald and co-workers (*20–22*) have shown that amines are hydrogen-bonded to water and that the rate constants for removal of this water are slow enough to compete with proton transfer and diffusional steps.

Methanol and trifluoroethanol show curvature in Figure 2, before the limiting rate constants are reached, rather than a sharp break at the diffusion-controlled limit. This finding may represent a requirement for solvation of the attacking alcohol to provide stabilization for the acidic proton of the attacking ROH in the transition state (*17*). Evidence exists for an increased reactivity of alcohol clusters in the reactions of alcohol with carbocations in dichloroethane (*23*).

Nonlinear Structure–Reactivity Correlations

Curved Brønsted plots or other structure–reactivity correlations are often taken as evidence for changes in transition-state structure with changing properties of the reactant that might be described by the Marcus equation (*24*) or other equations. However, it is important to evaluate other possible explanations for such curvature, including solvation effects that could decrease the reactivity of basic nucleophiles without any change in the structure of the transition state for nucleophilic attack. For example, solvation effects could provide a relatively simple explanation for the curvature of structure–reactivity correlations for reactions of basic oxygen anion nucleophiles with acyl compounds and carbon acids.

An example is shown in the Brønsted-type correlation of Figure 3, which shows rate constants for the reactions of substituted phenoxide and alkoxide ions with *p*-nitrophenyl thioacetate (*25*). Although considerable scatter of the rate constants for reactions of alkoxide ions occurs, a definite decrease in the slope for basic ions compared with those for less basic phenolate ions is evident. The slope of the correlation is 0.68 for the phenolate ions, and a line is drawn with a slope of 0.17 through points for alkoxide ions of similar geometric structure. Very similar curves are observed for

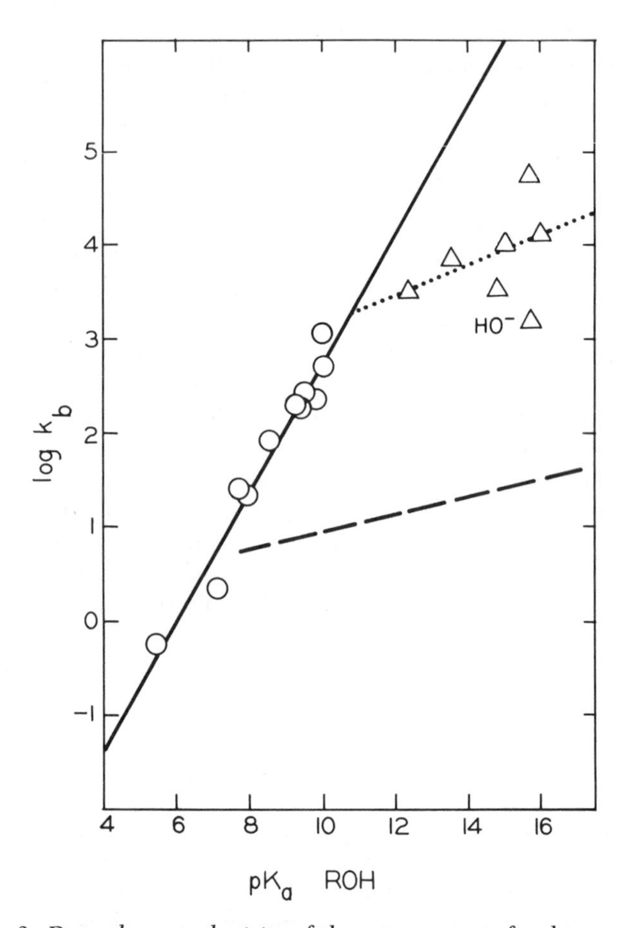

Figure 3. Dependence on basicity of the rate constants for the reactions of phenoxide, alkoxide, and hydroxide ions with p-nitrophenyl *thioacetate. The solid line for phenoxide ions (circles) has a least-squares slope of* β_{nuc} = *0.68, and a dotted line of slope 0.17 has been drawn through the points for ethoxide and trifluoroethoxide anions (triangles). The dashed line shows the expected rate constants if the full solvation energy change for phenolate ionization is added to* $\Delta G^{\ddagger}$ *(see the text). (Reproduced from reference 25. Copyright 1977 American Chemical Society).*

reactions of anionic oxygen nucleophiles with *p*-nitrophenyl acetate and other esters and for proton removal from carbon acids (*25–31*).

Kovach et al. (*32*) has suggested that the different behavior of the phenoxide and alkoxide ions results from resonance delocalization in the phenoxide anions. Although this delocalization may be a contributing factor, it is not the most important factor because increased slopes are also observed for weakly basic nucleophiles of aliphatic oxygen anions (*31*). A contribution of changing transition-state structure may also occur, but this contribution is not the principal cause of the change in slope because other measures of transition-state structure show no change or only a small change over a large range of nucleophile reactivity (*31*).

The strongest evidence that nonlinearity in reactions of oxygen anions does not arise from a change in transition-state structure comes from the demonstration by Pohl and Hupe (*30*) that the curvature is an intrinsic property of the nucleophilic reagents, not of $\Delta G°$ for the reaction. Figure 4 shows a plot of the rate constants for catalysis of proton removal from different carbon acids against the difference between the pK_a of the carbon acid and that of the conjugate acid of the catalyzing oxygen base. Figure 4 shows that the curvature is not a function of $\Delta G°$ for the reaction; instead, the curvature is a function of the absolute pK_a of the oxygen base. The observed curvature is sharp and would require an intrinsic barrier of 2.5 kcal mol^{-1} for the proton-transfer reaction, according to the Marcus equation. This value is much smaller than the intrinsic barrier of approximately 10 kcal mol^{-1} that was estimated from the change in Brønsted β values for the removal of a proton from a series of ketones and diketones of increasing acidity. The expected curvature for an intrinsic barrier of 10 kcal mol^{-1} is shown by the solid line in Figure 4. The gradual change in the primary deuterium isotope effect for hydron abstraction, shown by the solid line at the bottom of Figure 4, is also consistent with an intrinsic barrier of 10 kcal mol^{-1}. The dashed line for an intrinsic barrier of 2.5 kcal mol^{-1}, based on the observed curvature of the Brønsted plots, predicts a much sharper decrease in the isotope effect than is observed (*30*).

Nonlinearity in structure–reactivity correlations that arises from a solvent effect requires a change in the amount or the nature, not just the strength, of solvation. A change in the strength of solvation with increasing basicity should itself follow a linear structure–reactivity correlation. Consequently, a change in strength would change the slope, but not the linearity, of an observed structure–reactivity correlation, as in the desolvation of amines described previously.

Evidence exists that hydroxide and alkoxide ions are solvated by hydrogen bonding to three solvent molecules (*33, 34*). This solvation contributes to the observed equilibrium solvent isotope effects for the ionization of oxygen acids. However, these isotope effects decrease with decreasing basicity of the anion (*35*). The observed solvent deuterium isotope effect for the ionization of

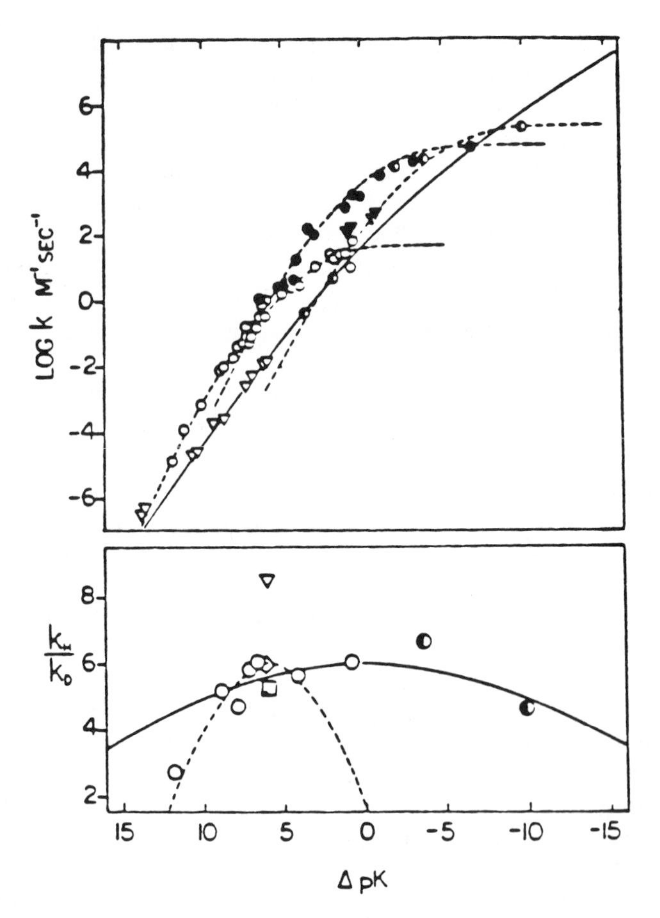

Figure 4. Brønsted plots and isotope effects, for proton abstraction from carbon acids, from Pohl and Hupe (30). *The upper plot shows rate data for oxyanion and thiol anion catalyzed proton abstraction from 4-(4-nitrophenoxy)-2-butanone* (○, ▽), *acetylacetone* (●, ▼), *and ethyl nitroacetate* (○). *The lower plot shows values of* k_H/k_D *for oxyanions* (○), *an amine* (◇), *and a thiol anion* (△) *with 4-(4-nitrophenoxyl)-2-butanone and oxyanions with ethyl nitroacetate* (○). *The solid line is the type of dependence expected for a large intrinsic barrier (10 kcal/mol). The dashed line is the type of dependence expected for a small intrinsic barrier (2.5 kcal/mol).*

carboxylic acids can be accounted for simply by the fractionation factor for H_30^+, without a significant contribution from solvation of the carboxylate anion (*31*). The low limiting rate constants for diffusion-controlled reactions of acetate anion with carbocations indicate that a kinetic barrier for removal of solvating water exists (*17*), but the strength of this binding is not enough to cause a significant change in the zero-point energy and isotope effect of the solvent. The decreasing isotope effects suggest that the strength of solvation decreases with decreasing pK_a of the anion. In the limit, water itself is not fully hydrogen-bonded; it is not ice.

Cohen and Jones (*36*) have shown that ortho substitution of two *tert*-butyl groups causes a large increase in the basicity of basic phenoxide ions. However, this effect becomes smaller with decreasing p*K* of the phenol and essentially disappears for phenols of $pK < 7$. This evidence indicates that the importance of solvation, and presumbly the number of solvating water molecules, decreases sharply for weakly basic substituted phenoxide ions.

The dashed line in Figure 3 shows the decrease in the rate constant that would be expected if the entire change in solvation energy that is brought about by two *tert*-butyl groups (*36*) were expressed to decrease the rate constant for nucleophilic attack by oxygen anions. This difference is much larger than is needed to explain the observed decreases in rate for the most basic anions.

The available data appear to be consistent with the notion that removal of one of the three water molecules from a basic alkoxide ion is necessary to make an electron pair available for nucleophilic attack. The requirement for this removal decreases the observed rate constants and β_{nuc} for basic oxygen anions that are solvated by three water molecules. This concept implies that partial solvation can be retained in the transition state. A somewhat similar situation has been calculated by Jorgensen and co-workers (*2*) for the attack of Cl^- on methyl chloride in water. With the less basic phenoxide ions, less solvation occurs initially and nucleophilic attack can occur with little or no loss of solvation energy.

An explanation of how these solvation effects change rate constants in a structure–reactivity correlation is not trivial. We cannot simply say that solvation decreases the rate constant, because we are considering a correlation with pK_a and solvation also decreases the basicity of the ion in the reference reaction. One way to illustrate this problem is shown in Figure 5. Consider a Brønsted correlation line that was constructed by using weakly basic nucleophiles, for which desolvation is not important. The lower left point on this line represents the *observed* pK_a of a more basic nucleophile that must lose a solvating water molecule, with the equilibrium constant K_D, before reaction and before addition of a proton according to equation 7. The pK_a of the desolvated ion is larger than the observed pK_a by $-\log K_D$, as shown by the lower horizontal arrow. However, this increase in p*K* causes a relatively small increase in reactivity of the desolvated ion, by the amount $-\beta \log K_D$, as shown by the right-hand arrow. Finally, the observed rate constant will be smaller than this value because the fraction of RO^- that exists as the reactive, desolvated ion is small. The observed $\log k$ is decreased by the amount of $-\log K_D$ (for small K_D), as shown by the left-hand arrow.

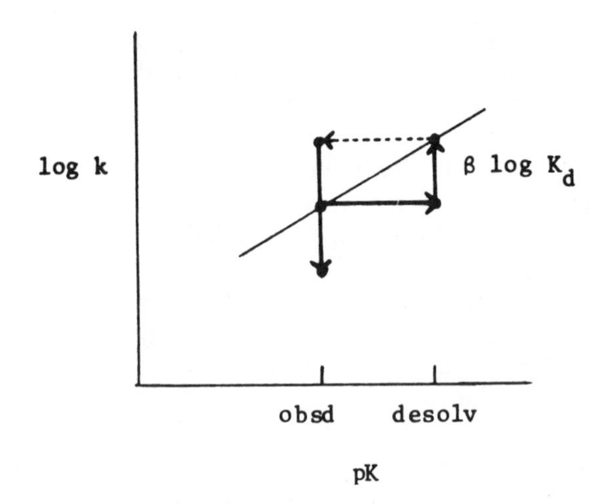

Figure 5. Schematic Brønsted plot to show how a requirement for partial desolvation can cause a negative deviation from a Brønsted plot by its effects on the observed pK, *reactivity, and concentration of solvated and desolvated ions* (31).

The final result is that the observed log k is smaller than the predicted log k by the amount of $(1 - \beta)$ log K_d. Thus, the effects of desolvation are expected to be largest when the value of β_{nuc} is small (*31*).

$$RO^-(HOH)_n \overset{K_D}{\rightleftharpoons} RO^-(HOH)_{n-1} + HOH \xrightarrow{k[X]} \text{product} \quad (7)$$

Literature Cited

1. Taft, R. W.; Abboud, J-L. M.; Kamlet, M. J.; Abraham, M. H. *J. Sol. Chem.* **1985,** *14,* 153–186, and references cited therein.
2. Chandrasekhar, J.; Smith, S. F.; Jorgensen, W. L. *J. Am. Chem. Soc.* **1985,** *107,* 154–163.
3. Benkovic, S. J.; Schray, K. J. In *Transition States of Biochemical Processes;* Gandour, R. D.; Schowen, R. L., Eds.; Plenum: New York, 1978; Chapter 13, p 493.
4. Westheimer, F. H. *Chem. Rev.* **1981,** *81,* 313–326.
5. Kirby, A. J.; Jencks, W. P. *J. Am. Chem. Soc.* **1965,** *87,* 3209–3216.
6. Buchwald, S. L.; Friedman, J. M.; Knowles, J. R. *J. Am. Chem. Soc.* **1984,** *106,* 4911–4916.
7. Skoog, M. T.; Jencks, W. P. *J. Am. Chem. Soc.* **1983,** *105,* 3356–3357.
8. Bourne, N.; Williams, A. *J. Am. Chem. Soc.* **1983,** *105,* 3357–3358.
9. Bourne, N.; Williams, A. *J. Am. Chem. Soc.* **1984,** *106,* 7591–7596.
10. Skoog, M. T.; Jencks, W. P. *J. Am. Chem. Soc.* **1984,** *106,* 7597–7606.
11. Kirby, A. J.; Varvoglis, A. G. *J. Chem. Soc. B.* **1968,** 135–141.
12. Jencks, W. P.; Haber, M. T.; Herschlag, D.; Nazaretian, K. L. *J. Am. Chem. Soc.* **1986,** *108,* 479–483.

13. Arnett, E. M.; Chawla, B. *J. Am. Chem. Soc.* **1979,** *101,* 7141–7145.
14. Arnett, E. M.; Jones, F. M., III; Taagepera M.; Henderson, W. G.; Beauchamp, J. L.; Holtz, D.; Taft, R. W. *J. Am. Chem. Soc.* **1972,** *94,* 4724–4726.
15. Taft, R. W. *Prog. Phys. Org. Chem.* **1983,** *14,* 328.
16. Taft, R. W.; Wolf, J. F.; Beauchamp, J. L.; Scorrano G.; Arnett, E. M. *J. Am. Chem. Soc.* **1978,** *100,* 1240–1249.
17. Richard, J. P.; Jencks, W. P. *J. Am. Chem. Soc.* **1984,** *106,* 1373–1383.
18. Bednar, R. A.; Jencks, W. P. *J. Am. Chem. Soc.* **1985,** *107,* 7117–7126.
19. Bednar, R. A.; Jencks, W. P. *J. Am. Chem. Soc.* **1985,** *107,* 7126–7138.
20. Grunwald, E.; Eustace, D. In *Proton-Transfer Reactions;* Caldin, E.; Gold, V., Eds.; Wiley: New York, 1975; pp 103–120.
21. Grunwald, E.; Ku, A. Y. *J. Am. Chem. Soc.* **1968,** *90,* 29–31.
22. Grunwald, E.; Ralph, E. K., III *Acc. Chem. Res.* **1971,** *4,* 107–113.
23. Sujdak, R. J.; Jones R. L.; Dorfman, L. M. *J. Am. Chem. Soc.* **1976,** *98,* 4875–4879.
24. Marcus, R. A. *J. Phys. Chem.* **1968,** *72,* 891–899.
25. Hupe, D. J.; Jencks, W. P. *J. Am. Chem. Soc.* **1977,** *99,* 451–464.
26. Jencks, W. P.; Gilchrist, M. *J. Am. Chem. Soc.* **1962,** *84,* 2910–2913.
27. Hupe, D. J.; Wu, D. *J. Am. Chem. Soc.* **1977,** *99,* 7653–7659.
28. Hupe, D. J.; Wu, D.; Shepperd, P. *J. Am. Chem. Soc.* **1977,** *99,* 7659–7662.
29. Pohl, E. R.; Wu, D.; Hupe, D. J. *J. Am. Chem. Soc.* **1980,** *102,* 2759–2763.
30. Pohl, E. R.; Hupe, D. J. *J. Am. Chem. Soc.* **1984,** *106,* 5634–5640.
31. Jencks, W. P.; Brant, S. R.; Gandler, J. R.; Fendrich, G.; Nakamura, C. *J. Am. Chem. Soc.* **1982,** *104,* 7045–7051.
32. Kovach, I. M.; Elrod, J. P.; Schowen, R. L. *J. Am. Chem. Soc.* **1980.** *102,* 7530–7534.
33. Gold, V.; Grist, S. *J. Chem. Soc., Perkin Trans. 2* **1972,** 89–95.
34. Gold, V.; Toullec, J. *J. Chem. Soc., Perkin Trans. 2* **1979,** 596–602.
35. Bell, R. P.; Kuhn, A. T. *Trans. Faraday Soc.* **1963,** *59,* 1789–1793.
36. Cohen, L. A., Jones, W. M. *J. Am. Chem. Soc.* **1963,** *85,* 3397–3402.

RECEIVED for review March 3, 1986. ACCEPTED July 11, 1986.

11

Rates and Equilibria of Electrophile–Nucleophile Combination Reactions

Calvin D. Ritchie

Department of Chemistry, State University of New York, Buffalo, NY 14214

The recently reported correlation of reactivities and one-electron oxidation potentials of nucleophiles is examined with new data for hydrazine in aqueous solution and several nucleophiles in $(CH_3)_2SO$ solution. The correlation fails to apply to these reactions. A thermodynamic cycle is utilized to estimate the free energies of ionization of pyronin–nucleophile adducts both in solution and in the solid state. A satisfying rationalization of the dichotomy of ionic and covalent crystals of these and similar compounds is obtained. The equilibrium constants for reactions of nucleophiles with several types of cations are examined as indicators of specific bonding effects such as steric and gem interactions.

ONE-STEP BOND-FORMING REACTIONS of an electrophile with a nucleophile, Lewis acid–base reactions, are among the most common elementary reactions in organic chemistry. An understanding of the factors determining the rates and equilibria of such reactions would constitute an understanding of much of the entire field of organic chemistry.

The task of finding systems of electrophiles and nucleophiles for which rates and equilibria of the simple combination reactions can be measured is not an easy one, and data have accumulated slowly. The efforts in my laboratory have focused primarily on carbocations, with pK_R values in the range measurable in dilute aqueous solution, reacting with common anionic and neutral nucleophiles. The pyronin cation [3,6-bis(dimethylamino)xanthylium cation] is particularly well suited for such studies (*1*). This cation has an unusually high pK_R of 11.5 and gives measurable rates and equilibria with a wide range of nucleophiles. Pyronin is also stable enough in dimethyl sulfoxide [$(CH_3)_2SO$] solution to allow studies in that solvent (*2*) for

0065-2393/87/0215-0169$06.00/0

comparison with reactions in water. The planar structure of the cation and the fact that the reactive site is a secondary carbon should minimize complications from steric effects.

Our most recent report (*3*) on reactions of pyronin cation showed a correlation between the rates of reactions of a variety of nucleophiles and the one-electron oxidation potentials of the nucleophiles in aqueous solution. The correlation included anionic, neutral, and "α-effect" nucleophiles; only thiolate ion nucleophiles showed significant deviations from the correlation.

The one-electron oxidation potentials of the nucleophiles in aqueous solution were obtained from a thermodynamic cycle involving the gas-phase bond-dissociation energies and the aqueous pK_a values of the H–nuc species. Also, the one-electron oxidation potentials, the standard free energies for the cation–nucleophile combination reaction, and the standard free energies for the homolytic bond dissociations of the adducts are interrelated by a thermodynamic cycle.

In this chapter, newly available data for hydrazine (*4*) and data for reactions of pyronin in $(CH_3)_2SO$ solution (*2*) will be used to further test this correlation, and the last-mentioned thermodynamic cycle will be applied to the estimation of equilibrium constants for cation–nucleophile combination reactions that cannot be measured directly. The equilibrium constants for reactions of nucleophiles with several types of cations will be compared.

Oxidation Potentials and Rates

The justification for equation 1 has been presented (*3*) where $\Delta G°_3(H_2O)$ is the standard free energy of dissociation (homolytic) in aqueous solution and $\Delta H°_3(gas)$ is the standard enthalpy of dissociation in the gas phase for an H–X molecule or $H–B^+$ ion.

$$\Delta G°_3(H_2O) = \Delta H°_3(gas) - 2 \text{ kcal/mol} \qquad (1)$$

The adiabatic ionization potential for hydrazine was recently determined (*4*) to be 187 kcal/mol. The proton affinity of hydrazine is −207 kcal/mol [adjusted to $A_p(NH_3)$ = −206 (*5*)]. Combined with the ionization potential of 313.6 kcal/mol for H· (*3*), the value of 80 kcal/mol for $\Delta H°_3(gas)$ of $H–NH_2NH_2^+$ is obtained.

Using the thermodynamic cycles presented earlier (*3*) with a value of 10.8 kcal/mol for the free energy of acid dissociation of hydrazinium ion in water, we obtain $\Delta G°_5$ = 16 kcal/mol for the standard oxidation potential of hydrazine in aqueous solution.

Table I shows the standard oxidation potentials and other related thermodynamic quantities along with the free energies of activation for reactions with pyronin cation in aqueous solution for all nucleophiles for which we have data. Most of the data were reported previously (*3*), but several errors in

Table I. Oxidation Potentials and Reactivities with Pyronin Cation in Water

Nucleophile	ΔH°_3(*Gas*)	ΔG°_2	ΔG°_5	ΔG°(*Soln*)	$\Delta G^\dagger$
HOO^-	88.6	16.0	20.5	−95	11.5
$C_6H_5O^-$	86.5	13.7	20.7	−74	—[a]
NO_2^-	78.0	4.4	21.5	—	—
$CH_3CH_2CH_2S^-$	88.6	14.8	21.7	−78	8.2
$C_6H_5S^-$	83.3	8.9	22.3	−68	6.7
CH_3O^-	102.0	22.6	27.3	−93	12.2
$CF_3CH_2O^-$	102.0	17.0	32.9	−84	13.8
N_3^-	92.5	6.4	34.0	−75	12.7
ClO^-	98.0	10.3	35.6	−74	13.1
CH_3COO^-	106.0	6.4	47.5	−67	—
CN^-	123.8	12.6	59.1	−74	17.3
OH^-	119.3	21.5	45.7	−105	15.1
NH_2NH_2	80.0	10.8	16.1	−72	13.2
Piperidine	94.0	15.2	26.7	−56	11.2
$CH_3CH_2CH_2NH_2$	103.0	14.4	36.5	−63	13.2
H_2O	141.0	0.0	89.3	−103	21.9

NOTE: Original references to the literature are given in reference 3, except for hydrazine, which is given in the present text. ΔG°_2 is the standard free energy of ionization of the conjugate acids of the nucleophiles. ΔG°_5 is the standard oxidation potential versus the normal hydrogen electrode. ΔG°(soln) is the standard free energy for transfer of the anionic nuleophile or of the conjugate acid of the neutral nucleophile from dilute gas to dilute aqueous solution.
[a]Not determined.

previous data have been corrected. (The values for ΔG°_1 in the previous report are all 2 kcal/mol too high, and the free energy of solution of acetate ion should be −67 kcal/mol, rather than −77 kcal/mol.) The free energy of activation for reaction of hydrazine with pyronin is estimated from the relative rates of reactions of hydrazine and *n*-butylamine with [*p*-(dimethylamino)phenyl]tropylium ion (*6*) and the measured rate of reaction of *n*-butylamine with pyronin cation (*1*).

If the correlation of ΔG°_5 and $\Delta G^\dagger$ is to make any sense, the correlation must be presumed to be between the $\Delta G^\dagger$ and the ΔG°_{ET} for the reaction

$$R^+ + X^- \xrightarrow{\Delta G^\circ_{ET}} R\cdot + X\cdot$$

where

$$\Delta G^\circ_{ET} = \Delta G^\circ_5(X^-) - \Delta G^\circ_5(R\cdot) \quad (2)$$

and $\Delta G^\circ_5(R\cdot)$ is the standard oxidation potential of the pyronin radical.

If we wish to compare reactions in different solvents, $\Delta G°_{ET}$ will change as a result of changes in the oxidation potentials of both X^- and of $R\cdot$. The general scheme (Scheme I)

$$\begin{array}{ccccccccc} R_w^+ & + & Nuc_w^- & \xrightarrow{\Delta G_{ET}{}^w} & R_w\cdot & + & X_w\cdot^+ \\ \Delta G_R{}^{+t}\downarrow & & \Delta G_X{}^{-t}\downarrow & & \Delta G_R\cdot^t\downarrow & & \Delta G_X{}^{t\cdot(+)}\downarrow \\ R_D^+ & + & Nuc_D^- & \xrightarrow{\Delta G^D{}_{ET}} & R_D\cdot & + & X_D\cdot^{(+)} \end{array}$$

Scheme I

gives the free energy of electron transfer in $(CH_3)_2SO$, $\Delta G_{ET}{}^D$, in terms of the quantity in water, $\Delta G_{ET}{}^w$, and the transfer free energies of the pertinent species(t):

$$\Delta G_{ET}{}^D = \Delta G_{ET}{}^w - \Delta G_R{}^{+t} + \Delta G_{R\cdot} - \Delta G_X{}^{-t} + \Delta G_{X\cdot}{}^{(+)t} \quad (3)$$

If we assume that the transfer free energy for $X\cdot$ is equal to that for HX, we obtain the last two terms of equation 3 by considering:

$$HX^+ \xrightarrow{K_a} H^+ + X^-$$

in water and $(CH_3)_2SO$:

$$\begin{aligned} \Delta G_X{}^{-t} - \Delta G_{HX}{}^{(+)t} &= 1.37\Delta pK_a - \Delta G_H{}^{+t} \\ &\simeq \Delta G_X{}^{-t} - \Delta G_{X\cdot}{}^{(+)t} \end{aligned} \quad (4)$$

where ΔpK_a is the pK_a in $(CH_3)_2SO$ minus that in water. The free energy of transfer of the proton from water to $(CH_3)_2SO$ has been estimated as −4.5 kcal/mol (7), and the pK_a values for a variety of acids in both solvents are known.

To estimate the transfer free energies of R^+ and $R\cdot$, we may consider the reaction

$$R^+ + n\text{-PrS}^- \xrightarrow{\Delta G°_c} n\text{-PrSR}$$

where R^+ is the pyronin cation. Equilibrium constants for this reaction in both water (*1*) and $(CH_3)_2SO$ (*2*) have been measured:

$$\Delta G_{PrSR}{}^t - \Delta G_{R+}{}^t - \Delta G_{PrS-}{}^t = -13.0 \text{ kcal/mol}$$

From the measured pK_a values of n-PrSH in water and $(CH_3)_2SO$ (8), we have

$$\Delta G_{PrS}{}^{-t} - \Delta G_{PrSH}{}^t = 13.0 \text{ kcal/mol}$$

where we have used −4.5 kcal/mol for the transfer free energy of the proton (7). Combining these last two equations

$$\Delta G_{\mathrm{PrSR}}{}^{t} - \Delta G_{\mathrm{PrSH}}{}^{t} = \Delta G_{\mathrm{R+}}{}^{t} \simeq \Delta G_{\mathrm{R\cdot}}{}^{t}$$

where the last approximate equality arises from the assumption that the difference in transfer free energies of PrSR and PrSH is due to a "group" contribution.

With these estimates, we finally have

$$\Delta G_{\mathrm{ET}}{}^{D} = \Delta G_{\mathrm{ET}}{}^{w} - 1.37\Delta pK_a(\mathrm{HX}) - 4.5 \tag{5}$$

Because $\Delta G_{\mathrm{ET}}{}^{w}$ is given by equation 2, obviously a plot of the quantity $\Delta G^{\circ}_5(\mathrm{X}^-) - 1.37\Delta pK_a(\mathrm{HX}) - 4.5$ versus $\Delta G^{\dagger}$, where the ΔG°_5 values are those for aqueous solution, in $(CH_3)_2SO$ should have the same slope and intercept as the plot of $\Delta G^{\circ}_5(\mathrm{X}^-)$ versus $\Delta G^{\dagger}$ in water if the correlation of $\Delta G^{\circ}_{\mathrm{ET}}$ with $\Delta G^{\dagger}$ holds. Pertinent data are given in Table II, and the plot of the data for both water and $(CH_3)_2SO$ is shown in Figure 1.

Table II. Data For Reactions in $(CH_3)_2SO$ Solution

Nucleophile	$-1.37\Delta pK_a - 4.5$	$\Delta G^{D}_{ET} + \Delta G^{\circ}_5(R)$	$\Delta G^{\dagger}$
n-$BuNH_2$	−5.2	31.3	10.2
$CH_3OCH_2CH_2NH_2$	−5.6	32.0	10.6
Piperidine	−3.8	23.0	8.9
n-PrS^-	−13.0	8.7	<4.7
CN^-	−9.7	49.0	8.2

The point for the reaction of hydrazine in water is badly off the correlation line for the other nucleophiles, and all of the data for reactions in $(CH_3)_2SO$ solution are not only off the correlation line but also fail to show any relationship when considered alone.

It is still surprising that so many of the reactions in water show the correlation between ΔG°_5 and $\Delta G^{\dagger}$. No theory of which we are aware anticipates such a correlation. The theory of curve crossings (*9–11;* S. Shaik, personal communication, 1985) ascribes the activation barriers for these reactions to an avoided crossing of surfaces, one for a diradical state and the other for an ionic state. The two surfaces, however, are related by "vertical" transitions, and solvation energies, particularly, make these far different from the "adiabatic" differences such as ΔG°_5 (S. Shaik, personal communication, 1985). Unfortunately, the fundamental data necessary to even begin to apply this theory to the present reactions are not available.

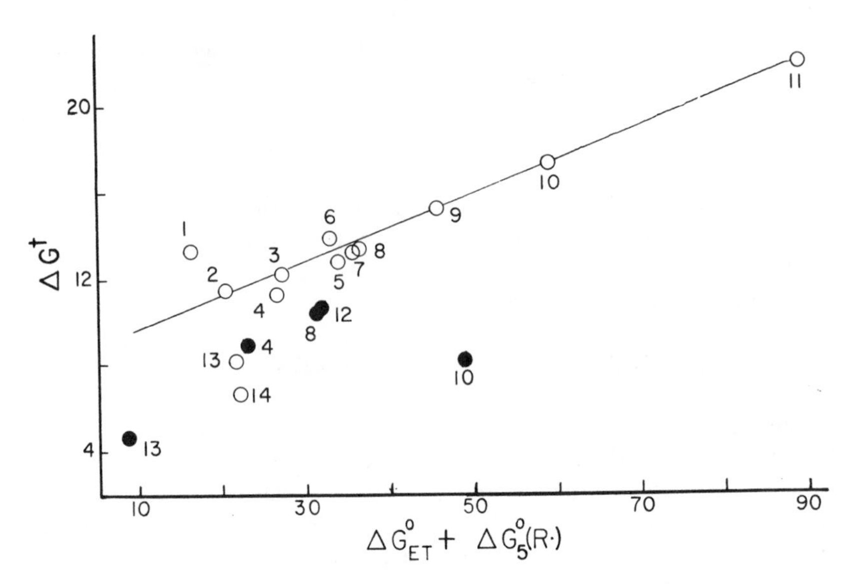

Figure 1. Reactivities versus oxidation potentials of nucleophiles. Open circles: aqueous solution. Closed circles: $(CH_3)_2SO$ solution. Nucleophiles are identified by numbers: 1, NH_2NH_2; 2, HOO^-; 3, CH_3O^-; 4, piperidine; 5, N_3^-; 6, $CF_3CH_2O^-$; 7, ClO^-; 8, n-$PrNH_2$; 9, HO^-; 10, CN^-; 11, H_2O; 12, $CH_3OCH_2CH_2NH_2$; 13, n-PrS^-; and 14, $C_6H_5S^-$.

Heterolytic and Homolytic Bond Dissociations of R–X

The heterolytic and homolytic dissociation of R–X species are related by a thermodynamic cycle to the electron transfer from X^- to R^+:

$$\Delta G°_{BDE} - \Delta G°_{ET} = \Delta G°_{ion} \tag{6}$$

where $\Delta G°_{BDE}$ is the free energy of dissociation of R–X into R· and X· and $\Delta G°_{ion}$ is the free energy of dissociation of R–X into R^+ and X^- (3). As already mentioned, $\Delta G°_5(R\cdot)$ is not known, so we cannot obtain absolute values for $\Delta G°_{ET}$. For a common R, however, differences in $\Delta G°_{ET}$ for different X are equal to differences in the $\Delta G°_5$ values, and we have

$$\Delta\Delta G°_{BDE} - \Delta\Delta G°_5 = \Delta\Delta G°_{ion} \tag{7}$$

where

$$\Delta\Delta G°_{BDE} = \Delta G°_{BDE}(RX) - \Delta G°_{BDE}(RY)$$
$$\Delta\Delta G°_5 = \Delta G°_5(X^-) - \Delta G°_5(Y^-)$$
$$\Delta\Delta G°_{ion} = \Delta G°_{ion}(RX) - \Delta G°_{ion}(RY)$$

The approximation that $\Delta\Delta G^\circ_{BDE}$ in solution is equal to that in the gas phase has already been justified (*3*), and equation 7 was used to obtain $\Delta\Delta G^\circ_{BDE}$ values from the other quantities. We will now obtain values for ΔG°_{ion} in cases where these values are not experimentally accessable.

The bond strengths of various C–X bonds, for cases where the other three groups bonded to carbon are either C or H, primarily reflect the stabilities of the C and X radicals formed on dissociation so long as steric effects are not serious. Thus, the difference in bond strength between an R–X and an R–Y in such cases is expected to be nearly constant for various R. A useful compilation of bond dissociation energies in the gas phase has been given by McMillen and Golden (*12*). In Table III, the differences in bond dissociation energies between R–X and R–OH, averaged for R representing methyl, ethyl, and benzyl where these substituents are available, are shown for a number of X groups. In Table III, the column headed $\Delta\Delta G^\circ_{ET}$(OH–X) shows the differences in ΔG°_5 values for OH and X from Table I.

If the $\Delta\Delta G^\circ_{BDE}$ values for R representing pyronin are approximately equal to those shown in Table III, the values for $\Delta\Delta G^\circ_{ion}$ can be calculated from equation 7. The value for ΔG°_{ion} for pyronin–OH is known (*1*) to be 3.4 kcal/mol, so values of ΔG°_{ion}, in the last column of Table III, can be obtained.

Table III. Estimates of ΔG_{ion}

X	D(*RX*) − D(*ROH*)	$\Delta\Delta G^\circ_{ET}(OH–X)$	ΔG°_{ion}
OH	0	0	3.4
OCH_3	−9.8	18.4	12
OC_6H_5	−28	25	0
CN	28	−13.4	18
NO_2	−31	24.2	−3
F	13	−36	−20
Cl	−10	−14	−21
O_2CCH_3	−12	−2	−11

If differential steric effects for reactions of different cations are not too serious, we may expect the differences in ΔG°_{ion} for various Xs to be the same as those in Table III. We (*1*) reported, for example, that the relative equilibrium constants for reactions of various Xs are the same for pyronin and [(dimethylamino)phenyl]tropylium cations. From the known pK_R, 11.5 (*1*), for pyronin and the estimates given in Table III, the maximum pK_R that a cation could have and still give measurable equilibria with the various X^-s in Table III can be estimated.

For the equilibrium constant for the cation–anion combination reaction to be measurable in reasonably dilute aqueous solution, it must be greater than approximately 10. We may estimate, then, that chloride and fluoride ions will give measurable equilibria only with cations having pK_R values less than approximately −5. Acetate ion should give measurable equilibria with cations having a pK_R value of less than approximately 2. Bunton and Huang (*13*) reported that acetate ion gives no reaction with tri-*p*-anisylmethyl cation, which has a pK_R of 0.8. This finding is not surprising because serious steric effects may be expected for the triarylmethyl systems (*1*).

Anyone who has worked with triarylmethyl or pyronin systems has probably been struck by the distinctions between the solid states of various derivatives. Malachite Green chloride, for example, is a highly colored crystalline ionic material, and the carbinol is a colorless, reasonably low melting [mp 163 °C (*14*)] solid. This distinction between ionic and covalent solids can be considered in another thermodynamic cycle:

$$\begin{array}{ccc} R_w^+ + X_w^- & \xrightarrow{-\Delta G^\circ_{ion}} & RX^w \\ \Delta G^\circ_{cryst} \downarrow & & \downarrow \Delta G^\circ_{ppt} \\ [R^+X^-]_{cryst} & \xrightarrow{\Delta G^\circ_{IC}} & [RX]_{cryst} \end{array}$$

For triarylmethyl and pyronin derivatives, the solubilities of covalent compounds, such as the carbinols, are on the order of 10^{-5} M in water. Thus, ΔG°_{ppt} must be approximately −7 kcal/mol.

The free energy of crystallization, ΔG°_{crys}, is equal to minus the solvation energies of the cation and anion plus the crystal energy. We have values for the solvation energies of various X^- (Table I). The solvation energies of the R^+ species are expected to be very close to −50 kcal/mol for these types of cations (*15*). A very rough estimate of the crystal energies can be obtained from an electrostatic calculation and an r_0^6 repulsion term. If we use r_0 = 3 Å and a Madelung constant of 1.6 for simple stacked ions with no interstack interactions are used, a value of −145 kcal/mol is calculated for the crystal energy. The free energy for formation of the crystal from the gas-phase ions involves the loss of entropy of two free particles. At 25 °C, this loss of entropy should contribute ca. +13 kcal/mol to the free energy.

The use of these rough estimates gives

$$\Delta G^\circ_{IC} = 75 + \Delta G^\circ_{soln}(X^-) - \Delta G^\circ_{ion} \tag{8}$$

Calculated values for pyronin derivatives and the maximum pK_R of the cation allowing formation of covalent solid are given in Table IV.

The facts that triphenylmethyl chloride (pK_R = −6.6) forms a covalent solid and that all known carbinols and methyl ethers are covalent indicate that the calculations are at least reasonable. Interestingly, the fluorides of most cations are predicted to form covalent solids. I do not know of data to support or contradict this prediction. Also, possibly an acetate can be found

Table IV. Ionic to Covalent Crystals

X	$-\Delta G^\circ_{soln}$	ΔG°_{ion}	ΔG°_{IC}	$pK_R(cov)$
OCH_3	93	12	−30	all
OC_6H_5	74	0	1	10
F	105	−20	−10	18
Cl	78	−21	17	−2
O_2CCH_3	67	−11	18	−2

for which an ionic crystal dissolves in water to give a covalent compound in solution because we found previously that ΔG°_{ion} should be positive for cations with a pK_R value below +2.

Steric and Other Specific Bonding Effects

Throughout much of the preceding discussions, we have assumed that steric and other specific bonding effects are absent in the compounds considered. Hine (*16*) focused attention on the equilibria

$$\text{R–OH} + \text{HX} \underset{RX}{\overset{K^{HX}}{\rightleftharpoons}} R\text{–}X + H\text{–}OH$$

as indicators of specific effects. Values of the equilibrium constants for these reactions can be calculated from those for the cation–anion combination reactions and the acidity constants for the H–X molecules where R^+ is one of the stable cations. From measured equilibrium constants for addition of HX to carbonyl compounds, $K_{RX}{}^{HX}$ for R^+ = α-hydroxy carbocations can be obtained. Data are shown for several systems in Table V.

All of the values in Table V, with the exception of those for trifluoroethoxide ion, are appreciably greater than zero. This finding indicates some special stability of an H–O bond or some unusual instability of a C–O bond. This same feature is shown for simpler carbon groups, such as methyl and ethyl, in the tabulations of bond strengths in the gas phase given by McMillen and Golden (*12*).

The other trends in values for the various Rs in Table V are consistent with the trends in pairwise interaction contributions to heats of formation evaluated by Allen (*19*) and discussed by Hine (*20*). For example, pyronin has a C–C–X pair interaction in place of an O–C–X interaction for pyridine-4-carboxaldehyde; for X = OH this difference favors the aldehyde adduct by approximately 7 kcal, although for other Xs, the difference is somewhat smaller. Thus, the values given for the aldehyde are smaller than those for pyronin. We have already pointed out that steric effects are expected in the triarylmethyl derivatives and that the comparison of these with pyronin derivatives are consistent with that expectation (*1*).

Table V. Values of log K_{RX}^{HX}

X	Ar_3C^{+}[a]	Pyr^{+}[b]	$RC^{+}H-OH$[c]	$ArC^{+}(CF_3)-OH$[d]
SO_3^{2-}	7.8	12.7	7.6	~2.9
CN^-	—[e]	17.2	6.0	~5.0
$HOCH_2CH_2S^-$	4.5	7.1	3.9	<3.0
HOO^-	—	4.1	2.9	~1.2
$CF_3CH_2O^-$	—	−0.3	<0.5	—
N_3^-	2.0	—	0.3	—
n-PrNH⁻	1.2	3.9	3.2	1.8
$O(CH_2CH_2)_2N^-$	—	3.6	3.1	—
CH_3ONH^-	4.2	5.1	4.4	—
H_2NNH^-	2.5	—	4.4	3.2

[a]Values for Malachite Green, Crystal Violet, or tri-*p*-anisylmethyl cation in water (*1*).
[b]Values for pyronin or [(dimethylamino)phenyl]tropylium cation in water (*1*).
[c]Values from additions to pyridine-4-carboxaldehyde (*17*).
[d]Values from additions to 1,1,1-trifluoroacetophenone (*18*).
[e]Not determined.

Some of the specific numbers in Table V are more difficult to understand, however. Peroxide, although an oxygen base like hydroxide, shows preference for bonding to carbon. Hydrazine and methoxylamine anions give the same value of K_{RX}^{HX} with the aldehyde but significantly different values with the triarylmethyl system. No explanations are apparent.

Acknowledgment

The work reported in this paper has been supported by Grant CHE 8205767 from the National Science Foundation.

Literature Cited

1. Ritchie, C. D.; Kubisty, C.; Ting, G. Y. *J. Am. Chem. Soc.* **1983,** *105,* 279
2. Ritchie, C. D. *J. Am. Chem. Soc.* **1983,** *105,* 3573.
3. Ritchie, C. D., *J. Am. Chem. Soc.* **1983,** *105,* 7313.
4. Mautner, M.; Nelsen, S. F.; Willi, M. R.; Frigo, T. B. *J. Am. Chem. Soc.* **1984,** *106,* 7384.
5. Moylan, C. R.; Brauman, J. I. *Annu. Rev. Phys. Chem.* **1983,** *34,* 187.
6. Ritchie, C. D.; Virtanen, P. O. I. *J. Am. Chem. Soc.* **1973,** *95,* 1882.
7. Cox, B. G.; Hedwig, G. R., Parker, A. J.; Watts, D. W. *Aust. J. Chem.* **1974,** *27,* 477.
8. Ritchie, C. D.; VanVerth, J. E.; Virtanen, P. O. I. *J. Am. Chem. Soc.* **1982,** *104,* 3491.
9. Warshel, A.; Weiss, R. M. *J. Am. Chem. Soc.* **1980,** *102,* 6218.
10. Pross, A.; Shaik, S. *Acc. Chem. Res.* **1983,** *16,* 363.

11. Shaik, S. *J. Am. Chem. Soc.* **1984,** *106,* 1227.
12. McMillen, D. F.; Golden, D. M. *Annu. Rev. Phys. Chem.,* **1982,** *33,* 493.
13. Bunton, C. A.; Huang, S. K. *J. Am. Chem. Soc.* **1972,** *94,* 3536.
14. Ritchie, C. D.; Sager, W. F.; Lewis, E. S. *J. Am. Chem. Soc.* **1962,** *84,* 2349.
15. Wolf, J. F.; Harch, P. G.; Taft, R. W. *J. Am. Chem. Soc.* **1975,** *97,* 2904.
16. Hine. J. *Structural Effects on Equilibria in Organic Chemistry;* Wiley-Interscience: New York, 1975; p 225.
17. Sander, E. G.; Jencks, W. P. *J. Am. Chem. Soc.* **1968,** *90,* 6154.
18. Ritchie, C. D. *J. Am. Chem. Soc.* **1984,** *106,* 7187.
19. Allen, T. L. *J. Chem. Phys.* **1959,** *31,* 1039.
20. Hine, J. *Structural Effects on Equilibria in Organic Chemistry;* Wiley-Interscience: New York, 1975; pp 8–10.

RECEIVED for review October 21, 1986. ACCEPTED January 6, 1986.

12

Nucleophilic Attacks on Low Lowest Unoccupied Molecular Orbital Compounds

Shmaryahu Hoz

Department of Chemistry, Bar-Ilan University, Ramat-Gan 52100, Israel

The nature of the transition state of nucleophilic reactions with LL [low lowest unoccupied molecular orbital (LUMO)] substrates is analyzed and reviewed. In cation–anion combination reactions, a partial radical character is developed on both the nucleophile and the substrate. Examination of a simple state diagram shows that this diradicaloid character is increased as the LUMO of the substrate is lowered. The model is further extended to other LL substrates such as carbonyl functions and activated olefins. Three empirical manifestations of the diradicaloid character of the transition state are discussed: (1) the correlation between the ionization potentials of the nucleophiles and their nucleophilicity toward LL substrates; (2) the α-effect phenomenon; and (3) the variations in the positional selectivity of 9-nitromethylenefluorene in nucleophilic reactions as a function of the solvent.

THE SWAIN–SCOTT EQUATION (*1*) is probably one of the most important empirical equations in the field of nucleophilic reactions:

$$\log (k/k_0) = sn \tag{1}$$

This equation relates the nucleophilicity of a nucleophile with its rate constant in a given S_N2 reaction. In this class of reactions, bond formation and bond cleavage are fused into a single transition state. A seemingly less complicated reaction is a reaction in which the status of only one bond is changed in the rate-determining step. Anion–cation combinations exemplify this type of reaction. The relationship between rate constants and nucleophilicity in these reactions is given by the Ritchie equation (*2, 3*):

0065-2393/87/0215-0181$06.00/0

$$\log (k/k_0) = N_+ \tag{2}$$

The two major differences between these two equations are as follows: (1) Unlike the Swain–Scott equation, the Ritchie equation lacks a selectivity parameter. Thus, a plot of log k versus N_+ always gives a unity slope. (2) Nucleophilicity ranking by the two scales is different. This difference is especially noticeable for the three nucleophiles, CN^-, CH_3O^-, and N_3^-. According to the N_+ scale, the nucleophilicity order is $N_3^- > CH_3O^- > CN^-$, whereas the reverse is true for the Swain–Scott n scale.

To test the generality of the Ritchie equation, we reacted the olefins 9-(dinitromethylene)fluorene (FDN), 9-(dicyanomethylene)fluorene (FDCN), and 9-(nitromethylene)fluorene (FN) with a series of nucleophiles (*4, 5*). The reaction of these three substrates resemble that of anion–cation combination reactions in the absence of leaving group departure at the transition state. Yet, in spite of the fact that these substrates are not positively charged, an excellent correlation was observed with the N_+ scale. Similar results were observed for other uncharged substrates such as carbonyl functions (*6*) and activated aryl halides (*7*). However, the three systems, FDN, FDCN, and FN, are unique in that the slopes of log k versus N_+ are significantly larger than 1. This necessitates the incorporation of a selectivity parameter (S_+) into the Ritchie equation.

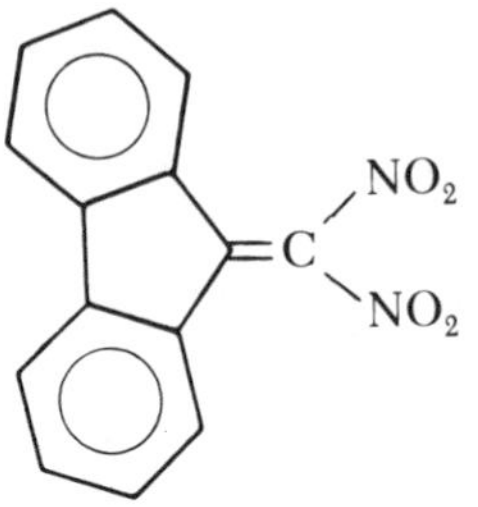

FDN 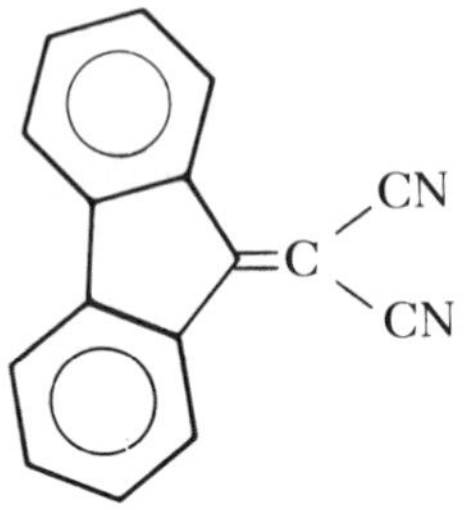 FDCN 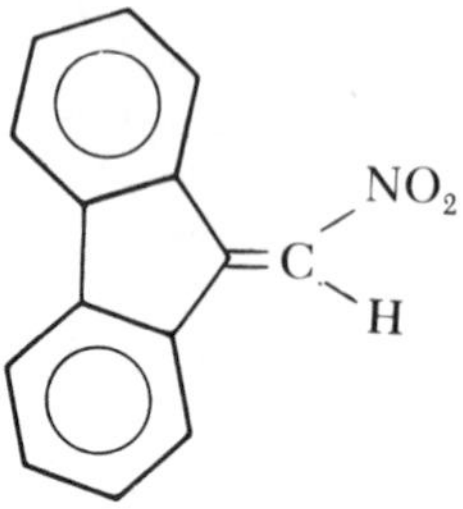 FN

Recently, we found (unpublished results) that the reactions of the diphenyl analogues of these three substrates do not follow the Ritchie equation and the data are highly scattered. One major difference between the two sets of substrates is the extent of steric crowding around the activated carbon. However, the mechanism by which the steric effect induces this change is not entirely clear.

Classification

Ritchie (*2*) suggested that the differences between the reactions that follow the Swain–Scott and the Ritchie equations stem from the coupling between bond formation and bond cleavage in the S_N2 process as opposed to only bond formation in the second class of reactions. We (5), on the other hand, suggested that the origin of the differences probably lies in the nature of the lowest unoccupied molecular orbital (LUMO) of the substrate. The LUMO of a typical S_N2 substrate being a σ^* orbital is of high energy, whereas cations, carbonyls, activated olefins, and other substrates that obey the Ritchie equation have a relatively low LUMO, usually π^*. The latter dichotomy is more in line with the generally accepted notion that highest occupied molecular orbital (HOMO)–LUMO interactions govern nucleophilic reaction (8). So that these two approaches can be distinguished, a system should be found that undergoes an S_N2 reaction and has a relatively low energy reactive LUMO. Because the vast majority of the substrates that possess both low LUMO and leaving group (e.g., aryl halides and esters) react with nucleophiles in a two-step process, an example that will meet the two demands is likely to be a borderline case in terms of classification.

Two available examples are nucleophilic reactions with organometallic cations and with activated bicyclobutanes. Sweigart and co-workers (9, *10*) showed that a linear correlation is obtained between the log k values for the reactions of nucleophiles (Nu) with various organic cations and the transition metal complexes of the same cations. The latter reactions resemble S_N2 reactions if the carbon–metal bond cleavage is considered as a departure of the leaving group:

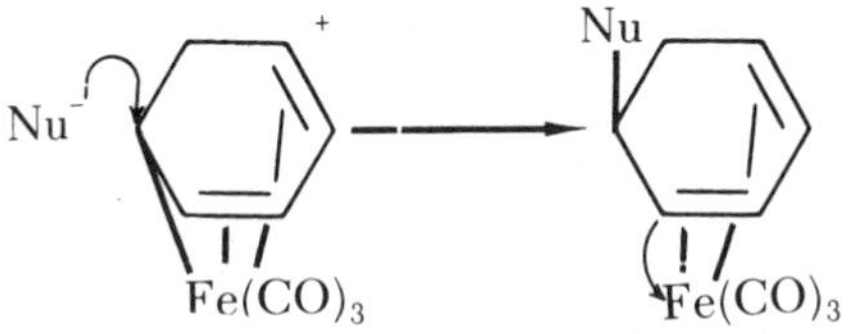

$$Nu^- + [C_6H_7Fe(CO)_3]^+ \longrightarrow Nu{-}C_6H_7Fe(CO)_3 \qquad (3)$$

The second example involves nucleophilic attacks on activated bicyclobutane. These attacks occur anti to the central bond, which is cleaved in the course of the reaction. The LUMO of this bond is lower than that of ethylene. Thus, if considered to be an S_N2 reaction, the reaction complies with the two aforementioned demands. Unfortunately, data are available (*11*) for the reaction of this system with CN^- and CH_3O^- only:

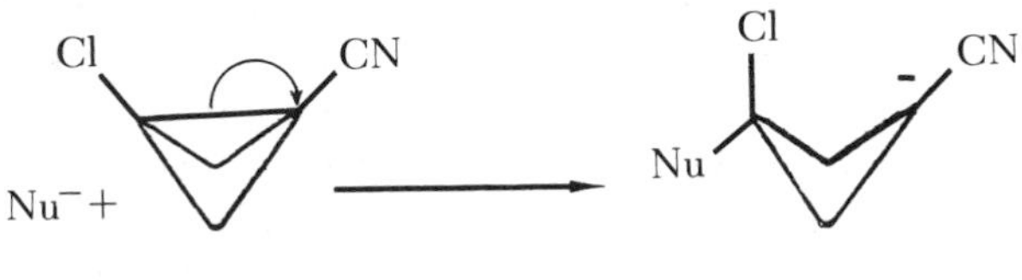

$$Nu^- + \text{Cl,CN-bicyclobutane} \longrightarrow \text{Nu,Cl-cyclobutyl carbanion (CN)} \qquad (4)$$

The order of reactivity of these two nucleophiles toward bicyclobutanecarbonitrile is $CH_3O^- > CN^-$. This order is for the nucleophilicity of the N_+ and not of the n scale. Thus, in addition to the fact that frontier orbitals are usually considered to control nucleophilic reactions, the last two examples lend more credence to the classification of the substrates according to the energies of their LUMOs as a possible origin of the differences between the two types of behavior. Therefore, in the following discussion we will refer to the substrates acording to their LUMO's energies, namely, LL (low LUMO) or HL (high LUMO) substrates.

Proposed Model

According to the model developed (*5, 12*), the transition state of the reaction of LL substrates with nucleophiles should be characterized not only by partial charges ($\delta\pm$) but also by a partial diradicaloid character. In other words, at the transition state, both the nucleophile and the substrate acquire a partial radical character ($\delta\cdot$). This finding emerges directly from the basic assumption of continuity between the two extreme situations that may take place upon a nucleophile–electrophile encounter. The first situation is the "normal" S_N2 reaction, whereas the second extreme is an electron-transfer reaction that results in the formation of a radical pair. The transition state of a reaction located in the continuity zone between these two extremes can therefore be described as a combination of the two extreme electronic configurations in variable proportions. For a given nucleophile, the amount of the diradicaloid character will increase with the proclivity of the substrate to undergo electron-transfer reactions. Thus, an LL substrate will in general show more of the diradicaloid nature at the transition state than HL substrates.

In the following discussion nucleophilic reactions with LL substrate will be analyzed by making use of basic principles of electronic states. We will show that (a) under certain circumstances the transition state of nucleophilic reactions must be diradicaloid and (b) the diradicaloid nature of the transition state will in general increase with the increase in the electrophilicity of the substrate.

To perform the analysis, we need to examine the classic state diagram (*13, 14*) shown in Figure 1. Curve a represents the ground state (S_0) of a molecule A–B, which correlates with the diradical state (D), and curve b is the excited state (S_1), which correlates with the zwitterionic state (Z) (*15, 16*). In general, for cations and anions such as those studied by Ritchie, that is triarylmethyl cations, tropilium cations, diazonium cations, and common anionic nucleophiles such as CH_3O^-, CN^-, and N_3^-, the zwitterionic state in the gas phase will always be of higher energy than the diradicaloid configuration (D). (In the case of A–B being Na–Cl, the separation between curves a and b at the plateau region ($IP_{Na} - EA_{Cl}$) is small (ca. 35 kcal/mol). However, the Coulombic bonding is larger than the covalent bonding, which leads to curve crossing, which places the ionic bond well below the covalent bond. On the other hand, in the case of A–B = CH_3–Cl, $IP_{CH_3} - EA_{Cl}$ is approximately 145 kcal/mol and the covalent bond is much stronger than the

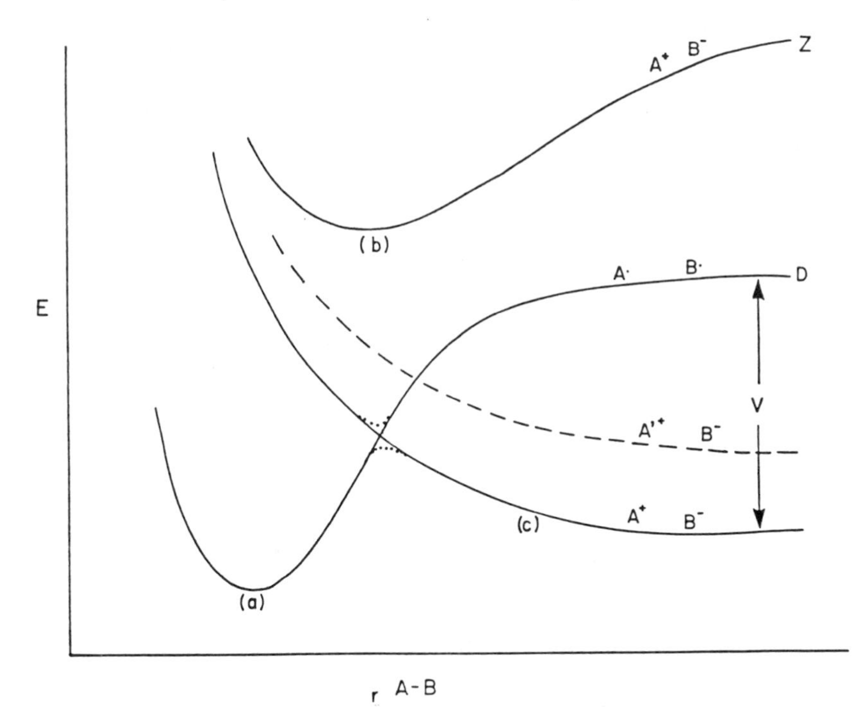

Figure 1. State diagram for the reaction A + B ⟶ AB. Curve a is the covalent diradicaloid state; curve b is the excited (S_1) zwitterionic state; curve c is the zwitterionic state in condensed phase. The dashed line is similar to curve c and is for the more electrophilic A'.

ionic one (using the Pauling equation, the covalent bond energy of Na–Cl can be estimated as 38 kcal/mol, whereas that of CH_3–Cl is approximately 80 kcal/mol). Therefore, for CH_3Cl and similar compounds, no curve crossing occurs in the gas phase. (For the relevant data, *see* references 17 and 18). In solution, to the first approximation the covalent diradicaloid curve a will not be significantly affected by the solvent, whereas curve b will be drastically lowered in energy (curve c) so as to appear in part below curve a. Thus, in solution, the A–B bond will be cleaved heterolytically by following curve a to the avoided crossing point with curve c and continuing along curve c to give $A^+ + B^-$ as is indeed observed in solvolytic reactions. The reaction in the reverse direction is nothing but the anion–cation combination reactions studied by Ritchie. For a covalent bond to be formed from the ionic species A^+ and B^-, the ions must go over a barrier that peaks in the region of the intersection (avoided crossing zone) of the potential ionic and diradical–covalent surfaces. More importantly, *upon going from A· + B· to the covalent bond A–B, a gradual decrease in the diradical character of the system occurs. At an intermediate point between the two extremes along curve a, the system, in a simple notation, will be described as* $\overset{\delta\cdot}{A} - \overset{\delta\cdot}{B}$. At the transition state, this configuration will be mixed with the ionic configuration in equal proportions to give the overall configuration, which using the same notation will be described as $\overset{\delta+\delta\cdot}{A} - \overset{\delta-\delta\cdot}{B}$. Thus, in this case, the transition state for anion–cation combination reaction is of a diradicaloid nature, and in order to reach this state, partial electron transfer (PET) from the nucleophile to substrate must take place.

This model may also be appropriate to other LL substrates, such as carbonyl groups (*6*), activated double bonds (*5, 19*), and aryl halides (*7*), that also obey the Ritchie equation. This seems appropriate because these LL substrates are highly polar and because particularly in solution the zwitterion **I** is a major contribution to their overall resonance structure:

$$C{=}X \rightleftarrows \underset{\textbf{I}}{C^{+}{-}X^{-}}$$

where X = O, $C(NO_2)_2$, $C(CN)_2$, $C(H)NO_2$, C(H)COPh, and so on. Thus, the carbon in **I** can be viewed as an equivalent to A^+ of the previous discussion.

This model is not applicable in cases where the transition state of the rate-limiting step cannot be identified with the point in which curves a and c intersect. Most of the data regarding the movement of a molecular system along the a–c combination path are obtained from solvolytic reactions. These studies indicate that curve c in Figure 1 is not necessarily a smooth line (as we have drawn for the sake of simplicity) but may have some energy minima along it that may correspond to intimate and solvent-separated ion pairs observed in the course of many solvolytic reactions. In cases where these two species are defined chemical entities, the species must be separated by a potential barrier. In many cases, the transfer from intimate to solvent-separated ion pairs is assumed to comprise the rate-determining step in the solvolytic reaction (*20*). Therefore, by microscopic reversibility, the rate-determining step for the same reaction in the reverse direction will not be the transition from intimate ion pairs to covalent compound but rather from the solvent separated to the intimate ion pair. Therefore, the transition state of the rate-determining step of this process will not contain the diradicaloid component, which is a crucial characteristic of the transition states of reactions obeying the Ritchie equation, and hence will not be accommodated by this equation.

The second point that is worth noting is the effect of the electrophilicity of A on the magnitude of the diradicaloid character of the transition state. Replacing A^+ by $A^{+\prime}$, which is of a higher electrophilicity, will diminish the separation (*V*) between curves a and c in Figure 1. Because in general the effect of such a change on the energy of the covalent bond as well as on the coulombic interaction will be secondary to the effect on *V*, the intersection point of curves a and c (and therefore the transition state) will move to the right (Figure 1). As we have previously pointed out, the diradicaloid nature of the A, B system decreases gradually upon moving from right to left (Figure 1, curve a). Thus, if the transition state is shifted to the right (an early transition state), it will acquire a larger diradicaloid character. Shifting the transition state further to the extreme right will result in a complete electron transfer that will precede the coupling of A· and B·. This finding, in addition to confirming our previous intuitive conclusion, also implies that the magnitude of the diradicaloid character at the transition states of a series of reactions in which either the substrate or the nucleophile is varied is not constant. At the intersection point, the electronic configurations represented by curves a and c are mixed in *equal proportions* to yield the final electronic configuration of the transition state.

Empirical Manifestation of the Model

In the following sections, three possible empirical manifestations of the proposed model are presented: (1) correlation between the nucleophilicity toward LL substrate and the ionization potentials of the nucleophiles, (2) the α effect, and (3) the positional selectivity of FN as a function of the solvent.

Nucleophilicity and Ionization Potentials. Previously, the suggestion that the differences in the nucleophilic ranking between the n and the N_+ scale may stem from a relatively large amount of electron transfer present in the transition state described by the N_+ scale was made. If this statement is true, then some correlation between the N_+ scale and the solution ionization potentials of the nucleophiles should exist. Because these data are not available, we have calculated (5) the energy associated with K_3 using the thermodynamic cycle shown in Scheme I.

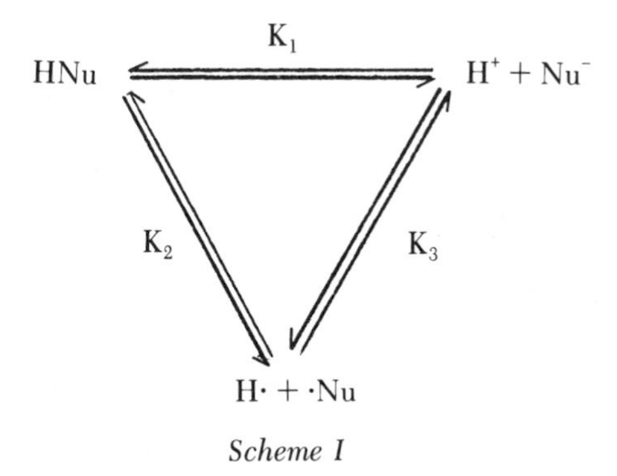

Scheme I

The energy associated with K_3 is the energy required to transfer an electron from the nucleophile to H^+. According to the data obtained this way, the ease with which an electron can be removed from the nucleophile is in the order $N_3^- > OH^- > CN^-$. As expected, this order is indeed the order of the N_+ scale; thereby, the suggested model is confirmed. Subsequent to this calculation, using the same thermodynamic cycle, Ritchie (*21*) [however, at the conference, Ritchie reported that this correlation holds only for aqueous solutions and not for reactions in $(CH_3)_2SO$] showed that activation energies for nucleophilic reactions of the pyronin cation correlate linearly with the solution ionization potentials of the nucleophiles. These two examples indicate that the transition state of the reactions of nucleophiles with LL substrates reflects in part features typical of electron-transfer processes.

α Effect. the α effect is defined as a positive deviation of an α nucleophile from a Brønsted-type plot (*22*). Several different origins exist for the α effect. The proposed model offers a consistent explanation of this effect based on transition-state stabilization (*12*). This effect manifests itself mainly with unsaturated substrates (*23*), which according to our terminology are LL substrates. In light of the previous discussion, the nucleophile in the transition state of these reactions acquires a radicaloid character. That radicals are relatively highly stabilized when located α to a lone pair of electrons is well-known (*24*). This statement is easily explained in terms of the molecular orbital (MO) diagram shown in Scheme II.

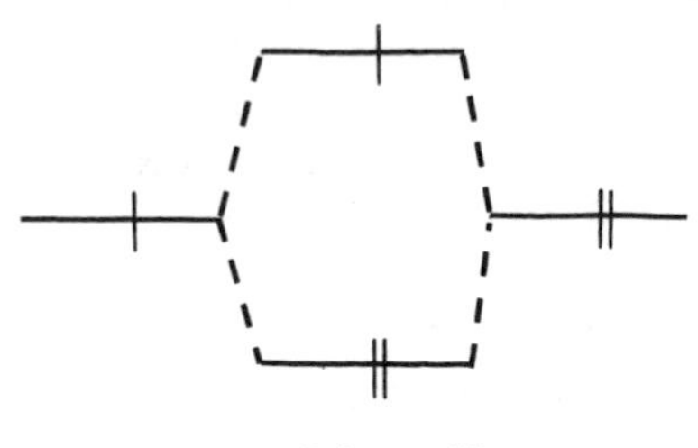

Scheme II

As can be seen from Scheme II, the fact that two electrons drop in energy whereas only one electron goes up leads to a net stabilization. This stabilization effect will be partly reflected in the transition state of the reaction of an α effect nucleophile with a LL substrate. Thus, an α effect is likely to be observed because this stabilization mechanism is unique to α nucleophiles and is unavailable to any "normal" nucleophile.

In principle, the magnitude of the α effect is expected to increase with the radicaloid character developed on the α nucleophile at the transition state. So that the validity of this hypothesis can be assessed, first, a probe for the degree of electron transfer at the transition state should be found. A possible probe in this case is the β_{nuc} value. As was pointed out by Bordwell and Clemens (*25*), it acquires values of 0.3–0.5 for S_N2 reactions and much higher values (1.1–1.5) for electron-transfer reactions. Thus, if the β_{nuc} value can indeed be correlated with the radical character in the nucleophile, a correlation between the α effect and β_{nuc} values should exist. Indeed, such a correlation was observed by Dixon and Bruice (*26*) in the reactions of hydrazines with a variety of substrates.

Further support for the hypothesis that β_{nuc} values can be used as a measure of the radical character of the nucleophile can be gained from Fukuzumi and Kochi's studies (*27*) on electrophilic aromatic substitution reactions:

$$\mathrm{ArH} + \mathrm{E}^+ \xrightarrow{k_e} \mathrm{ArE} + \mathrm{H}^+ \tag{5}$$

The ρ^+ value (equation 6) in these reactions was shown to reflect the mean separation of the reactants at the transition state (*27*):

$$\log k_e = \rho^+{}_e\sigma^+ + C \tag{6}$$

The proton affinity (pK_a) of arenes is also a linear function of σ:

$$\mathrm{ArH} + \mathrm{H}^+ \underset{K_a}{\rightleftharpoons} \mathrm{ArH_2}^+ \tag{7}$$

$$\Delta pK_a = \rho^+{}_a\sigma^+ \tag{8}$$

Thus, substituting σ^+ for ΔpK_a in equation 6 results in

$$\log k_e = (\rho^+{}_e/\rho^+{}_a)\Delta pK_a + C \tag{9}$$

The last equation is in fact a Brønsted-type equation in which β_{nuc} is replaced by $\rho^+{}_e/\rho^+{}_a$. Because $\rho^+{}_a$ is constant, β_{nuc}, like $\rho^+{}_e$, should also be interpreted as a measure of the mean distance between the reactants at the transition state.

It is clear from Figure 1 that as β_{nuc} increases, that is, as the separation of A and B at the transition state grows larger, the degree of electron transfer will also increase. This analysis predicts that for large enough β_{nuc} values (probably larger than 1), a complete electron transfer is expected. As noted, Bordwell found this situation indeed to be the case.

The proposed model does not imply that an α effect will be observed in the gas phase. Moving from right to left on curve b (Figure 1) gives an ionic bond that may later on decay to the ground state of the covalent compound. Because neither the exact mechanism nor the identity of the rate-limiting step is known for this process, no definite conclusion regarding the existence of the α effect in the gas phase can be derived on the basis of this model.

Positional Selectivity in FN. The reactions of nucleophiles with FN (for structures, *see* the introduction) present an interesting problem. In water, the nucleophile interacts with position 9 of the fluorene ring, whereas in dipolar aprotic solvents such as $(CH_3)_2SO$, the nucleophile attacks C-α;

this reaction leads ultimately to the formation of **II** (5), probably via a nucleophilic vinylic substitution mechanism (equation 10). In a solvent of intermediate polarity, for example, CH_3OH, both mechanisms are operative.

Nu^- + [fluorenylidene]=C(NO_2)(H) (α) $\xrightarrow[\text{or DMF}]{(CH_3)_2SO}$ [fluorenyl]–C(NO_2)(H)(Nu) $\xrightarrow{-NO_2^-}$ [fluorenylidene]=C(H)(Nu) (10)

II

In general, the reactivity of olefinic substrates is known to correlate with the electron-withdrawing power of the activating group (*28*). A good measure for the latter is the pK_a of the methano derivative of the activating group. When this criterion is employed for the two sites of FN, in water, the pK_a of CH_3NO_2 is 10.2 (*29*), whereas that of fluorene is 21–22 (*30*). Thus, in water, nucleophilic attack will take place at C-9 and the negative charge will be located on the nitromethide moiety. In $(CH_3)_2SO$, the pK_a of CH_3NO_2 rises to 17.2 and that of fluorene remains essentially unchanged (*30*). Prima facie, this result could have been taken as a satisfactory explanation for the observed shift in the site of the nucleophilic attack. However, a detailed structure–reactivity analysis shows this argument to be fallacious. This finding becomes apparent from a comparison of FDCN with FN. In spite of the fact that nitromethane is more acidic than malononitrile by approximately 1 pK_a unit (*29*), FN is less reactive than FDCN by 2 log units (*5*). On the other hand, a linear correlation does exist between the log k for the nucleophilic attacks on the three substrates and log k_{ion} (deprotonation reaction) for the corresponding methano derivatives. A similar correlation was reported by Bernasconi et al. (*31*) for the reactions of β-nitrostyrene, benzylidenemalononitrile, and Meldrum's acid derivative. Pearson and Dillon (*29*) have shown that a plot of log k_{ion} versus log K_a for many carbon acids in water is linear, with the marked exceptions of nitromethane and nitroethane, which deviate strongly from that line. Extrapolation using this plot shows that the kinetic acidity of nitromethane should be correlated with a pK_a of 17 rather than 10.2. Thus, the effective pK_a of nitromethane that should be used both in water and $(CH_3)_2SO$ is 17 (using this value in a plot of log k for nucleophilic reactions with FDN, FDCN, and FN in water versus pK_a values yields the expected linear correlation). Hence, the origin of the different positional selectivities in water and $(CH_3)_2SO$ cannot be explained on the grounds of variations of the pK_a of nitromethane.

A more suitable explanation can be suggested for this case, which makes use of the present model. Because at the transition state the substrate acquires a partial radical–anionic character, the radical anion of the substrate should be examined. The positional selectivity will most likely be determined by the location of the unpaired spin population in the model radical anion, toward which the radicaloid nucleophile will be attached in order to complete bond formation. A similar argument was invoked by Kochi (27) to explain the positional selectivities observed in electrophilic aromatic substitution.

To determine the spin population distribution, we (32) performed semiempirical (MNDO) calculations on the radical anion of FN with and without two water molecules hydrogen bonded to the nitro group. The optimized bond lengths are shown in Figures 2–4 (the geometry of the water molecules was not optimized and the NO · · · HOH separation was arbitrarily set to 2.0 Å). The MNDO coefficients of the SOMO on C-9 and C-α are given in Table I. Geometry optimization of FN (the neutral molecule) was performed with the aromatic portion of the molecule kept planar. The results show that the rest of the molecule is coplanar (within 1°) with the aromatic moiety. No

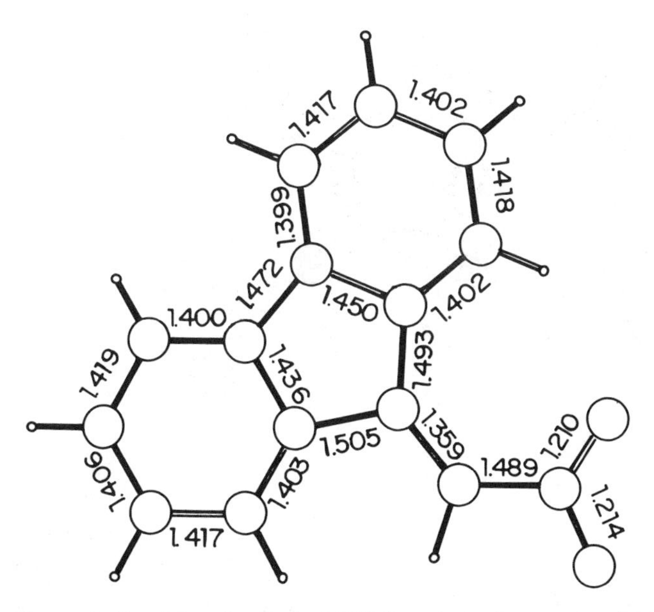

Figure 2. Bond lengths in the MNDO-optimized structure of FN.

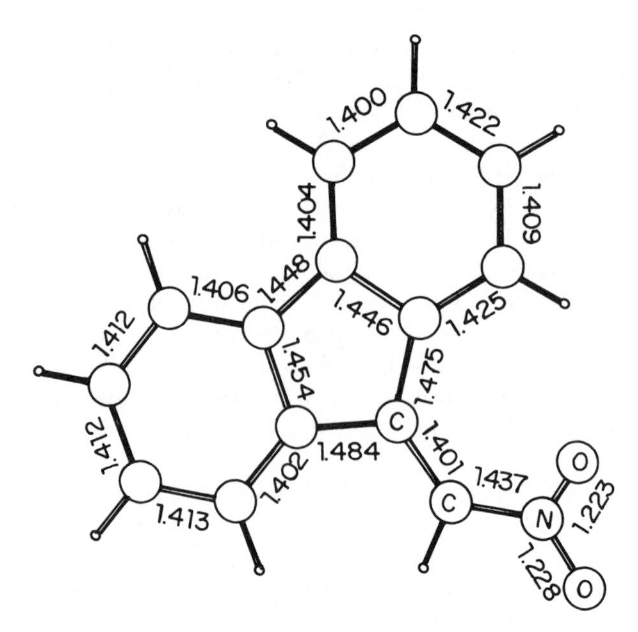

Figure 3. Bond lengths in the MNDO-optimized structure of FN⁻·.

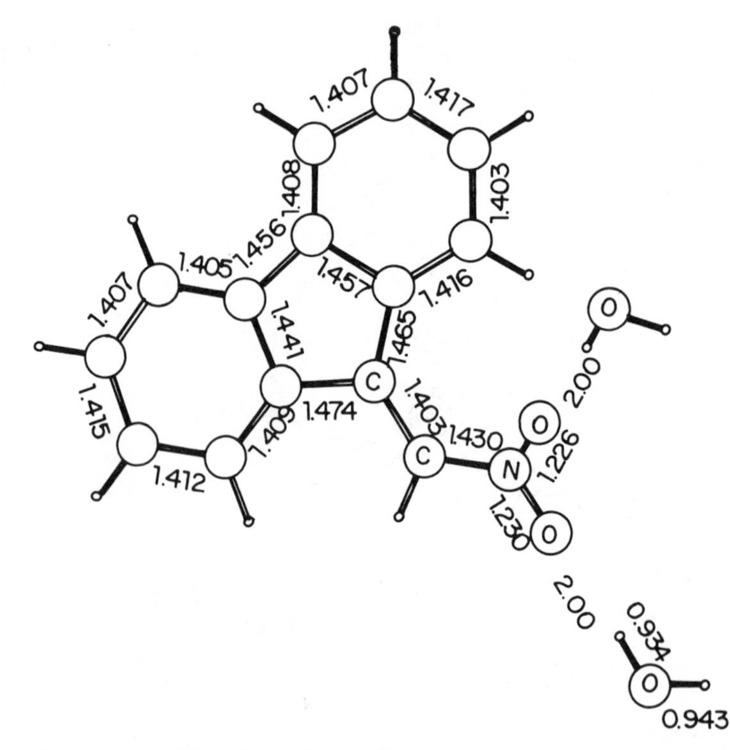

Figure 4. Bond lengths in the MNDO-optimized structure of FN⁻·$2H_2O$ (see the text).

Table I. Total Energies, MNDO, and RHF Open Shell STO–3G Coefficients of the SOMO on C-9 and C-α in FN·⁻

	FN·⁻			*FN·⁻2H₂O*		
Method	*Energy (eV)*	*C–9*	*C–α*	*Energy (eV)*	*C–9*	*C–α*
MNDO[a]	−2764.408	0.414	−0.443	−3466.816	−0.438	0.426
MNDO[b]	−2764.564	0.436	−0.404	−3467.133	−0.461	0.376
STO–3G[b]	−730.017[c]	0.458	−0.630	−879.958	−0.489	0.572

[a] Calculation performed on MNDO-optimized geometry of FN.
[b] Calculation performed on MNDO-optimized geometry of FN·⁻.
[c] Energy in au.

restrictions were imposed in the geometry optimization of the radical anion (FN·⁻). In its final structure, all carbon atoms are found essentially in a single plane. However, the dihedral angle around the C-9–C-α bond is 23°–26° with a slight pyramidalization about C-α. RHF open shell STO–3G (HONDO) (*33*, *34*) calculations were performed at the MNDO optimized geometries. The RHF coefficients at the SOMO are also given in Table I. Obviously, these calculations provide an indication of general trends rather than reliable absolute data. However, the calculations show that adding water molecules to the system indeed causes a shift of spin population from C-α to C-9. This shift in spin population is probably the governing factor in determining the positional selectivity in the reactions of FN with nucleophiles.

Acknowledgment

The assistance of D. Cohen in performing the MNDO calculation is gratefully acknowledged.

Literature Cited

1. Swain, G. C.; Scott, C. B. *J. Am. Chem. Soc.* **1953,** *75*, 141.
2. Ritchie, C. D. *Acc. Chem. Res.* **1972,** *5*, 348.
3. Ritchie, C. D.; Virtanen, P. O. I. *J. Am Chem. Soc.* **1972,** *94*, 4966.
4. Hoz, S.; Speizman, D. *Tetrahedron Lett.* **1978,** 1775.
5. Hoz, S.; Speizman, D. *J. Org. Chem.* **1983,** *48*, 2904.
6. Ritchie, C. D. *J. Am. Chem. Soc.* **1975,** *97*, 1170.
7. Ritchie, C. D.; Sawada, M. *J. Am. Chem. Soc.* **1977,** *99*, 3754.
8. Fujimoto, H.; Fukui, K. In *Chemical Reactivity and Reaction Paths;* Klopman, G., Ed.; Wiley: New York, 1974; Chapter 3.
9. Alavosus, T. J.; Sweigart, D. A. *J. Am. Chem. Soc.* **1985,** *107*, 985.
10. Kane-Maguire, L. A. P.; Honig, E. D.; Sweigart, D. A. *Chem. Rev.* **1984,** *84*, 525.
11. Hoz, S.; Aurbach, D. *J. Org. Chem.* **1984,** *49*, 4144.
12. Hoz, S. *J. Org. Chem.* **1982,** *47*, 3545.
13. Evans, M. G.; Polanyi, M. *Trans. Faraday Soc.* **1938,** *34*, 11.
14. Warshel, A.; Weiss, R. M. *J. Am. Chem. Soc.* **1982,** *102*, 6218.
15. Michl, J. *Top. Curr. Chem.* **1974,** *46*, 1.

16. Dauben, W. G.; Salem, L.; Turro, N. J. *Acc. Chem. Res.* **1975,** *8,* 41.
17. Pauling, L. *The Nature of the Chemical Bond;* Cornell University: Ithaca, NY, 1960.
18. Turner, D. W. *Adv. Phys. Org. Chem.* **1966,** *4,* 31.
19. Ritchie, C. D.; Kawasaki, A. *J. Org. Chem.* **1981,** *46,* 4704.
20. Harris, J. M. *Prog. Phys. Org. Chem.* **1974,** *11,* 89.
21. Ritchie, C. D. *J. Am. Chem. Soc.* **1983,** *105,* 7313.
22. Hoz, S.; Buncel, E. *Isr. J. Chem.* **1985,** *26,* 313.
23. Fina, N. J.; Edwards, J. O. *Int. J. Chem. Kinet.* **1973,** *5,* 1.
24. Nelson, S. F. In *Free Radicals;* Kochi, J. K., Ed.; Wiley: New York, 1973; Vol. 2, Chapter 21.
25. Bordwell, F. G.; Clemens, A. H. *J. Org. Chem.* **1981,** *46,* 1035.
26. Dixon, J. E.; Bruice, T. *J. Am. Chem. Soc.* **1972,** *94,* 2052.
27. Fukuzumi, S.; Kochi, J. K. *J. Am. Chem. Soc.* **1981,** *103,* 7240.
28. Patai, S.; Rappoport, Z. In *The Chemistry of Alkenes;* Patai, S.; Ed.; Wiley: New York, 1964; Chapter 8.
29. Pearson, R. G.; Dillon, R. L. *J. Am. Chem. Soc.* **1953,** *75,* 2439.
30. Matthews, W. S.; Bares, J. E.; Bartmess, J. E.; Bordwell, F. G.; Comforth, F. J.; Drucker, G. E.; Margolin, Z.; McCallum, R. I.; McCollum, G. J.; Vanier, N. R. *J. Am. Chem. Soc.* **1975,** *97,* 7006.
31. Bernasconi, C. F.; Fox, J. P.; Fornarini, S. *J. Am. Chem. Soc.* **1980,** *102,* 2810.
32. Dewar, M. J. S.; Thiel, W. *J. Am. Chem. Soc.* **1977,** *99,* 4899.
33. Dupuis, M.; King, H. F. *Int. J. Quantum Chem.* **1977,** *11,* 613.
34. Dupuis, M.; King, H. F. *J. Chem. Phys.* **1978,** *68,* 3998.

RECEIVED for review October 21, 1985. ACCEPTED February 10, 1986.

13

A Quasi-Thermodynamic Theory of Nucleophilic Reactivity

R. F. Hudson

Chemical Laboratory, University of Kent, Canterbury, Kent, England

Nucleophilic reactivity is related by a semiempirical equation to solvation energy and electron affinity of the nucleophile and the dissociation energy of the bond formed. A general equation, which shows that nucleophilicity is a function of bond formation in the transition state, reduces to the Swain and Brønsted equations under limiting conditions. The abnormally high reactivity of so-called α-nucleophiles is examined. Accordingly, the α effect is shown to be proportional to the extent of bond formation as given by the Brønsted coefficient β. A maximum α effect when β = 1, that is, when bond formation is complete, is supported by experimental data and by thermochemical calculations involving "normal" nucleophiles and α-nucleophiles. Calculations at the G–4.31 *and* G*–6.311 *levels shed further light on the origin of the α effect.*

The concept of nucleophilic reactivity is central to organic chemistry in that most reactions involve electron transfer in the transition state to some extent. The classical theory of Ingold et al. (*1*) invoked the idea by assuming that HO^- (very basic) was more nucleophilic than H_2O (weakly basic), without giving any clear-cut definition of nucleophilic reactivity. Indeed, it was known at the time, largely through the work of Conant and coworkers (*2*, *3*), that some ions, for example, I^- and Br^- were very nucleophilic but very weak bases. However, the idea persisted for some time, largely through the influence of the Brønsted theory (*4*) that nucleophilic reactivity (and also leaving-group ability) depended on proton basicity.

A breakthrough came when Swain and Scott (*5*) proposed a two-parameter equation that calibrated the reactivity of nucleophiles by reference to their rate of reaction toward methyl iodide, taken to be a "standard" electrophile. This equation was based, in effect, on the well-known order of reactivity in S_N2 reactions $I^- > Br^-$, $Cl^- > F^-$ and on the work of Foss (*6*),

0065-2393/87/0215-0195$06.00/0

which showed that reactivity toward divalent sulfur correlated closely with the redox potential of that nucleophile.

Although the Swain equation made a very important contribution, it led to some confusion because it was originally applied to all types of reactions including alkylation, acylation, and sulfonylation. The nucleophilic order is highly dependent on the nature of the substrate, and the equation was modified in the Swain–Edwards four-parameter equation (*7*).

The main drawback of these linear free-energy relationships is that they do not relate to reaction mechanisms. Curiously enough, like the original Brønsted equation, the main objective appeared to be the correlation of rate data rather than the interpretation of reaction mechanisms. This deficiency was partly remedied in the concept (*8*) of hard–soft acid–base (HSAB), which was in effect a qualitative extension of the Swain–Edwards equation but was more powerful in the sense that different types of reaction were related to the hard–soft classification, and the concept therefore has a wide application in organic synthesis and inorganic equilibria.

Our work evolved from early investigations into the mechanism of hydrolysis of acyl chlorides (*9–11*) and the reactivity of nucleophiles toward organophosphorus compounds (*12*). I was intrigued at that time by the fact that some nucleophiles rapidly dealkylated phosphate esters and hence were important in deprotection in nucleotide chemistry, and other nucleophiles were rapidly phosphorylated, and this finding is important in the search for antidotes for the nerve gases and also in predicting the reactivity of organophosphorus insecticides.

The orders of nucleophilic reactivity for alkylation and acylation were found to be quite different (*13, 14*) and in subsequent work (*15*) this finding was related to the extent of bond formation in the transition state as given empirically by the Brønsted coefficient, β. Previously, this difference was used to predict the position of bond fission in the alkaline hydrolysis of phosphinate, phosphonate, and phosphate esters (*12*). Jencks and Carriuolo (*16*) came to similar conclusions around the same time in outstanding work on the acylation of *p*-nitrophenyl acetate.

So that a simple interpretation of nucleophilic reactivity could be derived, (*17*), the approach developed by Polanyi and Evans (*18*) in their famous work on transition-state theory was used. The reactivity of a series of nucleophiles toward a given electrophile was represented by using recognizable energy terms and proceeding as follows.

First, the equilibrium for the addition of a nucleophile, N^-, to an electrophile, E^+, is

$$N^- + E^+ \overset{\Delta G^\circ}{\rightleftharpoons} N\text{–}E$$

The free-energy change, ΔG° is estimated in terms of experimental quantities by adopting the following thermodynamic cycle.

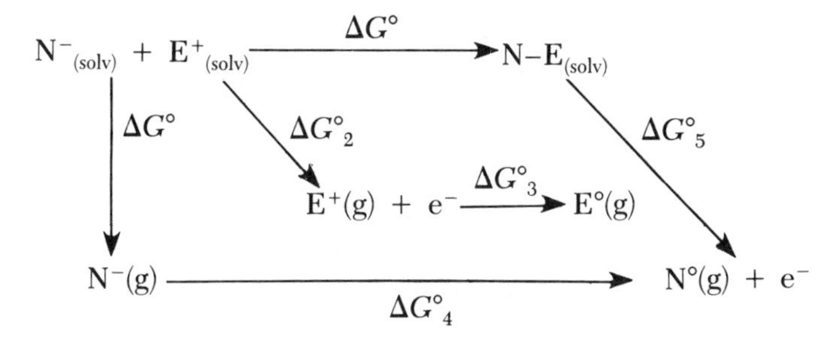

The following terms are involved: ΔG°_1 is the free energy of solution N^- $(-\Delta G_N{}^s)$, ΔG°_2 is the free energy of solution E^+ $(-\Delta G_E{}^s)$, ΔG°_3 is the electron affinity of E°, ΔG°_4 is the electron affinity of N° (E_N), and ΔG°_5 is the bond dissociation energy $(D_{N\text{-}E})$ and the associated entropy difference $(S_{N\text{-}E} - S_{N^\circ} - S_{E^\circ})$. Electron affinity is an enthalpy term, but the entropy differences $S_{N-} - S_e$ and $S_{E+} - S$ are negligible.

The required free energy change, ΔG°, is then given by $\Delta G^\circ = \Delta G_N{}^s + \Delta G_{E+}{}^s + \Delta G_{E_+ + e^- \rightarrow E^\circ} + E_N - T(S_{N^\circ} + S_{e-} - S_{N-}) - D_{N\text{-}E} - T[S_{N\text{-}E(solv)} - S_{N^\circ} - S_{E(g)}] = \Delta G_N{}^s + E_N - D_{N\text{-}E} - T[S_{N\text{-}E(solv)} - S_{N-(g)}]$.

Because ΔG° values for a series of nucleophiles, N^-, and a given electrophile (E) are being compared, terms involving E only are constant. Thus,

$$\Delta G^\circ = \Delta G_N{}^s + E_N - D_{N\text{-}E} - T[S_{N\text{-}E(solv)} - S_{N-}] + \text{constant}$$

In the derivation of all LFERs changes in $\delta_{N\text{-}E(solv)} - S_{N-(g)}$ can be assumed to be small as N^- is varied and hence

$$\Delta G^\circ \simeq \Delta G_N{}^s + E_N - D_{N\text{-}E} + \text{constant}$$

So that this approach can be adapted to the rate of analogous process, that is, the reaction between N^- and E–X, the bond is assumed to extend to the critical distance characteristic of the transition state. The nucleophile then interacts with the electrophilic center with transfer of charge *Ze*. This reaction leads to partial bond formation and partial desolvation involving free energy changes $\gamma D_{N\text{-}E}$ and $-\alpha \Delta G_N{}^s$, respectively.

The free energy of activation, ΔG^*, is then given by

$$\Delta G^* \simeq \alpha \Delta G_N{}^s + \beta E_N - \gamma D_{N\text{-}E} + \text{constant}$$

where βE_N is the energy change produced by the "partial" electron transfer *Ze*.

So that this equation can be applied to experimental data, relative magnitudes of α, β, and γ have to be found. As a first attempt to interrelate these coefficients, an electrostatic model was adopted for the desolvation and electron-transfer terms.

The free energy of transfer of an ion from gas phase to solution, that is, the free energy of solvation, is given by the Born equation (*19, 20*):

$$\Delta G_N{}^s = \frac{-Z^2e^2}{(2Dr)}$$

The equation can be applied with reasonable accuracy to changes in charge, Z, which are of concern in the present application, and to changes in ionic radius, *r*. The equation is difficult to apply when different solvents are employed because the value of the dielectric constant in the neighborhood of an ion is difficult to calculate.

The change in electrostatic energy on removal of charge *Ze* from N^- can be evaluated by using the energy required to remove this charge from a sphere (*21*), that is

$$\beta I_N = Z^2 E_N$$

Bond dissociation energies are given quite accurately by the Morse function (*22, 23*), but in view of its complexity it is assumed that

$$\gamma D_{N\text{-}E} \propto Z^n$$

With $n = 2$, equation 1 can be derived:

$$\Delta G^* \simeq (2Z - Z^2)\Delta G_N{}^s - Z^2\,(D_{N\text{-}E} - E_N) + \text{constant} \qquad (1)$$

This equation has been examined for several values of Z, but the equation does not appear to represent the experimental S_N2 reactivities particularly well (Table I).

In an alternative treatment (*17*), again desolvation is assumed to precede bond formation, and electron transfer and desolvation are assumed to be

Table I. Calculated Values of ΔE^* from Equations 1 and 2 and the Free Energy of Reaction, $\Delta G^\circ{}_C$, Compared to Swain's *N* Parameter

	$D_{CH_3\text{-}X}$	I_N	$\Delta G_N{}^s$	$\Delta G^\circ C$	$\Delta G^*{}_{calcd}{}^a$	$\Delta G^*{}_{calcd}{}^b$	n
F	112.0	83.5	96	67.5	(2.00)	(2.0)	2.00
Cl	84.6	88.2	64	67.6	3.35	3.6	3.05
Br	67.0	81.6	58	68.6	3.95	3.8	3.85
I	57.2	74.6	46	63.4	5.00	5.0	5.00
OH	92.6	35.0	105	47.4	4.50	3.9	4.25
SH	72.0	46.0	~73	47.0	5.60	5.5	5.10

NOTE: Energies are given as kcal m^{-1}.
[a] $-\Delta G^* = 0.07\,[(\Delta G_N{}^s + E_N) - 0.33 D_{N\text{-}C}]$ from equation 2 with $\alpha = 0.07$ and $\gamma = 0.023$.
[b] $-\Delta G^* = 0.17\,[\Delta G_n{}^s - 0.7\,(D_{N\text{-}C} - E_N)]$; cf. equation 1.

represented by a single parameter, that is, $\alpha \simeq \beta$. The equation then reduces to

$$\Delta G^* \simeq \alpha(\Delta G_N^s + E_N) - \gamma D_{N\text{-}E} + \text{constant} \quad (2)$$

where $1 > \alpha > \gamma$. This equation involves the ionization potential of the solvated nucleophile, $\Delta G_N^s + E_N$, and partial bond formation as opposing energy terms.

Several interesting limiting conditions occur for this equation. The first limiting condition occurs when the bond formation is small ($\gamma \longrightarrow 0$) and

$$\Delta G^* \simeq \alpha(\Delta G_N^+ + E_N) + \text{constant} \quad (3)$$

That is, the reactivity is related to the redox potential of N^-, and equation 3 reduces to the Swain equation (*5*):

$$\log (k/k^\circ) = sn$$

A second limiting condition occurs when the bond formation is large ($\gamma \longrightarrow 1$) and

$$\Delta G^* \simeq \Delta G_N^s + E_N^- - D_{N\text{-}E} + \text{constant} \quad (4)$$

That is, the reactivity is given by the affinity of the nucleophile toward the reaction center under consideration.

A third limiting condition occurs when similar nucleophiles are involved; for example, for a series of ions of the general structure AO^-, the bond energy term, $D_{N\text{-}E}$, may be treated as a constant (cf. equation 3).

Similarly, for a combination with a proton

$$\Delta G^H \cong (\Delta G_N^s + E_N) + \text{constant} \quad (5)$$

Combination of equations 3 and 5 gives

$$\Delta G^* \cong \alpha \Delta G^H + \text{constant} \quad (6)$$

That is, the extended Brønsted relation (*24*)

$$\log k = \beta_{nuc} pK_a + \text{constant}$$

Numerical reactivity constants can be derived from equation 2 by inserting empirical values of α and γ with the restriction that $\alpha > \gamma$ because γ/α is a measure of the extent of bond formation in the transition state. ($\gamma/\alpha = 1$ for complete bond formation.)

With values of $\alpha = 0.07$ and $\gamma = 0.023$, the calculated reactivity at a saturated carbon atom is compared with Swain's n values in Table I.

When the calculated values are scaled to 2.0 for F^-, a set of constants ΔG_{calcd} in reasonable agreement with the experimental values (n) is obtained. The quantitative use of this equation is, however, limited, but this usage does lead to qualitative conclusions of importance that are complementary to the well-known theory (8) of HSAB.

Thus, according to our general treatment, the relative reactivity of two nucleophiles is given by the relative magnitude of ($\Delta G_N^s + E_N$) and D_{N-E}. However, the relative reactivity is also given by the relative magnitude of α and γ; that is, the relative reactivity is a function of the extent of bond formation (and hence of the Brønsted β_{nuc}).

This result is shown diagrammatically in Figure 1. Accordingly, the relative reactivity of two nucleophiles, for example, I^- (soft) and OH^- (hard), increases as the selectivity increases. As the influence of the third term becomes significant with the increase in bond formation, a reversal occurs from the "soft" order $k_{I^-}/k_{OH^-} > 1$ to the "hard" order $K_{OH}/k_{RO^-} > 1$ ($\beta_{nuc} \simeq 0.15$). This effect is much more pronounced in reactions of hydrogen, for example, in the base-catalyzed elimination of Bu_tCl, k_{RS^-} $k_{RO^-} > 1$ ($\beta_{nuc} \simeq 0.15$), whereas in general, $k_{RO^-}/k_{RS^-} > 1$ for most β-eliminations.

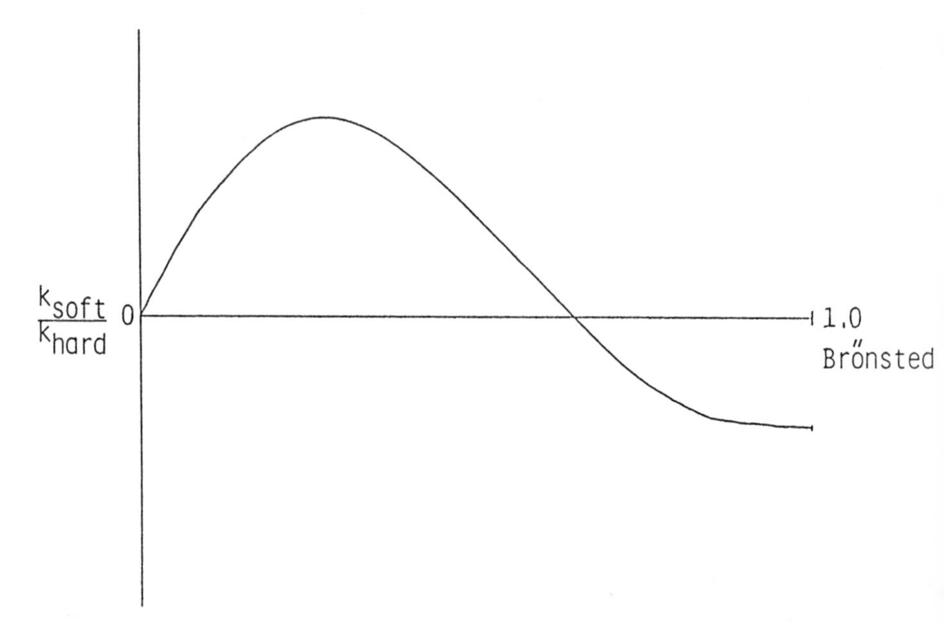

Figure 1. Change in selectivity with degree of bond formation in the transition state (β).

At a more theoretical level, nucleophilic reactivity can be treated by the polyelectron perturbation method (*25*), which in its simplest form gives the perturbation energy in terms of a first-order (Coulombic) term and a second-order term involving orbital energies α_j and α_k (Figure 2). Subsequently, this equation was adapted by Hudson and Filippini (*27*) and independently by Klopman et al. (*28*) to the problem of enhanced nucleophilic reactivity. This problem is particularly thorny, and it will now be discussed in some detail.

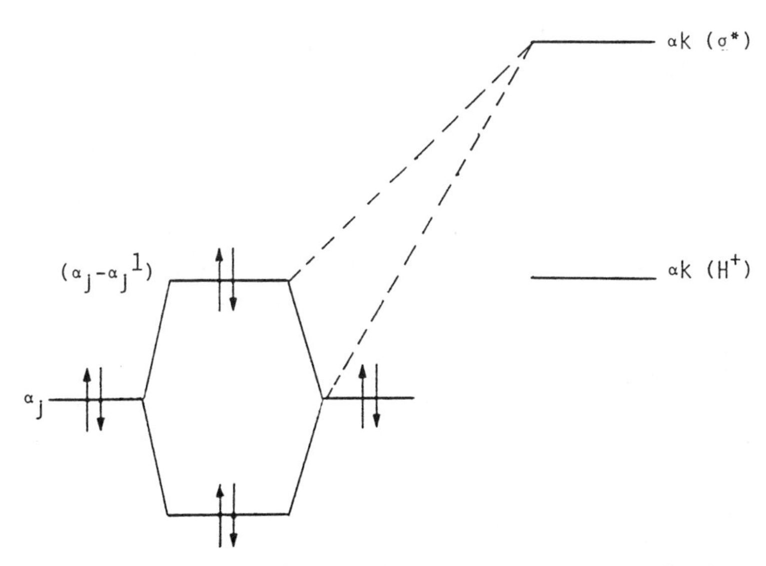

Figure 2. Perturbation treatment of the α effect. (Reproduced with permission from reference 48. Copyright 1973 Verlag Chemie).

According to my treatment, the α effect is attributed to electron-pair repulsions that cause orbital splitting (Figure 2), with a consequent increase in perturbation (i.e., stabilization) due to a decrease in the $\alpha_j - \alpha_k$ term. However, why this effect does not also affect the corresponding pK_a values to which the α effect is related is not clear. In fact, adjacent lone pairs produce large decreases in the rate of proton removal from pyridine and other nitrogen heterocycles (*29, 30*). Considering the reverse process, this effect should lead to an increase in pK_a. This explanation of the α effect has been criticized extensively in recent years (*31–33*).

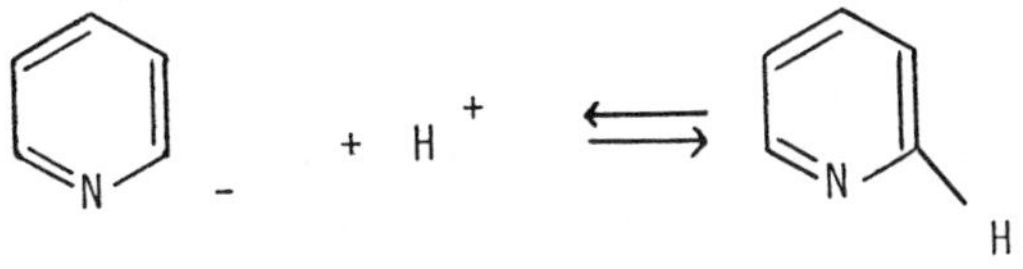

According to my treatment, the magnitude of the α effect increases with the extent of bond formation in the transition state as represented by the Brønsted, β_{nuc}, parameter. This finding is in accord with experimental observations, and Aubort and Hudson (*34*) showed some time ago (Table II) that the α effect for *p*-nitrophenyl acetate ($\beta_{nuc} \simeq 0.8$) is much greater than that for an alkyl bromide ($\beta_{nuc} \simeq 0.3$). Dixon and Bruice (*35*) showed a similar effect for the reaction of hydrazines with a wide range of electrophiles.

Table II. Comparison of the α Effect for Benzyl Bromide and *p*-Nitrobenzyl Bromide with the α Effect for *p*-Nitrophenyl Acetate (PNPA)

		ΔΔ *log* k		
Nucleophile, R	*Reference*	*RBr*	*PNPA*	*Ratio*
$CH_3COCHNO^-$	24	0.60	2.00	3.3
$RCONO(CH_3)O^-$	24	1.10	2.10	2.1
HO_2^-	25	1.70	3.50	1.9

The logical extension of this relationship is that the α effect is a maximum when β = 1.0, that is, when bond formation is complete. This relationship means that the α effect should be observed in equilibria as well as in kinetics.

This proposal was suggested by Hine and Weimar (*36*), who made thermochemical calculations of the affinities of some anions including HO_2^- and HO^- toward carbon and hydrogen for which an α effect of the order of 10^6 for the equilibrium was found. Moreover, Jencks and co-workers (*37, 38*) measured the equilibrium constant for the acylation of *N*-methylhydroxamic acids and obtained an experimental value of approximately 10^2 for the equilibrium α effect. This value is considerably less than the experimental value obtained from the rates of acylation. The apparent anomaly may be due to the two-stage process that is usually assumed for acylation:

$$AO^- + RCOX \underset{}{\overset{K_1}{\rightleftharpoons}} R\text{–}\underset{X}{\overset{O^-}{C}}\text{–}OA \overset{K_2}{\rightleftharpoons} R\text{–}CO\cdot OA + X^-$$

The kinetics refer to the formation and subsequent decomposition of a (charged) tetrahedral intermediate, whereas the affinities are given by the overall equilibrium constants K_1 and K_2. I suggest (vide infra) that the affinity of a nucleophile is highly dependent on the Coulombic interaction energy, and this dependence could lead to a larger α effect for the rate of acylation than for the equilibrium.

In recent work (*39*) on isothiocyanates, oximes were found to add readily to acyl isothiocyanates to give intermediates, for example

$$\text{(fluorenone oxime)}=N\text{–}OH + C_6H_5C(O)N{=}C{=}S \overset{k}{\rightleftharpoons} \text{(fluorenylidene)}=N\text{–}O\text{–}C(S)NHC(O)C_6H_5$$

where $k \gg 1$, whereas phenols of comparable pK_a show no tendency toward addition:

$$X\text{–}C_6H_4\text{–}OH + C_6H_5C(O)N{=}C{=}S \overset{k}{\rightleftharpoons} C_6H_5\text{–}OC(S)NHC(O)C_6H_5$$

where $k \ll 1$. To explain this remarkable difference in the behavior of oximes and phenols, a *hypothetical* series of processes was assumed and then the first law of thermodynamics was applied.

Thus, for the addition of an oxime to the isothiocyanate, the following processes can be postulated:

$$R_2C{=}N\text{–}OH \overset{K_H}{\rightleftharpoons} R_2C{=}N\text{–}O^- + H^+$$

$$PhCO\cdot NCS + H^+ \overset{K_1}{\rightleftharpoons} Ph\cdot CO\cdot NH\cdot C^+ {=}S$$

$$Ph\text{–}CO\cdot NH\cdot C^+ {=}S + R_2C{=}N\text{–}O^- \overset{K_2}{\rightleftharpoons} PhCO\cdot NH\cdot CS\cdot ON{=}CR_2$$

Although the purpose of this postulate is to explain the reaction affinity, the postulate is a reasonable reaction mechanism. In the presence of triethylamine, both oximes and phenols are benzoylated by reaction at the carbonyl group:

$$X\text{–}C_6H_4\text{–}O^- + C_6H_5C(O)N{=}C{=}S \rightarrow X\text{–}C_6H_4\text{–}OC(O)C_6H_5 + NCS^-$$

This example shows the regiospecificity of nucleophilic attack.

The observed equilibrium constant, K, is given by

$$K = K_H K_1 K_2$$

For two nucleophiles with similar pK_a values, for example, an oxime and phenol, K_H is constant, K_1 is common, and hence K is determined by K_2. In other words, the affinity of the oxime toward a carbonium center is greater than the affinity of a "normal" nucleophile of comparable pK_a.

To generalize, an α effect will be observed *in an equilibrium* when the relative proton affinity, ΔG°_H, is greater than the relative affinity toward the electrophile, ΔG_E. This statement is supported by the thermochemical calculations and ab initio calculations at the G-4.31 level (Table III). From the estimated values of the affinities toward carbon and hydrogen from thermochemical data, $\Delta G°_H \simeq \Delta G°_C$ for CH_3O^- and CHO_2^-, two normal nucleophiles of very different basicity. However, $\Delta G°_H - \Delta G°_C = 4.8$ kcal/mol when CH_3O^- and HO_2^- are compared; this result corresponds to an α effect of $10^{3.5}$ at 298 °C. This value may be compared with the experimental value of the α effect of $10^{2.4}$ for the α effect in the reaction of p-nitrophenyl acetate with HO_2^-. The value of β is found to be 0.65 from the log k–pK_a slope at pK_a–12, that is, for H_2O_2. Thus, for complete bond formation, a value of $10^{3.7}$ would be found for the α effect, if this effect is assumed to be proportional to β.

Table III. Highest Occupied Molecular Orbital (HOMO) Energies, Mulliken Charges, q_0, and Calculated and Experimental Proton and Methyl Cation Affinities of Some Oxyanions

	ΣHOMO		*Proton Affinity* $\Delta G°_H$		*Methyl Cation Affinity* $\Delta G°_C$		*Electronic Charge on Oxygen* q_0	
Oxyanion	*G*-4.31	*G**-6.311	*G*-4.31	Exptl	*G*-4.31	Exptl	*G*-4.31	*G**-6.311
HO^-	−0.0340	−0.0483	426.0	390.8	292.8	276.3	−1.144	−1.193
CH_3O^-	−0.0635	−0.0773	409.8	379.2	279.1	267.6	−0.949	−0.859
HO_2^-	−0.0868	−0.0942	387.4	379.4	259.6	272.6	−0.607	−0.614
FO^-	−0.1371	−0.1514	360.5		235.4		−0.471	−0.457
HCO_2^-	−0.1510		360.0	345.2	229.8	233.1	−0.788	

NOTE: HOMO energies are in hartrees. Mulliken charges are in units of electrons on the nucleophilic atom. At 0 K, the calculated affinities were uncorrected for zero-point vibration. The experimental affinities were at ΔH_{298} for the process $N^- + A^+ \rightarrow NX$ (A = H and C). The affinities were in units of kilocalories per mol.

When the origin of the enhanced affinity of the α-nucleophile toward the electrophilic center or, alternatively, the enhanced affinity of the normal nucleophile toward the proton is studied, the equilibria can be treated by a classical thermodynamic argument as follows (*40*). Consider the equilibria

$$N_1^- + E^+ \overset{K_1}{\rightleftharpoons} N_1\text{–}E \qquad N_1^- + H^+ \overset{K_1^H}{\rightleftharpoons} N_1\text{–}H$$

$$N_2^- + E^+ \overset{K_2}{\rightleftharpoons} N_2\text{–}E \qquad N_2^- + H^+ \overset{K_2^H}{\rightleftharpoons} N_2\text{–}H$$

Each process is imagined to involve (1) desolvation, (2) electron transfer, and (3) bond formation. Thus, according to

$$\Delta G° = -RT \ln K = \Delta H_N + E_N - D_{NE} + \Sigma E^+ + T\Sigma\Delta S \qquad (9)$$

where ΔH_N, E_N, and $D_{N\text{-}E}$ have already been defined (*see* equation 1); ΣE^+ represents energy terms for E that are constant when two nucleophiles are compared, and $\Sigma\Delta S$ is the net entropy change in the reaction, which can be neglected in the following treatment.

When this equation is applied to previously mentioned equilibria

$$RT \ln \frac{(K_2}{K_1)} - RT \ln \frac{K_2{}^{H}}{K_1{}^{H}} = (D_{H-N_1} - D_{H-N_2}) - (D_{E-N_1} - D_{E-N_2})$$
$$= (\Delta D_{H-N} - \Delta D_{E-N})$$

where N_1 represents a normal nucleophile and N_2 an α-nucleophile.

Thus, an α effect will be observed in an equilibrium when the difference in bond dissociation energies of a normal and an α-nucleophile joined to a proton is greater than the corresponding difference when the two nucleophiles are joined to the electrophilic center under consideration.

In a previous note presenting a perturbation treatment, the α effect was attributed to a combination of electron-pair repulsion and Coulombic interaction with the proton *and* the electrophilic center. This explanation can be examined further by reference to the molecular orbital data in Table III. A decrease in ΣHOMO (i.e., an increase in ionization potential) of a normal nucleophile is accompanied by a decrease in q_0, the charge on the nucleophilic atom. However, substitution at the nucleophilic atom to form an α-nucleophile, for example, HO_2^-, produces a large decrease in q_0 for a smaller decrease in ΣHOMO. For example, ΣHOMO for HCO_2^- is considerably *less* than ΣHOMO for HO_2^-, but the negative charge is *greater* for the HCO_2^- ion.

This relationship is shown diagrammatically in Figure 3, where the energy of the HOMO is plotted against the charge density, q_0, on the nucleophilic oxygen atom. If an extrapolation is made through the points for HCO_2^- and CH_3O^-, that is, the two normal nucleophiles, the HOMO energy is approximately 5.0 eV greater than that predicted for HO_2^- and approximately 5.7 eV greater than that predicted for FO^-. This finding means that the electron affinity of HO_2 is much less than expected on the basis of the nuclear charge.

These calculations suggest that the increased value of ΔD_{HN} relative to ΔD_{CN} may be explained in terms of bond polarities. The abnormally large decrease in negative charge, q_0, for HO_2^- reduces the Coulombic contributions to the energies of the H–O and C–O bonds. In view of the fact that H–O bonds are more polar than C–O bonds, a consequence partly of the reduced internuclear distance, this reduction in Coulombic energy is greater for interactions with the proton.

Therefore, OH^- (or CH_3O^-) has an abnormally high affinity for the proton, and as a consequence, HO_2^-, and other α-nucleophiles, has a greater relative affinity for carbon (and for other electrophilic centers). This finding appears to form the basis of the apparently high reactivity of α-nucleophiles, which is intrinsic to the reagent and should be observed in gas-phase equilibria (*see* Table III) as well as in solution kinetics.

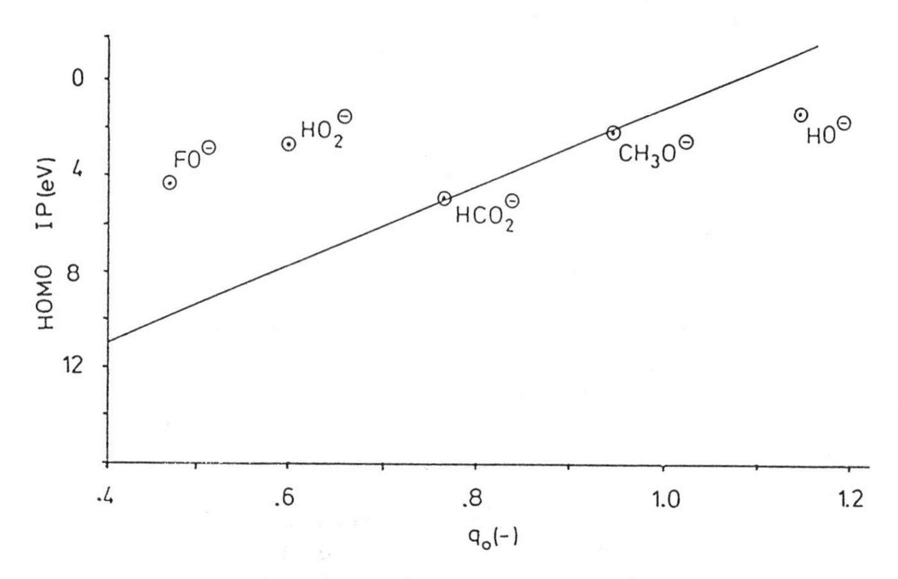

Figure 3. Relationship between the calculated HOMO energy and total charge on the terminal oxygen atom, q_0, *of several oxygen nucleophiles.*

Recent work (*32, 43*) has cast doubt on this conclusion. De Puy et al. have shown, by product analysis of reaction of HO_2^- with esters in the gas phase, the absence of a kinetic α effect.

These reactions are generally collision-controlled, and hence rate differences are due to probability factors. Under these conditions, reaction selectivity due to activation energy differences disappears, although as discussed previously the α effect will manifest itself in differences in the corresponding equilibria.

As a consequence, the α effect should be solvent independent or nearly so. However, recent work (*44, 45*) has shown appreciable increases in the α effect of oximes as the concentration of dimethyl sulfoxide increases. However, this increased α effect reaches a maximum at approximately 50 mol %, and the values in water and $(CH_3)_2SO$ appear to be very similar. In other words, on this limited evidence, the α effect is the same in pure solvents but may change appreciably in solvent mixtures. Liquid mixtures are notoriously complex, and hence we suggest that the increased α effects are due to changes in liquid structure, which are analogous to phase changes on a micro scale (cf. micelles).

In conclusion, the α effect is attributed to the *cooperative* effect of two terms: (a) orbital splitting, which determines the HOMO and the ionization potential of the nucleophile and (b) a Coulombic term, which produces bond polarity. The influence of these terms on the α effect has been analyzed recently by Laloi-Diard and Minot (*46*) by perturbation theory, which is an extension of our earlier treatment (*26*). The balance between these two terms, which is different for the combination of a nucleophile with a reaction center (electrophile) and a proton, produces the α effect.

A striking example of this theory is the large α effect (ca. 10^3) of pyridine *N*-oxides in contrast to the "normal" behavior of trimethylamine *N*-oxide and phenoxide (*47*). According to my explanation, P_π–P_π repulsion in the phenoxide and the inductive (Coulombic) effect in the aliphatic amine oxide alone do not produce an α effect. However, the combined action of the inductive effect and P_π–P_π repulsion in the aromatic *N*-oxide produces the observed α effect.

Acknowledgments

I thank S. Wolfe for his help and encouragement in the course of this work, which was developed during a stay at Queens University, Kingston, D. J. Mitchell for permission to use his computational data, and D. N. Hague and D. P. Hansell for helpful discussions.

Literature Cited

1. Gleave, J. L.; Hughes, E. D.; Ingold, C. K. *J. Chem. Soc.* **1935,** 236.
2. Conant, J. B.; Kirner, W. R. *J. Am. Chem. Soc.* **1924,** *44,* 232.
3. Conant, J. B.; Hussey, R. E. *J. Am. Chem. Soc.* **1925,** *47,* 476.
4. Brønsted, J. N. *Z. Phys. Chem.* **1922,** *102,* 169.
5. Swain, G. C.; Scott, C. B. *J. Am. Chem. Soc.* **1953,** *75,* 141.
6. Foss, O. *Acta. Chem. Scand.* **1947,** *1,* 307.
7. Edwards, J. O. *J. Am. Chem. Soc.* **1954,** *76,* 1540.
8. Pearson, R. G. *J. Am. Chem. Soc.* **1963,** *85,* 3533.
9. Archer, B. L.; Hudson, R. F. *J. Chem. Soc.* **1950,** 3259.
10. Brown, D. A.; Hudson, R. F. *J. Chem. Soc.* **1953,** 883, 3352.
11. Hudson, R. F. *Chimia* **1961,** *15,* 394.
12. Harper, D. C.; Hudson, R. F. *J. Chem. Soc.* **1958,** 1356.
13. Hudson, R. F.; Green, M. *Proc. Chem. Soc.* **1959,** 149.
14. Hudson, R. F.; Green, M. *J. Chem. Soc.* **1962,** 1055.
15. Hudson, R. F.; Loveday, G. W. *J. Chem. Soc.* **1962,** 1068.
16. Jencks, W. P.; Carriuolo, J. *J. Am. Chem. Soc.* **1960,** *82,* 1778.
17. Hudson, R. F. *Chimia* **1962,** *16,* 173.
18. Polanyi, M.; Evans, M. G. *Trans. Faraday Soc.* **1938,** *34,* 11.
19. Born, M.; *Z. Phys.;* **1920,** *1,* 45.
20. Feakins, D.; Watson, P. *J. Chem. Soc.* **1963,** 4734.
21. Ferreira, R. *Trans. Faraday Soc.,* **1963,** *59,* 1075.
22. Morse, P. M. *Phys. Rev.* **1929,** *34,* 57.
23. Glasstone, S.; Laidler, K. J.; Eyring, H. *The Theory of Rate Processes;* New York: McGraw-Hill; 1941, p 92.
24. Hudson, R. F. In *Chemical Reactivity and Reaction Paths;* Klopman, G., Ed. New York: Wiley; 1974, pp 167–252.
25. Hudson, R. F.; Klopman, G. *Theoret. Chim. Acta.* **1967,** *8,* 165.
26. Hudson, R. F.; Aubort, J. D. *Chem. Comm.* **1970,** 937.

27. Hudson, R. F.; Filippini, F. *Chem. Comm.* **1972,** 522.
28. Klopman, G.; Tsuda, K.; Louis, J. B.; Davis, R. E. *Tetrahedron* **1970,** *26,* 4549.
29. Zoltewicz, J. A.; Grahe, G.; Smith, C. L. *J. Am. Chem. Soc.* **1969,** *91,* 5501.
30. Kauffmann, T.; Wirthwein, R. *Angew. Chem. Intl. Ed.* **1971,** *10,* 20.
31. Liebman, J. F.; Pollack, R. M. *J. Org. Chem.* **1973,** *38,* 3444.
32. De Puy, C. H.; Della, E. W.; Filley, J.; Grabowski, J. J.; Bierbaum, V. M. *J. Am. Chem. Soc.* **1983,** *105,* 2482.
33. Wolfe, S.; Mitchell, D. J.; Schlegel, H. B. *J. Am. Chem. Soc.* **1981,** *103,* 7694.
34. Aubort, J. D.; Hudson, R. F. *Chem. Comm.* **1970,** 937.
35. Dixon, J. E.; Bruice, T. C. *J. Am. Chem. Soc.* **1972,** *94,* 2052.
36. Hine, J.; Weimar, R. D. *J. Am. Chem. Soc.* **1965,** *87,* 3387.
37. Gerstein, J.; Jencks, W. P. *J. Am. Chem. Soc.* **1964,** *86,* 4655.
38. Sander, E. G.; Jencks, W. P. *J. Am. Chem. Soc.* **1968,** *90,* 6154.
39. Hudson, R. F.; Hansell, D. P. *Chem. Comm.* **1985,** 1405.
40. Hudson, R. F.; Hansell, D. P.; Wolfe, S.; Mitchell, D. J.; *Chem. Comm.* **1985,** 1406.
41. Wolfe, S.; Mitchell, D. J.; Schlegel, H. B.; Minot, C.; Eisenstein, O. *Tetrahedron Lett.* **1982,** *23,* 615.
42. Laloi-Diard, M.; Verchere, J.-F.; Gosselin, P.; and Terrier, F. *Tetrahedron Lett.* **1984,** 1267.
43. Buncel, E.; Um Ik-Hwan *J. Chem. Soc. Chem. Commun.* **1986,** 595.
44. Laloi-Diard, M.; Minot, C. *Nouveau J. de Chimie* **1985,** *9,* 569.
45. Jencks, W. P.; Carrinolo, J. *J. Am. Chem. Soc.* **1960,** *82,* 1778.
46. Hudson, R. F. *Angew. Chem.* **1973,** *85,* 63.

RECEIVED for review October 21, 1985. ACCEPTED August 1, 1986.

14

Nucleophilicity and Distance

Fredric M. Menger

Department of Chemistry, Emory University, Atlanta, GA 30322

This paper focuses on the relationship between reactivity of a nucleophile with an electrophile and the distance separating the two species. The relationship, examined by means of rigid molecular frameworks bearing two functionalities at well-defined distances and angles, is shown to be extremely sensitive. Theoretical considerations also support the contention that distance is a key parameter in nucleophilic reactivity. Reactions in solution occur with enzyme-like rates when critical distances are achieved.

OPEN A TEXTBOOK ON ORGANIC CHEMISTRY, introductory or otherwise. You will find information on how nucleophilicity depends on basicity, polarizability, solvent, temperature, substituents, structure of electrophile, and so on. But you will *not* find much information on how nucleophilicity depends on geometric disposition. Too little is known about the subject. This lack of knowledge constitutes a serious gap in our understanding of chemical dynamics because, as will be shown, distance is critically important to nucleophilic reactivity. Distance information is required to fully characterize reaction pathways and to interpret structural data on enzymes. How, after all, can a particular arrangement of catalytic groups surrounding a substrate at an active site be evaluated without first understanding the connection between reactivity and alignment?

Our interest in nucleophilicity derives in large measure from the amazing velocities at which enzymatic nucleophiles attack bound substrates. Chymotrypsin, for example, has a serine hydroxyl that performs a nucleophilic attack on an amide carbonyl about 10^8 times faster than the laboratory rate at equivalent pH and temperature (*1*). A general-base catalysis by an imidazole ring accounts of 10–10^2 of this factor; at least 10^6 remains as a mystery. Why is the chymotrypsin hydroxyl such a powerful nucleophile? Few chemists would answer this question by invoking a mechanism unique to biology. No reason exists to suspect that enzymes operate by anything other than the principles familiar to every organic chemist. Yet the

0065–2393/87/0215–0209$06.00/0

interplay of effects that ultimately gives rise to enzymatic catalysis remains a matter of conjecture.

Enzyme-like nucleophilicities can be secured in the laboratory via (a) intramolecular systems and (b) judicious choice of solvent. (As will be argued later, these effects are not unrelated.) Two specific examples serve to illustrate the point.

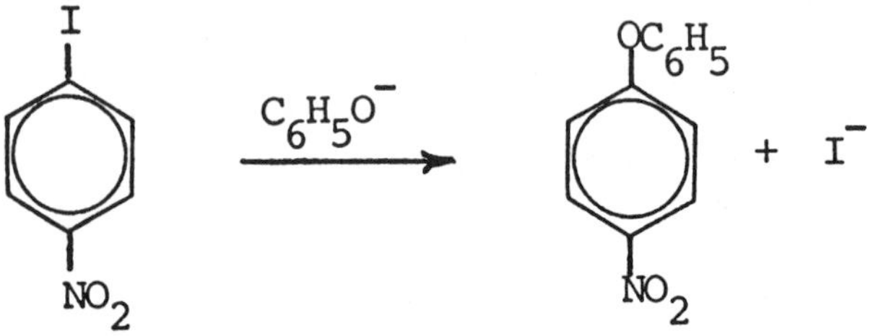

In the first reaction, intramolecular attack by the hydroxyl on the carboxyl carbonyl proceeds 5×10^8 times faster than the corresponding intermolecular process (2). In the second reaction, the rate increases 10^7-fold upon switching the solvent from methanol to dimethylformamide (3). Obviously, the huge rate increases in these organic systems do not necessarily prove that similar effects are at work in enzymes. But to be suspicious is quite natural, and many people, too numerous to mention, have pointed out the possible relationship between enzyme catalysis and intramolecularity or solvation effects.

The possibility that enzymatic reactivity is indeed related to intramolecularity leads next to an important question: *Why* are intramolecular reactions often very fast? Unfortunately, no good answer exists to this question. The difficulty is brought forth with a vengence in Kirby's scholarly compilation of effective molarity (EM) values (*4*) (where EM $= k_{intra}/k_{inter}$). EM values, which vary from very small (<0.3 M) to very large ($>10^{10}$ M), depend on ring size, substituents, solvent, and reaction type. No known theory can explain—let alone predict—these wild fluctuations. A person is reminded of the first law of sociology: "Some do, some don't". Kirby's compilation probably represents the largest and most variant body of unexplained data in physical organic chemistry.

Two recent articles from our laboratories delve into the question of intramolecularity (5, 6). The main postulate of reference 6 is that the rate of reaction between functionalities *A* and *B* is proportional to the time that *A* and *B* reside within a critical distance. Thus, fast intramolecular (or enzymatic) reactions are thought to occur when a carbon framework (or a protein structure) enforces residency at critical distances upon two reactive groups. Reference 6 contains (a) evidence supporting the postulate, (b) magnitudes of critical distances, and (c) examples of intramolecular reactions attaining enzyme-like rates when critical distances are imposed.

In the remainder of this chapter, important points that, for one reason or the other, could not be fully covered previously are discussed (in question–answer format).

Is It Reasonable to Postulate a Critical Distance?

A critical distance is both a reasonable and a time-honored concept. Consider the cases below in which critical distances have been invoked.

(a) The classical theory of Smoluchowski (7) assumes that a reaction between two hard-sphere molecules occurs only when the separation distance r reaches R, the "distance of closest approach" (equation 1). Note the presence in equation 1 of the "δ function", which equals either zero or unity: $\delta(r - R) = 0$ if $(r - R) > 0$; $\delta(r - R) = 1$ if $(r - R) = 0$.

$$k(r) = \frac{k_0\, \delta(r - R)}{4\pi r^2} \tag{1}$$

(b) Intermolecular triplet–triplet energy transfer is described by the Perrin equation (equation 2) (8). In this model, a donor is quenched if an acceptor at concentration C_A lies within a critical radius R_c. A typical donor–acceptor pair has an R_c of 13 Å; this value indicates that quenching abruptly ceases when the separation distance exceeds 13 Å.

$$R_c = \left(\frac{3000 \ln (I_o/I_a)}{4\pi N C_A}\right)^{1/3} \tag{2}$$

(c) In a statistical treatment of intramolecularity, Sisido (9) assumed that a reaction takes place only when the separation between functional groups at the ends of a chain becomes shorter than a distance r_0. Good fits between experimental and calculated cyclization constants were obtained with an $r_0 = 2.3$–2.7 Å.

(d) Electron transfer in any given system occurs, undoubtedly, over a range of encounter distances R. Because each separation distance has its own transfer probability, the rate constant must be obtained by integrating over the equilibrium distribution of distances [equation 3, where R_{ij} is the separation distance of the redox centers and $g(R_{ij})$ is the pair distribution function].

$$k_{ij} = c\int_0^\infty 4\pi R_{ij}^2 g(R_{ij}) k_{et}(R_{ij})\, dR_{ij} \tag{3}$$

Usually, however, the situation can be simplified by assuming an "effective" encounter distance; the observed rate then becomes a function of the rate at this particular separation distance and the work necessary to bring the reactants together (*10*). Although no rigorous meaning can be given to the effective encounter distance, this distance is often considered to be the "contact" separation of the reactants.

The terms "critical distance", "bonding distance", "encounter distance", "contact distance", and "van der Waals distance" can be used interchangeably. They all refer to the distance at which a nucleophile and electrophile initiate bond formation. The longer two species remain at this distance, the faster the rate.

How "Critical" Is the Critical Distance?

Four cases were just described that treat rate versus distance as if it were a step function. Surely this result is not accepted literally. But because physical–chemical data can often be fit to equations that incorporate a step function, the dependence of rate on distance must be very sharp if not exactly a step function. The situation is not unlike that of the "critical micelle concentration" or cmc. A surfactant in water forms large aggregates (micelles) when its concentration reaches the cmc. Micelles no doubt form over a range of concentrations, but the range is sufficiently small that for all practical purposes a precipitous, all-or-none behavior can be assumed. This situation is true for the reactivity–distance relationship. In reference 6, Menger cites several reactions whose rates manifest a severe dependence on distance (i.e., reactions in which a few tenths of an angstrom are worth many orders of magnitude in rate). Extremely fast rates are observed when two functional groups are held within a critical distance. Another particularly striking example is given in the next paragraph.

Intramolecular hydride transfer in equation 4 proceeds with an enzyme-like EM of 6.5×10^6 M. In other words, the intramolecular reaction is 6.5×10^6 times faster than the intermolecular counterpart at 1 M concentration (*11*). Davis et al. (*11*) argued that relief of strain *cannot* explain the fast rate because (a) the equilibrium constant in equation 4 is close to unity and (b) force-field calculations show that hydroxy ketone is only 1.7 kcal/mol more strained than the corresponding diketone, which lacks nonbonded H/C=O interactions. The extremely fast nucleophilic attack on the carbonyl is, however, expected from our "spatiotemporal" hypothesis. Because the mobile hydrogen is held rigidly only 2.35 Å away from the carbonyl carbon, well under the suspected critical distance of 2.8 Å (*6*), the conditions for an enzyme-like acceleration are met.

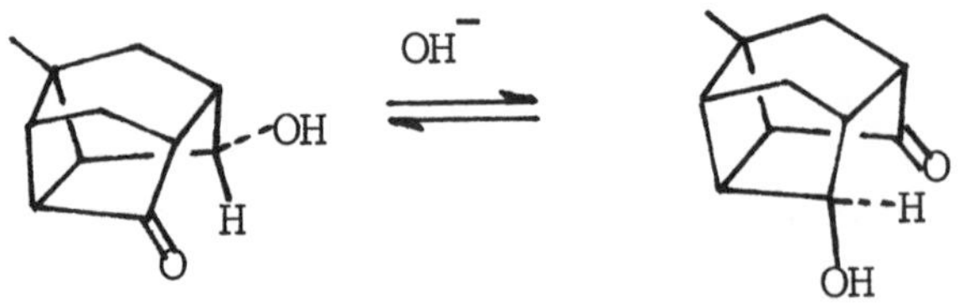

OH^- (4)

Why Is It Necessary To Invoke a New Postulate When Intramolecularity Has Already Been Explained by Entropic Factors?

The entropic theory of intramolecularity (*12*) states that nothing is remarkable about a nucleophile rapidly attacking an electrophile in the same molecule; this attack is a simple entropic consequence of converting a bimolecular reation into a unimolecular one. Our problems with this viewpoint, given in detail elsewhere (*6*), consist of the following: (a) Thermodynamics by its very nature cannot describe events on a molecular level. To say that a reaction is fast because of entropic factors is akin to saying that a day is hot because of climatic factors. Both statements may be correct, but neither contains a great deal of information. (b) Entropies of activation (which we have collected from the literature in large numbers) correlate poorly with intramolecular efficiencies. Thus, entropies of activation for reactions in solution can be used neither to rationalize nor to predict; the entropy theory is, in effect, a nontheory. (c) If intramolecular reactions are fast because placing both functionalities in the same molecule is entrophically beneficial, then minor structural variations in the intramolecular system should not impact greatly on the rate. The entropy school incorporates this conclusion into their widely quoted corollary: "Freezing the free rotation about a single bond linking two reactive functionalities improves the intramolecular rate by a factor of only 5". The corollary is illustrated schematically in equation 5.

CH_2–––(A) / CH_2–––(B) $\xrightarrow{k_{rel} = 1}$ CH_2–––(A) / CH_2–––(B) (5a)

CH–––(A) ‖ CH–––(B) $\xrightarrow{k_{rel} = 5}$ CH–––(A) ‖ CH–––(B) (5b)

However, in many instances, a single frozen rotation leads to a rate increase of $>10^4$, not 5. An example is shown below (*4*). What is the source of this huge discrepancy?

OH COOH EM $= 4 \times 10^4$

OH COOH EM $= 5 \times 10^8$

The entropy theory falters because it does not take into account the prodigious rate effects possible when critical distances are imposed. If two functional groups are connected by a long flexible chain, then inserting a cis double bond in the chain will indeed have only a minor effect on the rate.

Residual floppiness in the chain renders a critical distance unlikely. Although the entropy theory correctly predicts a trivial rate increase in such a system, the system itself is trivial. The really interesting rate increases, some of them reaching enzymatic levels, are not addressed by entropy theory.

One final point with regard to comparing the critical-distance and entropy theories is the following. Words like "true and false", "correct and incorrect", and "valid and invalid" have been avoided. Such descriptives have no place in discussions of chemical models that are, above all, fictitious. Models—one must never forget—are to be used, not believed. Thus, I do not claim the spatiotemporal hypothesis represents the "truth"; I merely claim that it is a valuable aid for thinking, especially in cases where entropic arguments fail to help.

Can the "Spatiotemporal" Postulate Be Useful in Explaining Nucleophilic Reactivity?

This question can be answered affirmatively by citing recent work of Breslow et al. (*13*), who were interested in the acylation of β-cyclodextrins (CD) by bound esters. When, for example, *m*-nitrophenyl acetate binds to the β-CD cavity, the ester transfers its acyl group 64 times faster than it hydrolyzes in water at the same pH:

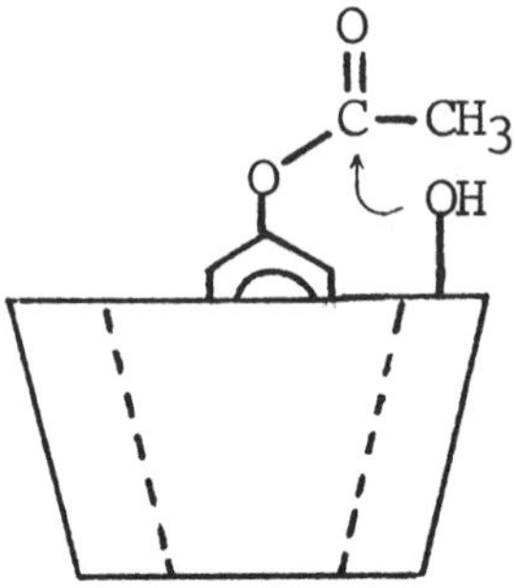

Because 64 is not a large acceleration, Breslow et al. began searching for more active substrates. They did this by first constructing molecular models of the tetrahedral intermediate for acylation by a variety of esters. They then assessed the quality of the "fit" within the cavity. This procedure led to the testing of *p*-nitrophenyl ferrocenylacrylate, a compound whose ferrocene system can enter the cavity and, with a slight tilt, rest its side chain above a β-cyclodextrin hydroxyl:

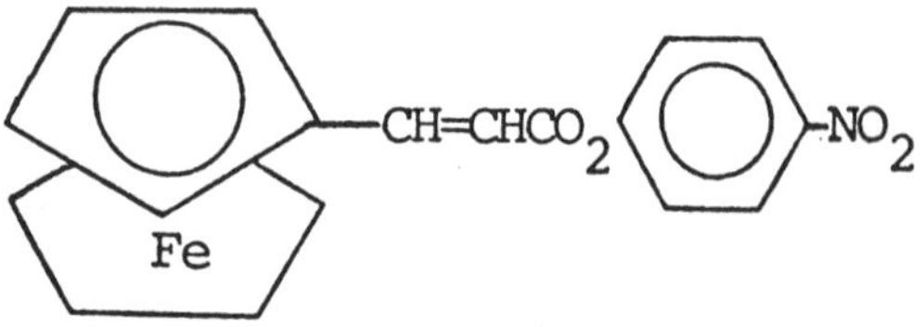

The resulting deacylation rate increases to 7.5×10^5 times faster than background, an impressive number. Note that the increase from 64 to 7.5×10^5 has nothing to do with the extent of binding because the association constant remains about the same. What in fact were Breslow et al. really doing when they were searching for improved "fits"? In terms of the postulate, they were searching for a substrate that would position its carbonyl at a critical distance from a hydroxyl a high percentage of the time. They succeeded in their search and achieved an enzyme-like rate. Although the cyclodextrin–ferrocene association constant is "normal" [in the words of Breslow et al. (*13*)], the conclusion cannot be made that this association requires little or no energy to attain the critical distance. Something is not gotten for nothing. Very likely the association constant would, in fact, be much *larger* than "normal" were it not for the enforced proximity of the carbonyl and hydroxyl. When binding energy is sacrificed, the reaction rate is enhanced.

What Is the Role of Solvent in the Spatiotemporal Postulate?

In reference 6, the critical distance for nucleophilic attack on a carbonyl is estimated roughly as 2.8 Å. This value is *less* than the diameter of a water molecule. The conclusion is inescapable: the nucleophile and electrophile must desolvate while forming a reactive complex. Once the complex is formed, however, the ensuing reaction can be extremely fast. Indeed, the step in which bonds are formed and broken may not even be rate-determining. I can hardly claim this idea is new. In 1952, Glew and Moelwyn-Hughes (*14*) suggested that the energy necessary to reorganize solvent molecules around reacting species comprises *almost all* the activation energy of some reactions. Dewar and Storch (*15*) recently reiterated the concept. The role of solvent is seen particularly clearly in the theoretical work of Chandrasekhar et al. (*16*) on the S_N2 reaction between Cl^- and CH_3Cl. The free energy of activation increases from 3.6 to 26.3 kcal/mol upon passing from the gas phase to water.

Note that nowhere in the spatiotemporal postulate is the term "transition state" used. This term is not used because the transition state, for reasons just mentioned, is considered peripheral to time and distance. Perhaps the love affair that physical organic chemists are having with "transition structure" is overly passionate. Diffusion theory, not quantum mechanics, may be the key to future progress.

Is the Time–Distance Concept Not Simply a "Strain" Theory?

The answer to this question depends on the definition of "strain". That large rate increases are possible when an intramolecular reaction relieves non-bonded tension elsewhere in the molecule is certainly true. A prime example is given (*4*):

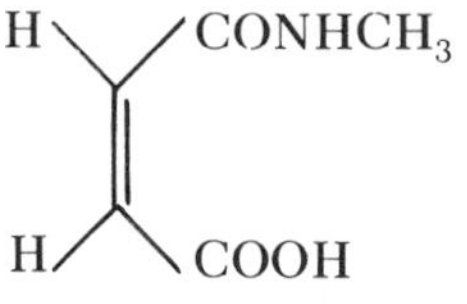

H, H / CONHCH$_3$, COOH — EM = 2×10^9

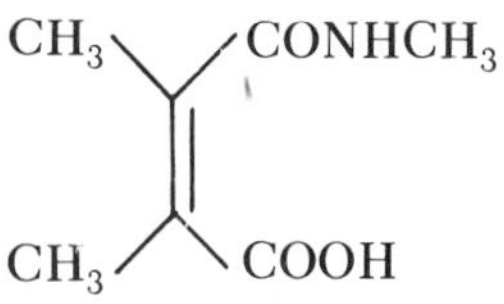

CH_3, CH_3 / CONHCH$_3$, COOH — EM = 3×10^{13}

Methyl–methyl interactions are reduced during cyclization of the molecule on the right. This sort of steric acceleration is different from the acceleration induced by holding two functionalities together. In fact, we take great care to avoid reactions where classical steric accelerations muddle the issue. Consider the following the intramolecular nucleophilic displacements (*17*):

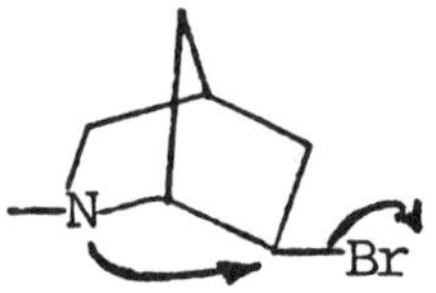

$k_{rel} = 2 \times 10^5$

$k_{rel} = 1$

Strain is *generated* in the cyclizations. Correction for strain effects would only *increase* the value of 2×10^5. The extremely fast rate of the azanorbornane derivative must, therefore, be attributed to an enforced residency at bonding distances.

Although the time–distance concept is *not* a traditional strain theory, a similarity exists between the two in that both invoke elevated ground-state energies. This concept requires energy to desolvate a nucleophile and an electrophile prior to holding them at bonding distances. The source of this energy depends on the system. In the azanorbornane derivative, the energy is "covalent" (i.e., imparted to the molecule during its synthesis). Enzymes, on the other hand, sacrifice binding energy to achieve proper geometries.

How Is the "Time" Component of the Spatiotemporal Hypothesis Treated?

Clearly, for two reactants simply to reach a contact distance is insufficient; they must also *retain* this disposition. The longer the time that two atoms spend poised in a position to react, the greater the probability of thermal activation and the faster the rate. Because the hypothesis involves, therefore, a *pair* of fluents (time and distance), we have found it useful to ignore the time by examining only rigid intramolecular systems. Unfortunately, this situation may not always be possible or desirable. In such cases, a "pre-association" constant relating solvent-separated species with van der Waals complexes must be evaluated (theoretically or experimentally). Data of this type are not generally available; this situation again leads to my opinion that less emphasis should be given to transition states and more to solution dynamics.

Is the Spatiotemporal Postulate Useful in Predicting Chemical Behavior?

If the reasonable assumption that the critical distance for nucleophilic attack on an ester (CD_e) is larger than that for attack on an amide (CD_a) is made, rigid carbon skeletons can then hold a nucleophile and an ester (or corresponding amide) at a distance D according to three possibilities: (a) $D > CD_e$ and CD_a, (b) $CD_a < D < CD_e$, and (c) $D < CD_e$ and CD_a. The kinetic consequences of the different geometries are shown in Table I. By far the most interesting case is the one in which the distance exceeds the needs of the amide but not the ester. In such an event, an intramolecular reaction should be fast only for the ester. Entropy theory, of course, would not make a distinction. In summary, EM values could conceivably vary widely with the nature of the functionalities in chemically similar reactions. I know of no other theory that can make this prediction nor of a study in which the possibility has been systematically tested. And if the spatiotemporal hypothesis stimulates an experiment, its formulation will have been worthwhile apart from whether the prediction (and others we are now testing) turn out to be correct or not.

Table I. Kinetic Consequences of Different Geometrics.

Distance	*EM of Ester*	*EM of Amide*
$D > CD_e$ and CD_a	small	small
$CD_a < D < CD_e$	large	small
$D < CD_e$ and CD_a	large	large

Acknowledgment

Support by the National Science Foundation is greatly appreciated.

Literature Cited

1. Bender, M. L.; Kézdy, F. J.; Gunter, C. R. *J. Am. Chem. Soc.* **1964,** *86*, 3714.
2. Hershfield, R.; Schmir, G. L. *J. Am. Chem. Soc.* **1973,** *95*, 7539.
3. Parker, A. J. *Chem. Rev.* **1969,** *69*, 1.
4. Kirby, A. J. *Adv. Phys. Org. Chem.* **1980,** *17*, 183.
5. Menger, F. M. *Tetrahedron* **1983,** *39*, 1013.
6. Menger, F. M. *Acc. Chem. Res.* **1985,** *18*, 128.
7. Steiger, U. R.; Keizer, J. *J. Chem. Phys.* **1982,** *77*, 777.

8. Cowan, D. O.; Drisko, R. L. *Elements of Organic Photochemistry;* Plenum: New York, 1976; p 286.
9. Sisido, M. *Macromolecules* **1971,** *4,* 737.
10. Pispisa, B.; Palleschi, A.; Barteri, M.; Nardini, S. *J. Phys. Chem.* **1985,** *89,* 1767.
11. Davis, A. M.; Page, M. I.; Mason, S. C.; Watt, I. *J. Chem. Soc., Chem. Commun.* **1984,** 1671.
12. Page, M. I.; Jencks, W. P. *Proc. Natl, Acad. Sci. U.S.A.* **1971,** *68,* 1678.
13. Breslow, R.; Czarniecki, M. F.; Emert, J.; Hamaguchi, H. *J. Am. Chem. Soc.* **1980,** *102,* 762.
14. Glew, D. N.; Moelwyn-Hughes, E. A. *Proc. R. Soc. London, Ser. A* **1952,** *211,* 254.
15. Dewar, M. J. S.; Storch, D. M. *J. Chem. Soc., Chem. Commun.* **1985,** 94.
16. Chandrasekhar, J.; Smith, S. F.; Jorgensen, W. L. *J. Am. Chem. Soc.* **1985,** *107,* 154.
17. Hutchins, R. O.; Rua, L. *J. Org. Chem.* **1975,** *40,* 2567.

RECEIVED for review October 21, 1985. ACCEPTED June 17, 1986.

15

Bond-Cleavage Reactions with Hard Acid and Soft Nucleophile Systems

Kaoru Fuji

Institute for Chemical Research, Kyoto University, Uji, Koyoto 611, Japan

Carbon – heteroatom bonds can be cleaved by an appropriate combination of a hard acid and a soft nucleophile. Synthetically useful selective C–O bond cleavage in the presence of other C–O bond(s) is described. Reductive dehalogenation of α-haloketones is presented as an example that illustrates the concept of hard–soft affinity inversion. Finally, regio- and stereoselective functionalization of 1,3-dienes is demonstrated by the thienium cation Diels–Alder cyclization involving the C–S bond cleavage.

EVERY CHEMICAL BOND IS REGARDED as composed of a combination of a Lewis acid and a Lewis base. Moreover, the bond has hard–soft dissymmetry, unless the bonded groups are identical. Thus, most chemical bonds are divided into two groups. One group includes bonds consisting of a soft acid and a hard base, and the other group includes those bonds consisting of a hard acid and a soft base. As carbon–heteroatom bonds belong to the former group, they can be cleaved by combinations of a hard acid and a soft nucleophile according to the hard and soft acids and bases (HSAB) principle (*1*, *2*). Here we describe several combinations consisting of a hard Lewis acid and a soft nucleophile that are useful for the cleavage of carbon–heteroatom bonds.

Carbon–Oxygen Bond Cleavage

The methyl ether is an ideal protecting group of hydroxyl groups in terms of stability. However, chemical stability has been a main drawback of this protecting group because of the lack of deblocking methodology. Recently, a number of reagents have been developed for demethylation. These reagents may fall into three categories (Scheme I). In the reaction of type 1, as the nucleophilic attack of Y^- controls the reaction, demethylation occurs with

0065-2393/87/0215-0219$06.00/0

Type 1

$$\text{R-O-Me} + \text{Y}^- \longrightarrow \text{R-O}^- + \text{MeY} \qquad \text{Y} = \text{RS}^-\ (3),\ \text{I}^-\ (4,5)$$

Type 2

$$\text{R-O-Me} + \text{Z-Y} \longrightarrow \text{R-}\overset{+}{\underset{\text{Z}}{\text{O}}}\text{-Me} + \text{Y}^- \longrightarrow \text{R-O-Z} + \text{MeY}$$

$$\text{YZ} = \text{BBr}_3\ (6),\ \text{TMSI}\ (7,8)$$

Type 3

$$\text{R-O-Me} + \text{Z} \longrightarrow \text{R-}\overset{+}{\underset{\text{Z}^-}{\text{O}}}\text{-Me} + \text{Y} \longrightarrow \text{R-O-Z}^- + \text{MeY}^+$$

Scheme I. Three types of reagents for demethylation.

the substrates originally containing an electron-withdrawing group like the aromatic ring or the carbonyl group as R. Thiolate (*3*) and iodide (*4, 5*) have been widely used as nucleophiles.

In the type 2 reaction, as represented by boron tribromide (*6*) or trimethylsilyl iodide (*7, 8*), the reagent involves the intramolecular combination of a hard acid and a soft base. The soft base (Y^-) is generated in the same amount as the substrate reacting with ZY, because the stoichiometric reaction of ZY with a substrate affords a stoichiometric amount of nucleophile Y^- Reaction temperature, reaction time, and the solvent system are the main factors controlling the reaction in this case. On the other hand, as a hard acid and a soft nucleophile are supplied from separate sources in the type 3 reactions, the ratio of a substrate, a hard acid, and a soft nucleophile can be freely changed to modify the reactivity. Numerous combinations along these lines have been developed for C–O bond cleavage reactions (*9*). Among the combination systems we developed (*10–13*), examples of the selective cleavage of a particular C–O bond in the presence of another C–O bond are given in Schemes II and III and Chart I. All of the reactions proceeded through the attack of a soft nucleophile in the S_N2 sense at the soft carbon activated by the coordination of the adjacent oxygen atom with a hard acid as illustrated in Scheme IV.

Deblocking of a methyl ether with a combination of a hard Lewis acid and a thiol has been successfully applied in the total synthesis of natural products such as lythranidine (*14*) and quassin (*15*) (Scheme V).

Carbon–Halogen Bond Cleavage: Hard–Soft Affinity Inversion

All chemical bonds, except for symmetrically substituted ones, are inherently dissymmetric in two respects: charge dissymmetry and hard–soft dissymmetry. The reversal of charge dissymmetry by modifying the structural unit (umpolung) has been well documented and has become one of the important principles in synthetic organic chemistry (*16*). On the other hand,

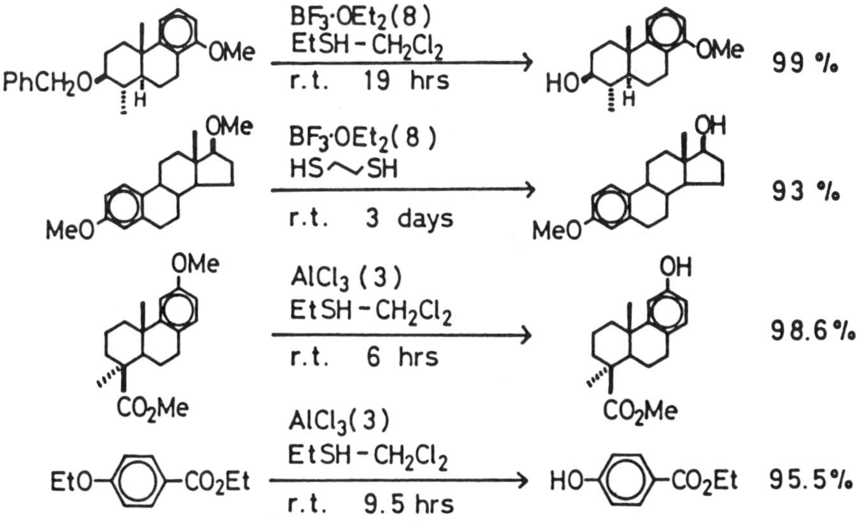

PhCH$_2$O … OMe → $BF_3 \cdot OEt_2$(8), EtSH - CH_2Cl_2, r.t. 19 hrs → HO … OMe 99 %

MeO … OMe → $BF_3 \cdot OEt_2$(8), HS∼SH, r.t. 3 days → MeO … OH 93 %

OMe … CO_2Me → $AlCl_3$ (3), EtSH - CH_2Cl_2, r.t. 6 hrs → OH … CO_2Me 98.6 %

EtO–C$_6$H$_4$–CO_2Et → $AlCl_3$(3), EtSH - CH_2Cl_2, r.t. 9.5 hrs → HO–C$_6$H$_4$–CO_2Et 95.5 %

Scheme II. Selective dealkylation of ethers.

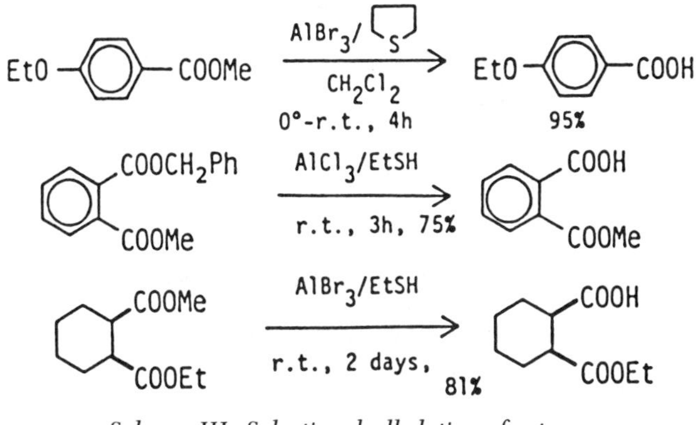

EtO–C$_6$H$_4$–COOMe → $AlBr_3$/tetrahydrothiophene, CH_2Cl_2, 0°-r.t., 4h → EtO–C$_6$H$_4$–COOH 95%

C$_6$H$_4$(COOCH$_2$Ph)(COOMe) → $AlCl_3$/EtSH, r.t., 3h, 75% → C$_6$H$_4$(COOH)(COOMe)

C$_6$H$_{10}$(COOMe)(COOEt) → $AlBr_3$/EtSH, r.t., 2 days, 81% → C$_6$H$_{10}$(COOH)(COOEt)

Scheme III. Selective dealkylation of esters.

the reversal of hard–soft dissymmetry, which may be called "hard–soft affinity inversion" (Scheme VI) has never been claimed as a method of choice in synthetic organic chemistry. Realization of this principle includes dehalogenation of α-chloro- and α-fluoroketones with a combination system of aluminum chloride and ethanethiol (*17*).

As shown in Table I, halogen atoms are reductively removed to afford the corresponding dithioacetal in almost all cases. Clearly, the favorable interaction based on the HSAB principle as shown in **I** operates in deiodination and debromination. However, the electron shift shown in **I** is hardly acceptable in the case of defluorination and dechlorination because F^+ and

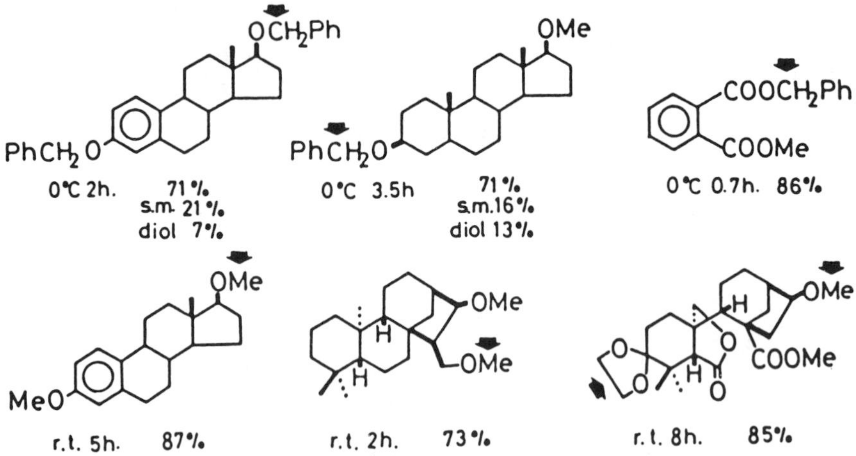

Chart I. Selective dealkylation with $AlCl_3$–NaI–CH_3CN system. The arrow denotes the position of dealkylation.

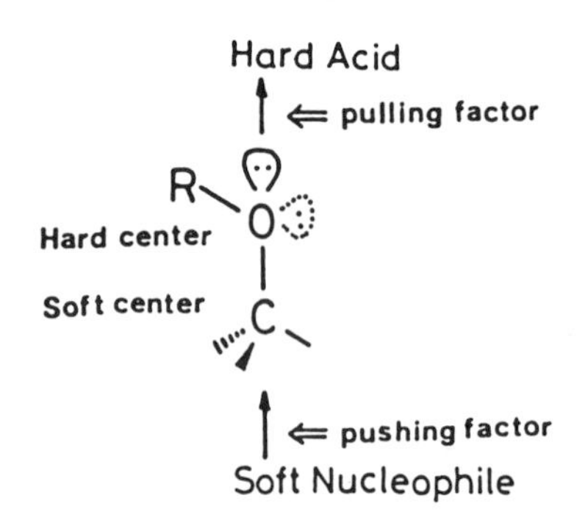

Scheme IV. Attack of a soft nucleophile.

Cl^+ are much harder than Br^+ and I^+. Product distribution from α-bromoacetophenones was quite different from those from α-chloro- and α-fluoroacetophenones under the extremely mild conditions (Scheme VII).

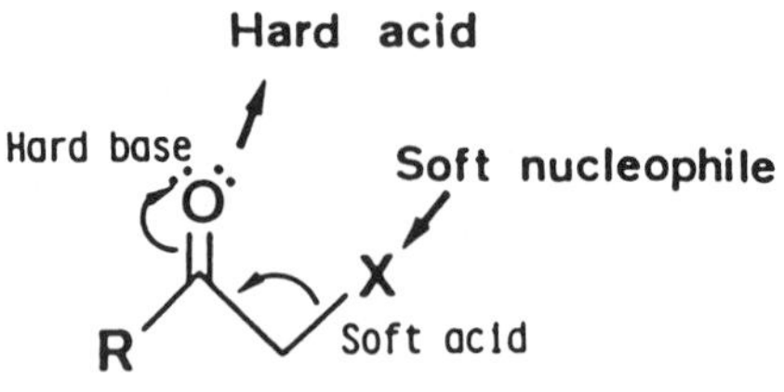

Remarkably, 1,2-diethylthiostyrene was obtained in more than 50% yield from the latter two haloketones. Iodine and bromine were removed with

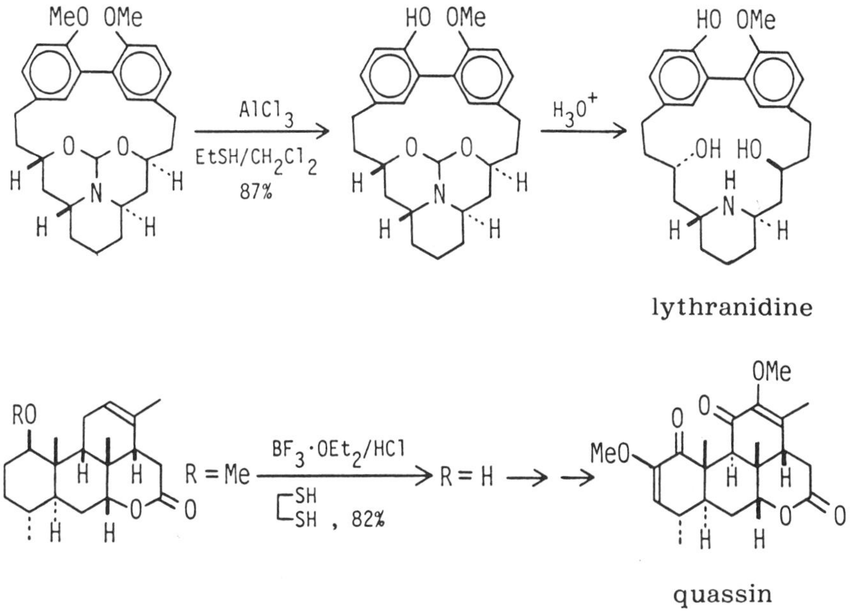

lythranidine

quassin

Scheme V. Deblocking of a methyl ether.

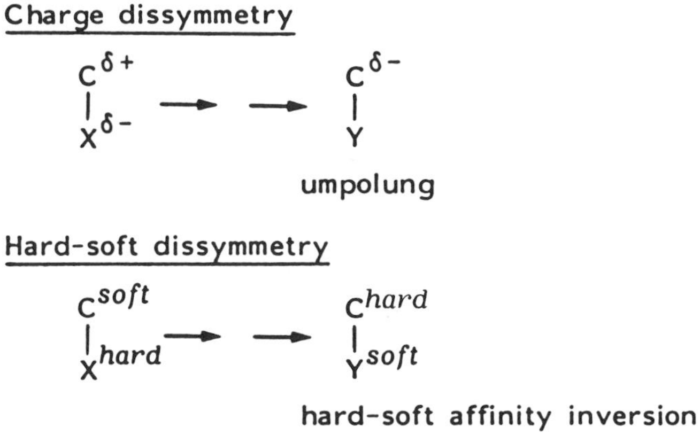

Charge dissymmetry

$C^{\delta+}-X^{\delta-} \rightarrow \rightarrow C^{\delta-}-Y$

umpolung

Hard-soft dissymmetry

$C^{soft}-X^{hard} \rightarrow \rightarrow C^{hard}-Y^{soft}$

hard-soft affinity inversion

Scheme VI. Two types of dissymmetry in C–X bonds in the activated S_N2 reaction.

dimethyl sulfide as a soft nucleophile instead of ethanethiol, while fluorine and chlorine remained almost intact (Scheme VIII). This result suggests that the initial formation of dithioacetal is indispensable for defluorination and dechlorination. Thus, hard–soft affinity inversion from the hard carbonyl oxygen atom to the soft sulfur atom plays a crucial role in defluorination and dechlorination (**II**). α-Halodithioacetal undergoes either direct dehalogena-

Table I. Dehalogenation of α-Haloketones with $AlCl_3$ and EtSH

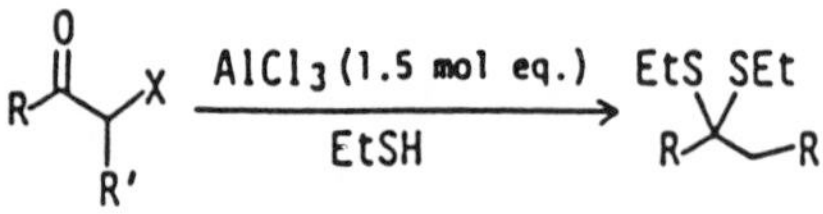

$$RCOCH(R')X \xrightarrow[\text{EtSH}]{AlCl_3\ (1.5\ \text{mol eq.})} RC(SEt)_2CH_2R'$$

Substrate R	Substrate R'	Substrate X	Time min	Product Yield, %
C_6H_5	H	I	5	>99 [a]
C_6H_5	H	Br	20	98
p-Br-C_6H_4	H	Br	30	86
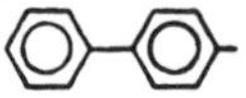	H	Br	20	75
-CH_2-$(CH_2)_3$-CH_2-		Br	15	56
-CH_2-$(CH_2)_4$-CH_2-		Br	15	56
C_6H_5	H	Cl	10	77
-CH_2 $(CH_2)_3$-CH_2-		Cl	15	69
C_6H_5	H	F	20	67
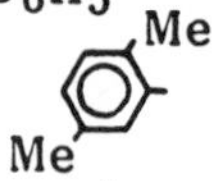 Me, Me	H	F	15	72 [b]
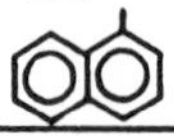	H	F	20	87 [c]

NOTE: All reactions were run at 0° C.
[a] Acetophenone was obtained.
[b] A mixture of **a** (53%), **b** (8%), and **c** (11%).
[c] A mixture of **d** (26%) and **e** (61%).

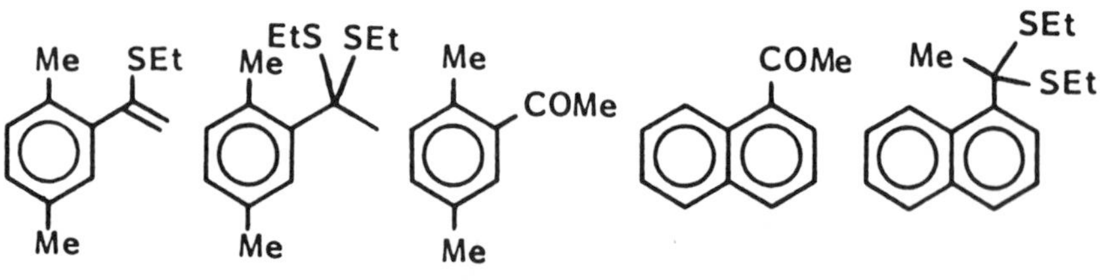

a: Me, SEt, Me
b: EtS, SEt, Me, Me
c: Me, COMe, Me
d: COMe
e: Me, SEt, SEt

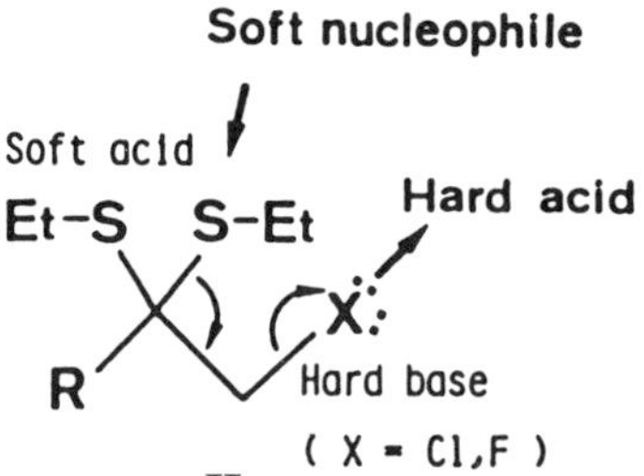

Soft nucleophile

Soft acid

Et–S S–Et

Hard acid

R

X:

Hard base

(X = Cl,F)

II

tion (path a) or 1,2-migration of the sulfur function (path b) followed by desulfurization (Scheme IX). Existence of the path b may allow the 1,2-transposition of the carbonyl group, when α-chloro- or α-fluorodibenzyl ketone is dehalogenated. Results listed in Scheme X clearly indicate that both pathways a and b are operative to remove fluorine and chlorine.

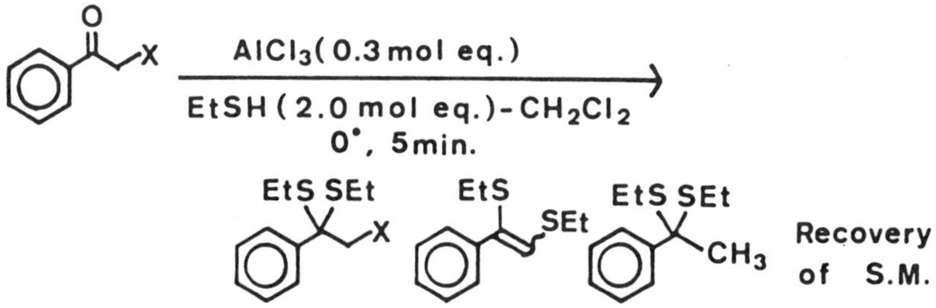

$AlCl_3$(0.3 mol eq.), EtSH (2.0 mol eq.)-CH_2Cl_2, 0°, 5min.

Starting Material	Yields(%)			
	EtS SEt (X)	EtS / SEt	EtS SEt (CH_3)	Recovery of S.M.
X = Br	0	4	14	54
Cl	3	53	8	18
F	10	55	6	27

Scheme VII. Partial dehalogenation of α-haloacetophenones.

$AlCl_3$ / Et_2S

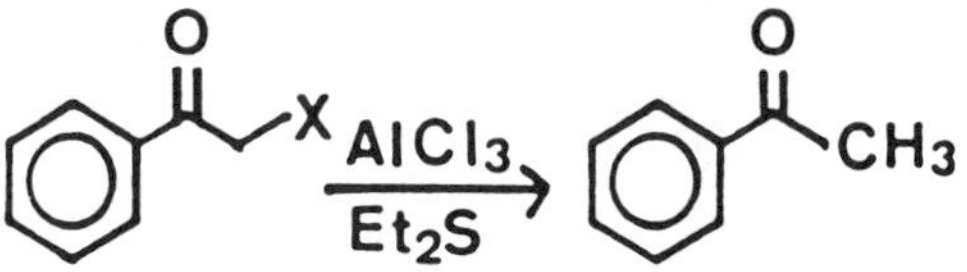

X =		
I	60%	(5 min)
Br	73%	(20 min)
Cl	6%	(100 min)
F	0%	(80 min)

Scheme VIII. Dehalogenation of α-haloacetophenones with $(CH_3)_2S$.

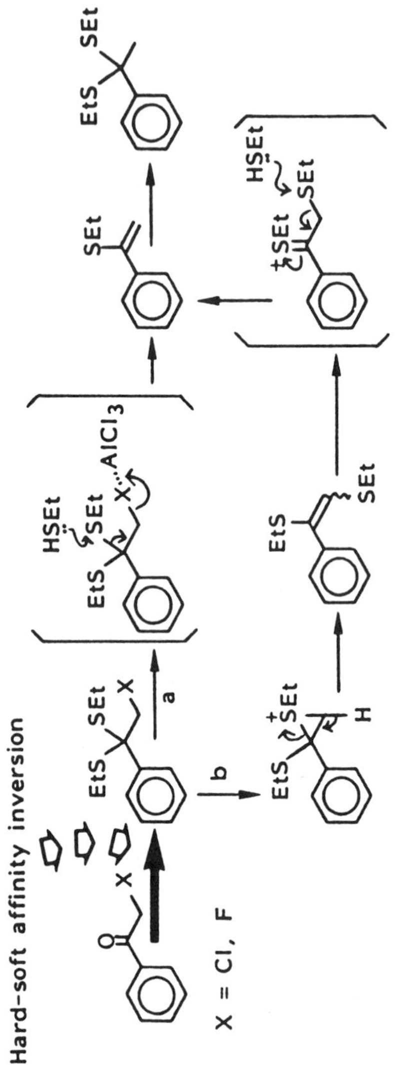

Scheme IX. Hard–soft affinity inversion.

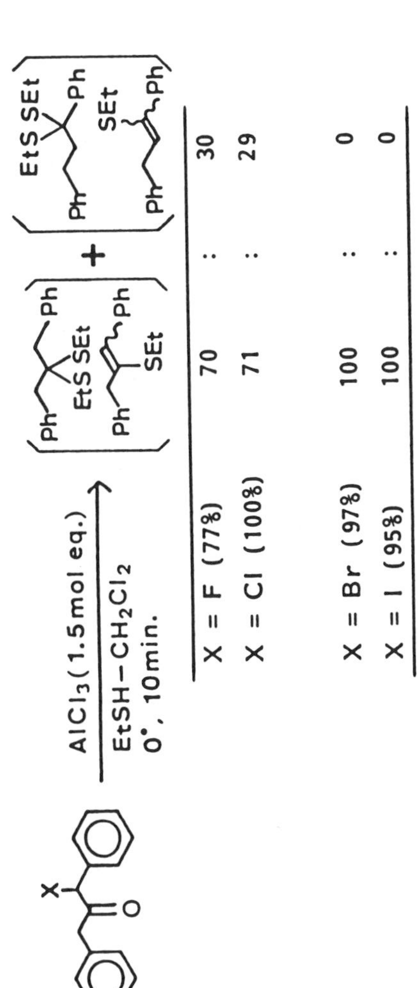

Scheme X. Dehalogenation of α-halodibenzyl ketones.

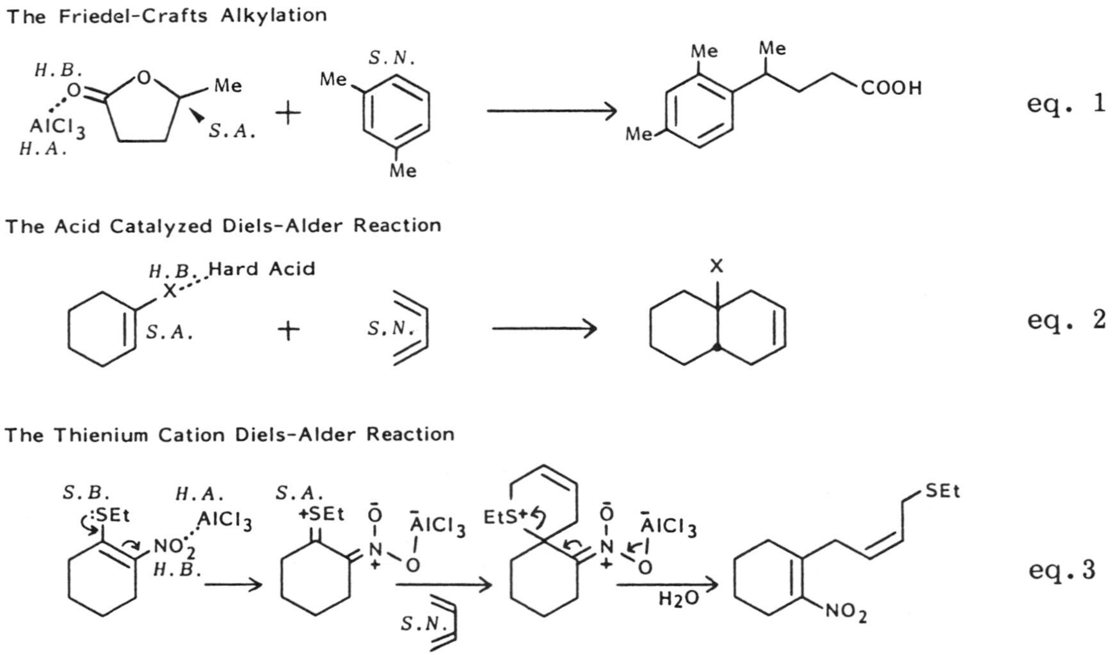

The Friedel-Crafts Alkylation

H.B.
O
AlCl$_3$
H.A.
Me
S.A.
+
S.N.
Me
Me
Me
Me
Me
COOH

eq. 1

The Acid Catalyzed Diels-Alder Reaction

H.B. Hard Acid
X
S.A.
+
S.N.
X

eq. 2

The Thienium Cation Diels-Alder Reaction

S.B.
:SEt
NO$_2$
H.B.
H.A.
AlCl$_3$
S.A.
+SEt
$\bar{O}$
N
+
O
$\bar{A}$lCl$_3$
S.N.
EtS+
$\bar{O}$
N
+
O
$\bar{A}$lCl$_3$
H$_2$O
SEt
NO$_2$

eq. 3

Scheme XI. Friedel-Crafts alkylation and Diels–Alder reactions. H.A. is hard acid, S.A. is soft acid, H.B. is hard base, S.B. is soft base, and S.N. is soft nucleophile.

Table II. Reaction of Nitroolefins with Dienes

Entry	Diene	Nitroolefin	Product[a,b)]	Isolated yield %
1	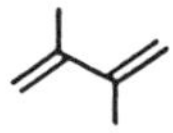	f	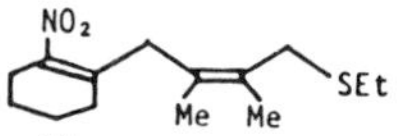 NO_2, SEt, Me, Me	84
2	//	g	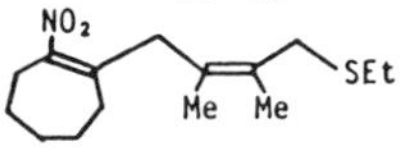 NO_2, SEt, Me, Me	86
3	//	h	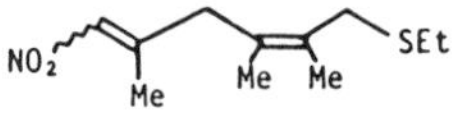 NO_2, Me, SEt, Me, Me	65
4	//	i	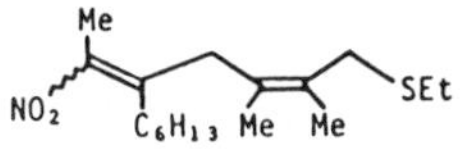 Me, NO_2, C_6H_{13}, SEt, Me, Me	84(93)[c)]
5		f	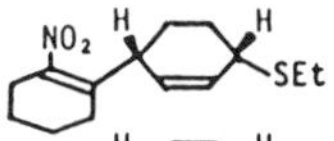 NO_2, H, H, SEt	54
6	//	g	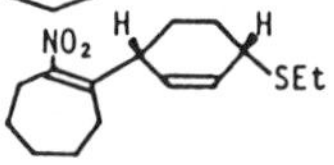 NO_2, H, H, SEt	58
7	//	h	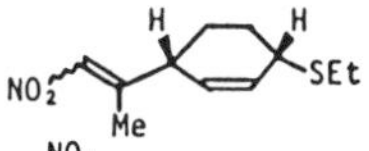 NO_2, Me, H, H, SEt	71
8	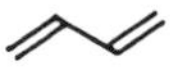	f	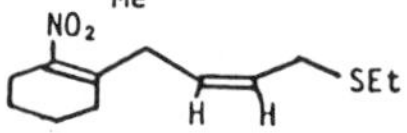 NO_2, SEt, H, H	79
9	//	i	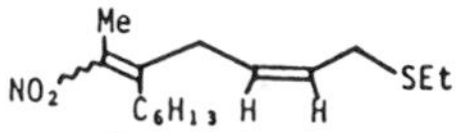 Me, NO_2, C_6H_{13}, SEt, H, H	58
10		f	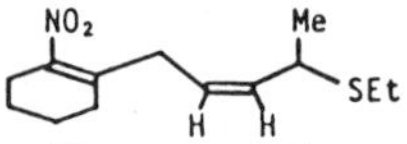 NO_2, Me, SEt, H, H	61(83)[c)]
11	//	g	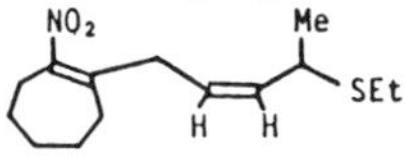 NO_2, Me, SEt, H, H	82
12	//	i	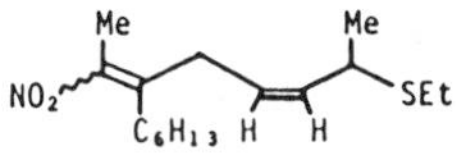 Me, Me, NO_2, SEt, C_6H_{13}, H, H	63(73)[c)]
13	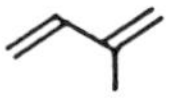	f	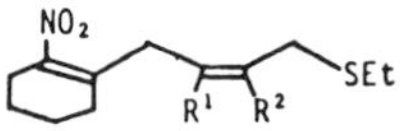 NO_2, SEt, R^1, R^2	68[d)]

a) All new compounds have been fully characterized by spectral means and have satisfactory combustion analysis or high-resolution peak matching. b) Geometry represented by a wavy line was not determined. c) Numbers in parentheses are the yields based on the consumed starting material. d) A mixture of two isomers (R^1 = Me, R^2 = H and R^1 = H, R^2 = Me) in a 6:1 ratio.

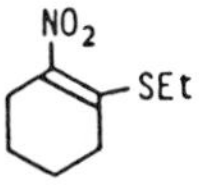

f: NO_2, SEt

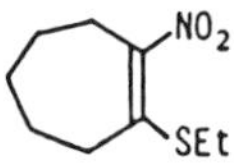

g: NO_2, SEt

h: NO_2, H, SEt, Me

i: NO_2, Me, SEt, C_6H_{13}

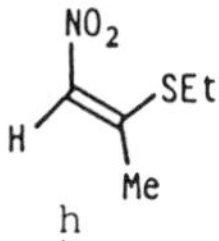

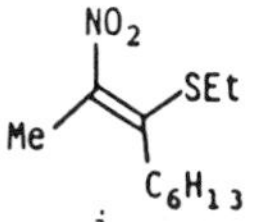

Carbon–Sulfur Bond Cleavage: The Thienium Cation Diels–Alder Reaction

The carbon–carbon bond may be formed when a carbon nucleophile is used in the combination system. The most representative examples include the Friedel-Crafts-type alkylation of aromatics (Scheme XI, equation 1) (*18*) and the acid-catalyzed Diels–Alder reaction (Scheme XI, equation 2). The reaction of a combination system consisting of aluminum chloride and 1,3-dienes leading to regio- and stereoselective functionalization of 1,3-dienes via the thienium cation Diels–Alder reaction (*19*) (Scheme XI, equation 3) is described here.

The reaction of 2-ethylthionitroolefins with aluminum chloride furnishes the thienium cation due to the favorable hard–hard interaction between the nitro group and aluminum chloride, whereby the sulfur atom is changed from a soft base to a soft acid. Trapping of the soft thienium cation with a diene, which is a soft nucleophile, affords the hetero Diels–Alder product that undergoes the fragmentation during the workup procedure to give rise to (Z)-olefin selectively (Scheme XI, equation 3). This reaction results in the regio- and stereoselective 1,4-functionalization of 1,3-dienes. Results are listed in Table II.

Cyclohexadiene provided 1,4-cis adducts stereoselectively (entries 5–7 in Table II). With 1,3-pentadiene, products carrying ethylthio group at the more substituted end was obtained with complete regioselection (entries 10–12 in Table II). Predominant formation of one of the two possible isomers was observed with 2-methyl-1,3-butadiene, though regioselectivity was not as high as in the case of 1,3-pentadiene (entry 13 in Table II). Intervention of a Diels–Alder-type adduct accounts for these selectivities.

Though thioketones have been used as dienophiles in hetero Diels–Alder reactions (*20*), only one example has been reported for the use of a thienium cation as a dienophile (*21*).

Acknowledgment

This work is partially supported by Grant 58570871 from the Ministry of Education, Science and Culture, Japan. Acknowledgment is made to coworkers in the papers cited in this review.

Literature Cited

1. Pearson, R. G. *Hard and Soft Acids and Bases;* Dowden, Hutchinson, & Ross: Strousberg, 1977.
2. Saville, B. *Angew. Chem. Int. Ed. Engl.* **1967,** *6*, 928.
3. Bartlett, P. A.; Johnson, W. S. *Tetrahedron Lett.* **1970,** 4459.
4. Mustafa, A.; Sidky, M. M.; Mahram, M. R. *Ann. Chem.* **1967,** *704*, 182.
5. Bennett, C. R.; Cambie, R. C. *Tetrahedron* **1967,** *23*, 927.

6. McOmie, J. F. W.; Watts, M. L.; West, D. E. *Tetrahedron* **1968,** *24*, 2289.
7. Ho, T.-L.; Olah, G. A. *Angew. Chem. Int. Ed. Engl.* **1976,** *15*, 774.
8. Jung, M. E.; Lyster, M. A. *J. Am. Chem. Soc.* **1977,** *99*, 968.
9. For an extensive review, see: Bhatt, M. V.; Kulkarni, S. U. *Synthesis* **1983,** 249.
10. Fuji, K.; Ichikawa, K.; Node, M.; Fujita, E. *J. Org. Chem.* **1979,** *44*, 1661.
11. Node, M.; Nishide, K.; Fuji, K.; Fujita, E. *J. Org. Chem.* **1980,** *45*, 4275.
12. Node, M.; Nishide, K.; Sai, M.; Fuji, K.; Fujita, E. *J. Org. Chem.* **1981,** *46*, 1991.
13. Node, M.; Ohta, K.; Kajimoto, T.; Nishide, K.; Fujita, E.; Fuji, K.; *Chem. Pharm. Bull.* **1983,** *31*, 4178.
14. Fuji, K.; Ichikawa, K.; Fujita, E. *J. Chem. Soc., Perkin Trans. 1* **1980,** 1066.
15. Grieco, P. A.; Ferrino, S.; Vidari, G. *J. Am. Chem. Soc.* **1980,** *102*, 7586.
16. Seebach, D. *Angew. Chem. Int. Ed. Engl.* **1979,** *18*, 239.
17. Fuji, K.; Node, M.; Kawabata, T.; Fujimoto, M.; *Chem. Lett.* **1984,** 1153.
18. Mosby, W. L. *J. Am. Chem. Soc.* 1952, *74*, 2564.
19. Fuji, K.; Khanapure, S. P.; Node, M.; Kawabata, T.; Ito, A. *Tetrahedron Lett.* **1985,** *26*, 779.
20. Weinreb, S. M.; Staib, R. R. *Tetrahedron,* **1982,** *38*, 3087.
21. Corey, E. J.; Walinsky, S. W. *J. Am. Chem. Soc.* **1972,** *94*, 8932.

RECEIVED for review October 21, 1985. ACCEPTED January 24, 1986.

16

Basicity and Nucleophilicity of Transition Metal Complexes

Ralph G. Pearson

Department of Chemistry, University of California, Santa Barbara, CA 93106

Rates of reaction of transition metal nucleophiles correlate both with oxidation potentials for ML_n^- *and with the* pK_a *values of the corresponding acids,* HML_n*. Therefore, the two parameters,* E° *and* H, *in the Edwards equation are not independent parameters. The same result is found for other nucleophiles, if the donor atom is C, N, O, or F. However, for bases with heavier donor atoms,* E° *and* H *are not as correlated with each other. For transition metal complexes, soft ligands, L, increase acidity and decrease nucleophilic reactivity. Hard ligands have the opposite effect.*

TRANSITION METAL COMPLEXES, either as anions or as neutral molecules, are often very good nucleophiles for alkyl halides and sulfonates. Some of them, such as vitamin B_{12s}, a cobalt (I) species, and the solvent-separated ion pair, $Na^+:S:Fe(CO)_4^{2-}$, where S is *N*-methylpyrrolidinone, are among the most reactive known (*1*, *2*). Reactions are of several types, for example

$$RX + ML_n \rightarrow RML_n^+ + X^- \quad (1)$$

$$RX + ML_n \rightarrow RMXL_n \quad (2)$$

where M is the metal atom and L_n stands for *n* ligands bound to M.

A variety of mechanisms have been found (*3*). Simple S_N2 substitution mechanisms are most common, followed by free-radical pathways. The latter occur by two mechanisms: (a) removal of X as an atom, followed by reactions of R·; and (b) single-electron transfer from ML_n to RX, to give R·. Hydride ion transfer reactions are also known (*4*). Reactions believed to be simple S_N2 processes in which RX is methyl iodide are discussed here.

$$RX + HML_n \rightarrow RH + X^- + ML_n^+ \quad (3)$$

0065-2393/87/0215-0233$06.00/0

Characteristically, transition metal nucleophiles react much faster with methyl iodide than with methyl tosylate. Rate constant ratios ranging from 30 to 3×10^5 have been found (3). Such behavior qualifies transition metal complexes to be called supersoft nucleophiles (5). Even larger ratios are found for reagents such as $Co(CN)_5^{3-}$, up to 10^9. Such large ratios are found only for free-radical pathways (6) and may be used as a mechanistic probe.

Nucleophilic Reactivity and Redox Potential

A pioneering study of transition metal nucleophiles was made by Dessy et al. (7). These workers measured not only rates of reaction of various ML_n^- with CH_3I but also the oxidation potentials at a platinum electrode. A good linear relation was found when log k_2 was plotted against $E_{1/2}$ for various nucleophiles.

$$ML_n \longrightarrow ML_n + e \qquad E_{1/2} \tag{4}$$

Such a finding was not unexpected because Edwards (8) had already put forward his equation

$$\log (k/k_0) = \alpha E^\circ + \beta H \tag{5}$$

relating nucleophilicity to two properties of the nucleophile: the ease of electron loss, as given by E°, the redox potential, and the proton basicity, as given by H ($H = pK_a + 1.74$). Soft nucleophiles would be expected to have large values of α and small values of β.

An even better reason exists to expect that the nucleophilicity of a transition metal complex would depend strongly on its ease of oxidation. Reaction with an alkyl halide, according to either reaction 1 or 2, is an example of oxidative addition. The oxidation state of the metal increases by two units. Taking a definite example, we find that the change is more than a formal one. The reactant is a typical d^{10} complex, whereas the product is a definite d^8 complex. The product has the right coordination number, square-planar structure, and visible–UV spectra found for similar Pt(II) complexes.

$$Pt^0[P(CH_3CH_2)_3]_3 + CH_3Br \longrightarrow CH_3Pt^{II}[P(CH_3CH_2)_3]_3^+ + Br^- \tag{6}$$

The more readily the metal atom in the complex ML_n can be oxidized, the more rapidly it can react with alkyl halides. Consideration of a large amount of experimental data on oxidative additions, in general, leads to a metal ordering: $Os^0 > Ru^0 > Fe^0 \gg Ir^I > Rh^I > Co^I \gg Pt^{II} > Pd^{II} \gg Ni^{II}$ (9). For any one metal, $M^{-I} > M^0 > M^I$ and so on.

The associated ligands can play an equally important role to that of the metal. Ligands that stabilize the reduced form of a metal will deactivate the metal complex toward nucleophilic behavior. Such ligands are CO, PR_3, C_2H_4, and other soft bases. Hard bases such as amines and alcohols will activate the complex by stabilizing the oxidized state. A complex such as $Rh^{I}Cl(CO)[P(C_6H_5)_3]_2$ is a very poor nucleophile, but $Rh^{I}(C_2DOBF_2)$ is extremely reactive (*10*). C_2DOBF_2 is a complex ligand that has four nitrogen donor atoms.

These conclusions are somewhat unexpected. Soft ligands are commonly thought to put more negative charge density on the metal atom than hard ligands and thus make the metal atom a better electron donor. However, Kubota (*11*) made a detailed study of the reaction

$$IrX(CO)[P(C_6H_5)_3]_2 + CH_3I \longrightarrow CH_3IrIX(CO)[P(C_6H_5)_3]_2 \qquad (7)$$

The order of rates found for different Xs was $F^- > N_3^- > Cl^- > Br^- > NCO^- > I^- > NCS^-$ with hard F^- reacting 100 times faster than soft (S-bonded) NCS^-.

Nucleophilicity and Brønsted Basicity

At the time of the study by Dessy et al. (*7*), little was known about the Brønsted basicity of ML_n^- in a quantitative sense. Therefore, what role, if any, was played by the parameter H in the Edwards equation was unclear. Recently, a number of pK_a values were measured for transition metal hydrides, both neutral, HML_n, and cationic, HML_n^+. The solvents used were chiefly methanol (*12*) and acetonitrile (*13*; J. R. Norton and J. Sullivan, personal communication).

$$ML_n^- + H^+ \longrightarrow HML_n \qquad pK_a \qquad (8)$$

Table I gives a listing of relative rate constants for reaction with CH_3I for a number of transition metal bases. The values are normalized to that for $Co(CO)_4^-$ set equal to unity and are valid for solvents such as tetrahydrofuran (THF) or glyme. Also given are the pK_a values for the conjugate acids in acetonitrile. In some cases, pK_a values in methanol were adjusted to acetonitrile by using references of the same charge type.

A very strong correlation between nucleophilic reactivity and Brønsted basicity occurs. The correlation is not perfect, but it cannot be expected to be, because a variety of types of nucleophiles are represented, and the solvents also vary. Clearly, the dependence on H in equation 5 is as great as the dependence on $E°$.

Some data seem to contradict this conclusion. The basicity of $Pt[P(CH_3CH_2)_3]_3$ is much greater than that of $Pd[P(CH_3CH_2)_3]_3$ or $Ni[P(CH_3CH_2)_3]_3$ (*14*). The rate constants for reaction with CH_3I are exactly opposite (*3*). However, these reactions are quite complicated. The actual

Table I. Nucleophilic Reactivity and Basicity of a Series of Transitional Metal Bases

Base	*Relative Rate*[a]	*p*K_a*(*CH_3CN*)*[b]
$C_5H_5Fe(CO)_2^-$	70,000,000	19.4
$C_5H_5Ru(CO)_2^-$	7,500,000	20.2
$Co(dmgH)(P[CH_3(CH_2)_3]_3)^-$	57,000	17.5[c]
$Re(CO)_5^-$	25,000	21.0
$Rh(dmgH)[P(C_6H_5)_3]^-$	1,700	16.5[c]
$C_5H_5W(CO)_3^-$	500	16.1
$Mn(CO)_5^-$	77	15.2
$C_5H_5Mo(CO)_3^-$	67	13.9
$C_5H_5Cr(CO)_3^-$	4	13.3
$Co(CO)_4^-$	1	8.5
$IrCl(CO)[As(C_6H_5)_3]_2$	0.10	9.3[d]
$IrCl(CO)[P(C_6H_5)_3]_2$	0.07	9.0[d]
$IrBr(CO)[P(C_6H_5)_3]_2$	0.03	8.6[d]
$V(CO)_6^-$	—[e]	—[f]

[a] See references 3 and 12.
[b] Reference 13 and J. R. Norton and J. Sullivan (personal communication).
[c] Estimated from data in 50% CH_3OH–H_2O; dmgH$^-$ is monoanion of dimethylglyoxime.
[d] Estimated from data in 100% CH_3OH and results for the pK_a of $M[P(OCH_3)_3]_4$ in CH_3OH and CH_3CN (M = Ni, Pd, or Pt).
[e] No reaction
[f] Very strong; reference 29.

reactants are probably $Pt[P(CH_3CH_2)_3]_2$ or $Pd[P(CH_3CH_2)_3]_2$, formed by dissociation of a ligand (*15*). Also, these phosphine complexes are prone to react by free-radical mechanisms (*16*).

A kinetic study of reaction 3, where RX is *n*-butyl bromide, gave the reactivity order (*4*) $HW(CO)_4[P(OCH_3)_3]^- > HCr(CO)_4[P(OCH_3)_3]^- > HW(CO)_5^- > CpV(CO)_3H^- > HCr(CO)_5^- > HRu(CO)_4^- > HFe(CO)_3[P(OCH_{33}]^- \gg HFe(CO)_4^-$. This order is also that of decreasing basicity for these anions. Whether the mechanism for reaction 3 is a direct H^- transfer to RX or oxidative addition of RX occurs first, followed by reductive elimination of RH, is unclear.

Although the available data leave some questions, another good reason to believe that a strong correlation exists between nucleophilic reactivity and basicity follows. The protonation reaction 8 for ML_n^-, or ML_n, is also an oxidative addition. For example

$$Pt^0\,[P(CH_3CH_2)_3]_3 + H^+ \rightarrow HPt^{II}\,[P(CH_3CH_2)_3]_3^+ \qquad (9)$$

Therefore, any predictions about the effect of changing the metal, or the ligands, L, are exactly the same for reaction 1 or 2 and reaction 9.

Accordingly, we predict that soft bases, such as CO, $P(OCH_3)_3$, and C_2H_4 will increase acidity, and hard bases, such as NR_3 and R_2O, will decrease acidity. These predictions are borne out very well (*12*). Soft bases will stabilize the lower oxidation state, for example

$$H_2Fe^{II}(CO)_4 \rightarrow HFe^{0}(CO)_4^- + H^+ \qquad (10)$$

Alternatively, π-bonding ligands, such as CO, stabilize the anion by delocalizing the lone electron pair.

A striking example of the role of the ligands is shown by $Rh^{I}(bipy)_2^+$, studied by Sutin and co-workers (*17*). The protonated form, $Rh^{III}(bipy)_2H^{2+}$ has a pK_a of 7.3 in water. In contrast, $Rh^{III}(NH_3)_5H^{2+}$ has a p$K_a > 14.0$ and $Rh(CNR)_4H^{2+}$ has a p$K_a < 0$ (*18*). The isocyanide ligand is similar to carbon monoxide. The high basicity of $Rh(bipy)_2^+$ correlates with the high nucleophilic reactivity of $Rh(CDOBF_2)$.

Relationship between E° *and* H

The arguments given state that easily oxidized bases will be good nucleophiles toward alkyl halides and also strong bases toward the proton. This statement means that $E°$ and H in the Edwards equation are no longer independent parameters but essentially one and the same property for transition metal bases. For bases where the donor atom is a nonmetal, this correlation is normally not the case. The fluoride ion is a stronger base than the iodide ion but is more difficult to oxidize. Still, examination of bases of the representative elements more closely to see if $E°$ and H are truly independent is worthwhile. Actually $E°$, the redox potential in water, is not the best parameter to use. $E°$ is measurable only for a few nucleophiles and is complicated by the nature of the products formed. For example, in

$$2I^- = I_2 + 2e \qquad E° \qquad (11)$$

the iodine–iodine bond strength as a factor is not desirable.

A much better parameter to use is the ionization potential (IP) of the nucleophile (EA is electron affinity).

$$I(g)^- = I(g) + e \qquad EA \qquad (12a)$$

$$NH_3(g) = NH_3^+(g) + e \qquad IP \qquad (12b)$$

Ritchie (*20*) recently emphasized this point and showed how the corresponding values can be obtained in aqueous solution

$$I^-(aq) = I(aq) + e \qquad E^{\circ\prime} \tag{13a}$$

$$NH_3(aq) = NH_3^+\ (aq) + e \qquad E^{\circ\prime} \tag{13b}$$

For the present purposes, a method equivalent to Ritchie's may be used.

For a series of anionic nucleophiles, the energies of the two gas-phase reactions are compared (*PA* is proton affinity):

$$X^-(g) = X(g) + e \qquad EA \tag{14}$$

$$H^+(g) + X^-(g) = HX(g) \qquad PA \tag{15}$$

For a series of neutral nucleophiles, the corresponding reactions are

$$B(g) = B^+(g) + e \qquad IP \tag{16}$$

$$H^+(g) + B(g) = BH^+(g) \qquad PA \tag{17}$$

Any linear relationship between equations 14 and 15, or between equations 16 and 17 will also be found in solution. Solvation energies and entropies will either be constant, as for the H^+ in equation 15, or will cancel as for X^- in equations 14 and 15, or nearly cancel, as for B^+ and BH^+ in equations 16 and 17. The reasoning involved in making these statements, and supporting literature citations, were given by Ritchie (*19*).

Figure 1 shows a plot of the gas-phase electron affinities of a number of anions plotted against the gas-phase proton affinities. The circles refer to anions where the donor atom is a second-row element, C,N, O, or F. The straight line is a least-squares fit to these values only. The slope is -0.87 and the correlation coefficient is -0.923. In spite of the scatter, clearly a strong negative correlation between the electron affinity of a radical and the proton affinity of its anion occurs. A strong base, such as CH_3^-, loses its electron readily, whereas PO_3^- (metaphosphate ion) is a weak base and is difficult to oxidize.

The crosses refer to anions where the donor atom is from the third, fourth, or fifth row. These data points fall regularly below the line: $I^- > Br^-$ Cl^-; $SeH^- > SH^-$; $AsH_2^- > PH_2^-$; and $GeH_3^- > SiH_3^-$. These nucleophiles are all weaker bases than their ease of oxidation would suggest. This behavior is easily traced to low values of the homolytic bond energy of HX.

If the homolytic bond energy (D_0) was constant for all HX

$$HX(g) = H\cdot(g) + X\cdot(g) \qquad D_0 \tag{18}$$

then a perfect linear relationship would occur in Figure 1, and the negative slope would be unity. However, D_0 does not have to be constant to obtain a straight line. For simple HX molecules, D_0 and EA values are as follows:

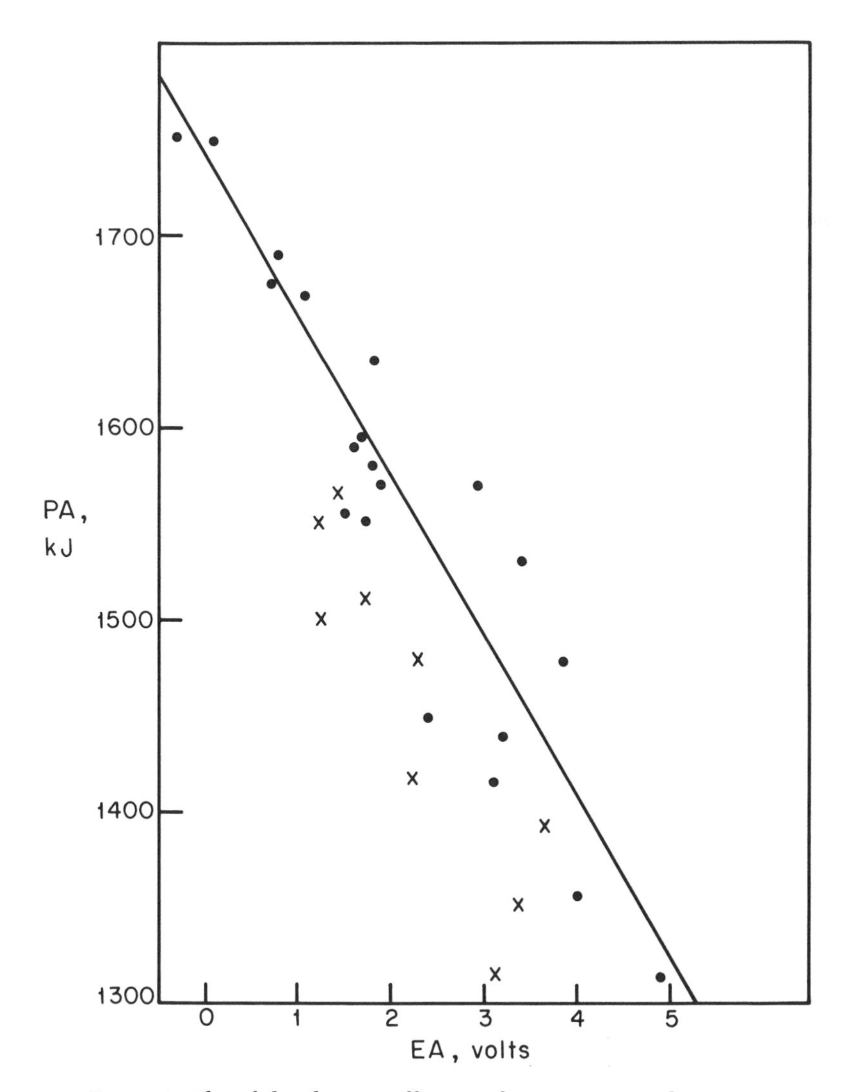

Figure 1. Plot of the electron affinities of anions against their proton affinities. Circles are for C, N, O, and F donor atoms, and crosses are for heavier donor atoms. Data are from reference 32 for proton affinities and references 20, 31, and 32 for electron affinities.

Molecule	D_0 of HX (kcal)	EA of X· (eV)
HF	136	3.80
H_2O	119	1.83
NH_3	107	0.74
CH_4	105	0.08

The resulting opposite trends of D_0 and PA are responsible for the negative slope of Figure 1 being less than unity. Also, the scatter of the anions of the second-row elements is due to variations in D_0 not related to the electron affinity.

Figure 2 is a plot of the ionization potential against proton affinity for a number of neutral molecules, where again the donor atom is C, N, O, or F. The molecules include HF, H_2O, NH_3, N_2, CO, singlet CH_2, and a sampling of ethers, alcohols, amines, esters, amides, nitriles, and other organic molecules. The latter compounds are considered to represent several hundred such molecules for which both IP and PA are known (*20–21*). The straight line drawn has a slope of -0.55, and the correlation coefficient is 0.939.

For these neutral bases where the donor atom is a second-row element, again a strong correlation between the basicity and the ease of losing an electron exists. The larger scatter in Figure 1 is due to random variations in the homolytic bond energy.

$$BH^+(g) = B\cdot(g) + H\cdot(g) \qquad D_0 \tag{19}$$

Such scatter is expected in view of the wide variety of molecules. If a series of closely related molecules, such as substituted anilines, had been chosen, then a good straight line would be obtained with a negative slope close to 1 (*22*).

If molecules in which the donor atom is in the third, fourth or fifth row of the periodic table are added to Figure 2, then similar results to those of Figure 1 are found. The heavier donor atoms lie below the line, corresponding to weak homolytic bond energies. Exceptions occur for alkyl phosphines and phosphites, which lie above the line. These exceptions are related to the fact that the ionization potentials of these molecules are anomalously large (*23*).

As explained earlier, Figures 1 and 2 will not be greatly changed if data in solution are used. Individual bases will move up or down, parallel to the lines shown. Therefore, under the normal conditions for nucleophilic substitution, ease of oxidation ($E^{\circ\prime}$) and Brønsted basicity are not independent parameters for bases where the donor atom is a second-row element. Either parameter may be used as a measure of nucleophilic reactivity.

For heavier donor atoms, the basicity is much lower than the $E^{\circ\prime}$ values would suggest. Nucleophilic reactivity would probably correlate with the $E^{\circ\prime}$ values, but not with pK_a. This statement is based on the assumption that

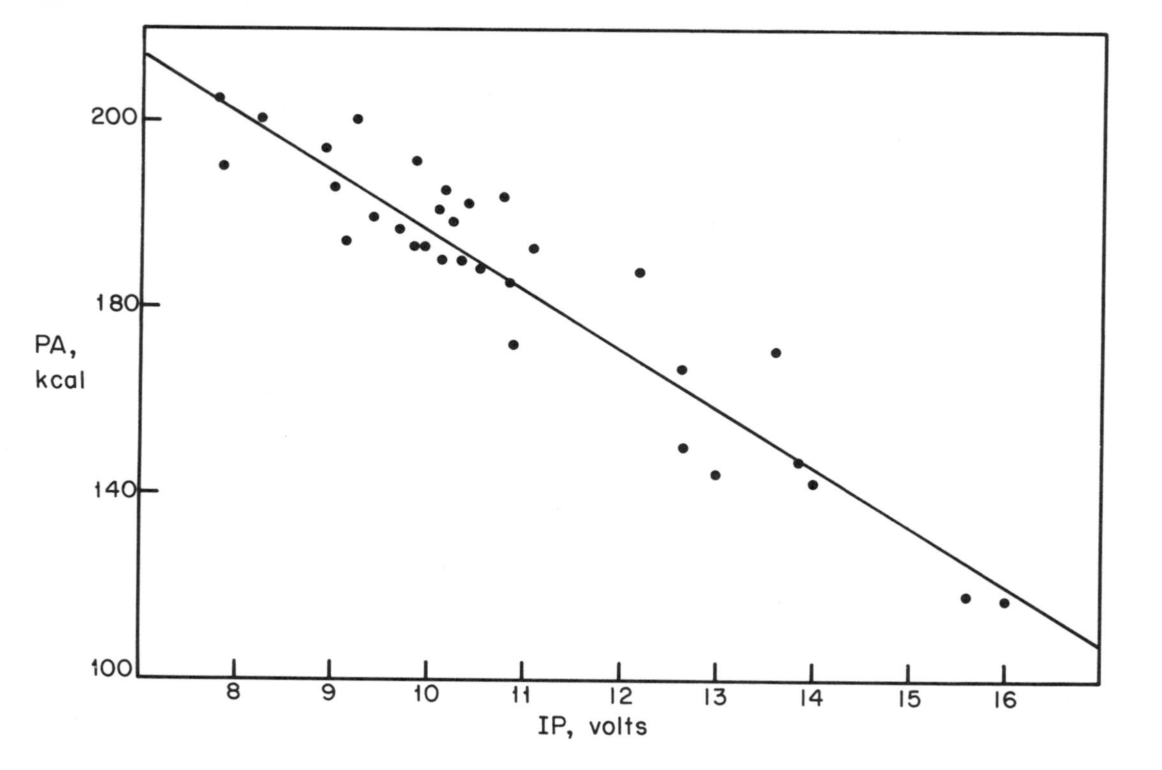

Figure 2. Plot of ionization potentials of neutral molecules against their proton affinities. Molecules are C, N, O, and F donors. Data are from references 20 and 21.

partial electron transfer to the substrate from the nucleophile is substantial in the transition state. This situation would be the case for reactions with CH_3I and most other alkyl halides. Thus, for substrates of this nature, the ease of electron loss, or ease of oxidation, is the important property determining nucleophilic reactivity.

This discussion helps to explain some interesting results recently reported by Bordwell and Hughes (*24*). Large differences in rate constants are found for reactions of carbanions, nitranions, and oxanions with benzyl chlorides. But these differences are largely eliminated if comparisons are made at the same proton basicities. The remaining differences are rather small and are substrate-dependent. The order is $C^- > O^- > N^-$ for a chloride leaving group but changes to $O^- > N^- > C^-$ for a tosylate leaving group (*25*).

That basicity is the dominant factor in determining nucleophilic reactivity in these cases cannot be concluded from these results. Because the nucleophiles are structurally similar, very probably $E^{\circ\prime}$ would vary in the same way as pK_a does for all of them. If the oxygen atom is replaced by sulfur, the resultant ion is 10^3–10^5 times more reactive, even though the sulfanion is much less basic. This result strongly suggests that electron transfer plays the dominant role.

The correlation between proton affinity and electron affinity, as in reactions 14 and 15, has its origin in two factors. Within each ion, a certain distribution of charges exists. The potential of these charges will affect the removal of an electron or the addition of a proton almost equally. This relationship is a classical electrostatic effect. The partial transfer of a pair of bonding electrons from the nucleophile to the proton leads to covalent bonding between the two. Easy electron loss means strong proton binding, if other quantum mechanical requirements are met, such as good orbital overlap.

The failure of the heavier donor atoms to follow the same pattern as the lighter ones must be largely an atom size effect. The diffuse orbital of an iodide ion will not overlap well with that of a proton, at least in comparison to that of fluoride ion. As for nucleophilic reactivity, the relative reactivity of F^- and I^- will be substrate-dependent. If covalent bonding in the transition state is large, and if the accepting orbital of the electrophile is diffuse, then I^- will do very well. Alkyl halides are usually in this category.

For other electrophiles, such as esters, much less covalent bonding in the transition state may occur and ionic bonding may be large. The accepting orbitals may be concentrated in space, similar to that of the proton. Clearly, I^- will not be a good nucleophile, compared to F^- (for an interesting analysis of nucleophile–electrophile interactions that does not use orbital theory, see reference 26).

Returning to transition metal nucleophiles, the experimental evidence already cited indicates that very likely pK_a and $E^{\circ\prime}$ are strongly correlated.

Some additional evidence is available, at least for anionic ML_n^-. The homolytic bond energy, D_0, is known for a number of HML_n molecules (*12*). The D_0 values all fall within a narrow range, between 54 and 73 kcal/mol. Furthermore, the variations that exist are related to pK_a. Molecules that are strong acids have weak M–H bonds (*12*). These two observations mean that a good negative correlation between EA and PA will certainly exist.

For neutral ML_n bases, the situation is not so clear. A number of gas-phase proton affinities have been measured for neutral complexes where ionization potentials are also known (*27*). A larger range of homolytic bond energies is found, between 53 and 87 kcal/mol. A plot of PA versus IP resembles Figure 2, but the correlation coefficient is only $^-$0.694. However, in some cases gas-phase protonation occurs on the ligand and not on the metal. In such cases, the homolytic bond energy is not that of M–H bond.

Compared to the representative elements, the transition metals are remarkable in that little variation in atomic sizes occurs in going from the first to the second and third series. Orbital sizes do not change greatly, and the strength of covalent bonds to ligands remains much more constant (*28*). This statement means that the atom size factor, which separates out the heavier donor atoms from C, N, O, and F, will not be present for the transition metals.

Acknowledgment

Acknowledgment is made to the donors of the Petroleum Research Fund, administered by the American Chemical Society, for support of this work.

Literature Cited

1. Schrauzer, G. L.; Deutsch, E.; Windgassen, R. J. *J. Am. Chem. Soc.* **1968,** *90,* 2441.
2. Collman, J. P.; Finke, R. G.; Cawse, J. N.; Brauman, J. I. *J. Am. Chem. Soc.* **1977,** *99,* 2515.
3. Pearson, R. G.; Figdore, P. E. *J. Am. Chem. Soc.* **1980,** *102,* 1541.
4. Kao, S. C.; Spillett, C.; Ash, C.; Lusk, R.; Park, Y. K.; Darensbourg, M. Y. *Organometallics* **1985,** *4,* 83.
5. Pearson, R. G.; Songstad, J. *J. Org. Chem.* **1967,** *32,* 2899.
6. Pearson, R. G.; Gregory, C. D. *J. Am. Chem. Soc.* **1976,** *98,* 4098.
7. Dessy, R. E.; Pohl, R. L.; King, R. B. *J. Am. Chem. Soc.* **1966,** *88,* 5121.
8. Edwards, J. O. *J. Am. Chem. Soc.* **1954,** *76,* 1540.
9. Lukehart, C. M. *Fundamental Transition Metal Organometallic Chemistry;* Brooks-Cole: Monterey, CA; 1985; p 228.
10. Collman, J. P.; Mac Laury, M. R. *J. Am. Chem. Soc.* **1974,** *96,* 3019.
11. Kubota, M. *Inorg. Chim. Acta* **1973,** *7,* 195.
12. Pearson, R. G. *Chem. Rev.* **1985,** *85,* 41.
13. Norton, J. R.; Jordan, R. F. *J. Am. Chem. Soc.* **1982,** *104,* 1255.
14. Yoshida, T.; Matsuda, T.; Okano, T.; Tetsumi, K; Otsuka, S. *J. Am. Chem. Soc.* **1979,** *101,* 2027.

15. Pearson, R. G.; Jayaraman, R. *Inorg. Chem.* **1974,** *13,* 246.
16. Kochi, J. K. *Organometallic Mechanisms and Catalysis;* Academic: New York, 1978; Chapter 7.
17. Chou, M.; Creutz, C.; Mahajan, D.; Sutin, N.; Zipp, A. P. *Inorg. Chem.* **1982,** *21,* 3989.
18. Sigal, I. S.; Gray, H. B. *J. Am. Chem. Soc.* **1981,** *103,* 2220.
19. Ritchie, C. D. *J. Am. Chem. Soc.* **1983,** *105,* 7313.
20. Rosenstock, H. M.; Draxl, K.; Steiner, B. W.; Herron, J. T. *J. Phys. Chem. Ref. Data* **1977,** *6, Suppl. No. 1.*
21. Lias, S. G.; Liebman, J. F.; Levin, R. D. *J Phys. Chem. Ref. Data* **1984,** *13,* 695.
22. Lias, S. G.; Ausloos, P. *Ion-Molecule Reactions: Their Role in Radiation Chemistry;* American Chemical Society: Washington, DC, 1975; p 91.
23. Elbel, S.; Bergmann, H.; Ensslin, W. *J. Chem. Soc., Faraday Trans. 2* **1974,** 555.
24. Bordwell, F. G.; Hughes, D. L. *J. Am. Chem. Soc.* **1984,** *106,* 3234.
25. Bordwell, F. G.; Hughes, D. L. *J. Org. Chem.* **1982,** 47, 3224.
26. Bader, R. F. W.; MacDougall, P. J.; Lau, C. D. H. *J. Am. Chem. Soc.* **1984,** *106,* 1594.
27. Stevens, A. E.; Beauchamp, J. L. *J. Am. Chem. Soc.* **1981,** *103,* 190.
28. Pearson, R. G. *Inorg. Chem.* **1984,** *23,* 4675.
29. Calderazzo, F. In *Catalytic Transition Metal Hydrides;* Slocum, D. W.; Moser, W. R., Eds.; New York Academy of Sciences: New York, 1983; p 37.
30. Bartmess, John E.; Scott, J. A.; McIver, R. T., Jr. *J. Am. Chem. Soc.* **1979,** *101,* 6046.
31. DePuy, C. H.; Bierbaum, V. M.; Damrauer, R. *J. Am. Chem. Soc.* **1984,** *106,* 4051.
32. Henchman, M.; Viggiano, A. A; Paulson, J. F.; Freedman, A.; Wormhoudt, J. *J. Am. Chem. Soc.* **1985,** *107,* 1453.

RECEIVED for review October 21, 1985. ACCEPTED January 27, 1986.

LINEAR FREE-ENERGY RELATIONSHIPS FOR SOLVENT NUCLEOPHILICITY

17

Electrophilic Interference in Methods for Estimating Nucleophilic Assistance in Solvolyses

J. Milton Harris[1], Samuel P. McManus[1], M. R. Sedaghat-Herati[1], N. Neamati-Mazraeh[1], M. J. Kamlet[2], R. M. Doherty[2], R. W. Taft[3], and M. H. Abraham[4]

[1] Department of Chemistry, University of Alabama, Huntsville, AL 35899
[2] Naval Surface Weapons Center, White Oak Laboratory, Silver Spring, MD 20910
[3] Department of Chemistry, University of California, Irvine, CA 92717
[4] Department of Chemistry, University of Surrey, Guildford, Surrey, GU2 5XH, England

Methods for estimating the extent of nucleophilic solvent assistance (NSA) in solvolyses generally ignore electrophilic solvent assistance (ESA), or they assume that variation in ESA covaries with solvent ionizing power during solvent variation. In the present work, we argue that these assumptions are incorrect and that they lead to overestimation of NSA. The solvatochromic equation is applied to the solvolyses of tert-*butyl chloride, 1-adamantyl chloride, and a mustard derivative to permit estimation of the sensitivity of these substrates to ESA. The results show that ESA can vary dramatically even when the leaving group is unchanged. Finally, we examine a recent failure of the EtOH–TFE (trifluoroethanol) approach for estimating NSA and conclude that failure is due to the unusually high sensitivity of the model compound, 1-adamantyl chloride, to ESA.*

SOLVOLYTIC DISPLACEMENT REACTIONS can be affected by solvents in several ways, including nucleophilic solvent assistance (NSA) and electrophilic solvent assistance (ESA). NSA can be defined as electron donation from solvent to the developing positive dipole of a reacting C–X bond, and ESA can be defined as electron acceptance by the solvent from the leaving group, **I**.

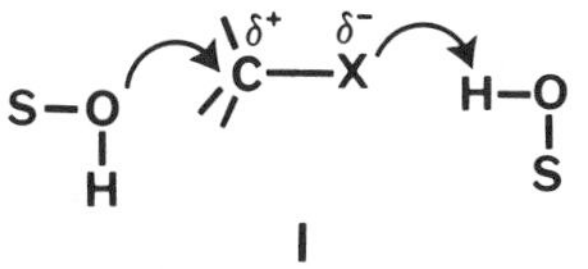

S–O(H) ↷ $\overset{\delta+}{C}$—$\overset{\delta-}{X}$ ↷ H–O(S)

I

0065-2393/87/0215-0247$06.00/0

Several methods are available for estimating the extent of NSA in solvolyses (*1–6*). Generally, either these methods ignore ESA or they assume that variation in ESA is minor during solvent variation or that ESA covaries with solvent ionizing power during solvent variation. The goal of this paper is to demonstrate that these assumptions regarding ESA lead to overestimation of NSA.

To illustrate this point, we will examine the ethanol–trifluoroethanol (EtOH–TFE) method of Raber et al. (*3*). The EtOH–TFE method is based on the observation that plots of log *k* for solvolytic substrates against 1-adamantyl chloride or bromide (1-AdCl or 1-AdBr) in a series of aqueous ethanols and TFEs fall into two general types, one type in which the plot is linear and a second type in which the plot is nonlinear. In the second case, the TFE points are found below the line defined by the aqueous ethanols (Figure 1). According to the Raber et al. interpretation, 1-AdCl solvolysis provides a measure of solvent ionizing power *and solvent electrophilicity*. The linear plot is then interpreted as being the result of the substrate in question reacting without NSA, as does the 1-adamantyl model. In the nonlinear plot, then, the conclusion is that nonlinearity results because the substrate in question has NSA acting as an additional factor that is not modeled by 1-AdCl. The TFE points appear to be "too slow" because NSA is absent in this weakly nucleophilic solvent.

Thus, Figure 1 is considered to be evidence that the mustard compound $CH_3SCH_2CH_2Cl$, **II**, reacts with NSA in solvents such as aqueous ethanol. However, other experimental methods show that **II** does not receive NSA (*6*). For example, the hydrolysis products of deuterium-labeled **II** are completely scrambled as would be expected for reaction through the sulfonium ion. We contend that the EtOH–TFE method fails in this case because it ignores ESA. Specifically, failure results because of large differences in the susceptibility of the substrate (**II**) and 1-AdCl to solvent electrophilicity. In other words, the TFE points fall below the ethanol line because 1-AdCl is receiving additional ESA in the more highly electrophilic TFE and not because NSA is absent in reaction of the mustard. This same assumption of the EtOH–TFE method, that rate of reaction of 1-AdCl (or any single substrate) can provide a combined measure of solvent ionizing power and solvent electrophilicity, is made in several other methods for measuring NSA (*7*). Our position is that this general assumption is incorrect.

Solvatochromic Method

To investigate the possibility of variable ESA, we have used the solvatochromic method of Kamlet and co-workers (*7–10*). According to this approach, a solvent-dependent phenomenon (in this case log solvolysis rate *k*) is a function of four solvent properties: solvent hydrogen-bond donor ability or electrophilicity, α; solvent hydrogen-bond acceptor ability or nu-

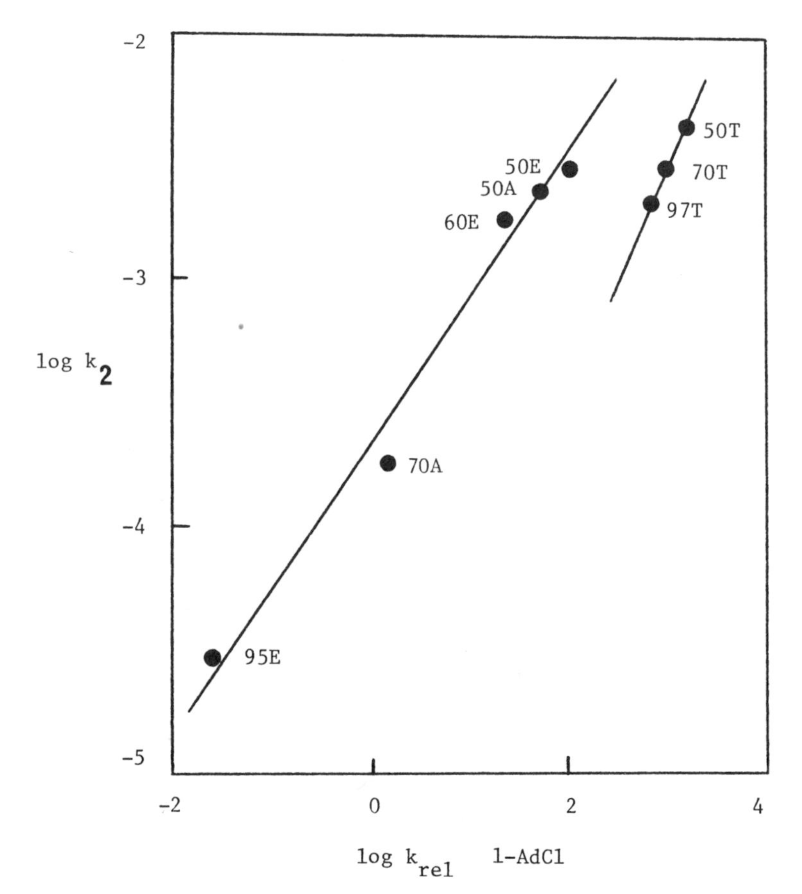

Figure 1. A typical nonlinear EtOH–TFE plot. Here the substrate indicated to react with NSA is $CH_3SCH_2CH_2Cl$, II (6). Rates are in reciprocal seconds and are relative rates for 1-adamantyl chloride. In solvent abbreviations, for example, 97T represents 97% aqueous trifluoroethanol, 95E represents 95% aqueous ethanol, and 70A represents 70% aqueous acetone, and so on.

cleophilicity, β; solvent dipolarity–polarizability, π*; and a measure of the cavity term δ_H^2.

$$\log k = \log k_0 + a\alpha + b\beta + s\pi^* + h\delta_H^2/100 \quad (1)$$

Three of the four solvent parameters (the exception being the cavity term) have been empirically determined from several quite different experiments. These averaged values of the parameters have proven to be useful for predicting a diverse collection of phenomena ranging from solubility to toxicity to chromatographic retention times (*8–10*). The fourth parameter

(δ_H^2) is the Hildebrand solubility parameter, which is calculated from the molar heat of vaporization. The coefficients of the independent variables are understood in the present case to represent the sensitivity of the rate to the associated variables.

We first applied the solvatochromic equation (SCE) to solvolysis of *tert*-butyl chloride (*t*-BuCl) to determine if the method could give a reasonable result for this much-studied reaction (*7*). Abraham et al. (*11*) had previously attempted correlation of these rates with the SCE without the cavity term, but as Bentley and Carter (*12*) have noted, an unsatisfactory result was achieved (*7*). First, TFE and hexafluoroisopropyl alcohol (HFIP) did not fit the correlation. Second, no rate dependence on solvent nucleophilicity β was found, despite other works indicating a weak dependence on this parameter (*12, 13*). Also, different correlations were observed for hydroxylic and nonhydroxylic solvents; Bentley considered this finding to indicate that the dehydrohalogenation transition state (in nonhydroxylic solvents) and the solvolysis transition state (in hydroxylic solvents) were significantly different and thus concluded that the two types of reactions should not be included in the same correlation.

Our decision to attempt correlation of *t*-BuCl rates with the SCE was based on the availability of an expanded data set (21 versus 15 solvents) and a recent appreciation that the cavity term should be included in the SCE (*14*). Application of the four-parameter SCE, equation 1, to correlation of the full 21-solvent data set gives an excellent correlation that removes the objections to the previous study (r = 0.9973 and sd = 0.24) (*7*):

$$\log k = -14.58 + 0.48\delta_H^2/100 + 5.09\,\pi^* + 4.17\alpha + 0.71\beta \quad (2)$$

Points for TFE and HFIP, as well as hydroxylic and nonhydroxylic solvents, fit the correlation nicely. Also, a weak, but statistically significant, dependency on solvent nucleophilicity β exists. Also noteworthy is the large dependence of the reaction rate on solvent electrophilicity. These results indicate that application of the SCE to reaction rates is legitimate.

We are currently applying further kinetic tests to the SCE to determine the range of its applicability to kinetic phenomena. One such test, for example, is to determine the sensitivity of reaction rate of sulfonium salt solvolysis to solvent electrophilicity (i.e. its a value). Because the leaving group is neutral in this case, such a reaction would be expected to have a very weak dependence on electrophilicity.

Variable Electrophilicity and Failure of the EtOH–TFE Method

If we assume that the SCE can be used to correlate kinetic processes, then a means of testing our earlier conclusions regarding failure of the EtOH–TFE method is provided. According to our proposal, a nonlinear EtOH–TFE plot can result from enhanced ESA for 1-AdCl and from NSA for the test substrate. Examining the EtOH–TFE plot for *t*-BuCl provides an illustration of these two factors.

Figure 2 shows the typical nonlinear EtOH–TFE plot that had previously been assumed to be typical for a k_s substrate reacting with NSA. In this case of *t*-BuCl solvolysis, however, we can use equation 1 to calculate the magnitude of NSA for *t*-BuCl to see if this assistance will account for the nonlinearity. An example is for 40% ethanol and 97% TFE. According to the original assumption of Raber et al. (*3*), these two solvents differ only in nucleophilicity because the rate for 1-AdCl is unchanged. Thus, the difference in rates of approximately 1 log unit in these two solvents would, according to this assumption, result from the $b\beta$ term for *t*-BuCl ($b = 0$ for 1-AdCl). We estimate that β for 40% ethanol is very unlikely to be higher than 0.5, and because $b = 0.71$, only 0.4 log unit can be attributed to NSA for *t*-BuCl solvolysis. According to our proposal, the remaining 0.6 log unit results from enhanced sensitivity to electrophilicity (the *a* value) on the part of 1-AdCl.

To look at the nonlinearity as resulting from ESA to 1-AdCl, we can compare the difference in rates (1.4 log units) between 60% ethanol and 97%

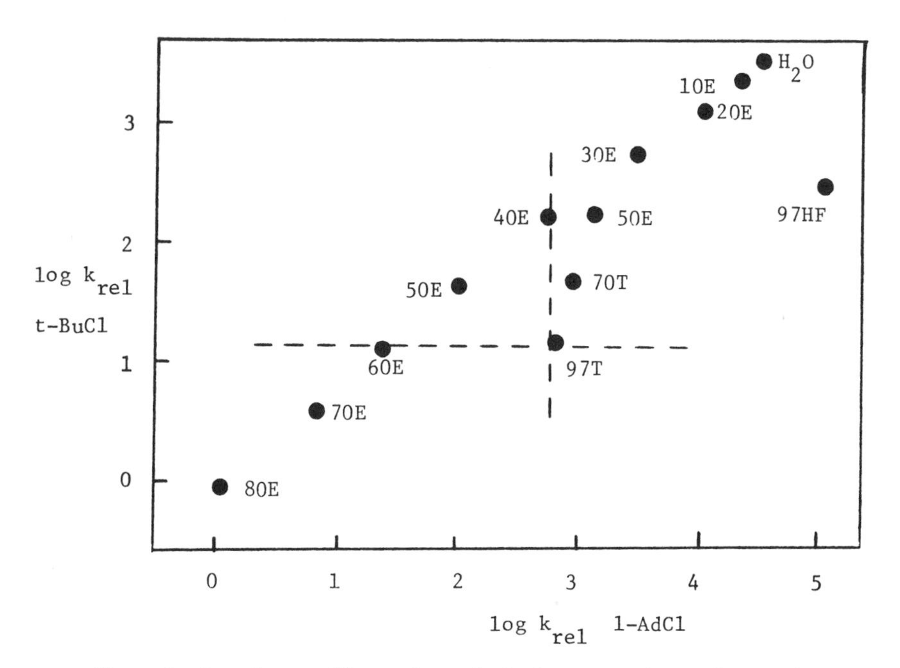

Figure 2. Correlation of logarithms of solvolysis rates for 1-adamantyl chloride versus tert-*butyl chloride at 25 °C (7). E represents ethanol, T represents trifluoroethanol, and HF represents hexafluoro-2-propanol.*

TFE (Figure 2). Assuming that the β value for 60% ethanol is 0.5, we calculate (from $b\beta$) that 0.4 log unit of the 1.4 log units results from NSA to *tert*-butyl. Thus, we see that ESA must account for 1 log unit, and because the difference in α values for 60% EtOH and TFE is 0.5, we calculate

$$(\Delta a)(\Delta\alpha) = \text{ESA} \tag{3}$$

$$(\Delta a)(0.5) = 1.0$$

$$\Delta a = 2$$

Because the a value for t-BuCl equals 4.17, this analysis indicates that the a value for 1-AdCl must be approximately 6.

Accurate determination of the a for 1-AdCl requires solution of the full SCE equation. Unfortunately, this solution is difficult to accomplish because determining rates in weakly nucleophilic, nonhydroxylic solvents is a complex matter. An example is the study of Kevill and Kim (*15*) on the solvolysis of 1-adamantyl arenesulfonates in acetonitrile. These workers found that the reaction proceeded to give an early equilibrium between the reactant and the nitrilium ion formed by acetonitrile acting as nucleophile. So that the equilibrium could be removed and good kinetics obtained, trapping of the cation with azide ion to give a tetrazole product was necessary. Presumably, other solvolyses of 1-Ad derivatives in other weakly nucleophilic solvents would be similarly complex. Having rates in these nonhydroxylic solvents is critical if the complete equation is to be solved because of colinearity between certain of the independent variables in hydroxylic solvents.

To circumvent this problem, and to permit estimation of a values for substrates such as 1-AdCl, we have developed "the method of double differences" or MODD. This approach is based on the observation that the differences in δ_H^2 and π^* values for the two pairs of solvents ethanol–methanol and TFE–HFIP are quite similar and that the differences between the differences for the pairs (i.e., the double differences) are therefore quite small: $\Delta\pi^*_{\text{TFE–HFIP}} = 0.06$, $\Delta\pi^*_{\text{EtOH–MeOH}} = 0.08$, $\Delta\Delta\pi^* = -0.02$, $\Delta\Delta\delta_H^2/100 = -0.05$, $\Delta\Delta\alpha = 0.55$, and $\Delta\Delta\beta = -0.15$. If the dipolarity–polarizability and cavity terms are assumed to be negligible, then the rate difference for these pairs can be reduced to

$$\Delta\Delta \log k = a(0.55) + b(-0.15) \tag{4}$$

Because b is 0 for 1-AdCl, the equation in this case is further simplified:

$$\Delta\Delta \log k = 0.55a \tag{5}$$

Substituting the proper values for the rates gives an a for 1-AdCl of 6.47. Applying the MODD equation, equation 4, to t-BuCl solvolysis gives an a of

4.29, which is close to the actual value of 4.17 obtained from solution of the full SCE.

The a value from the MODD of 6.47 is consistent with the value of 6 calculated previously from the EtOH–TFE plot for *t*-BuCl where the plot nonlinearity was assumed to result from a combination of NSA for *t*-BuCl and ESA for 1-AdCl. A consistent picture emerges from these preliminary results: further support is provided for the conclusion that the SCE is suitable for treating kinetic results, and apparently the EtOH–TFE method for detecting NSA overestimates the importance of nucleophilicity by ignoring variations in substrate susceptibility to ESA.

Electrophilicity and Mustard Solvolysis

To return to the earlier question of failure of the EtOH–TFE method for solvolysis of **II,** we can apply the MODD to determine the a value for reaction of this compound; again, an a value much smaller than the value of 6 for 1-AdCl should be found. Examination of Figure 1 shows that a 2 log unit difference between 60% ethanol and 97% TFE exists. Because $b = 0$ for mustard, all this difference must come from differences in a values. Application of equation 3 gives $(\Delta a)(\Delta \alpha) = \text{ESA} = 2$, $(\Delta a)(0.5) = 2$, and $\Delta a = 4$ or $a_{\text{II}} = 2$. Application of the MODD, equation 5, gives an a value of 1.6 for **II.**

Again, reasonable agreement exists between the a value necessary to account for the nonlinearity of the EtOH–TFE plot and the a value calculated by the approximate MODD approach derived from the SCE.

Acknowledgments

The work at the University of Alabama was supported by The Army Research Office (DAAG29–82–K–0181). R. M. Doherty and M. J. Kamlet received support from the Naval Surface Weapons Center Foundational Research Program.

Literature Cited

1. Bentley, T. W.; Schleyer, P. v. R. *Adv. Phys. Org. Chem.* **1977,** *14,* 1.
2. Harris, J. M. *Prog. Phys. Org. Chem.* **1974,** *11,* 89.
3. Raber, D. J.; Neal, W. C., Jr.; Dukes, M. D.; Harris, J. M.; Mount, D. L. *J. Am. Chem. Soc.* **1978,** *100,* 8137.
4. Harris, J. M.; McManus, S. P. *J. Am. Chem. Soc.* **1974,** *96,* 4693.
4. Streitwieser, A., Jr. *Solvolytic Displacement Reactions;* McGraw-Hill: New York, 1962.
6. McManus, S. P.; Neamati-Mazraeh, N.; Hovanes, B. A.; Harris, J. M. *J. Am. Chem. Soc.* **1985,** *107,* 3393.
7. Harris, J. M.; Taft, R. W.; Abraham, M. H.; Doherty, R. M.; Kamlet, M. J. *J. Chem. Soc., Perkin Trans. 2,* in press.

8. Kamlet, M. J.; Abboud, J. L. M.; Taft, R. W. *Prog. Phys. Org. Chem.* **1981,** *14,* 485.
9. Taft, R. W.; Abraham, M. H.; Doherty, R. M.; Kamlet, M. J. *Nature (London)* **1985,** 313, 384.
10. Taft, R. W.; Abraham, M. H.; Doherty, R. M.; Kamlet, M. J. *J. Am. Chem. Soc.* **1985,** *107,* 3105.
11. Abraham, M. H.; Taft, R. W.; Kamlet, M. J. *J. Org. Chem.* **1981,** *46,* 3053.
12. Bentley, T. W.; Carter, G. E. *J. Org. Chem.* **1983,** *48,* 579.
13. Kevill, D. N.; Kamil, W. A.; Anderson, S. W. *Tetrahedron Lett.* **1982,** *23,* 4635.
14. Kamlet, M. J.; Doherty, R. M.; Abraham, M. H.; Taft, R. W.; Harris, J. M. *J. Chem. Soc., Perkin Trans. 2,* in press.
15. Kevill, D. N.; Kim. C. B. *J. Org. Chem.* **1974,** *39,* 3085.

RECEIVED for review November 19, 1985. ACCEPTED June 30, 1986.

18

Nucleophilicities of Aqueous, Alcoholic, and Acidic Media

T. W. Bentley

Department of Chemistry, University College of Swansea, Singleton Park, Swansea SA2 8PP, Wales

The development of quantitative scales of solvent nucleophilicity based on solvolysis reactions is reviewed. Effects of solvent nucleophilicity are illustrated by product studies, by correlations of kinetic data, and by quantitative estimates of competing nucleophilic pathways, including competing solvent-assisted and anchimerically assisted pathways. The problem of separating quantitatively the nucleophilic and electrophilic solvent contributions to reactivity is discussed. Recent results on the nucleophilicities of aqueous sulfuric acid mixtures are presented.

AQUEOUS MEDIA PROVIDE ECONOMICAL SOLVENTS for a wide range of organic reactions. Organic cosolvents (alcohols, acetone, and dioxane) may be added to aid solubility, and acids or bases may also be present. Reactivity in pure water is of interest in both chemistry and biology. Reactions in pure organic solvents (alcohols and carboxylic acids) are also important, and the reactions discussed here will all involve solvolysis (equation 1) of a substrate (RX) with a large excess of solvent (SOH).

$$\mathrm{RX + SOH \longrightarrow ROS\ (or\ alkene) + HX} \qquad (1)$$

These reactions are also of considerable academic importance because studies (*1*, *2*) of solvolytic reactions have led to important contributions to the mechanistic principles of organic chemistry.

Experimental Section

Rates and products of solvolytic reactions (equation 1) were obtained by standard NMR or high-performance liquid chromatographic (HPLC) methods. Rates of reactions were also monitored from the appearance of acid (HX, equation 1), determined by the change in conductance of the solution or by titrimetric methods. For anions X^- containing a suitable chromophore, reaction rates were monitored by the change in UV spectrum of the solution. The chemicals required for this study were either

0065-2393/87/0215-0255$06.00/0

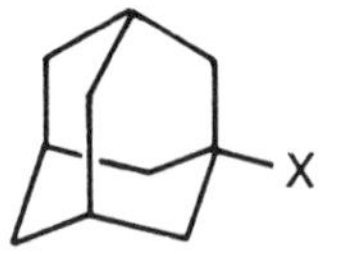

(I) 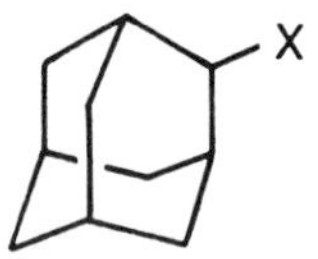 (II)

commercially available or prepared by standard methods. Particular attention was given to solvolyses of 1- and 2-adamantyl substrates (**I** and **II**, respectively).

Results and Discussion

Solvent Dependence of Reactivity. Solvolysis reactions were investigated to obtain structure–reactivity relationships, but these studies were complicated by the solvent dependence of relative rates (Table I). These results show a 10^{10} variation in relative rates of solvolyses of methyl and 2-adamantyl tosylates (2-AdOTs) in trifluoroacetic acid (TFA) compared with those of ethanolysis. Even for two secondary systems, relative rates for 2-AdOTs–$(CH_3)_2CHOTs$ vary from 36 in trifluoroacetic acid to 0.0011 in ethanol (*4*). Hence, separate intrinsic structural effects must be separated from solvent-induced effects.

Table I. Relative Rates of Solvolyses of Alkyl Tosylates

	k_{rel} *for Alkyl Tosylates*		
Solvent	*CH_3OTs*	*$(CH_3)_2CHOTs$*	*2-AdOTs*[a]
Ethanol	1	1	1
Water	21	1.1×10^3	7.1×10^5
Trifluoroacetic acid	2.2×10^{-4}	64	2.1×10^6

NOTE: Kinetic data are from references 3 and 4.
[a] 2-Adamantyl tosylate (**II**, X = OTs).

The initial objective of our work was to quantify solvent effects (particularly solvent nucleophilicity) by adapting the Grunwald–Winstein equation (2) (*5*). In equation 2, k is the rate of solvolysis of a substrate (RX) in any solvent relative to 80% v/v ethanol–water (k_0) and Y is the solvent ionizing power defined by $m = 1.000$ for solvolyses of *tert*-butyl chloride at 25 °C. In this chapter, a discussion of equation 2 and similar free-energy relationships is presented. At the time our work began (1969), in collaboration with Schleyer, mechanisms of solvolytic reactions were close to a "high" in controversy (*6–8*). More recent mechanistic developments (*9–13*) are not reviewed in detail here, but increased recognition of the importance of nucleophilic solvent assistance should be noted.

$$\log (k/k_0)_{RX} = mY \quad (2)$$

Two particularly significant new sources of experimental data have been developed during the past 20 years or so. Solvents of low nucleophilicity were examined, for example, TFA (*14*), trifluoroethanol (TFE) (*15*), and hexafluoroisopropyl alcohol (HFIP) (*16, 17*). Also, the field of gas-phase ion chemistry has expanded rapidly, and data on intrinsic gas-phase reactivity provide important supporting evidence for our interpretations of the kinetic data.

Reactivity in Weakly Nucleophilic Media. In this section, evidence illustrating the unusual reactivity of the fluorinated alcohols TFE and HFIP will be discussed. Various physical properties have been summarized (*18*), including pK_a values in water: for TFE, $pK_a = 12.37$; and for HFIP, $pK_a = 9.30$. HFIP is about 7 pK_a units more acidic than isopropyl alcohol (*18*), presumably because of strong electron withdrawal by the CF_3 groups (*19*). This electronic effect also reduces the nucleophilicity or cation-solvating power of HFIP, and the reduced availability of the oxygen lone pair is reflected in the high first ionization energy of HFIP (12.35 eV) (*10*). The first ionization energy of TFE (11.7 eV) is 1.2 eV higher than that of ethanol (*10*). Probably, for any solvent, SOH, an effect reducing the nucleophilicity of the oxygen lone pair will also increase the tendency to ionize to SO^- and H^+ or at least to form a hydrogen bond to a suitable acceptor. This relationship leads to complications in attempts to separate nucleophilic and electrophilic contributions by solvents.

An analysis of the transport properties of alkali metal halides in TFE led to association constants, which were interpreted in terms of greater anion solvation by TFE than by ethanol and also less cation solvation by TFE (*19*). Despite evidence that chloride ion is strongly solvated by TFE, HCl is about 15 times more soluble in ethanol than in TFE; this result emphasizes the weak solvation of protons by TFE (*19*). Continuing this trend, alkali metal salts are very sparingly soluble in HFIP (*18*).

For organic reactions in which positive charge develops on carbon, structural effects on reactivity are enhanced in weakly nucleophilic solvents (*14*). The effect of alkyl substitution on S_N reactions can be quantified by the substituent parameter ρ^*, which is much more negative in weakly nucleophilic media [e.g., TFA (*14, 20*), HFIP (*4*), or TFE (*4*)] than in "normal" alcohol solvents. Apparently in weakly cation solvating media, the electron demand for stabilization of positive charge is met from within the molecule [e.g., by enhanced electron donation from adjacent alkyl groups (*4*)].

In carbocation chemistry, attack by solvent on a reactive intermediate (R^+ or R^+X^-) will lead to alcohols, ethers, or esters (equation 3, carbon attack) and to alkenes (hydrogen attack). If two solvents are in competition, attack by the more nucleophilic solvent might be expected to predominate.

However, some of these trapping reactions are very unselective, even for competition between azide ion and water (*21*). Apparently, solvent electrophilicity and bulkiness are significant at least when the selectivity is relatively low (*22–24*). Substantial (>10-fold) differences in selectivity favoring water or ethanol over TFE are observed for a wide variety of solvolyses: for example, for *p, p'*-dichlorobenzhydryl chloride (*25*), benzoyl chloride (*26*), 1-anisyl-2-methylpropen-1-yl tosylate (*23*), and some D-glucopyranosyl systems (*27*). To know the ethanol–TFE selectivity for strongly solvent assisted S_N2 processes (e.g., for solvolyses of methyl substrates) would be interesting, but these data are currently unavailable.

$$R^+X^- + SOH \longrightarrow ROS + HX \tag{3}$$

In some cases, nucleophilic attack by solvent may be an undesired reaction pathway, but a protic solvent of high ionizing power may be required to provide adequate heterolytic reactivity or electrophilic catalysis. Examples include many cyclizations and reactions involving neighboring group participation (k_Δ processes) (*28, 29*). Suitable solvents include TFE (*30–37*), HFIP (*35–38*), aqueous sulfuric acid (*39*), and carboxylic acids (*28, 29*).

Suppression of competing nucleophilic pathways, by changing from normal solvents to fluorinated alcohols, provided evidence for homoallylic participation (*40*). For solvolyses of cyclohexen-4-yl tosylate (equation 4), the substitution product (**IV**) is important, but the bicyclic product (**V**) is formed in significant quantities in HFIP (Table II). The stereochemistry of the substitution product (**IV**; Table II) supports the interpretation that a displacement of solvent with inversion of stereochemistry (k_s process) occurs in nucleophilic media, changing to a homoallylic (k_Δ) process with retention of stereochemistry in HFIP. In acetic acid, the k_Δ process is just beginning, and in formic acid, the k_s and k_Δ processes occur about equally.

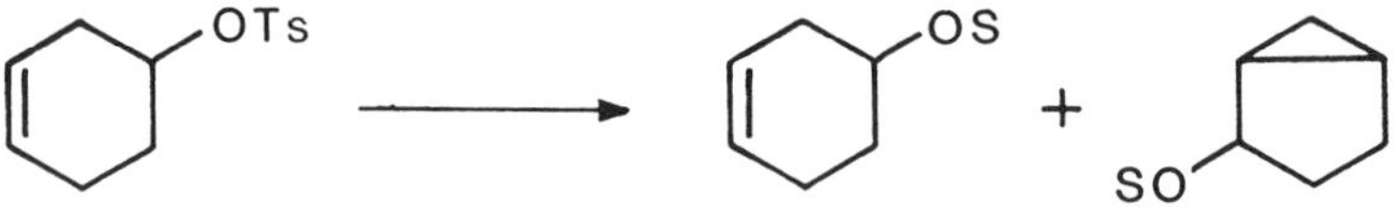

OTs (III) ⟶ OS (IV) + SO (V) (4)

Product data cited previously (*26, 35–37, 41*) support independent evidence from kinetic studies (Table III) that HFIP is even less nucleophilic than TFE. Solubilities of alkali metal salts show the same trend in cation solvating power (*18, 19*). These diverse results warrant emphasis because Abraham et al. have implied (*42*) that the solvatochromic parameter β is a measure of solvent nucleophilicity. However, the β values for TFE and HFIP are both zero (*43*), and the relationship between β and solvent nucleophilicity is therefore questionable.

Table II. Yield (Percent) and Stereochemistry of Products of Solvolyses of Cyclohexen-4-yl Tosylate (III) in Solvents of Various Nucleophilicity and Ionizing Power

Solvent	*Cyclization, V*	*Substitution, IV*	*(% Retention)*
70% Dioxane–water	—[a]	—[a]	(0)
Acetic acid[b]	<3	20	(17)
Formic acid	<3	58	(40)
HFIP	10	65	(100)

NOTE: Data are from reference 40.
[a] Quantitative data were not obtained.
[b] The major products are alkenes.

Table III. Comparison of Scales of Solvent Nucleophilicity

Solvent	N_{OTs}[a]	$\log (k/k_0)_{Et_3O^+PF_6^-}$[b]	N_{KL}[b]
CH_3CH_2OH	0.00	0.55	0.46
80% CH_3CH_2OH	0.00[c]	0.00[c]	0.00
CH_3OH	−0.04	0.64	0.58
H_2O	−0.44[d]	−1.01	−0.87
CH_3CO_2H	−2.35	−1.42	−1.34
HCO_2H	−2.35	−1.71	−1.61
97% TFE	−2.79	−2.28	−2.22
97% HFIP	−4.27	—[e]	—[e]
TFA	−5.56	—[e]	—[e]

[a] From reference 3.
[b] From reference 60.
[c] By definition.
[d] From reference 61.
[e] Not determined.

Quantitative Scales of Solvent Nucleophilicity. Solvolytic studies in solvents of low nucleophilicity led to renewed interest in quantitative measures of solvent nucleophilicity. Peterson and Waller (*44*) derived a scale of solvent nucleophilicity (N_{PW}) from the rates of displacement by solvent of tetramethylenehalonium ions (**VI**) in liquid sulfur dioxide. The reaction is approximately half-order in carboxylic acid, possibly because dimer-monomer preequilibrium occurs (*44*). More recently, hydrolysis of the iodonium salt (**VIII**) in competition with anionic or solvent nucleophiles was studied. A scale of nucleophilicity relative to water was obtained by quan-

(VI) + RCO_2H ⟶ (VII)

$$\text{(VIII)} \xrightarrow{H_2O + SOH} \text{(IX)} + \text{(X)}$$

titative analysis of the products (**IX** and **X**) (*45*). These values varied 1 order of magnitude in the order $CH_3CH_2OH > HCO_2H > (CH_3)_2CHOH > H_2O > CH_3CO_2H > (CH_3)_3COH$ and are not directly comparable with those defined for pure solvents.

Our approach was developed from a plot of logarithms of rate constants for methyl tosylate versus Y values. This method can be illustrated by the plot (Figure 1) versus Y_{OTs}, a scale of solvent ionizing power based on solvolyses of 2-adamantyl tosylate (**II**, X = OTs) (*3*). The normal alcohols, water, and alcohol–water mixtures form a line of similar slope to a line joining the data points for acetic and formic acids (Figure 1). The slope of these lines is approximately 0.3, which was assumed (*3, 46*) to be the sensitivity of solvolyses of methyl tosylate to solvent ionizing power (m value, equation 1).

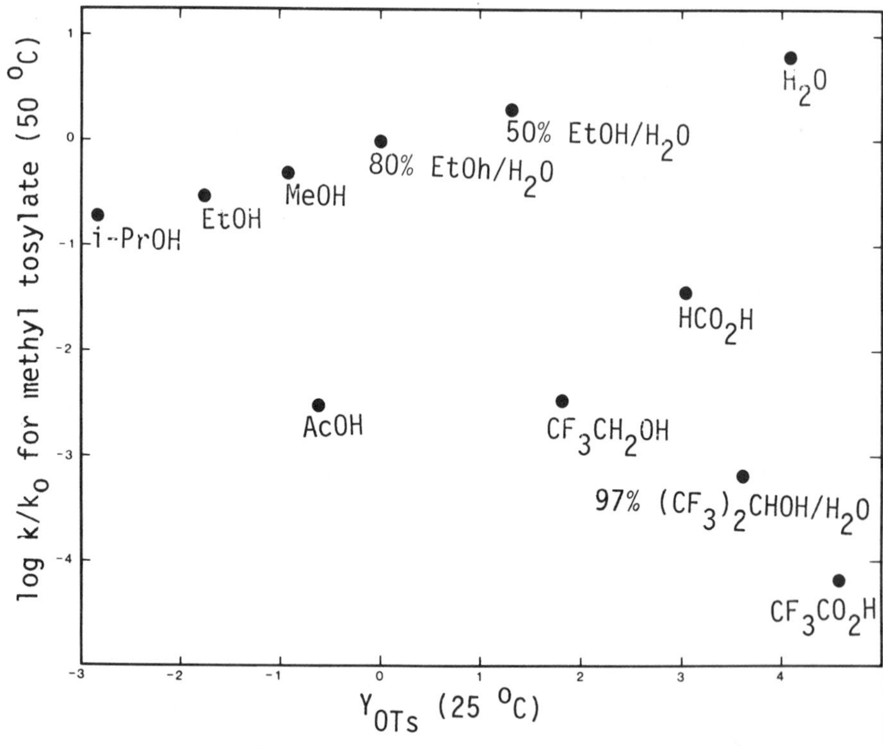

Figure 1. Correlation of logarithms of rate constants for solvolyses of methyl tosylate versus Y_{OTs} (data from reference 3).

This conclusion was supported by Peterson and Waller's data (*44*) showing that acetic and formic acids reacted at about the same rate with the chloronium ion (**VI**). Solvent nucleophilicity (N) could then be obtained by rearrangement of the extended Grunwald–Winstein equation (*47*):

$$\log (k/k_0)_{RX} = lN + mY \tag{5}$$

Defining $l = 1$ for the sensitivity of solvolyses of methyl tosylate to solvent nucleophilicity gave

$$N = \log (k/k_0)_{CH_3OTs} - 0.3Y \tag{6}$$

The modern definition of N values is based on Y_{OTs} values (3) and is referred to as N_{OTs}:

$$N_{OTs} = \log (k/k_0)_{CH_3OTs} - 0.3Y_{OTs} \tag{7}$$

The evolutionary process to equation 7 was not so straightforward as implied previously. When the work on N values began (1970), we were still using the original Grunwald–Winstein Y values (5), because evidence justifying their continued use for *most* solvents had just appeared (*48*). The N values we reported in 1972 were referred to as N_{BS} and were calculated from equation 6 with original Y values except for CF_3CO_2H (*46*). In the 1976 paper (3), N_{OTs} values (equation 7) were reported along with an additional set of N values calculated by using Y values (equation 6). We now consider that Y values are unsuitable because solvolyses of *tert*-butyl chloride are weakly (but significantly) sensitive to solvent nucleophilicity (*49–52*). Consequently, only N_{OTs} values are now recommended, and only these values have been used in our subsequent correlations (3, *49*, *73*). Another advantage of Y_{OTs} is that solvent effects attributed to the tosylate leaving group are incorporated into the Y_{OTs} values.

Scales of solvent ionizing power for other leaving groups are currently being developed (*49*, *53–58*), but whether different N scales are required for different leaving groups is not yet clear. A shortage of experimental data for solvolyses of methyl halides in a wide range of solvents exists—these data are difficult to obtain because of the low reactivity of methyl substrates. Solvolyses of acid chlorides (e.g., benzoyl chloride and *p*-nitrobenzoyl chloride) show greater reactivity and greater sensitivity to solvent nucleophilicity (*59*), but acid chlorides are less satisfactory model compounds because they appear to react by at least two competing mechanisms (*26*).

An alternative set of N values was reported by Kevill and Lin (N_{KL} values) (*60*). As with N_{OTs}, the lN/mY equation (5) is employed, but the model substrates required to define the parameters are positively charged. Solvolyses of triethyloxonium hexafluorophosphate and *tert*-butylsulfonium ion define N_{KL} values:

$$\log (k/k_0)_{(CH_3)CS^+(CH_3)_2} = Y^+ \qquad (8)$$

$$N_{KL} = \log (k/k_0)_{(CH_3CH_2)_3O^+PF_6^-} - 0.55Y^+ \qquad (9)$$

For these solvolyses, the Y^+ scale (equation 8) is compressed relative to the Y scale (equation 10) (*59*), so the mY correction term (e.g., in equations 6 and 7) is greatly reduced.

$$Y^+ = -0.09Y \qquad (10)$$

The m value of 0.55 (equation 9) was obtained from preliminary N_{KL} data and equation 11. Hence, the N_{KL} scale is based on $l = 1$ for solvolyses of ethyl substrates, whereas the N_{OTs} scale is based on $l = 1$ for solvolyses of methyl substrates. For this reason alone, the N_{KL} scale would show a lower range of values than N_{OTs} (*see* Table III).

$$\log (k/k_0)_{CH_3CH_2OTs} = N_{KL} + mY \qquad (11)$$

Comparisons between N_{KL} and N_{OTs} are further complicated by the choice of m values and by leaving group effects. Kevill and Rissmann (*62*) suggested that the value of $m = 0.3$ (equations 6 and 7) is too low and that differences between N_{KL} and N_{OTs} values could be reconciled if a higher m value (e.g., 0.55) were used. In contrast, the m value of 0.55 quoted in equation 9 is not critical because of the low range of Y^+ values. Also, the Y^+ scale partly reflects nucleophilic solvent participation (*51*), and removal of the $0.55Y^+$ term from equation 9 may be preferable (*62*). The similarity of values of N_{KL} and $\log (k/k_0)_{(CH_3CH_2)_3O^+PF_6^-}$ can be seen from Table III.

Some surprising consequences and limitations of this evidence (*60, 62*) should be considered. Product data (*22–27, 30–37*) do not support the implications from N_{KL} values (Table III) that the difference in nucleophilicity between CH_3CH_2OH and H_2O is the same as that between H_2O and 97% TFE, although nucleophilicities of binary aqueous mixtures are only approximate guides to the nucleophilicities of the pure solvents. The revised m value, tentatively suggested for methyl [0.5 (*60*) or 0.55 (*62*)], is surprisingly close to the unequivocal m value for solvolyses of 2-*endo*-norbornyl tosylate [**XI**, $m = 0.69 \pm 0.02$ (*4*)]. Replacement of $m = 0.30$ by $m = 0.55$ (equation

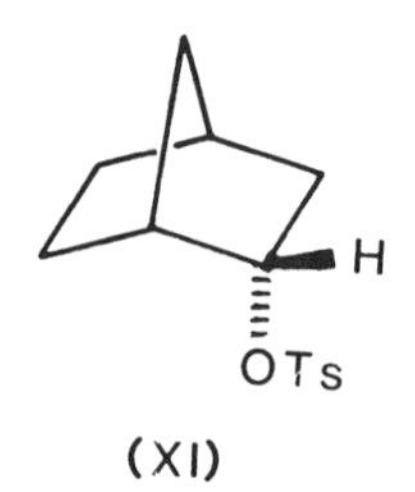

(XI)

6) substantially reduces the N values for 97% HFIP and TFA (revised N_{OTs} values would then be -5.2 and -6.7, respectively). These values could be compared with those derived from data for positively charged leaving groups (*63*).

Separation of Nucleophilic and Electrophilic Contributions to Solvation Effects. The discussion on the appropriate choice of m value for the N_{OTs} scale (equation 7) is part of a more general problem of dissecting the nucleophilic contributions (corresponding to the lN term in equation 5) and electrophilic contributions (included in the mY term in equation 5) to solvent effects. When a substrate reacts by a nucleophilically solvent assisted pathway, m decreases and l increases, often in a uniform manner (equation 12). Schadt et al. proposed (*3*) that an increase in nucleophilic assistance (increase in l) caused delocalization of positive charge, which led to a decrease in m. All of the deviations from the rates expected for S_N1 (k_c) reactivity were attributed to nucleophilic solvent assistance (*3*); Schadt et al. (*3*) assumed that m values for k_c processes would be the same as for 2-adamantyl (**II**), and more recent data for 1-adamantyl (**I**), 1-adamantylmethylcarbinyl, and 1-bicyclo[2.2.2]octyl tosylates support this assumption (*4, 53*).

$$l = (1 - m)/0.7 \tag{12}$$

A more complex alternative explanation can be illustrated by assuming that Y values should be dissected into two terms, π^* for dipolarity-polarizability and α for hydrogen-bond donation. Therefore, S_N1 substrates with the same leaving group may differ in sensitivity to α (i.e., to electrophilic assistance) (*64*). We now have sufficient data to test this proposal. The fluorinated alcohols (TFE and HFIP) have high α values, but these alcohols do not show anomalous behavior in plots of Y_{Cl}, Y_{Br}, Y_I, Y_{OClO_3}, Y_{OTs}, and $Y_{picrate}$ (all based on solvolyses of either **I** or **II**) (*54, 58*). Probably, differences in electrophilic assistance for adamantyl substrates over this wide range of leaving groups would be greater than differences in electrophilic assistance for two substrates having the *same* leaving group. Therefore, apparently electrophilic assistance *and* general electrostatic effects can be satisfactorily included in the mY term of equation 5. This conclusion is particularly important for the extension of our solvolytic studies to aqueous acidic media.

To avoid oversimplification, we note that changes in m values can occur independently of changes in l values when charge delocalization occurs either within the carbocation [e.g., solvolyses of benzyl substrates (*3, 65, 66*)] or within the anionic leaving group [e.g., solvolyses of picrates (*58*)]. (It has recently been shown (*74*) that solvolyses of benzyl tosylate do fit equation 12 satisfactorily.) Such effects appear to be general electrostatic as opposed to specific electrophilic or nucleophilic. These effects can be observed in a range of methanol–water mixtures for which changes in electrophilic assistance are less likely than for TFE and HFIP [the α values (*43*) are for CH_3OH 0.93, for H_2O 1.17, for TFE 1.51, and for HFIP 1.96].

Steric Effects on Solvent Nucleophilicity. The N_{OTs} values (Table III) show that TFA is, by this measure, less nucleophilic than HFIP. As the rate ratios 2-AdOTs:$(CH_3)_2CHOTs$ in TFA and 97% HFIP were very similar, we initially proposed (*17*) that solvolyses of $(CH_3)_2CHOTs$ in these two weakly nucleophilic solvents were k_c processes. More detailed studies (*4*, *50*) later showed that solvolyses of secondary tosylates in 100% HFIP were closer to limiting (k_c) than those in TFA. This apparent reverse of the order of nucleophilicity could be caused by steric effects (*50*), which could also contribute to the lower precision of the lN/mY equation (5) in correlations of rate data for secondary substrates; for example, for 2-propyl tosylate, solvolyses in TFE and HFIP are predicted to be slower than observed, whereas reaction in CF_3CO_2H is predicted to be faster than observed (*3*).

Although steric effects on nucleophilicity are difficult to quantify, the following two examples further illustrate results that appear to be caused primarily by steric effects on nucleophilicity. Rearrangement of ^{15}N-labeled benzenediazonium tetrafluoroborate (equation 13) occurs in competition with dediazoniation. The overall rate of dediazoniation shows a low sensitivity to changes in solvent, even from water and fluorinated alcohols to FSO_3H (*67*). As might be expected, dediazoniation is slightly faster in TFE and HFIP than in FSO_3H (*68*). Remarkably, the extent of rearrangement after 70% dediazoniation follows the order HFIP > TFE > FSO_3H > H_2O, and probably steric effects contribute significantly to this order (*68*). Steric factors also might explain the preference for equatorial attack in the order HFIP > TFE > HCO_2H, following 1,3-hydride shifts in substituted cyclohexyl cations (*69*). Steric effects were emphasized by McManus (*70*), who compared ionization potentials (*IP*) and *N* values with proton affinites (*PA*) in the series H_2O, CH_3OH, CH_3CH_2OH, $(CH_3)_2CHOH$, and $(CH_3)_3COH$. A plot of *IP* versus *PA* was linear, but a similar plot of *N* values showed deviations (rate retardation) for the more sterically hindered alcohols [$(CH_3)_2CHOH$ and $(CH_3)_3COH$].

$$C_6H_5{-}N^+{\equiv}{}^{15}N \longrightarrow C_6H_5{-}{}^{15}N^+{\equiv}N \qquad (13)$$

Solvent Effects on Competing Nucleophilically Solvent Assisted (k_s) and Anchimerically Assisted (k_Δ) Processes. Schleyer et al.'s work (*8*) on solvolyses of 2-adamantyl (**II**) arose from earlier studies (*71*) of the competition between k_s and k_Δ processes in β-arylalkyl systems (**XII** and **XIII**). Apparently, no crossover occurred between the two processes, because rate–product correlations were observed. Hence, any cationic intermediates in the k_s process must be sufficiently strongly solvated to prevent attack by the β-aryl group. This work led subsequently to a substantial upward revision in estimates of the extent of nucleophilic assistance in acetolysis and formolysis, two of the classical solvolytic processes.

$ArCH_2CH_2X$ **XII** $ArCH_2CHXCH_3$ **XIII**

Competition between k_s and k_Δ processes is strongly dependent on solvent nucleophilicity. Comparing solvolyses of **XII** with the nonparticipating model ethyl tosylate shows a 7500-fold change in relative rates from ethanol to TFA; the corresponding change for **XIII** compared with 2-propyl tosylate is 100-fold. The k_s and k_Δ processes were first separated by plotting the observed rates k_t versus Hammett σ values for a wide range of aryl substituents. Deactivated systems reacted exclusively by the k_s pathway. Although activated systems reacted extensively by the k_Δ pathway, the contribution of the k_s pathway could be obtained by an iterative procedure based on a linear extrapolation of the data for deactivated systems (72).

After the k_s and k_Δ contributions in each solvent were separated, the kinetic data for **XII** could then be correlated to give equations 14 and 15.

$$\log (k_s/k_s°) = 0.33Y + 0.78N \quad (14)$$

$$\log (k_\Delta/k_\Delta°) = 0.67Y \quad (15)$$

Inclusion of an lN term in equation 15 gave a small negative l value (−0.05) with only a slight improvement in the correlation. Similarly, for the secondary system (**XIII**), equations 16 and 17 were obtained.

$$\log (k_s/k_s°) = 0.50Y + 0.46N \quad (16)$$

$$\log (k_\Delta/k_\Delta°) = 0.82Y \quad (17)$$

Surprisingly, inclusion of an lN term in equation 17 gave a sizable negative l value (−0.23), although again no great improvement in the correlation occurred. As these results are based on data for only five solvents and considering the need to dissect the k_s and k_Δ contributions, the negative l value may not be significant. The equations for the k_s processes (14 and 16) give m and l values similar to those for solvolyses of ethyl tosylate and 2-propyl tosylate in the same solvents (72). These equations (14–17) imply that both increases in Y and decreases in N will favor k_Δ more than k_s processes, a situation attainable with aqueous sulfuric acid (39). Other dissections of k_s more than k_Δ processes have also been reported (32).

Solvolyses in Aqueous Sulfuric Acid. Kinetic data (73) for solvolyses of CH_3OTs and **XI** and derived values of N_{OTs} and Y_{OTs} are shown in Table IV.

Table IV. Kinetic Data for Aqueous Sulfuric Acid

	Rate Constants, 10^4k(s^{-1})			
% H_2SO_4[a]	CH_3OTs[b]	XI[c]	N_{OTs}[d]	Y_{OTs}[e]
0	1.39	1.19	−0.44	4.10
30	0.64	2.65	−0.87	4.43
50	0.18	6.4	−1.57	4.90
60	0.08	15	−2.02	5.29
70	0.05	62	−2.5	6.1

[a] Percent w/w H_2SO_4–H_2O.
[b] At 50 °C.
[c] Mesylate at 25 °C.
[d] Calculated from equation 7.
[e] Calculated by assuming log (k/k_0) for **XI** $= 0.72Y_{OTs} + 0.05N_{OTs}$.

These results show substantial increases in Y_{OTs} and decreases in N_{OTs} as the concentration of sulfuric acid increases. Quantitive dissections of k_Δ and k_s have been attempted (*73*).

Acknowledgments

This work was initiated in collaboration with P. v. R. Schleyer, whose continued interest is appreciated. Exchanges of information with D. N. Kevill and financial support from the Science and Engineering Research Council are gratefully acknowledged.

Literature Cited

1. Streitwieser, A., Jr. *Chem. Rev.* **1956,** *56,* 571.
2. Lowry, T. H.; Richardson, K. S. *Mechanism and Theory in Organic Chemistry,* 2nd ed.; Harper and Row: New York, 1981; Chapter 4.
3. Schadt, F. L.; Bentley, T. W.; Schleyer, P. v R. *J. Am. Chem. Soc.* **1976,** *98,* 7667.
4. Bentley, T. W.; Bowen, C. T.; Morten, D. H.; Schleyer, P. v. R. *J. Am. Chem. Soc.* **1981,** *103,* 5466.
5. Grunwald, E.; Winstein, S. *J. Am. Chem. Soc.* **1948,** *70,* 846.
6. Shiner, V. J., Jr.; Fisher, R. D.; Dowd, W. *J. Am. Chem. Soc.* **1969,** *91,* 7748.
7. Sneen, R. A.; Larsen, J. W. *J. Am. Chem. Soc.* **1969,** *91,* 6031.
8. Schleyer, P. v. R.; Fry, J. L.; Lam, L. K. M.; Lancelot, C. J. *J. Am. Chem. Soc.* **1970,** *92,* 2542.
9. Harris, J. M. *Prog. Phys. Org. Chem.* **1974,** *11,* 89.
10. Bentley, T. W.; Schleyer, P. v. R. *Adv. Phys. Org. Chem.* **1977,** *14,* 50.
11. Bentley, T. W.; Schleyer, P. v. R. *Adv. Phys. Org. Chem.* **1977,** *14,* 51–58.
12. Bentley, T. W.; Schleyer, P. v. R. *Adv. Phys. Org. Chem.* **1977,** *14,* 40.
13. Jencks, W. P. *Acc. Chem. Res.* **1980,** *13,* 161.
14. Peterson, P. E.; Kelley, R. E., Jr.; Belloli, R.; Sipp, K. A. *J. Am. Chem. Soc.* **1965,** *87,* 5169.
15. Scott, F. L. *Chem. Ind. (London)* **1959,** 224.

16. Sunko, D. E.; Szele, I. *Tetrahedron Lett.* **1972**, 3617.
17. Schadt, F. L.; Schleyer, P. v. R.; Bentley, T. W. *Tetrahedron Lett.* **1974**, 2335.
18. Matesich, M. A.; Knoefel, J.; Feldman, H.; Evans, D. F. *J. Phys. Chem.* **1973**, *77*, 366.
19. Evans, D. F.; Nadas, J. A.; Matesich, M. A. *J. Phys. Chem.* **1971**, *75*, 1708.
20. Katritzky, A. R.; Marquet, J.; Lopez-Rodriquez, M. L. *J. Chem. Soc., Perkin Trans. 2* **1983**, 1443.
21. Ta-Shma, R.; Rappoport, Z. *J. Am. Chem. Soc.* **1983**, *105*, 6082.
22. Ando, T.; Tsukamoto, S. *Tetrahedron Lett.* **1977**, 2775.
23. Kaspi, J.; Rappoport, Z. *J. Am. Chem. Soc.* **1980**, *102*, 3829.
24. McManus, S. P.; Zutaut, S. E. *Tetrahedron Lett.* **1984**, 2859.
25. Rappoport, Z.; Ben-Yacov, H.; Kaspi, J. *J. Org. Chem.* **1978**, *43*, 3678.
26. Bentley, T. W.; Carter, G. E.; Harris, H. C. *J. Chem. Soc., Perkin Trans. 2* **1985**, 983.
27. Sinnott, M. L.; Jencks, W. P. *J. Am. Chem. Soc.* **1980**, *102*, 2026.
28. Johnson, W. S. *Acc. Chem. Res.* **1968**, *1*, 1.
29. Van Tamelen, E. E. *Acc. Chem. Res.* **1968**, *1*, 111.
30. Trahanovsky, W. S.; Doyle, M. P. *Tetrahedron Lett.* **1968**, 2155.
31. Raber, D. J.; Dukes, M. D.; Gregory, J. *Tetrahedron Lett.* **1974**, 667.
32. Hanack, M.; Collins, C. J.; Stutz, H.; Benjamin, B. M. *J. Am. Chem. Soc.* **1981**, *103*, 2356.
33. Collins, C. J.; Hanack, M.; Stutz, H.; Auchter, G.; Schoberth, W. *J. Org. Chem.* **1983**, *48*, 5260.
34. Lambert, J. B.; Finzel, R. B. *J. Am. Chem. Soc.* **1982**, *104*, 2020.
35. Ferber, P. H.; Gream, G. E. *Aust. J. Chem.* **1981**, *34*, 2217.
36. Ferber, P. H.; Gream, G. E. *Aust. J. Chem.* **1981**, *34*, 1051.
37. Ferber, P. H.; Gream, G. E.; Wagner, R. D. *Aust. J. Chem.* **1980**, *33*, 1569.
38. Allard, B.; Casadevall, A.; Casadevall, E.; Largeau, C. *Nouv. J. Chim.* **1980**, *4*, 539.
39. Peterson, P. E.; Vidrine, D. W. *J. Org. Chem.* **1979**, *44*, 891.
40. Lambert, J. B.; Featherman, S. I. *J. Am. Chem. Soc.* **1977**, *99*, 1542.
41. Bruzik, K.; Stec, W. J. *J. Org. Chem.* **1981**, *46*, 1618.
42. Abraham, M. H.; Taft, R. W.; Kamlet, M. J. *J. Org. Chem.* **1981**, *46*, 3053.
43. Kamlet, M. J.; Abboud, J.-L. M. Abraham, M. H.; Taft, R. W. *J. Org. Chem.* **1983**, *48*, 2877.
44. Peterson, P. E.; Waller, F.J. *J. Am. Chem. Soc.* **1972**, *94*, 991.
45. Peterson, P. E.; Vidrine, D. W.; Waller, F. J. Henrichs, P. M.; Magaha, S.; Stevens, B. *J. Am. Chem. Soc.* **1977**, *99*, 7968.
46. Bentley, T. W.; Schadt, F. L.; Schleyer, P. v. R. *J. Am. Chem. Soc.* **1972**, *94*, 992.
47. Winstein, S.; Fainberg, A. H.; Grunwald, E. *J. Am. Chem. Soc.* **1957**, *79*, 4146.
48. Raber, D. J.; Bingham, R. C.; Harris, J. M.; Fry, J. L.; Schleyer, P. v. R. *J. Am. Chem. Soc.* **1970**, *92*, 5977.
49. Bentley, T. W.; Carter, G. E. *J. Am. Chem. Soc.* **1982**, *104*, 5741.
50. Bentley, T. W.; Bowen, C. T.; Parker, W.; Watt, C. I. F. *J. Am. Chem. Soc.* **1979**, *101*, 2486.
51. Kevill, D. N.; Kamil, W. A.; Anderson, S. W. *Tetrahedron Lett.* **1982**, 4635.
52. Bunton, C. A.; Mhala, M. M.; Moffatt, J. R. *J. Org. Chem.* **1984**, *49*, 3639.
53. Bentley, T. W.; Carter, G. E. *J. Org. Chem.* **1983**, *48*, 579.
54. Bentley, T. W.; Carter, G. E.; Roberts, K. *J. Org. Chem.* **1984**, *49*, 5183.
55. Kevill, D. N.; Bahari, M. S.; Anderson, S. W. *J. Am. Chem. Soc.* **1984**, *106*, 2895.

56. Creary, X.; McDonald, S. R. *J. Org. Chem.* **1985,** *50,* 474.
57. Kevill, D. N.; Anderson, S. W. *J. Org. Chem.* **1985,** *50,* 3330.
58. Bentley, T. W.; Roberts, K. *J. Org. Chem.* **1985,** *50,* 4821.
59. Swain, C. G.; Mosely, R. B.; Bown, D. E. *J. Am. Chem. Soc.* **1955,** *77,* 3731.
60. Kevill, D. N.; Lin, G. M. L. *J. Am. Chem. Soc.* **1979,** *101,* 3916.
61. Bentley, T. W.; Bowen, C. T. *J. Chem. Soc., Perkin Trans. 2* **1978,** 557.
62. Kevill, D. N.; Rissmann, T. J. *J. Org. Chem.* **1985,** *50,* 3062.
63. Kevill, D. N. Presented at the 190th National Meeting of the American Chemical Society, Chicago, IL, Sept 1985; paper ORGN 213.
64. Kamlet, M. J.; Abboud, J. L. M.; Taft, R. W. *Prog. Phys. Org. Chem.* **1981,** *13,* 485.
65. Young, P. R.; Jencks, W. P. *J. Am. Chem. Soc.* **1979,** *101,* 3288.
66. Harris, J. M.; Shafer, S. G.; Moffatt, J. R.; Becker, A. R. *J. Am. Chem. Soc.* **1979,** *101,* 3295.
67. Zollinger, H. *Angew Chem., Int. Ed. Engl.* **1978,** *17,* 141.
68. Szele, I.; Zollinger, H. *J. Am. Chem. Soc.* **1978,** *100,* 2811.
69. Schneider, H.-J.; Busch, R. *J. Org. Chem.* **1982,** *47,* 1766.
70. McManus, S. P. *J. Org. Chem.* **1981,** *46,* 635.
71. Lancelot, C. J.; Schleyer, P. v. R. *J. Am. Chem. Soc.* **1969,** *91,* 4291.
72. Schadt, F. L.; Lancelot, C. J.; Schleyer, P. v. R. *J. Am. Chem. Soc.* **1978,** *100,* 228.
73. Bentley, T. W.; Jackson, S. J.; Roberts, K.; Williams, D. J. *J. Chem. Soc., Perkin Trans. 2,* in press.
74. Kevill, D. N.; Rissman, T. J. *J. Chem. Res.* **1986,** 252.

RECEIVED for review October 21, 1985. ACCEPTED January 27, 1986.

19

Nucleophilicity Studies of Reactions in Which the Displaced Group Is a Neutral Molecule

Dennis N. Kevill, Steven W. Anderson, and Edward K. Fujimoto

Department of Chemistry, Northern Illinois University, DeKalb, IL 60115

A comparison of the rates of solvolysis of the tert-*butyldimethylsulfonium ion and the 1-adamantyldimethylsulfonium ion presents strong evidence that the solvent dependence of the* tert-*butyldimethylsulfonium ion solvolysis rates is governed primarily by solvent nucleophilicity effects. Leaving-group contributions based upon 1-adamantyldimethylsulfonium ion solvolyses are better incorporated into the establishment of the solvent nucleophilicity scale based upon triethyloxonium ion solvolysis. Alternative solvent nucleophilicity scales based upon the solvolysis of S-methylbenzothiophenium ions are discussed. Analyses of the extent of nucleophilic participation by the solvent in the solvolyses of methyldiphenylsulfonium and benzhydryldimethylsulfonium ion will be presented. The relative nucleophilicities of various anionic and neutral nucleophiles toward the triethyloxonium ion in ethanol have been determined.*

THEORETICAL TREATMENTS OF S_N2 REACTIONS, in the presence of solvent molecules, using statistical mechanics simulations involving a large number of water molecules (*1, 2*) or quantum mechanics calculations involving a few water molecules (*3, 4*), are in their infancy. As an alternative, the roles of solvent nucleophilicity (N) and solvent ionizing power (Y) during solvolysis reactions can be treated in terms of a linear free energy relationship (LFER). In 1951, Winstein et al. (*5*) put forward equation 1 for the correlation of solvolysis rates of neutral substrates. In equation 1, k and k_0 represent the specific rates of solvolysis of a substrate in a given solvent and in the standard solvent (80% ethanol), and l and m are measures of the sensitivity of the

0065-2393/87/0215-0269$06.00/0

solvolysis of a given substrate toward changes in N and Y values. This equation can be considered as an extension of an earlier equation (without the lN term) put forward (*6*) to correlate the rates of unimolecular solvolyses.

$$\log (k/k_0) = lN + mY \tag{1}$$

Recently, Swain et al. (*7*) have put forward equation 2, where $\Delta\log k$ is the incremented $\log k$ relative to the lowest value in the series, A and B are measures of solvent electrophilicity and nucleophilicity, and a, b, and c are determined by details of the solvolysis under consideration. Although this equation appears to have an advantage over equation 1 in that A and B values can be obtained without the study of solvolysis reactions, use of such values frequently leads to unreasonable predictions (*8*).

$$\Delta\log k = aA + bB + c \tag{2}$$

For the analysis of S_N1 solvolyses, Abraham et al. (*9*) have proposed an equation (equation 3) based on sensitivities toward solvatochromatic properties. In equation 3, π^* is a measure of solvent dipolarity–polarization, α is a measure of solvent hydrogen bond donor acidity, and β is a measure of solvent hydrogen bond acceptor basicity. More recently, a term governing cavity effects has been added, and this term is considered to represent an important contribution (*10, 11*). The cavity term can be directly related to the square of the Hildebrand solubility parameter (*10–12*). A similar analysis by Koppel and Palm (*13, 14*) involves terms governed by solvent polarity, solvent polarizability, electrophilic solvation ability, and nucleophilic solvation ability. Recently, a cavity term has also been added to this analysis (*12*).

$$\log k = \log k_0 + s\pi^* + a\alpha + b\beta \tag{3}$$

For situations where solvent nucleophilicity may be a factor, Kevill (*8*) favors the use of the extended Grunwald–Winstein equation (equation 1). Scales of N_{OTs} and Y_{OTs} values based upon the use of methyl tosylate and 2-adamantyl tosylate as model S_N2- and S_N1-reacting substrates have been developed (*15, 16*). Also Y scales have been developed for other anionic leaving groups using 1-adamantyl or 2-adamantyl derivatives (*17–19*), where S_N2 reaction is impossible or severely hindered.

Solvolyses of tert-*Butyl Derivatives*

When a scale of Y_{Cl} values (*17*) was used together with N_{OTs} values within equation 2, an l value of 0.30 was determined for *tert*-butyl chloride solvolysis. An S_N2 (intermediate) mechanism, involving both an intermediate and covalent interaction by a nucleophilic solvent molecule, was proposed

(*17*). Abraham et al. (*9*) and Dvorko et al. (*12*) suggested that the data can be alternatively rationalized by assuming that the moderate differences in the response of specific rates to solvent variation between 1-adamantyl chloride and *tert*-butyl chloride arise not from a higher response to changes in solvent nucleophilicity for *tert*-butyl chloride but from a higher response to changes in solvent electrophilicity for 1-adamantyl chloride. The basis for this alternative explanation is the covarience between the N and Y scales usually observed for a binary solvent system. However, the proportionality factor involved in the covarience varies widely with the system, even changing from negative to positive, and by a judicious choice of standard solvents, including solvents from different systems, problems associated with covarience can be avoided (*15, 20, 21*).

In an analysis of the solvolysis of *tert*-butyl chloride in terms of equation 3, Abraham et al. (*9*) found that the specific rates could be correlated against π^* and α, without the need to include a term relating to basicity (β) values. However, a new analysis with inclusion of the cavity term has indicated (*11*) a minor contribution from the $b\beta$ term. Conversely, an analysis (*12*) in terms of the Koppel–Palm equation, with inclusion of a cavity term, did not find any dependence on nucleophilicity (basicity).

The situation regarding the analysis of *tert*-butyl chloride solvolyses in terms of a LFER is clearly confused, a general acceptance of the beliefs that, at most, only a minor contribution arises from the term governed by solvent nucleophilicity and that the dominant contribution is from a combination of solvent dipolarity–polarizability, and that electrophilic solvation exists.

We reasoned that the best approach toward estimating the extent of nucleophilic assistance during solvolysis of *tert*-butyl derivatives was not to try to refine the existing LFER treatments (*7, 9, 11–13, 17*) by use of more sophisticated parameters or by addition of further parameters. Instead, the type of substrate studied should be changed in a manner that would remove, or greatly reduce in magnitude, the factors that dominate when the leaving group is anionic. An excellent substrate for such a study is the *tert*-butyldimethylsulfonium ion (t-BuSMe$_2^+$), conveniently used in conjunction with the essentially nonnucleophilic (*22*) trifluoromethanesulfonate (triflate) anion. Solvolysis of this cation will proceed with charge dispersal at the transition state and with ejection of a neutral molecule. This ion had been studied in a variety of solvents (*23*), and with the notable exception of the acetolysis, an approximately linear relationship, showing small rate reductions with increasing Y values, was found ($m \sim -0.09$). The data were interpreted in terms of the Hughes–Ingold concept, with increased solvent polarity preferentially stabilizing the more intensely charged initial state, which led to a reduction in the rate. All of the binary systems studied were of the type where an increase in the Y value is accompanied by a decrease in the N value. We extended the study to 2,2,2-trifluoroethanol (TFE)-containing solvents (*24, 25*), and in particular, we studied TFE–H_2O mixtures, which are unusual in that both N and Y values increase together (*26*). Kevill et al. (*25*) found for TFE–H_2O mixtures that the rate *increased* with increasing Y values (Figure 1). A rationalization of the variation in rates with solvent composition, consistent with all of the data, is that the specific rates increase with increasing solvent nucleophilicity. Consistent with this idea, acetic and formic acids, having very similar N values (*15, 27*) but very different Y values (*15, 28*), gave specific rates of solvolysis of t-$BuSMe_2^+$ that were very similar (8.10×10^{-6} s^{-1} for acetolysis and 6.78×10^{-6} s^{-1} for formolysis, at 50.0 °C). When specific solvolysis rates were plotted logarithmically against N_{KL} values (*29*), some lateral dispersion was found but the slopes of the plots (*25*) for a given binary system were around 0.35, consistent with the value estimated for *tert*-butyl chloride solvolyses (*17*).

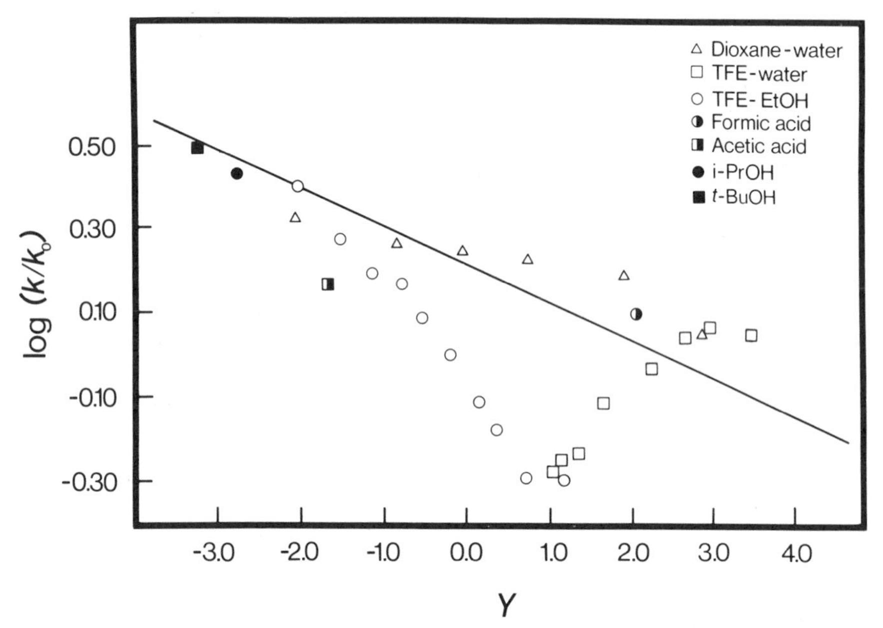

Figure 1. Grunwald–Winstein plot for the solvolysis of the t-$BuSMe_2^+$ *ion. The line represents the best fit to the data of reference 23 (excluding acetolysis). The points represent our data for the solvolysis of* t-$BuSMe_2^+OTf^-$ *at 50.0 °C.*

Solvolyses of the 1-Adamantyldimethylsulfonium Ion

If the relative rates of t-$BuSMe_2^+$ solvolyses are indeed governed by solvent nucleophilicity, then a study of 1-adamantyldimethylsulfonium ion (1-$AdSMe_2^+$) solvolyses (equation 4) should show quite different characteristics, because specific nucleophile solvation from the rear is prevented by the cage structure.

$$1\text{-AdSMe}_2^+ \; SO_3CF_3^- \xrightarrow{ROH} 1\text{-AdOR} + SMe_2 + ROH_2^+ + SO_3CF_3^- \quad (4)$$

(1-AdSMe_2^+)

The specific rates of solvolysis have been found to vary very little with the solvent. For 41 solvent systems, the specific rates at 70.4 °C varied only by a factor of less than 7, from $1.09 \times 10^{-6}\ s^{-1}$ in *tert*-butyl alcohol to $7.09 \times 10^{-6}\ s^{-1}$ in 97% hexafluoroisopropyl alcohol (HFIP). This lack of variation is even more remarkable considering the large temperature dependence, as reflected in enthalpies of activation of approximately 35 kcal/mol. Reasonable rates are observed at moderate temperatures only because of entropies of activation of approximately +18 eu.

For each binary mixture within which both $t\text{-BuSMe}_2^+$ and 1-AdSMe_2^+ were studied (TFE–EtOH, TFE–H_2O, EtOH–H_2O, MeOH–H_2O, acetone–H_2O, and dioxane–H_2O), the specific rates of solvolysis of 1-AdSMe_2^+ always showed a variation with change in solvent composition in the opposite direction to that for $t\text{-BuSMe}_2^+$. Further, the rate variations were more modest for 1-AdSMe_2^+. These observations can be rationalized in terms of an effect related to the Hughes–Ingold concept but with solvent nucleophilicity substituted for the more generalized concept of solvent polarity. For both substrates, the effect will favor a reduction in rate with increased solvent nucleophilicity because of a larger stabilizing effect at the more intensely charged initial state. However, for $t\text{-BuSMe}_2^+$ solvolyses, this effect is outweighed by an additional effect involving specific nucleophilic solvation of the developing carbenium ion, an effect prevented for 1-AdSMe_2^+ solvolyses by the cage structure. For positively charged *tert*-butyl substrates, nucleophilic assistance from the solvent is the dominant influence in determining solvent composition induced changes in specific rates, whereas nucleophilic assistance was only a minor factor for neutral *tert*-butyl substrates.

Another feature supporting the importance of nucleophilicity is the observation that the ratio of specific solvolysis rates ($k_{t\text{-BuX}}/k_{1\text{-AdX}}$) in 97% HFIP gives a ratio of 2.56 (at 25 °C) for X = Br (*30*) and 2.60 (at 70.4 °C) and 2.71 (at 49.7 °C) for X = SMe_2^+. Electrophilic solvation would not be expected to be a factor for a positively charged leaving group, and the similarity of the specific rate ratio for X = SMe_2^+ or X = Br suggests that it is also not an important factor when X is a halogen; that is, *t*-BuCl and 1-AdCl have similar responses to solvent electrophilicity. The value of approximately 2.6 can then be considered as a base point, perhaps reflecting a slightly better *nonspecific* nucleophilic solvation of the *t*-BuX relative to that from 1-AdX. The ratio then increases for either X = SMe_2^+ or Cl as the nucleophilicity of the solvent increases, because of the incursion of specific nucleophilic assistance for t-BuX substrates, but not for 1-AdX substrates. If, as an alternative, the low ratio in 97% HFIP is ascribed as being due to a higher response to solvent electrophilicity for 1-AdCl relative to t-BuCl (*9, 12*), then the similar ratio observed for the comparison with the *tert*-alkyldimethylsulfonium salts is difficult to explain.

Solvent Nucleophilicity Scales Based upon R–X⁺ Substrates

Kevill and Lin (*29*) pointed out several years ago that solvent nucleophilicity scales are best developed by using a positively charged substrate, because the modified form required for the extended Grunwald–Winstein equation (equation 5) has solvent ionizing power (Y) replaced by a measure (Y^+) of the influence of a given solvent upon the leaving-group ability of an initially positively charged leaving group. Because the range of Y^+ values is much narrower than that of Y values, any errors in solvent nucleophilicity scales are reduced due to an incorrect estimate being made of the m value.

Kevill and Lin developed (*29*) an N_{KL} scale of solvent nucleophilicities using equation 6.

$$\log (k/k_0) = lN + mY^+ \tag{5}$$

$$N_{KL} = \log (k/k_0)_{Et_3O^+} - 0.55Y^+_{t\text{-BuSMe}_2^+} \tag{6}$$

Two reasons led to the choice of t-$BuSMe_2^+$: the analogy with the use of *tert*-butyl chloride to establish the traditional Y scale (*6*) and the observation that several of the required values were available in the literature (*23*). Clearly, this choice was not good, and, indeed, the mY^+ corrections (fortunately small) were in the wrong direction. The required Y^+ scale is better based on 1-$AdSMe_2^+$ solvolyses, and we have developed a revised scale (N_{KL}'), defined as in equation 7. The $0.55Y^+$ term is, however, now so small that to a very good degree of approximation $\log (k/k_0)_{Et_3O^+}$values can be directly used as a solvent nucleophilicity scale (designated $N_{Et_3O^+}$). Indeed, N_{KL}' and $N_{Et_3O^+}$ values, for 34 solvents (Table I), correlate very well ($r = 0.9958$) with each other:

$$N_{KL}' = \log (k/k_0)_{Et_3O^+} - 0.55\ Y^+_{1-\text{AdSMe}_2^+} \tag{7}$$

$$N_{KL}' = 1.06N_{Et_3O^+} - 0.042 \tag{8}$$

Either the N_{KL}' or the N_{Et_3O+} scale can do an excellent job of correlating the rates of both positively charged and neutral substrates. The logarithms of the specific rates of solvolysis of the N-(methoxymethyl)-N,N-dimethyl-m-nitroanilinium ion correlate linearly with $N_{Et_3O}{}^+$ values in aqueous ethanol and aqueous TFE with a value of 0.75 for the slope (*31*). An analysis has

Table I. Solvent Nucleophilicity Values Based upon the Solvolysis of the Triethyloxonium Ion at 0.0 °C.

Solvent[a,b]	$N_{Et_3O^+}$[c]	N_{KL}'[d]	*Solvent*[a,b]	$N_{Et_3O^+}$[c]	N_{KL}'[d]
100% EtOH	0.55	0.56	80% dioxane	−0.10	−0.04
80% EtOH	0.00	0.00	70% dioxane	−0.24	−0.22
60% EtOH	−0.35	−0.38	60% dioxane	−0.34	−0.34
40% EtOH	−0.64	−0.70	40% dioxane	−0.60	−0.60
20% EtOH	−0.95	−1.03	20% dioxane	−0.80	−0.84
100% MeOH	0.64	0.62	100% TFE	−2.17	−2.42
80% MeOH	0.19	0.14	97% TFE	−2.28	−2.50
60% MeOH	−0.25	−0.30	90% TFE	−1.73	−1.93
40% MeOH	−0.49	−0.56	70% TFE	−1.28	−1.47
20% MeOH	−0.88	−0.96	50% TFE	−1.05	−1.21
95% acetone	−0.23	−0.31	80% T–20% E	−0.92	−1.10
90% acetone	−0.12	−0.20	60% T–40% E	−0.42	−0.52
80% acetone	−0.22	−0.29	40% T–60% E	−0.05	−0.12
70% acetone	−0.36	−0.42	20% T–80% E	0.41	0.38
60% acetone	−0.46	−0.52	100% H_2O	−1.01	−1.15
40% acetone	−0.71	−0.78	100% AcOH	−1.42	−1.46
20% acetone	−0.86	−0.96	100% HCO_2H	−1.71	−1.73

[a] On a volume–volume basis (at 25.0 °C), except for TFE–H_2O mixtures, which are on a weight percentage basis.
[b] Other component water, except for TFE–EtOH (T–E) mixtures.
[c] Decimal logarithm of the specific rate of solvolysis of Et_3O^+ at 0.0 °C relative to the specific rate of solvolysis in 80% aqueous ethanol (values from references 24 and 29).
[d] Defined according to equation 7, where Y^+ is the decimal logarithm of the specific rate of solvolysis of 1-AdSMe_2^+ at 70.4 °C relative to the specific rate of solvolysis in 80% aqueous ethanol.

recently been published comparing the use of $N_{Et_3O^+}$ and N_{OTs}, in conjunction with Y_{OTs}, for the correlation of the specific rates of solvolysis of allyl arenesulfonates (*21*). A minor source of irritation is that confusion can arise because l values based on the use of N_{KL}' or $N_{Et_3O^+}$ values must be scaled before they can be compared with l values based on the use of N_{OTs} values. The need for scaling can be avoided by using a positively charged methyl derivative as the standard substrate. This usage will also have the advantages of maximizing the range of N values and minimizing steric interactions.

We have recently completed a study, at 25.4 °C, of the solvolyses of S-methyldibenzothiophenium trifluoromethanesulfonate (MeDBTh$^+$OTf$^-$) in 37 solvents (equation 9): the 33 solvents studied for Et_3O^+ solvolysis plus four HFIP-containing solvents. A plot (Figure 2) shows that, for the 33 solvents studied with both substrates, a good linear plot is obtained when $\log (k/k_0)_{Et_3O^+}$

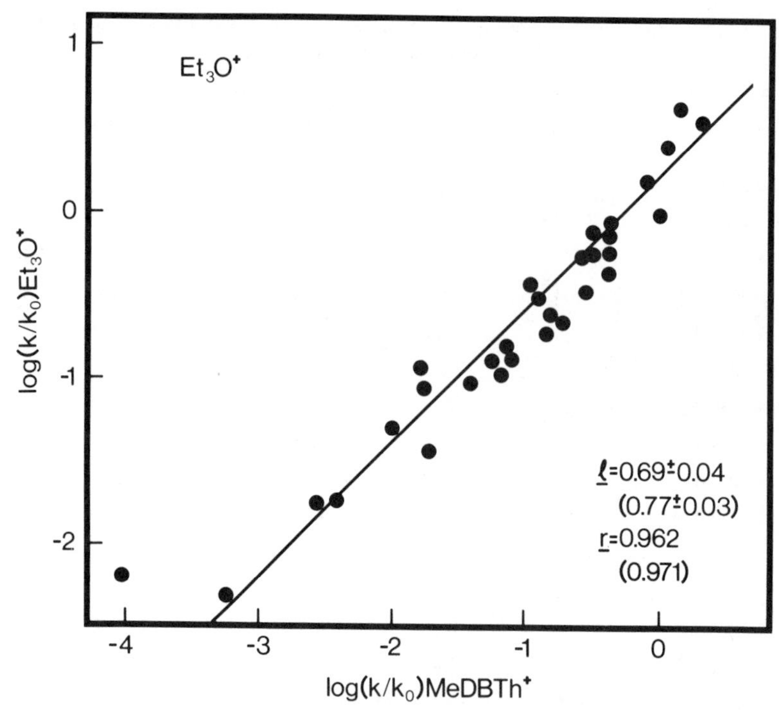

Figure 2. A plot of log $(k/k_0)_{Et_3O^+}$ *(at 0 °C) against log* $(k/k_0)_{MeDBTh^+}$ *(at 25.4 °C).*

is plotted against log $(k/k_0)_{MeDBTh^+}$. The only large deviation is for the study in 100% TFE, which has log (k/k_0) values of -2.17 for Et_3O^+ and of -4.25 for $MeDBTh^+$. The cause of this deviation is not known. For all solvents, the slope (l value) is 0.69 ± 0.04 ($r = 0.962$) and, with omission of the data for 100% TFE, the l value is 0.77 ± 0.03 ($r = 0.971$).

$$\text{MeDBTh}^+\ (\text{S-Me},\ SO_3CF_3^-) \xrightarrow{\text{ROH}} \text{dibenzothiophene} + MeOR + ROH_2^+ + SO_3CF_3^- \qquad (9)$$

Other S_N2-reacting RX^+-type substrates have also been correlated against log $(k/k_0)_{MeDBTh^+}$ (termed N_{MeDBTh^+}). The specific rates of solvolysis of the methyldiphenylsulfonium ion at 50.0 °C in 26 solvents led to an l value of 0.86 ± 0.04 ($r = 0.968$). The specific rates of solvolysis of the S-methylbenzothiophenium ion, at 25.4 °C, in all of the 37 solvents studied for the $MeDBTh^+$ ion led to an excellent correlation, with an l value of 0.93 ± 0.02

(r = 0.996). The *S*-ethyldibenzothiophenium ion, at 25.4 °C, in 11 solvents gave a linear correlation, with an l value of 0.97 ± 0.03 (r = 0.997). The data for the solvolysis of the *N*-(methoxymethyl)-*N*,*N*-dimethyl-*m*-nitroanilinium ion (*31*) in six solvents gave, with inclusion of 100% TFE, an l value of 0.46 ± 0.05 (r = 0.983) and, with exclusion of 100% TFE, an l value of 0.55 ± 0.04 (r = 0.992).

The lack of any need to include the mY^+ term in correlations of the solvolyses of RX^+ substrates is nicely illustrated by the analyses presented in Table II. For each substrate, the l values and the correlation coefficients (r) are of essentially the same values, irrespective of whether log $(k/k_0)_{RX}{}^+$ is correlated with both N and Y^+ or only with N. Further, the m values follow no rational pattern and have extremely large associated standard deviations. The obvious conclusion is that the rates are adequately correlated by use of equation 10, where N can be represented by N_{KL}', $N_{Et_3O^+}$, N_{Th^+} or N_{MeDBTh^+}. ($N_{Th}{}^+$ is defined in the same manner as N_{KL}', but with $MeDBTh^+$ as the standard substrate).

$$\log\ (k/k_0)_{RX^+} = lN \qquad (10)$$

Table II. Correlation of Specific Solvolysis Rates for Thiophenium Ions at 25.4 °C.

Ion	n[a]	*Correlation Parameters*	*Constant(s)*	*Estimate*	r[b]
S, Me^+	33	N_{KL}', Y^+	l	1.25 ± 0.06	0.967[c]
			l	1.19 ± 0.09	0.968[d]
			m	−0.35 ± 0.46	
S, Me^+	33	N_{KL}', Y^+	l	1.23 ± 0.05	0.9729[c]
			l	1.25 ± 0.08	0.9730[d]
			m	0.17 ± 0.41	
S, Et^+	11	N_{Th^+}, Y^+	l	0.93 ± 0.03	0.996[c]
			l	0.97 ± 0.04	0.997[d]
			m	0.46 ± 0.30	

[a] Number of solvents.
[b] Correlation coefficient.
[c] Using log $(k/k_0) = lN$.
[d] Using log $(k/k_0) = lN + mY$.

Solvolyses of the Benzhydryldimethylsulfonium Ion

Solvolyses of the benzhydryldimethylsulfonium ion have several features in common with solvolyses of the 1-$AdSMe_2^+$ ion. For 38 solvents, the specific rates at 25.4 °C vary only by a factor of 16 (from 1.02×10^{-4} s^{-1} in 80% acetone to 16.4×10^{-4} s^{-1} in 70% HFIP). This variation is only slightly greater than the factor of seven observed for 1-$AdSMe_2^+$ solvolyses. Also, the entropies of activation are very high, a value of +15.3 eu for ethanolysis can be compared to values of +20.3 eu for the S_N1 ethanolysis of 1-$AdSMe_2^+$ and −13.4 eu for the S_N2 ethanolysis of the methyldiphenylsulfonium ion.

A feature highly suggestive of S_N1 reaction, but which was not observed for the corresponding 1-$AdSMe_2^+$ solvolyses, is the observation that additions of dimethyl sulfide to the solvolyses in 95% acetone lead to common-molecule rate depression. For example, addition of 0.1 M dimethyl sulfide to the solvolysis at 25.6 °C led to a reduction in the specific rate from 3.39×10^{-4} s^{-1} to 2.28×10^{-4} s^{-1}. These findings are consistent with those for the corresponding halides. The adamantyl carbocation formed from the bromide is not captured by external nucleophiles during solvolysis (*32*, *33*), but common-ion return is observed during solvolyses of benzhydryl chloride (*34*). Consistent with this difference in the nature of the capture of the two carbocations, the 1-$AdSMe_2^+$ solvolyses in 96–60% aqueous ethanol show a slight preference for product formation by reaction with water (selectivity value S of 1.35 ± 0.11), but the corresponding solvolyses of the benzhydryldimethylsulfonium ion, at 24.9 °C, show a modest preference for reaction with the more nucleophilic ethanol (S value of 0.413 ± 0.032). Preference for reaction with the less nucleophilic water is usually taken to imply product formation at the solvent-separated ion pair stage, and a modest preference for reaction with ethanol would be consistent with capture at the free ion stage, as indicated by the common-molecule rate depression.

Although the range of specific rates as the solvent is varied is only slightly larger than for 1-$AdSMe_2^+$, the detailed nature of the variations within a given binary mixed solvent system is more complex. In aqueous methanol, the rates decrease slightly with increasing water content. In aqueous acetone and aqueous ethanol, shallow rate minima are observed. Bunton et al. (*35*) have suggested that the dispersions in Grunwald–Winstein plots for solvolyses of benzhydryl chloride may arise because of differences between alkyl and arylalkyl derivatives in initial-state interactions. These effects are probably less well canceled in going to the transition state for solvolysis of arylalkyl derivatives relative to alkyl derivatives because of conformational constraints applied at the transition state by resonance effects. This finding would also imply that use of benzhydryl halides as standards for the analysis of solvation effects (*12*) might not, if adopted, lead to the desired reduction in the number of factors that must be considered.

Nucleophilicity of Addenda toward the Triethyloxonium Ion in Ethanol

Kevill and Lin's (*27*) earlier study of the ethanolysis of the triethyloxonium ion, at 0.0 °C, has been extended to a consideration of the competition between the solvent and added nucleophile, either anionic or neutral, for reaction with the substrate (equation 11). This study is related to earlier studies, at 25.0 °C, of competition between water and added nucleophile for reaction with methyl bromide (*36*), methyl iodide (*37*), or the cyclic pentamethyleneiodonium ion (*38*) and of competition between methanol and added nucleophile for reaction with methyl iodide (*39*) or *trans*-$Pt(py)_2Cl_2$ (*39*). The nucleophilicities are usually expressed, relative to the solvent, in terms of the Swain–Scott equation (*36*) (equation 12). In equation 12, k and k_0 are second-order rate coefficients for reaction of a substrate with the added nucleophile and with the solvent, n is a measure of the nucleophilicity of the added nucleophile, and s is a measure of the sensitivity of the substrate toward changes in nucleophilicity. The value of s is taken as unity for the standard substrate.

$$Et_3O^+ \xrightarrow{EtOH,\ k_0} 2Et_2O + H^+ \qquad\qquad (11)$$

$$Et_3O^+ \xrightarrow{X^-,\ k} EtX + Et_2O$$

$$\log (k/k_0) = sn \qquad\qquad (12)$$

When the substrate is positively charged, as with the Et_3O^+ ion, the situation for attack by anions is complicated by k being a function of ionic strength (*40, 41*). We found empirically that a plot of log (k/k_0), as determined from the product ratio, against concentration of added tetra-*n*-butylammonium salt was essentially linear over a range of 0.004–0.02 M. Five values measured within this range were extrapolated to give values at extremely low added salt concentration (but still with 0.0038 M $Et_3O^+PF_6^-$). These values are compared with literature values in Table III. The n values can be obtained for nitrate ion [previously estimated (*39*)] and *p*-toluenesulfonate ion (not previously measured). The range of values is somewhat less than for the same anions determined under other standard conditions; this result possibly reflects the high reactivity of the Et_3O^+ ion. However, the present study differs from each of the previous studies in four aspects: temperature, electrophilic site, leaving group, and solvent. In particular, the desolvation of the anions is a critical factor and large variations of relative nucleophilicities

Table III. Nucleophilicity Values (n)[a] for Various Anions

	Nucleophilicity Scales					
Anion	*SS*[b]	*PSS(1)*[c]	*PSS(2)*[d]	*S*[e]	*P*[f]	*KF*[g]
I^-	5.08	7.42	5.46	5.41	4.11	4.59
NCS^-	4.77	6.70	5.75	5.42	4.06	4.21
Br^-	3.89	5.79	4.18	4.48	3.19	4.21
Cl^-	3.04	4.37	3.04	3.36	2.40	3.67
NO_3^-	—[h]	(1.5)	—	—	—	2.91
TsO^-	—	—	—	—	—	2.23

[a] From the Swain–Scott equation (equation 12) with an s value of unity.
[b] Reference 36; $CH_3Br + H_2O$ at 25 °C.
[c] Reference 39; $CH_3I + CH_3OH$ at 25 °C.
[d] Reference 39; $Pt(py)_2Cl_2 + CH_3OH$ at 25 °C.
[e] Reference 37; $CH_3I + H_2O$ at 25 °C.
[f] Reference 38; $C_4H_8I^+ + H_2O$ at 25 °C.
[g] This work; $Et_3O^+ + EtOH$ at 0 °C.
[h] Not determined.

with change of solvent are usually observed (*42*). Any attempt at a detailed comparison could not at this stage be justified. Variation only of the nature of the electrophilic site and of the identity of a halide ion leaving group has led to very poor correlations between the resultant n scales (*39*).

For neutral nucleophiles, we have utilized a series of ring-substituted *N,N*-dimethylanilines. The second-order rate coefficients should now be independent of nucleophile concentration, and this was confirmed by showing that log (k/k_0), obtained from the product ratios, was independent of the amine concentration for 0.008 to 0.08 M *N,N*-dimethyl-*p*-toluidine. The log (k/k_0) values could also be conveniently determined for *m*-CH_3-, H-, *p*-Br-, and *m*-Cl-substituted derivatives (equation 13). For the *m*-NO_2 derivative, even at 0.32 M, the dominant reaction is solvolysis and only an approximate value for log (k/k_0) could be obtained. A Hammett plot against the tabulated σ values (*43*) (omitting the approximate *m*-NO_2 data) led to a linear plot and a slope (ρ value) of -2.77 ± 0.15 ($r = -0.996$). This value is similar to values for reaction with other ethyl derivatives, derived from kinetically determined k values: -3.60 for reaction with ethyl iodide in nitrobenzene at 25.0 °C (*44*) and -2.86 for reaction with ethyl perchlorate in benzene at 25.0 °C (*45*).

$$Et_3O^+ \xrightarrow{XC_6H_4NMe_2,\ k} EtMe_2N^+C_6H_4X + Et_2O$$
$$Et_3O^+ \xrightarrow{EtOH,\ k_0} 2Et_2O + H^+ \qquad (13)$$

Role of the Counterion in Silver Ion Assisted Reactions of Ethyl Iodide

The study with tetra-*n*-butylammonium nitrate in competition with ethanol has been extended to the parallel reaction with silver nitrate. Also, the reaction of silver nitrate with ethyl iodide (at 20.0 °C) has been investigated. A preliminary account of these experiments has appeared (*46*). In a turn of the century study of the reaction of ethyl iodide with silver nitrate in ethanol, the rate was claimed to increase with an increase in nitrate ion concentration but without any change in the division of product between ethyl ether and nitrate ester. This observation had been used as evidence for the intermediate formation of $NO_3^-R^+$ or $NO_3^-R^+X^-Ag^+$ intermediates (*47*). Indeed, at the outset of our investigation, this finding was the only convincing evidence remaining for these intermediates. However, for all three systems, the product ratio responded very similarly toward changes in the concentration of the nitrate salt (Figure 3). Apparently, the previous claim of a constant

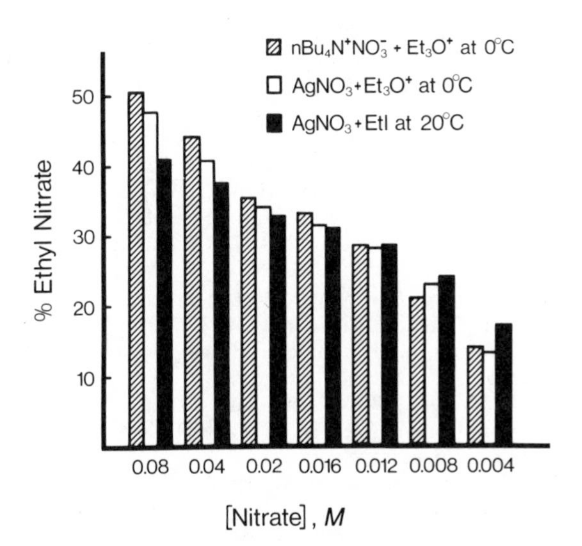

Figure 3. Percentage of reaction diverted to nitrate ester as a function of the concentration of added nitrate salt in the ethanolysis of 0.0038 M triethyloxonium hexafluorophosphate or ethyl iodide.

product ratio in the reaction of ethyl iodide with silver nitrate in ethanol is erroneous. We conclude that silver ion assisted nucleophilic substitution reactions of alkyl halides can be considered as S_N1 or S_N2 reactions, involving an equilibrium concentration of a positively charged substrate. For S_N2 reactions involving attack by an anion (equation 14), the kinetics are complicated both by the presence of the equilibrium and by the reaction being between oppositely charged species. Other possible complications include aggregation of the silver salt (*48*) or of the intermediate and counter ion.

$$RX + Ag^+ \rightleftharpoons (RXAg)^+ \quad (14a)$$

$$Y^- + RXAg^+ \rightarrow YR + X^-Ag^+ \quad (14b)$$

Acknowledgments

Abstracted, in part, from the Ph.D. Thesis of S. W. Anderson, Northern Illinois University, August 1985, and the M.S. Thesis of E. K. Fujimoto, Northern Illinois University, August 1984.

Literature Cited

1. Jorgensen, W. L.; Chandrasekhar, J. Abstracts of Papers, 190th National Meeting of the American Chemical Society, Sept. 8–13, 1986; Abstract ORGN 89.
2. Chandrasekhar, J.; Smith, S. F.; Jorgensen, W. L. *J. Am. Chem. Soc.* **1984,** *106*, 3049.
3. Cremaschi, P.; Gamba, A.; Simonetta, M. *Theor. Chim. Acta.* **1972,** *25*, 237.
4. Morokuma, K. *J. Am. Chem. Soc.* **1982,** *104*, 3732.
5. Winstein, S.; Grunwald, E.; Jones, H. W. *J. Am. Chem. Soc.* **1951,** *73*, 2700.
6. Grunwald, E.; Winstein, S. *J. Am. Chem. Soc.* **1948,** *70*, 846.
7. Swain, C. G.; Swain, M. S.; Powell, A. L.; Alunni, S. *J. Am. Chem. Soc.* **1983,** *105*, 502.
8. Kevill, D. N. *J. Chem. Res. Synop.* **1984,** 86.
9. Abraham, M. H.; Taft, R. W.; Kamlet, M. J. *J. Org. Chem.* **1981,** *46*, 3053.
10. Taft, R. W.; Abraham, M. H.; Doherty, R. M.; Kamlet, M. J. *J. Am. Chem. Soc.* **1985,** *107*, 3105.
11. Kamlet, M. J.; Doherty, R. M.; Abraham, M. H.; Taft, R. W.; Harris, J. M., submitted for publication. We wish to thank Dr. Kamlet for providing us with a preprint of this manuscript.
12. Dvorko, G. F.; Ponomareva, E. A.; Kulik, N. I. *Russ. Chem. Rev. (Engl. Transl.)* **1984,** *53*, 948.
13. Koppel, I. A.; Palm, V. A. *Reakts. Sposobn. Org. Soedin,* **1971,** *8*, 296.
14. Koppel, I. A.; Palm, V. A. In *Advances in Linear Free Energy Relationships;* Chapman, N. B.; Shorter, J., Eds.; Plenum: London, 1972; p 203.
15. Schadt, F. L.; Bentley, T. W.; Schleyer, P.v.R. *J. Am. Chem. Soc.* **1976,** *98*, 7667.
16. Bentley, T. W.; Bowen, C. T.; Brown, H. C.; Chloupek, F. J. *J. Org. Chem.* **1981,** *46*, 38.
17. Bentley, T. W.; Carter, G. E. *J. Am. Chem. Soc.* **1982,** *104*, 5741.
18. Kevill, D. N.; Bahari, M. S.; Anderson, S. W. *J. Am. Chem. Soc.* **1984,** *106*, 2895.
19. Bentley, T. W.; Carter, G. E.; Roberts, K. *J. Org. Chem.* **1984,** *49*, 5183.
20. Kevill, D. N.; Rissmann, T. J. *J. Chem. Soc., Perkin Trans. 2* **1984,** 717.
21. Kevill, D. N.; Rissman, T. J. *J. Org. Chem.* **1985,** *50*, 3062.
22. Stang, P. J.; Hanack, M.; Subramanian, L. R. *Synthesis* **1982,** 85.

23. Swain, C. G.; Kaiser, L. E.; Knee, T. E. C. *J. Am. Chem. Soc.* **1958,** *80,* 4092.
24. Kevill, D. N.; Kamil, W. A. *J. Org. Chem.* **1982,** *47,* 3785.
25. Kevill, D. N.; Kamil, W. A.; Anderson, S. W. *Tetrahedron Lett.* **1982,** *23,* 4635.
26. Kaspi, J.; Rappoport, Z. *J. Am. Chem. Soc.* **1980,** *102,* 3829.
27. Peterson, P. E.; Waller, F. J. *J. Am. Chem. Soc.* **1972,** *94,* 991.
28. Fainberg, A. H.; Winstein, S. *J. Am. Chem. Soc.* **1956,** *78,* 2770.
29. Kevill, D. N.; Lin, G. M. L. *J. Am. Chem. Soc.* **1979,** *101,* 3916.
30. Bentley, T. W.; Bowen, C. T.; Parker, W.; Watt, C. I. F.; *J. Am. Chem. Soc.* **1979,** *101,* 2486.
31. Knier, B. L.; Jencks, W. P. *J. Am. Chem. Soc.* **1980,** *102,* 6789.
32. MacMillan, J.; Pryce, R. J. *J. Chem. Soc. B.* **1970,** 337.
33. Harris, J. M.; Raber, D. J.; Hall, R. E.; Schleyer, P.v.R. *J. Am. Chem. Soc.* **1970,** *92,* 5729.
34. Bailey, T. H.; Jackson, E.; Kohnstam, G.; Queen, A. *Chem. Commun.* **1966,** 122.
35. Bunton, C. A.; Mhala, M. M.; Moffatt, J. L. *J. Org. Chem.* **1984,** *49,* 3639.
36. Swain, C. G.; Scott, C. B. *J. Am. Chem. Soc.* **1953,** *75,* 141.
37. Scott, J. M. W. *Can. J. Chem.* **1970,** *48,* 3807.
38. Peterson, P. E.; Vidrine, D. W.; Walker, F. J.; Henrichs, P. M.; Magaha, S.; Stevens, B. *J. Am. Chem. Soc.* **1977,** *99,* 7968.
39. Pearson, R. G.; Sobel, H.; Songstad, J. *J. Am. Chem. Soc.* **1968,** *90,* 319.
40. Beringer, F. M.; Gindler, E. M. *J. Am. Chem. Soc.* **1955,** *77,* 3200.
41. Pocker, Y.; Parker, A. J. *J. Org. Chem.* **1966,** *31,* 1526.
42. Lowry, T. H.; Richardson, R. S. *Mechanism and Theory in Organic Chemistry,* 2nd Ed., Harper and Row: New York, 1981; p 337.
43. Jaffe, H. H. *Chem. Rev.* **1953,** *53,* 191.
44. Rossell, J. B. *J. Chem. Soc.* **1963,** 5183.
45. Kevill, D. N.; Shen, B. W. *J. Am. Chem. Soc.* **1981,** *103,* 4515.
46. Kevill, D. N.; Fujimoto, E. K. *J. Chem. Soc., Chem. Commun.* **1983,** 1149.
47. For a review, with references, see Kevill, D. N., in *The Chemistry of Functional Groups,* Supplement D; Patai, S.; Rappoport, Z., Eds.; Wiley: New York, 1983; pp 939–948.
48. Pocker, Y.; Wong, W. H. *J. Am. Chem. Soc.* **1975,** *97,* 7097.

RECEIVED for review October 21, 1985. ACCEPTED July 31, 1986.

Nucleophilicity, Advances in Chemistry Series 215
Changes and Corrections to Chapter 20

Please note that the following changes and corrections are necessary because of errors introduced by ACS staff or outside typesetters.

The correct spelling of the author's name is Fărcaşiu, not "Fărcaşciu" as in the running heads at the top of each page.

On page 286, in paragraph 2 under "Solvent Nucleophilicity", line 2, "(*20a*) and the counterion was derived from antimony pentafluoride (*20b*) in sulfur dioxide" should read "(20a)—the counterion was derived from antimony pentafluoride (*20b*)—in sulfur dioxide".

On page 287, in paragraph 1 under "Method for Measuring Solvent Basicity", on line 2, "we used" should read "we had to use".

On page 288, in the next to the last line, "although AA" should read "whereas AA".

On page 289, in paragraph 2, line 1, "Although in the first set" should read "Whereas in the first set".

On page 290, in the second complete paragraph, on the last line, references "(*30–35*)" should read "(*29, 33*)".

On page 292, paragraph 2, line 6, "from the effect of the II" should read "from the effect of the 2NP". In line 7 of the same paragraph, "solution where" should read "solution, where".

On page 295, in the Acknowledgments, "co-workers; the assistance of Gaye" should read "co-worker, Miss Gaye". In line 4 of the same paragraph, "with Herbert" should read "with Professors Herbert".

20

Basicity–Nucleophilicity–Electrophilicity in the Protonation of Aromatics

Dan Fărcaşiu[1]

Corporate Research Science Laboratories, Exxon Research and Engineering Company, Annandale, NJ 08801

The relative strengths of weakly basic solvents are evaluated from the extent of protonation of hexamethylbenzene by trifluoromethanesulfonic acid (TFMSA) in those solvents or from the effect of added base on the same protonation in solution in trifluoroacetic acid (TFA), the weakest base investigated. The basicity TFA < difluoroacetic acid < dichloroacetic acid (DCA) < chloroacetic acid < acetic acid parallels the nucleophilicity. 2-Nitropropane appears to be a significantly stronger base than DCA by the first approach, although in the second type of measurement, the two have essentially equal basicity. The discrepancy is due to an interaction, possible for hydroxylic solvents such as DCA, with the anion of TFMSA. This "anion stabilization" is a determining factor of carbocationic reactivity in chemical reactions, including solvolysis. A distinction is made between carbocation stability, determined by structure, and persistence (existence at equilibrium, e.g., in superacids), determined by environment, that is, by anion stabilization.

Nucleophilicity and Basicity

ALL NUCLEOPHILES ARE BASES (*1*). In fact, within the definition of Lewis (*2*), nucleophilicity is basicity. Following Ingold (*3*), however, physical–organic chemists have normally used the Brønsted–Lowry definition (*4*, *5*) of bases as affinity for protons. Likewise, nucleophilicity referred to affinity for nuclei of other elements, most often carbon (*3*). Another classification reserves basicity, and its counterpart acidity, for equilibrium measurements, while nucleophilicity and its counterpart electrophilicity refer to rate measurements (*6*). The terms "carbon basicity" and "hydrogen nucleophilicity" have been employed (*7–9*). This classification does not seem to have gained much acceptance.

[1]Current address: Department of Chemistry, Clarkson University, Potsdam, NY 13676

0065-2393/87/0215-0285$06.00/0

In everyday practice, a restricted view is normally taken: basicity refers to the proton and is determined in equilibrium processes, although one can and does speak of kinetic basicity and acidity (*10*), which do not necessarily parallel the corresponding thermodynamic, or equilibrium, quantities (*11*). Likewise, nucleophilicity refers to all nuclei other than the proton and is a kinetic parameter, although, again, one can speak of equilibrium nucleophilicity (*12*). These practical meanings of basicity and nucleophilicity are used in this paper.

A quantitative, or at least semiquantitative, relationship between basicity and nucleophilicity is intuitively expected and was sometimes observed (*13*). In other cases, however, not even a "limited orderly behavior" was seen (*12*). The present work was undertaken as a further study of the relationship between basicity and nucleophilicity.

Solvent Nucleophilicity

Nucleophilicity has been discussed by many investigators as one of the properties of the reaction medium that determine the reactivities in carbocationic solvolysis reactions (*6*, *14–18*). The nucleophilicity parameter of a solvent, N (*15*, *16*), was determined for most systems by dissection of solvent effects into solvent ionizing power (described by the parameter Y) and nucleophilicity (N), by a correlation technique (*19*) that could raise doubts as to whether the relevant solvent properties were fully separated into the two parameters.

Other sets of N values were determined more directly from the rates of reaction between nucleophiles and tetramethylchloronium ions (*20a*) and the counterion was derived from antimony pentafluoride (*20b*) in sulfur dioxide solutions (*20a*) and of reaction between triethyloxonium fluorophosphate and nucleophiles used as solvents (*21*). The first approach (*20a*) has been criticized (*21*) on the ground that nucleophilicity of individual molecules or small clusters might be different from that of the compound as a solvent. The second approach (*21*) might be questioned for the assumption that solvent electrophilicity (*18*) has no effect on the measured rates.

Although criticisms leveled at each of the experimental approaches employed to determine nucleophilicities are valid, we must start from the realization that these three scales of N parameters (*19*, *20a*, *21*) are all we have so far. Even though the numbers might not be quite correct, the ordering of solvents is in general as predicted from their structures. Moreover, at least a rough parallelism among the three scales of N values exists (*19*, *20a*, *21*).

Measurements of Superacidity

We previously reported on a method of superacidic strength evaluation based on the measurement by ^{13}C-NMR spectroscopy of the degree of protonation of aromatic hydrocarbons (*22*). Using benzene as the indicator base (equation 1), Fărcaşiu and co-workers (*23–25*) established the acidity order HF–SbF_5 and HBr–$AlBr_3$ > HF–TaF_5.

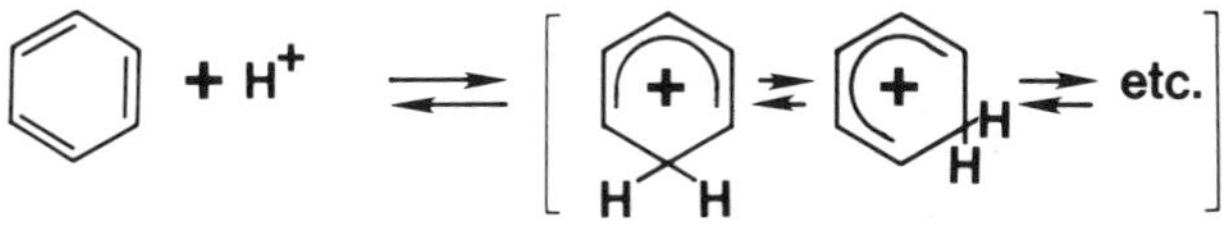

$$C_6H_6 + H^+ \rightleftharpoons [\ldots \rightleftharpoons \ldots \rightleftharpoons \text{etc.}] \quad (1)$$

The high strength of the hydrogen bromide–aluminum bromide system has invalidated an entire scheme of classification of superacids (*26*).

When the hydrocarbon base was hexamethylbenzene (**I**), which is estimated as 10^{10} times more basic than benzene (*27*), the strength of acids such as trifluoromethanesulfonic acid (triflic acid) could be examined (*28*, *29*). The position of the protonation equilibrium (**I** ⇄ **II**, equation 2) was also determined by ^{13}C-NMR spectroscopy.

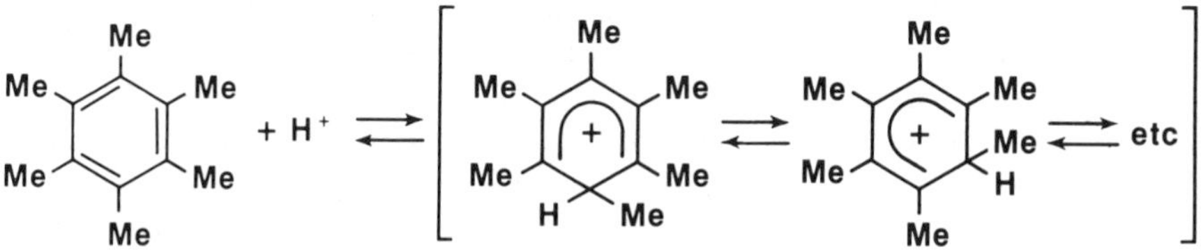

$$\textbf{I} + H^+ \rightleftharpoons [\textbf{II} \rightleftharpoons \ldots \rightleftharpoons \text{etc}]$$

Equation 2

The Hammett acidity function (*30*, *31*) for triflic acid was reported as H_0 = −13 for an "aged" acid (*32a*) presumably containing small amounts of water and −14.2 for an acid to which a little of its anhydride had been added (*32b*). We found that hexamethylbenzene is fully protonated to **II** when dissolved in triflic acid at a concentration of 0.5 M (acid-to-hydrocarbon ratio of about 22) (*28*, *29*).

Method for Measuring Solvent Basicity

At smaller ratios of acid to aromatic, partial protonation of the aromatic was observed. To vary the ratio systematically, from zero up, we used a solvent that is nonacidic toward hexamethylbenzene. If the solvent had some basicity toward triflic acid, another protonation equilibrium was established (equation 3, in which B is the basic solvent):

$$B + H^+ \rightleftharpoons BH^+ \quad (3)$$

Protonation of hexamethylbenzene in such a solvent is effected both by the triflic acid (symbolized by H^+ in equations 2 and 3) and by the conjugate acid of the solvent, BH^+. The degree of protonation (percentage of **II** in the mixture with **I**) at given stoichiometric concentrations of triflic acid and hydrocarbon is determined by the actual concentration of triflic acid as determined by equation 3 and by the acidic strength of BH^+ (both dictated by the basic strength of B). For instance, if in two solvents, B and B′, triflic acid is converted to BH^+ and $B'H^+$ to the extent of 99.0% and 99.9%, respectively, essentially all the observed protonation of the aromatic is effected by the conjugate acid of the solvent in each case. The protonating ability of the former solution (in B) is about 1 order of magnitude higher than that of the latter solution (in B′), because of the higher acid strength of BH^+. The amount of **II** at equilibrium, which is easily determined, is then a measure of the basic strength of the solvent.

Relative Strengths of Weakly Basic Solvents

Using this approach, we measured the relative base strengths of solvents of low basicity. Our intention was to compare the values obtained with the corresponding nucleophilicity parameters, *N* (*19, 20a, 21*), to determine whether a proportionality, or at least a parallelism, exists between the two properties. The solvents investigated were acetic acid (AA), chloroacetic acid (CA), dichloroacetic acid (DCA), difluoroacetic acid (DFA), trifluoroacetic acid (TFA), 2-nitropropane (2NP), and acetonitrile. To assure solubilization of the unprotonated aromatic, **I**, and to provide an internal NMR standard, we normally added 20–25% of chloroform to the samples. (Each sample contained approximately 0.4 mmol of hexamethylbenzene/mL of solvent.) Insufficient solubilities in nitromethane and tetrahydrothiophene 1,1-dioxide precluded their investigation. All other solvents, including CA, which is a solid at room temperature, gave samples that were homogeneous at 60 °C, the temperature at which all the proton shifts and transfers of equation 2 are fast on the ^{13}C-NMR (22.5-MHz) time scale. Clean ^{13}C-NMR spectra were obtained; these spectra contained, besides the signals of solvent and of acid, one signal for the aromatic and one for the methyl carbon of hexamethylbenzene. The exception was the acetonitrile solutions, for which some small impurity peaks developed in the aliphatic region (16–20 ppm) upon heating at 60 °C. Consequently, acetonitrile was dropped from the study.

The degree of protonation of hexamethylbenzene as a function of the acid-to-hydrocarbon ratio in some of the solvents is shown in Figure 1. The relative basicity could be assessed only for the least basic compounds in the series, namely, TFA < DFA < DCA. Protonation to a small extent might occur in CA at acid-to-hydrocarbon ratios of 10 and higher. The inherent error of our measurements, however, does not allow any confidence for levels of protonation of 5% or less. No protonation occurs at the acid-to-hydrocarbon ratio of 10 in 2NP, although AA, certainly more basic than CA, was not tested.

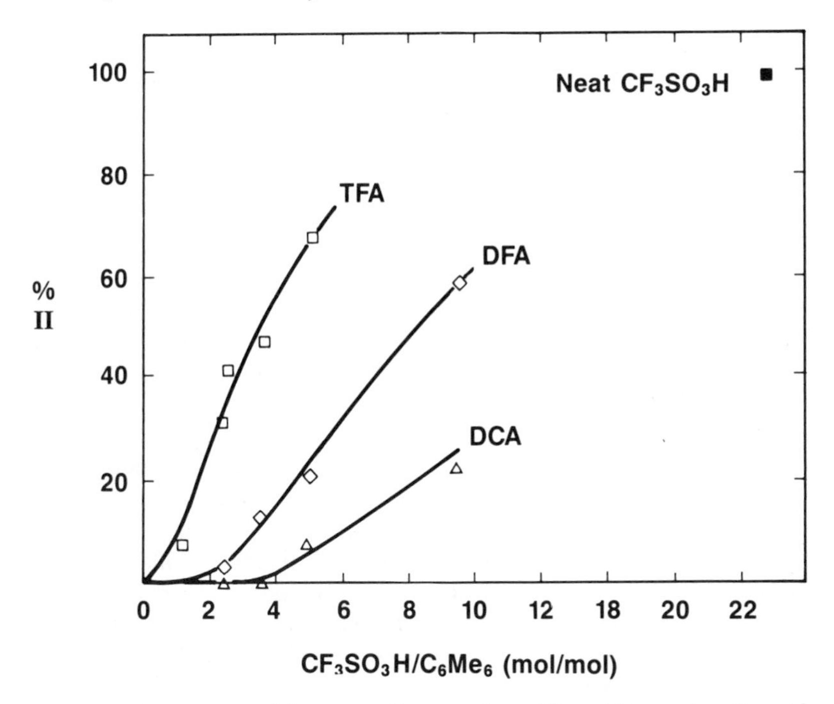

Figure 1. Protonation of hexamethylbenzene by triflic acid in carboxylic acid solvents.

To differentiate among the more basic solvents of our group, we conducted another series of experiments, in which the base was added in small increments to solutions containing fixed quantities of triflic acid and hexamethylbenzene in the least basic solvent of the series, TFA. The relative basicity could then be evaluated by the displacement of the protonation equilibrium of equation 2 to the left (less **II**) induced by B.

Although in the first set of experiments two reagents (hexamethylbenzene and triflic acid) had to be measured accurately in small quantities for each sample, three reagents (hexamethylbenzene, triflic acid, and base) had to be likewise dosed for each sample in the second set of experiments; this situation led to higher errors. We resorted, therefore, to pairwise comparisons of bases. A batch solution containing 15 mmol of triflic acid and 10 mL of solvent (75:25 TFA–chloroform) was made and used to prepare four samples containing one base and four samples containing another base (0.5 mmol of hexamethylbenzene, variable quantities of base, and 1 mL of batch acid solution). The samples were prepared in a dry box (*24*); the acid was added to the other components in an NMR tube that had been cooled to below −70 °C. The degree of protonation in each sample was measured and plotted on the same graph for the samples made with the same batch solution of acid (comparison of two bases). In view of the reported "aging" of triflic acid (*32*), a batch solution of acid was used all at once, so each pair of bases was compared in experiments employing acid of the same "age". The comparisons are shown in Figures 2–4.

Combination of data from Figures 1–3 leads to a basicity sequence (TFA < DFA < DCA < CA < AA) that parallels the nucleophilicity sequence found by Peterson and Waller (*20*).

On the other hand, Figure 4 shows that little difference between basicities of DCA and 2NP exists. In fact, the difference between the experimental points is within the combined uncertainty of the respective measurements. This result was not expected on the basis of the experiments run with the two bases as solvents. Indeed, 22% protonation to give **II** was observed for an acid-to-hydrocarbon ratio of 9.5 with DCA as the solvent (Figure 1), while no protonation was seen with 2NP as the solvent for an acid-to-hydrocarbon ratio of 10. These observations are in line with the concept of anion stabilization (*30–35*).

Stabilization of ions in solution occurs by solvation, which has a non-specific, "polarity", component and a specific component involving a direct (quasi-chemical) interaction between ions and solvent. The specific interaction between a carbocation and a solvent is of a kind that destroys the carbocationic state. Thus, the protonation reaction of equation 2 is correctly written in the form of equation 4, in which XH is the acid:

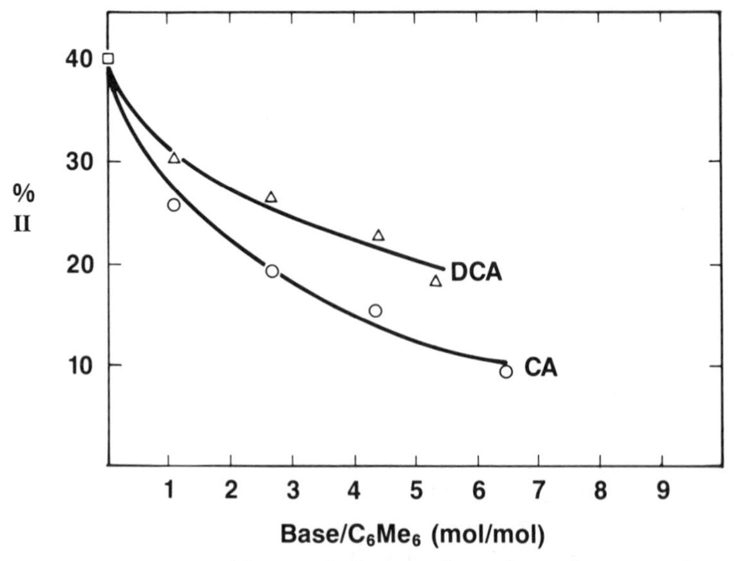

Figure 2. Protonation of hexamethylbenzene by triflic acid in TFA solution, in the presence of DCA or CA.

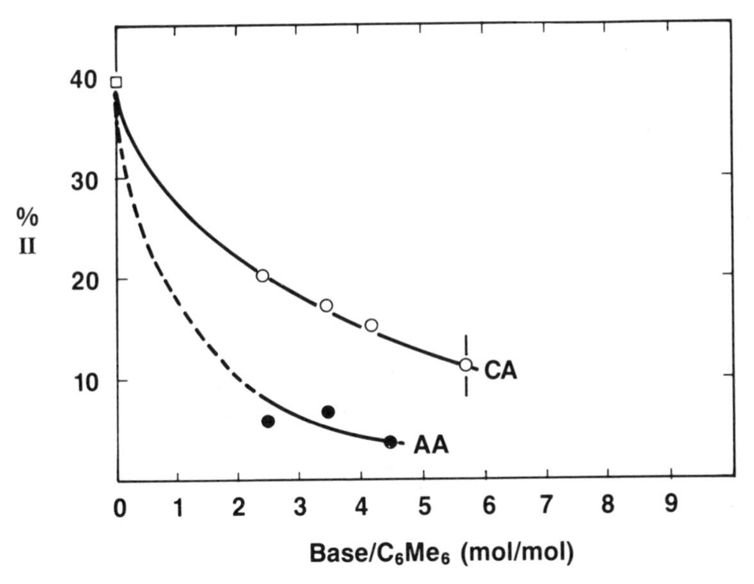

Figure 3. Protonation of hexamethylbenzene by triflic acid in TFA solution, in the presence of CA or AA.

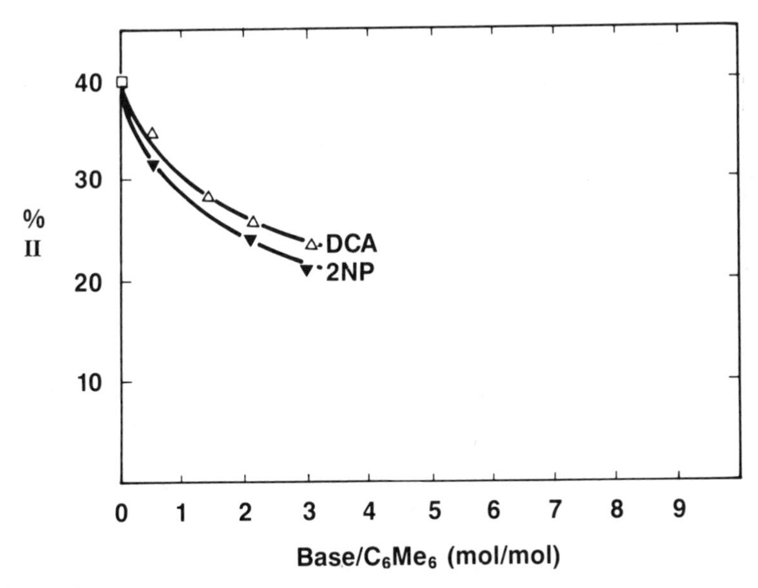

Figure 4. Protonation of hexamethylbenzene by triflic acid in TFA solution, in the presence of DCA or 2NP.

$$XH + \mathbf{I} \rightleftarrows X^- + \mathbf{II} \quad (4)$$

Any molecule that could specifically stabilize the cation **II** will be a base and combine with the proton; therefore, the equilibrium is displaced away from **II**. For other carbocations (such as the alkyl cations), the solvents interacting specifically with the positively charged species will be nucleophiles and combine with the carbocations. This relationship does not mean that carbocations cannot be nucleophilically solvated as discussed recently (*34, 35*). Interestingly, though, apparently n donors interact preferentially with hydrogen atoms rather than the cationic carbon (*36*), thus reacting as bases rather than nucleophiles (*37*). Also, Sharma et al. (*34*) concluded that nucleophilic stabilization is not the dominant solvation for carbocations in solution. In any event, nucleophilic solvation decreases rather than enhances carbocationic character. On the contrary, any specific interaction between the solvent (SOH) and the anion, such as formation of hydrogen bonds (equation 5), displaces the equilibrium of equation 4 toward formation of carbocations:

$$X^- + n\mathrm{SOH} \rightleftarrows (\mathrm{SOH}\ \cdots)_n X^- \quad (5)$$

(Hydrogen bonding is a quasi-chemical interaction because this bonding has a preferred, if not unique, space arrangement and a preferred, if not unique, stoichiometry.)

This representation explains why a carboxylic acid such as DCA, which forms hydrogen bonds with the anion, is a good solvent for carbocationic processes, while 2NP, which has a similarly low basicity but cannot act as a hydrogen bond donor to anions, does not provide a detectable amount of protonated hexamethylbenzene (**II**). The actual (low) basicity of 2NP can be assessed from the effect of the **II** on the position of the protonation equilibrium (equation 2 or 4) in TFA solution where anion stabilization is provided by the solvent.

Stable Carbocations and Persistent Carbocations

In superacid systems consisting of a composite of a Brønsted acid and a Lewis acid, the anion interacts chemically with one or more molecules of the latter. Thus, up to 4 equiv of SbF_5 has been recommended for conversion of alkyl halides to the corresponding carbocations (*38*). For those systems, the anion stabilization is accomplished by chemical bonding between the Lewis acid and the anion. The existence of carbocations in such solutions has been demonstrated by various physical methods.

The often used phrase "stable carbocations in super acids" is misleading and should be abandoned. We "see" the carbocations by various spectroscopic techniques because we remove from solution anything that can react with them. One species that cannot be removed is the anion; charge neutralization by combination of the anion and cation is the normal occurrence even for "nonnucleophilic" anions (*39*). So that the carbocation can be "seen", the anion has to be specially stabilized. The cation is then "bare" and very reactive, rather than "stabilized" in solution. Indeed, as the solution (solvent and Lewis acid) is changed to make the carbocation more "stable" as judged by NMR spectroscopy, the reactivity of the carbocation increases in reactions such as carbon monoxide trapping or hydride transfer.

A similar discussion (*40*) has been presented for free radicals, for which the distinction between *stable* and *persistent* has been introduced. The same distinction should be made for carbocations, stability being determined by structure and persistence by environment (degree of anion stabilization).

While simple, nonstabilized ions like *tert*-butyl are generated only in strong superacids—e.g., with SbF_5 (*41, 42*), $AlBr_3$ (*43, 44*), TaF_5 (*45*), $GaBr_3$ (*46*), or $AlCl_3$ (*47*) as Lewis acids—more stable species, such as **II**, can be obtained under milder conditions, in which the anion is stabilized by hydrogen bonding with the excess of acid [cooperative effect (*29, 33*)] or with the solvent (equation 5). The term "cooperative effect" for the property of an acid to form $X_nH_{n-1}^-$ clusters is preferred over the older term "homoconjugation" (*48*), because today homoconjugation is used to describe π-electron conjugation in systems with an interruption in the σ skeleton; cf. homoaromaticity (*49*). The solvents that interact with anions and as a consequence favor carbocationic processes are called anion-stabilizing solvents. The anion-stabilizing ability makes, for example, methanol a better solvent for carbocationic solvolysis than acetonitrile or dimethylformamide, although the latter solvents are more polar (*50a, 50b*).

Solvation Effects in Solvolysis Reactions

On the basis of our findings, the standard representation of the ionization of an organic substrate in solution (*51*) should be amended as shown in Scheme I. The solvent is involved in the transition state (**III**) of ionization. In the general formulation (**III**), the solvent interacts nucleophilically with the organic moiety and electrophilically with the leaving group (*52*). For solvent mixtures, solvent sorting is predicted, so that the most nucleophilic component interacts with $R^{\delta+}$, and the most electrophilic component interacts with $X^{\delta-}$ in **III**. Steric hindrance limits the nucleophilic participation to one solvent molecule (unless interaction with hydrogen atoms rather than the cationic carbon is considered), although several solvent molecules stabilize the departing anion by hydrogen bonding. Indeed, a combination of ab initio molecular orbital and statistical mechanics calculations indicates that about four water molecules are hydrogen-bonded to the entering nucleophile at the transition state of the S_N2 reaction between methyl chloride and the chloride anion (*53*); the same number of solvent molecules are bound to the structurally identical chloride leaving group (*53*). As the mechanism of the nucleophilic substitution changes gradually from S_N2 to S_N1, the hydrogen bonds between the solvent and chloride leaving group become stronger and stronger. The specific interaction of the leaving group with the solvent (anion-stabilizing effect) is the main driving force of the heterolysis leading to the ion pair (**IV**). The polarity of solvent plays a lesser role. On the other hand, for ionization of tertiary alkyl substrates in solvents as nucleophilic as the aqueous alcohols, the nucleophilic interaction between solvent and the incipient carbocation in **III**, if present, should be much smaller than the electrophilic solvent interaction with the leaving group (*18*).

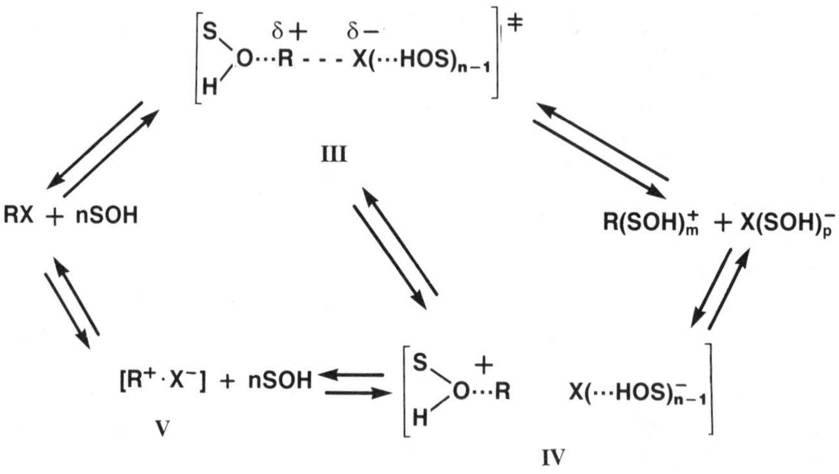

$$\left[\begin{matrix} S \\ H \end{matrix} \!\! > \! \overset{\delta+}{O \cdots R} - - - \overset{\delta-}{X} (\cdots HOS)_{n-1} \right]^{\ddagger}$$

III

$RX + nSOH$ $\rightleftharpoons$ III $\rightleftharpoons$ $R(SOH)_m^+ + X(SOH)_p^-$

$RX + nSOH \rightleftharpoons [R^+ \cdot X^-] + nSOH$ (V)

$[R^+ \cdot X^-] + nSOH \rightleftharpoons \left[\begin{matrix} S \\ H \end{matrix} \!\! > \! \overset{+}{O} \cdots R \quad X(\cdots HOS)_{n-1}^- \right]$ (IV)

III $\rightleftharpoons$ IV $\rightleftharpoons$ $R(SOH)_m^+ + X(SOH)_p^-$

Scheme I

A mechanistic model involving nucleophilic assistance, but not taking into account the variable electrophilic assistance in different solvents, has been proposed (*54, 55*) for the solvolysis of *tert*-butyl halides. The analysis was based on a comparison of solvent effects on the solvolysis rates of *tert*-butyl and adamantyl substrates. The solvent properties were analyzed in terms of parameters *N* and *Y*; the electrophilic assistance was incorporated into *Y* (*54, 56*). Such an approximation had been acceptable in the original work (*14–16*), which dealt mostly with aqueous alcohols as solvents. This approximation is no longer permissible when materials like TFA and fluorinated alcohols are used as solvents. In fact, Fainberg and Winstein (*56*) pointed out that different solvent mixtures could not be placed on the same correlation line.

The difference in behavior between *tert*-butyl and adamantyl substrates, observed essentially in solvents of low nucleophilicity but high anion-stabilizing power, is most probably due to the higher susceptibility of the "cage" substrate to the electrophilic assistance. In reference 18, we presented other known examples of reactions in which the increased sensitivity of adamantyl substrates to anion stabilization (electrophilic assistance) by the solvent is manifested.

No treatment of solvent effects on carbocationic reactivity that does not treat explicitly the anion-stabilizing effect should be acceptable (*18*), even for processes in which the rate-determining step is solvent attack on an ion pair. Probably, backside anion stabilization by the solvent intervenes even in the contact ion pair. In fact, rotation of the anion in this pair can deliver a molecule of solvent attached to the anion to the cation from the front; thus, the predominant retention of configuration occasionally observed in the solvolysis products is explained.

The process represented by **V** ⟶ **IV** occurs, however, in the solvolysis of a preionized material, such as triethyloxonium fluorophosphate, which was employed by Kevill and Lin (*21*) to establish a scale of nucleophilicity parameters, *N*. Because the effect of variable anion stabilization by solvent was not subtracted, whether their *N* values measure true and only nucleophilic reactivities is uncertain. This doubt would be dispelled, however, by an experiment in which the reagent would be treated with small amounts of nucleophile in a better anion-stabilizing solvent, such as TFA, or even sulfuric acid.

Acknowledgments

The experimental contribution of my co-workers; the assistance of Gaye Marino; the collaboration of my colleague, Dr. Rodney V. Kastrup, and his staff, who performed the NMR measurement; and most helpful discussions with Herbert C. Brown and Alan R. Katritzky are gratefully acknowledged. Heartfelt thanks are addressed also to a reviewer who divided the original text into sections and also provided headings and subheadings.

Literature Cited

1. Ingold, C. K. *Structure and Mechanism in Organic Chemistry*, 2nd ed.; Cornell University: Ithaca, NY, 1969; pp 236–237.
2. Lewis, G. N. *Valence and the Structure of Atoms and Molecules;* Chemical Catalog: New York, 1923; p 141.
3. Ingold, C. K. *Structure and Mechanism in Organic Chemistry*, 2nd ed.; Cornell University: Ithaca, NY, 1969; p 240.
4. Brønsted, J. N. *Recl. Trav. Chim. Pays-Bas* **1923,** *42*, 718.
5. Lowry, T. M. *J. Soc. Chem. Ind., London* **1923,** *42*, 43.
6. Swain, C. G.; Scott, C. B. *J. Am. Chem. Soc.* **1953,** *75*, 141.
7. Parker, A. J. *Proc. Chem. Soc., London* **1961,** 371.
8. Hine, J.; Weimar, R. D., Jr. *J. Am. Chem. Soc.* **1965,** *87*, 3387.
9. Hudson, R. F. In *Chemical Reactivity and Reaction Paths;* Klopman, G., Ed.; Wiley: New York, 1974; p 184.
10. Ingold, C. K. *Structure and Mechanism in Organic Chemistry*, 2nd ed.; Cornell University: Ithaca, NY, 1969; p 450.
11. Eigen, M. *Angew. Chem., Int. Ed. Engl.* **1964,** *3*, 1.

12. Ritchie, C. D. *J. Am. Chem. Soc.* **1983,** *105,* 3573, and other papers in the series.
13. Bordwell, F. G.; Hughes, D. L. *J. Org. Chem.* **1983,** *48,* 2206.
14. Grunwald, E.; Winstein, S. *J. Am. Chem. Soc.* **1948,** *70,* 846.
15. Winstein, S.; Grunwald, E.; Jones, H. W. *J. Am. Chem. Soc.* **1951,** *73,* 2700.
16. Winstein, S.; Fainberg, A. M.; Grunwald, E. *J. Am. Chem. Soc.* **1957,** *79,* 4146.
17. Harris, J. M. *Prog. Phys. Org. Chem.* **1974,** *11,* 89, and references cited therein.
18. Fărcaşiu, D.; Jähme, J.; Rüchardt, C. *J. Am. Chem. Soc.* **1985,** *107,* 5717.
19. Bentley, T. W.; Schleyer, P. v. R. *Adv. Phys. Org. Chem.* **1977,** *14,* 1, and references cited therein.
20a. Peterson, P. E.; Waller, F. J. *J. Am. Chem. Soc.* **1972,** *94,* 991.
20b. Peterson, P. E., personal communication.
21. Kevill, D. N.; Lin, G. M. L. *J. Am. Chem. Soc.* **1979,** *101,* 3916.
22. Fărcaşiu, D. *Abstracts of Papers,* 173rd National Meeting of the American Chemical Society, New Orleans, LA; American Chemical Society: Washington, DC, 1977; ORGN 188.
23. Fărcaşiu, D.; Fisk, S. L.; Melchior, M. T. Presented at the International Symposium on Chemistry of Carbocations, Royal Society of Chemistry, Bangor, Wales, Sept 10, 1981.
24. Fărcaşiu, D.; Fisk, S. L.; Melchior, M. T.; Rose, K. D. *J. Org. Chem.* **1982,** *47,* 453.
25. Fărcaşiu, D. *Acc. Chem. Res* **1982,** *15,* 46.
26. Olah, G. A.; Prakash, G. K. S.; Sommer, J. *Science (Washington, D.C.)* **1979,** *206,* 13.
27. Mackor, E. L.; Hofstra, A.; van der Waals, J. H. *Trans. Faraday Soc.* **1958,** *54,* 186.
28. Fărcaşiu, D.; Marino, G.; Kastrup, R. V.; Rose, K. D. *Abstracts of Papers,* 185th National Meeting of the American Chemical Society, Seattle, WA; American Chemical Society: Washington, DC, 1983; ORGN 160.
29. Fărcaşiu, D. Presented at the EUCHEM Conference on Superacidic and Superbasic Media (Liquid and Solid), Cirencester, England, Sept 13, 1984.
30. Hammett, L. P. *Physical Organic Chemistry,* 2nd ed.; McGraw-Hill: New York, 1970; p 278.
31. Rochester, C. *Acidity Functions;* Blomquist, A. T., Ed.; Organic Chemistry Monographs; Academic: New York, 1970; Vol. 17.
32a. Kramer, G. M. *J. Org. Chem.* **1975,** *40,* 302.
32b. Kramer, G. M., personal communication.
33. Fărcaşiu, D. Seminar at the Faculty of Engineering, Kyoto University, Kyoto, Japan, March 22, 1985.
34. Sharma, R. B.; Sen Sharma, D. K.; Hiraoka, K.; Kebarle, P. *J. Am. Chem. Soc.* **1985,** *107,* 3747.
35. Sen Sharma, D. K.; Meza de Höjer, S.; Kebarle, P. *J. Am. Chem. Soc.* **1985,** *107,* 3757.
36. Meot-Ner (Mautner), M.; Ross, M. M.; Campana, J. E. *J. Am. Chem. Soc.* **1985,** *107,* 4839.
37. Henchman, M.; Hierl, P. M.; Paulson, J. F. *J. Am. Chem. Soc.* **1985,** *107,* 2812.
38. Kelly, D. P.; Brown, H. C. *Aust. J. Chem.* **1976,** *29,* 957.
39. Zefirov, N. S.; Koz'min, A. S. *Acc. Chem. Res.* **1986,** *18,* 154.
40. Griller, D.; Ingold, K. U. *Acc. Chem. Res.* **1976,** *9,* 13.
41. Olah, G. A.; Tolgyesi, W. S.; Kuhn, S. J.; Moffatt, M. E.; Bastien, I. J.; Baker, E. B. *J. Am. Chem. Soc.* **1963,** *85,* 1328.

42. Brouwer, D. M.; Mackor, E. L. *Proc. Chem. Soc. London* **1964,** 147.
43. Kramer, G. M. *Int. J. Mass Spectrom. Ion Phys.* **1976,** *19,* 139.
44. Kramer, G. M. *J. Org. Chem.* **1979,** *44,* 2616.
45. Fărcaşiu, D. *J. Org. Chem.* **1979,** *44,* 2103.
46. Jensen, F. R.; Beck, B. H. *Tetrahedron Lett.* **1966,** 4287.
47. Kalchschmid, F.; Mayer, E. *Angew. Chem.* **1976,** *88,* 849.
48. Kolthoff, I. M.; Chantooni, M. K., Jr. *J. Am. Chem. Soc.* **1965,** *87,* 1004.
49. Winstein, S. *J. Am. Chem. Soc.* **1959,** *81,* 6524.
50a. Abraham, M. H.; Taft, R. W.; Kamlet, M. J. *J. Org. Chem.* **1981,** *46,* 3053.
50b. Wiseman, J. R., personal communication.
51. Winstein, S.; Appel, B.; Baker R.; Diaz, A. *Spec. Publ.—Chem. Soc.* **1965,** *No. 19,* 109.
52. Bentley, T. W.; Bowen, C. T.; Morten, D. H.; Schleyer, P. v. R. *J. Am. Chem. Soc.* **1981,** *103,* 5466.
53. Chandrasekhar, J.; Smith, S. F.; Jorgensen, W. L. *J. Am. Chem. Soc.* **1985,** *107,* 154.
54. Bentley, T. W.; Bowen, C. T.; Parker, W.; Watt, C. I. F. *J. Am. Chem. Soc.* **1979,** *101,* 2486.
55. Bentley, T. W.; Carter, G. E. *J. Am. Chem. Soc.* **1982,** *104,* 5741.
56. Fainberg, A. H.; Winstein, S. *J. Am. Chem. Soc.* **1957,** *79,* 1602.

RECEIVED for review October 21, 1985. ACCEPTED JANUARY 24, 1986.

21

Equivalent Scales for Correlation Using Two Solvent Parameters

Paul E. Peterson

Department of Chemistry, University of South Carolina, Columbia, SC 29208

Conversion of the Swain A *and* B *solvent parameter scales to nucleophilicity,* N, *and ionizing power,* Y, *is described. Results using* Y *values based on* tert-*butyl chloride rates are given, and the results of the comparable transformation using* Y *values based on adamantyl tosylate rates are presented. For hydroxylic "reaction" solvents and a few others, the converted values resemble published values or are reasonable. For other "nonreaction" solvents (the majority), we assign no meaning to the converted values, although we list them for others to examine. The nucleophilicity ratio of two solvents, chosen to be acetic and formic acid, may be arbitrarily chosen to give equivalent scales. The ratio may be adjusted to agree with chemical experience independent of the effect of solvent variation. The nucleophilicity of solvent component molecules in a constant solvent is an example of such independent chemical experience.*

THE GREAT TENDENCY OF CYCLOOCTANE OXIDE to react by transannular hydrogen shift in the solvent trifluoroacetic acid was ascribed to the successful competition of internal hydrogen nucleophiles with relatively nonnucleophilic solvent (*1*). Subsequently, Peterson et al. (*2*) extended studies of neighboring-group participation in trifluroroacetic acid to other major types of solvolytic reactions, including tosylate solvolyses. 1,4-Halogen participation was clearly revealed through halogen shifts and kinetic evidence (*3a*), whereas only limited evidence for such participation had been found by Winstein and co-workers [discussed in Peterson (*36*)] working in more nucleophilic solvents.

With tosylate solvolysis rates available for the first time in trifluoroacetic acid, Peterson and Waller (*4*) noted that plots of log *k* versus the ionizing power, *Y*, for the solvolysis of methyl, ethyl and secondary tosylates showed deviations from linearity that could be attributed to solvent nucleophilicity.

0065-2393/87/0215-0299$06.00/0

From the deviations, a scale of nucleophilicity was derived. Halogenated acetic acids were included, on the basis of reactivities with halonium ions. Other scales appeared from the Schleyer group (*5*, *6*) at about the same time. The various nucleophilicity scales were used to correlate solvolysis rates by now familiar four-parameter equations $\Delta G = N + mY$ or $\Delta G = sN + mY$. (G = free energy; N = solvent nucleophilicity; Y = solvent ionizing power; s = sensitivity; m = sensitivity.) Previously, parameters for such equations had not been determined.

The availability of the above-mentioned scales and equations led Peterson et al. (*7*) to reexamine an existing correlation of reaction rates by an equation involving two solvent parameters—the Swain–Moseley–Bown equation (*8*):

$$\Delta G = c_1 d_1 + c_2 d_2 \quad (1)$$

Here, the *d*'s are solvent parameters and the *c*'s are sensitivities to them. We were quite interested to find that a suitable transformation revealed an equivalent pair of solvent parameters that were interpretable as nucleophilicity and ionizing power. The parameters had been listed in a number of physical organic textbooks, but their significance had not been clear.

Transformation of the A *and* B *Parameters*

Recently, Swain et al. (*9*) introduced another set of solvent and reaction parameters for the correlation of free energies. Further commentary regarding them has appeared (*10*, *11*). The free energies are given by equation 1.

$$G = aA + bB + C \quad (2)$$

In equation 2, A and B are parameters characteristic of the solvent. The parameters a and b are characteristic of the reaction. I have now converted these parameters to obtain N and Y values. The conversion is described in the Appendix.

Kevill (*12*) reported a related transformation of A and B to obtain N and Y that utilizes the assumption that methyl tosylate solvolysis rates obey the equation $N_{\mathrm{OTs}} = \log (k/k_0) - 0.3Y_{\mathrm{OTs}}$ to determine the proportions of N and Y in A and B. Kevill tabulated values of N and Y for hydroxylic solvents. In common with Kevill's treatment, these solvents and a few others that we designate as reaction solvents are the only ones for which we have interpreted the converted values in terms of independent nucleophilicity and ionizing power parameters.

In our conversion, the assumed nucleophilicity ratio for two chosen solvents is used to determine the proportions of N and Y in A and B. We have given the formulas to determine sensitivity parameters, which were not considered by Kevill (*12*). Although we give converted N values for all solvents in the Swain data set, we note that values for solvents that do not serve as nucleophiles in the Swain data set are presently regarded as arbitrary numbers arising from the assumptions made. We shall see that these N parameters for nonreaction solvents tend, in fact, to be linearly related to the Y parameters. We have not determined whether the modest improvement in correlations presumably made possible by Swain's determination of two solvent parameters for the nonreaction solvents may be traced to any solvent property that is describable in familiar terms.

Here, we give an abbreviated description of our conversion, which is outlined fully in the Appendix. We assume that the parameters are separable into components as follows:

$$A_{\mathrm{hep}}{}^{\mathrm{S1}} = n_A N_{\mathrm{hep}}{}^{\mathrm{S1}} + y_A Y_{\mathrm{hep}}{}^{\mathrm{S1}} \quad (3)$$

$$B_{\mathrm{hep}}{}^{\mathrm{S1}} = n_B N_{\mathrm{hep}}{}^{\mathrm{S1}} + y_B Y_{\mathrm{hep}}{}^{\mathrm{S1}} \quad (4)$$

The superscripts and subscripts are solvent designations. The term $A_{\mathrm{hep}}{}^{\mathrm{S1}}$ refers to the difference in A values between heptane, a zero point on the Swain scale, and solvent S1, for example. In our first investigated conversion, we base the Y values on the rates of solvolysis of *tert*-butyl chloride, as calculated from the A, B, a, and b parameters. As shown in the Appendix, making the additional assumption that the nucleophilicities of two solvents, Sref1 and Sref2, have the ratio R that allows the calculation of the constants needed to apply equations 3 and 4. The results are

$$y_A = \frac{RA_{\mathrm{hep}}{}^{\mathrm{Sref2}} - A_{\mathrm{hep}}{}^{\mathrm{Sref1}}}{RY_{\mathrm{hep}}{}^{\mathrm{Sref2}} - Y_{\mathrm{hep}}{}^{\mathrm{Sref1}}} \quad (5)$$

$$\frac{y_A}{y_B} = \frac{BA_{\mathrm{hep}}{}^{\mathrm{Sref2}} - A_{\mathrm{hep}}{}^{\mathrm{Sref1}}}{RB_{\mathrm{hep}}{}^{\mathrm{Sref2}} - Y_{\mathrm{hep}}{}^{\mathrm{Sref1}}} = f(R) \quad (6)$$

$$\frac{n_A}{n_B} = \frac{BA_{\mathrm{hep}}{}^{\mathrm{S2}} - A_{\mathrm{hep}}{}^{\mathrm{S1}}}{RB_{\mathrm{hep}}{}^{\mathrm{S2}} - Y_{\mathrm{hep}}{}^{\mathrm{S1}}} = \text{constant} \quad (7)$$

$$s_{\mathrm{RX}} = a_{\mathrm{RX}} n_A + b_{\mathrm{RX}} n_B \quad (8)$$

$$m_{\mathrm{RX}} = a_{\mathrm{RX}} y_A + b_{\mathrm{RX}} y_B \quad (9)$$

Values of n_AN or n_BN for all solvents were obtained from equations 3 or 4, because the y value was available from equation 5, when the nucleophilicity ratio, R, for the reference solvents acetic and formic acid was chosen to be 1. The nN values for solvolysis solvents were compared with the Schadt–Bentley–Schleyer proposed set of nucleophilicity values. The lin-

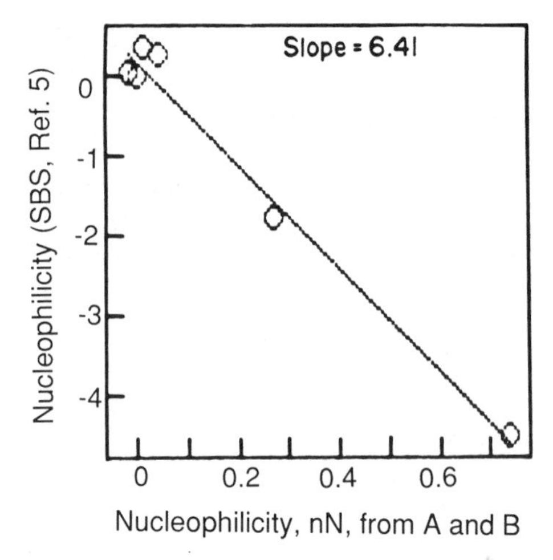

Figure 1. Plot of nucleophilicities of Bentley et al. (6) *versus the unscaled nucleophilicities,* n_{AN}, *from transformation of the* A *and* B *parameters. Solvents were ethanol, methanol, 50% ethanol in water, water, acetic acid, formic acid (superimposed), and trifluoroacetic acid.*

earity of the plot (Figure 1) shows that the A and B values are indeed interpretable in terms of familiar parameters. The slope of the plot was used to define the magnitude of n_A. The numerical values used in our conversion are as follows (parameter, value): y_A, 0.06274; y_B, 0.09534; n_A, -0.15595; and n_B, 0.20386 (all data are devised from data having two to three significant numbers; the additional numbers are supplied to facilitate the reproduction of our values). The equations involving them are summarized in the Box. In Tables I and II, the "*tert*-butyl chloride based" A and B values and s and m values, respectively, are given.

Summary of Equations

$$s = 0.15595a + 0.20386b$$

$$m = 0.06274a + 0.09534b$$

$$A_{S1}{}^{S2} = -0.15595N_{S1}{}^{S2} + 0.06274Y_{S1}{}^{S2}$$

$$B_{S1}{}^{S2} = 0.20386N_{S1}{}^{S2} + 0.09534Y_{S1}{}^{S2}$$

$$N_{S1}{}^{S2} = (A_{S1}{}^{S2} - 0.06274Y_{S1}{}^{S2})/-0.15595$$

$$Y_{S1}{}^{S2} = 7.37A_{S1}{}^{S2} + 5.65B_{S1}{}^{S2}$$

Table I. Solvent Nucleophilicity (*N*) and Ionizing Power (*Y*) Obtained by Transformation of *A* and *B* Parameters: *tert*-Butyl Chloride Based

Solvent j from Ref 9	*Structure*	N *from* A *and* B	Y *from* A *and* B	*Solvent j from Ref 9*	*Structure*	N *from* A *and* B	Y *from* A *and* B
1	CCl_4	1.64	−6.61	32	$(CH)_5N$	2.53	−2.01
2	$CHCl_3$	1.39	−1.98	33	PhBr	1.92	−3.85
3	CH_2Cl_2	1.86	−2.25	34	PhCl	1.96	−4.05
4	HCOOH	−1.73	2.38	35	$PhNO_2$	2.00	−4.76
5	$HCONH_2$	1.15	1.25	36	PhH	2.00	−4.76
6	$MeNO_2$	1.92	−1.13	37	$PhNH_2$	2.64	0.17
7	MeOH	−0.27	−0.85	38	$(CH_2)_5CO$	2.11	−2.90
8	CS_2	1.70	−6.31	39	$(CH_2)_6$	1.25	−8.71
9	Cl_2CCCl_2	1.40	−7.05	40	$Me(CH_2)_4Me$	1.12	−9.18
10	$ClCHCCl_2$	1.85	−4.97	41	Et_3N	1.33	−7.53
11	CF_3COOH	−4.75	3.48	42	$(Me_2N)_3PO$	3.61	−3.16
12	MeCN	1.85	−1.62	43	PhCN	2.12	−2.08
13	$ClCH_2CH_2Cl$	2.01	−2.36	44	PhMe	1.96	−5.19
14	MeCOOH	−1.73	−1.61	45	PhOMe	2.12	−3.47
15	EtOH	−0.08	−1.79	46	PhNHMe	2.23	−0.21
16	MeSOMe	2.46	−0.60	47	$2.6\text{-}C_5H_3NMe_2$	2.40	−3.30
17	$HOCH_2CH_2OH$	0.40	1.29	48	$Me(CH_2)_5Me$	1.18	−9.19
18	MeCOMe	2.15	−2.78	49	PhCOMe	2.43	−2.42
19	$HCONMe_2$	2.25	−1.74	50	$o\text{-}C_6H_4Me_2$	2.17	−5.76
20	$CH_3CH_2CH_2OH$	0.01	−2.07	51	$m\text{-}C_6H_4Me_2$	1.18	−9.19
21	Me_2CHOH	0.14	−2.36	52	$p\text{-}C_6H_4Me_2$	2.11	−5.93
22	MeCOEt	2.07	−3.32	53	$Me_3CCH_2CHMe_2$	1.08	−9.29
23	$(CH_2)_4O$	2.11	−4.16	54	Bu_2O	1.61	−7.17
24	MeCOOEt	1.79	−4.32	55	H_2O	0.00	3.82
25	$O(CH_2CH_2)_2O$	2.04	−4.01	56	96% MeOH	−0.06	−0.15
26	$MeCONMe_2$	2.45	−1.73	57	80% EtOH	0.07	0.00
27	BuOH	0.05	−2.27	58	60% EtOH	0.17	1.05
28	EtOEt	1.54	−6.39	59	50% EtOH	0.17	1.36
29	Me_3COH	0.76	−3.06	60	80% MeCOMe	0.63	−0.68
30	$MeOC_2H_4OMe$	1.59	−4.83	61	70% MeCOMe	0.58	−0.16
31	$BuNH_2$	3.32	−1.49				

Examination of the A, B, N, *and* Y *Parameters*

Examination of the properties of only the hydroxylic solvents was facilitated by a plotting program I wrote for the IBM personal computer. Data from two separate files may be read to an array for plotting. A third file is read to discriminate points for plotting as large or small circles.

As Taft et al. (*10*) noted, a large proportion of the solvents considered by Swain have *A* roughly linearly related to *B*. As expected, the *N* and *Y* values in our conversion exhibit a similar phenomenon (Figure 2), as do related plots (Figures 3–6). Amines and amides, some of which served as nucleophiles in the Swain data set, also deviate from the linearity of the plots for nonreaction solvents.

Table II. Sensitivities of Nucleophilicity(*s*) and Sensitivity to Ionizing Power (*m*) Obtained by Transformation of *a* and *b*: *tert*-Butyl Chloride Based

i *from Ref 9*	*Reactant*	s *from a and* b	m *from a and* b	i *from Ref 9*	*Reactant*	s *from a and* b	m *from a and* b
1	MeBr	1.36	0.06	39	5-methylfurfural	0.26	0.18
2	MeOTs	1.05	0.23	40	1-nitroso-2-naphthol	0.47	0.23
3	BuBr	0.97	0.29	41	2-nitroso-1-naphthol	0.26	0.25
4	$PhCH_2Cl$	0.98	0.38	42	$Et_4N + I^-$	0.08	1.59
5	Me_2CHOTs	0.22	0.57	43	Kosower Z	−5.36	3.65
6	*cyclo*-C_5H_9OTs	0.10	0.67	44	MPI	5.30	−3.53
7	*cyclo*-$C_6H_{11}OTs$	0.03	0.76	45	$3\text{-}MeOC_5H_4N + O^-$ (1)	−1.24	0.53
8	*endo*-$C_7H_{11}OTs$	−0.13	0.71	46	$3\text{-}MeOC_5H_4N + O^-$ (2)	−2.23	0.56
9	*exo*-$C_7H_{11}OTs$	−0.10	0.86	47	$3\text{-}MeOC_5H_4N + O^-$ (3)	−2.13	0.49
10	Ph_2CHCl	−0.20	1.67	48	$PhNO_2$	0.27	0.21
11	2-AdOTs	−0.32	0.92	49	$4\text{-}MeOPhNO_2$	0.36	0.25
12	Me_3CCl, Y	0.00	1.00	50	$4\text{-}Et_2NPhNO_2$	0.30	0.36
13	Me_3CBr	0.37	0.94	51	Ph_2CO	−0.30	0.11
14	$PhCMe_2O_2COPh$	−0.46	0.43	52	pyrimidine	−0.35	0.12
15	Ph_3CF	−1.95	1.11	53	pyridazine	−0.53	0.23
16	Ph_3COAc	−1.39	0.75	54	pyrroline oxide	−0.47	0.24
17	$MeI + (EtCH_2)_3N$	0.85	0.47	55	iron imine	−0.45	0.23
18	$MeI + PhNMe_2$	0.78	0.50	56	oximate	−3.15	1.61
19	$MeI + 3\text{-}ClPhNMe_2$	0.77	0.57	57	sulfoxide	0.57	0.17
20	$MeI + 4\text{-}ClPhNMe_2$	0.78	0.57	58	Dimroth E_T30	−3.83	2.33
21	$MeI + 3\text{-}MePhNMe_2$	0.78	0.49	59	Dimroth E_T26	−3.70	1.89
22	$MeI + 4\text{-}MeOPhNMe_2$	0.78	0.47	60	$Brooker_{XR}$	0.19	0.99
23	$EtI + Et_3N$	0.75	0.48	61	Davis *A*	−1.87	0.50
24	$EtO_2CCH_2Br + Et_3N$	−0.08	0.46	62	Davis *B*	−1.20	0.22
25	$EtO_2CCH_2I + Et_3N$	−0.02	0.42	63	Davis E_{Ct}	−0.63	0.10
26	$4\text{-}O_2NPhF + Et_4N + N_{3-}$	−1.52	0.60	64	$HCONMe_2$	−2.29	2.93
				65	$POCl_3$	1.02	2.06
27	$PhSO_2Cl + PhNH_2$	−1.11	−0.11	66	Me_2CHCH_2Cl, trans	2.49	1.42
28	$ClSO_2NCO$ + hexene	1.26	0.34	67	Me_2HPO, band 1	−4.14	3.63
29	TCNE + 4-methoxy-styrene	0.52	0.65	68	$(Me_3C)_2NO$, N	−0.19	0.10
				69	piperidyloxy, N	−0.17	0.09
30	Br_2 + 1-pentene	0.17	1.06	70	pyrrolinyloxy, N	−0.18	0.10
31	$Br_2 + Me_4Sn$	0.17	1.04	71	$4\text{-}AcC_5H_4NMe$, 2-H	−0.17	0.13
32	$2\text{-}PhSPhCO_3CMe_3$	−0.30	0.30	72	$4\text{-}AcC_5H_4NMe$, 3-H	−0.20	0.12
33	Berson Ω	−0.05	0.04	73	$4\text{-}AcC_5H_4NMe$, 5-H	−0.20	0.12
34	sulfoxide rearrangement	−0.13	0.11	74	$4\text{-}AcC_5H_4NMe$, 6-H	−0.20	0.14
35	$PhCO_2H$	0.61	0.18	75	$4\text{-}AcC_5H_4NMe$, Ac-H	−0.61	0.26
36	$2\text{-}O_2NPhOH$	0.50	0.12	76	2-fluoropicoline, F	−1.09	0.33
37	picramic acid	1.41	0.23	77	Et_3PO, P	−9.62	3.60
38	*o*-vanillin	0.32	0.22				

If the nonreaction solvents (the majority of the Swain solvents) exhibit linearity between two solvent parameters, the solvent properties can be represented by only one parameter. For the nonreaction solvents, the converted *N* values are independent of the converted *Y* values only to the extent that small deviations from linearity exist in the plot of *N* versus *Y*. The scaling and zero points must arise from Swain's assumptions of the values of certain parameters, but we have not investigated this aspect. We list all of the converted values for the convenience of others who may discern meaning in the small deviations from linearity for the nonreaction solvents or for those who simply want to use the converted parameters for correlations.

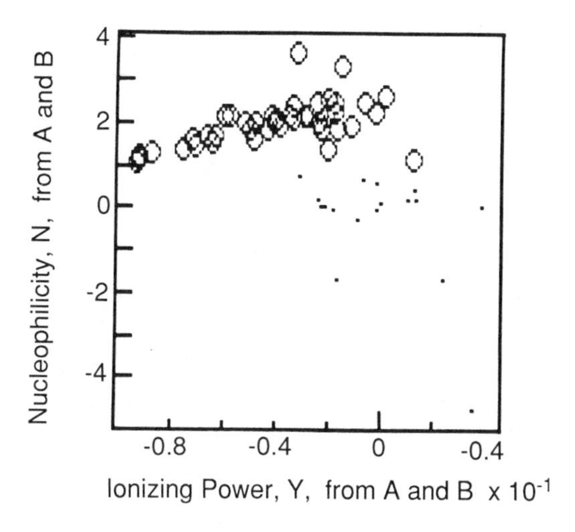

Figure 2. Plot of N *from* A *and* B *versus* Y *from* A *and* B (t-*BuCl-based*). *Hydroxylic solvents are shown as dots.*

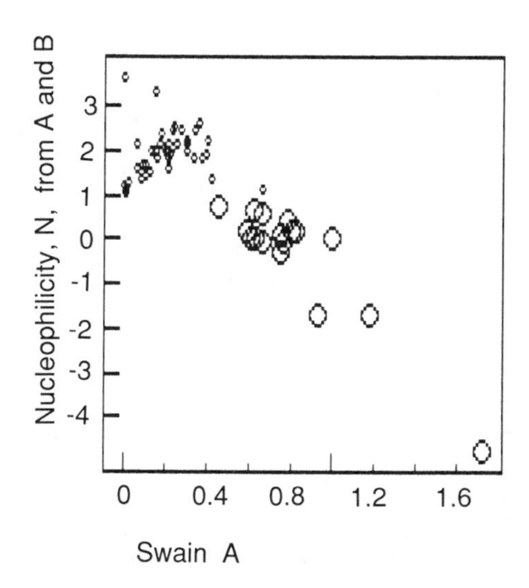

Figure 3. Plot of N *from* A *and* B (t-*BuCl-based*) *versus* A. *Hydroxylic solvents are shown as the larger circles.*

We now ask whether our converted N and Y parameters for the hydroxylic reaction solvents resemble the A or B values. This resemblance is not easy to judge from the data of Table I because the range of the various parameters affects the size of the multipliers, y_A, y_B, n_A, and n_B. Plots (Figures 3 and 4) where only the hydroxylic solvents (large circles) are considered significant show that our derived nucleophilicity, N, is roughly

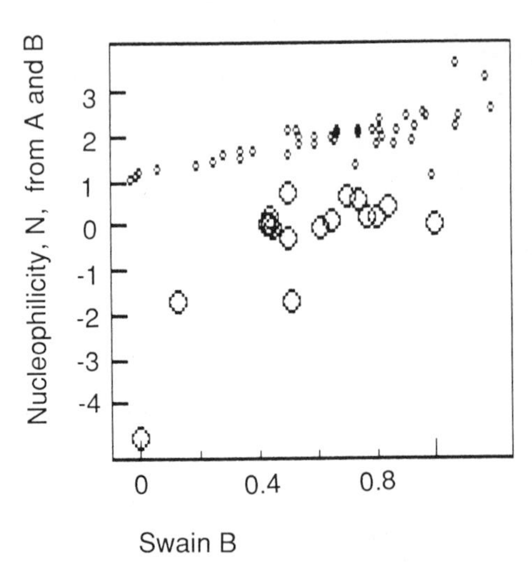

Figure 4. Plot of N *from* A *and* B *(t-BuCl-based) versus* B.

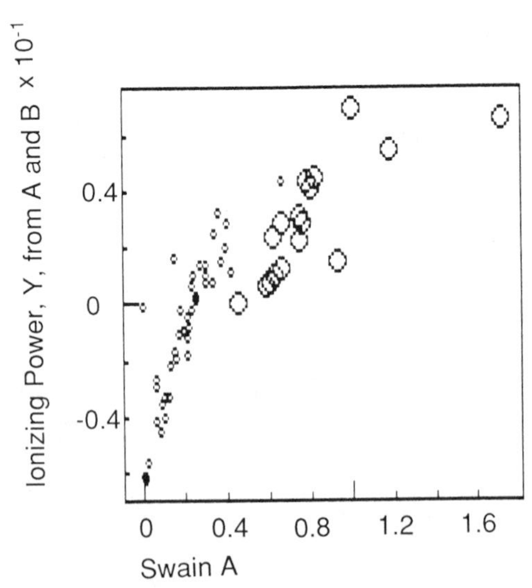

Figure 5. Plot of Y *from* A *and* B *(t-BuCl-based) versus* A.

linearly related to the Swain *A* value (with a negative coefficient), whereas the *B* value shows less correlation with *N*.

Although it may seem surprising that the "electrophilic" *A* parameter correlates with $-N$, we surmise that the *N* parameter represents both bond-forming "true nucleophilicity" and electrophilicity (the ability to promote the ionization of oxygen- or fluorine-containing leaving groups in solvents of low *N*). This hypothesis has been previously mentioned in connection with our conversion of Swain's *d* parameters. (7). Apparently, neither *A* nor *B* is closely correlated with *Y* (Figures 5 and 6).

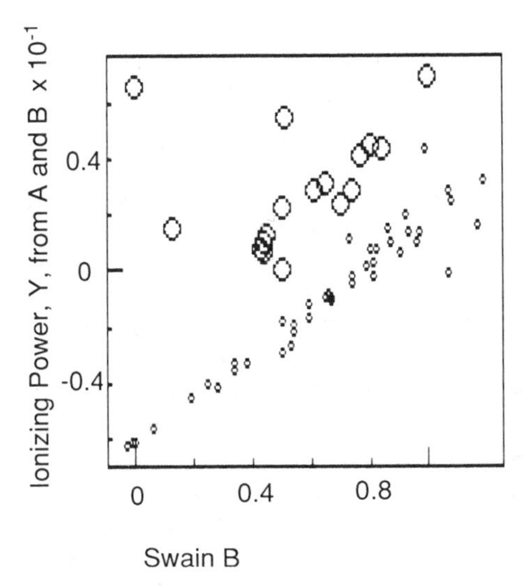

Figure 6. Plot of Y *from* A *and* B *(t-BuCl-based) versus* B.

Examination of R

We next investigate the effect of the ratio *R* in equation 5. In the original nucleophilicity scales, the nucleophilicities of acetic and formic acids were set equal, based on interpretations of solvolysis data in the literature and on the nucleophilicities of these in SO_2 solvent, a property perhaps somewhat distantly related to the nucleophilicity in a pure solvent. We have now calculated y_A as a function of *R*; y_A changes sign at an *R* value near 1.2 (Figure 7). Equation 3 shows that *N* values become scaled *A* values when y_A becomes 0 at an *R* value near 1.2. We note that changing *R* is merely a roundabout way to vary the proportion of *A* and *B* in the *N* and *Y* parameters.

For purposes of calculating the free energy change for a change of solvent in a reaction, whether we use solvent parameters based on $R = 1.0$ or 1.2 or any value is irrelevant. All parameters are mathematically equiv-

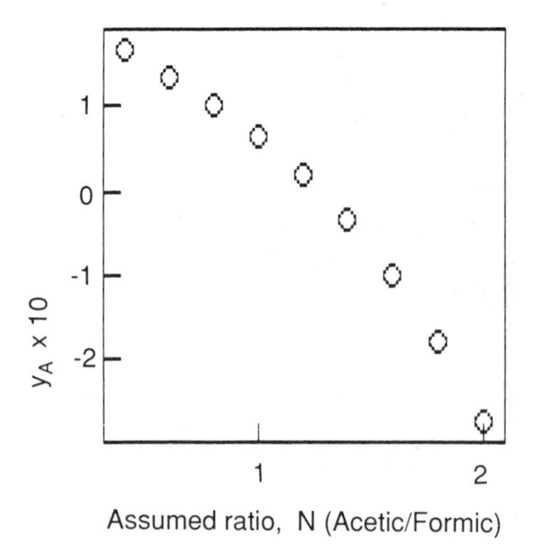

Figure 7. Plot of y_A *from equation 5 versus the chosen* N *ratio,* R, *for acetic and formic acid.*

alent! For computational purposes, whether acetic and formic acid are assigned the same nucleophilicity is irrelevant. Accordingly, concerns that the assignment of $R = 1$ has led to erroneous scales can be forgotten. Another way to view the situation is that N scales work when various proportions of the Y solvent property are added or subtracted, because any amount of the Y property may be resubtracted or added to N in choosing the proportion and sign of the two solvent parameters to represent a free energy change.

What property *independent* of solvent effect correlation might be used to select a chemically reasonable N scale? One such possibility is to set the relative nucleophilicities of the reference solvents, acetic and formic acid in the example discussed here, to be the same as those values found for the molecules acting as nucleophiles in a *constant solvent*. That task, in fact, was done for acetic acid and formic acid reacting with halonium ions in SO_2. This assignment led to a nucleophilicity scale having additional properties in agreement with the properties found in the constant solvent. These properties are the lower nucleophilicity of trifluoroacetic acid and the higher nucleophilicity of alcohols. We see that the N scale can be chosen to be chemically reasonable and preserve an accuracy of calculation identical with that of the AB scale (subject only to the two-significant-figure accuracy available in making the conversion).

Examination of the Sensitivities

We note that the sensitivities to nucleophilicity for methyl, primary, secondary, and tertiary halides resemble those in constant solvent in our converted

"s" scale. This observation provides another reason to prefer our converted values in comparison to the original values, in which sensitivities were not readily interpretable in terms of independent chemical experience. The ethyl tosylate to methyl bromide s ratio in the Swain–Scott study is 0.66, compared to the butyl bromide to methyl bromide s ratio of 0.71 obtained from equation 8. The latter equation also gives a reasonable isopropyl tosylate to methyl tosylate s ratio (0.21). As Swain noted, the b parameters, discussed as comparable to s, have no familiar order. Both a and b have large magnitudes for *tert*-butyl chloride reactions, for example, whereas the sensitivity of 0, set in the present conversion, leads to the familiar s order methyl > primary > secondary > tertiary.

The s sensitivities are determined by the choice of a reaction whose sensitivity to N is set equal to 0 (*see* equation 8). If we explore parameter conversions that provide a closer match of sensitivities to those in constant solvent than those based on *tert*-butyl chloride rates, we can base Y on another standard reaction, or even an imaginary reaction correlated by any desired proportion of A and B.

Other Standard Reactions

Because the solvolysis of adamantyl chloride has been used as a standard reaction, we list here the results of applying the conversion equations to this standard, using the equation given by Swain to represent the adamantyl rates. R of equation 5 was again set equal to 1 for acetic and formic acids. The y_A value (equation 5) was found to be 0.04684, compared with 0.0627 when *tert*-butyl chloride solvolysis was the standard reaction for obtaining Y.

The $n_A N$ values from solving equation 3 for nN, like those obtained by using Y values based on *tert*-butyl chloride, needed to be multiplied by a scale factor to obtain N values comparable to those in the Schadt–Bentley–Schleyer scale. For simplicity, we used the previously used scale factor $n_A = 6.51$ to derive the new adamantyl-based N scale. The spread of N values is slightly greater for the adamantyl-derived values obtained in this way. A plot (not given here) showed that the "$Y_{tert\text{-butyl based}}$" and "$Y_{\text{adamantyl based}}$" N scales are very similar. The Y scales themselves as obtained from the A and B parameters showed slightly greater differences. The adamantyl-based N and Y scales are given in Table III. Sensitivity parameters are given in Table IV.

Conclusion

Perhaps a main benefit of converting A and B parameters and sensitivities to them to other scales has been the clarification of how two-parameter equations work as applied to solvent effects. Swain's statistical method, which avoids the assignment of solvent parameters based on any one reaction, is an

Table III. Solvent Nucleophilicity (*N*) and Ionizing Power (*Y*) Obtained by Transformation of A and B Parameters: Adamantyl Tosylate Based

Solvent j from Ref 9	*Structure*	N *from* A *and* B	Y *from* A *and* B	*Solvent j from Ref 9*	*Structure*	N *from* A *and* B	Y *from* A *and* B
1	CCl_4	1.41	−6.56	32	$(CH)_5N$	2.18	−2.62
2	$CHCl_3$	1.19	−2.23	33	PhBr	1.92	4.12
3	CH_2Cl_2	1.60	−2.63	34	PhCl	1.69	−4.32
4	HCOOH	−1.49	2.75	35	$PhNO_2$	1.72	−4.98
5	$HCONH_2$	0.99	0.81	36	PhH	1.72	−4.98
6	$MeNO_2$	1.65	−1.62	37	$PhNH_2$	2.27	−0.66
7	MeOH	−0.23	−0.67	38	$(CH_2)_5CO$	1.82	−3.30
8	CS_2	1.46	−6.30	39	$(CH_2)_6$	1.07	−8.36
9	Cl_2CCCl_2	1.21	−6.88	40	$Me(CH_2)_4Me$	0.97	−9.75
10	$ClCHCCl_2$	1.85	−5.12	41	Et_3N	1.15	−7.31
11	CF_3COOH	−4.09	4.72	42	$(Me_2N)_3PO$	3.10	−4.02
12	MeCN	1.60	−2.05	43	PhCN	1.82	−2.55
13	$ClCH_2CH_2Cl$	1.73	−2.78	44	PhMe	1.68	−5.36
14	MeCOOH	−1.49	−0.90	45	PhOMe	1.84	−3.84
15	EtOH	−0.06	−1.60	46	PhNHMe	1.92	−0.88
16	MeSOMe	2.11	−1.31	47	$2,6\text{-}C_5H_3NMe_2$	2.06	−3.76
17	$HOCH_2CH_2OH$	0.34	1.08	48	$Me(CH_2)_5Me$	1.01	−9.78
18	MeCOMe	1.85	−3.21	49	PhCOMe	2.09	−2.97
19	$HCONMe_2$	1.94	−2.29	50	$o\text{-}C_6H_4Me_2$	1.87	−5.95
20	$CH_3CH_2CH_2OH$	0.00	−1.88	51	$m\text{-}C_6H_4Me_2$	1.01	−9.78
21	Me_2CHOH	1.12	−2.19	52	$p\text{-}C_6H_4Me_2$	1.81	—[a]
22	MeCOEt	1.78	−3.68	53	$Me_3CCH_2CHMe_2$	0.93	−9.29
23	$(CH_2)_4O$	1.82	−4.46	54	Bu_2O	1.38	−7.17
24	MeCOOEt	1.54	−4.51	55	H_2O	0.00	3.82
25	$O(CH_2CH_2)_2O$	1.76	−4.31	56	96% MeOH	−0.05	−0.15
26	$MeCONMe_2$	2.11	−2.34	57	80% EtOH	0.06	0.00
27	BuOH	0.04	−2.08	58	60% EtOH	0.14	1.05
28	EtOEt	1.32	−6.39	59	50% EtOH	0.14	1.36
29	Me_3COH	0.66	−3.02	60	80% MeCOMe	0.54	−0.68
30	$MeOC_2H_4OMe$	1.37	−4.91	61	70% MeCOMe	0.50	−0.16
31	$BuNH_2$	2.85	−2.40				

[a] Not determined.

interesting one. The *A* and *B* parameters may be transformed to mathematically equivalent ones by choosing any reaction to represent one of the new solvent parameters, provided the reaction is not one of those that actually requires only one parameter for a reasonable correlation. The other new solvent parameter is, by definition, not present in the correlation equation for the chosen reaction. This observation leads us to the relative amounts of the second parameter to be used in the transformation (equation 7).

An infinity of scales for the second new solvent parameter are still available, because the reaction data offer no way to discern whether some of the first solvent property is admixed with the second. The chemist may elect to use one of the scales for the second parameter, which reflects his chemical experience under different circumstances—for example, experience with reactions in constant solvent.

Table IV. Sensitivities of Nucleophilicity (*s*) and Sensitivity to Ionizing Power (*m*): Adamantyl Tosylate Based

Solvent i	*Reactant*	s *from* a *and* b	m *from* a *and* b	*Solvent* i	*Reactant*	s *from* a *and* b	m *from* a *and* b
1	MeBr	1.61	0.07	**40**	1-nitroso-2-naphthol	0.64	0.25
2	MeOTs	1.32	0.26	**41**	2-nitroso-1-naphthol	0.41	0.28
3	BuBr	1.24	0.32	**42**	Et_4N + I^-	0.73	1.73
4	$PhCH_2Cl$	1.30	0.42	**43**	Kosower Z	−4.76	3.98
5	Me_2CHOTs	0.49	0.63	**44**	MPI	−4.73	3.85
6	*cyclo*-C_5H_9OTs	0.39	0.73	**45**	3-MeOC_5H_4N + O^- (1)	−1.23	0.58
7	*cyclo*-$C_6H_{11}OTs$	0.34	0.83	**46**	3-MeOC_5H_4N + O^- (2)	−2.37	0.62
8	*endo*-$C_7H_{11}OTs$	0.14	0.78	**47**	3-MeOC_5H_4N + O^- (3)	−2.28	0.53
9	*exo*-$C_7H_{11}OTs$	0.23	0.94	**48**	$PhNO_2$	0.40	0.21
10	Ph_2CHCl	0.45	1.82	**49**	4-MeOPhNO_2	0.52	0.27
11	2-AdOTs	−0.00	1.00	**50**	4-Et_2NPhNO_2	0.49	0.40
12	Me_3CCl, Y	0.40	1.00	**51**	Ph_2CO	−0.31	0.12
13	Me_3CBr	0.81	0.02	**52**	pyrimidine	−0.36	0.13
14	$PhCMe_2O_2COPh$	−0.36	0.47	**53**	pyridazine	−0.53	0.26
15	Ph_3CF	−1.82	1.21	**54**	pyrroline oxide	−0.47	0.26
16	Ph_3COAc	−1.31	0.81	**55**	iron imine	−0.43	0.25
17	MeI + $(EtCH_2)_3N$	1.18	0.52	**56**	oximate	−3.01	1.75
18	MeI + $PhNMe_2$	1.10	0.54	**57**	sulfoxide	0.60	0.18
19	MeI + 3-ClPhNMe_2	1.12	0.62	**58**	Dimroth E_T30	−3.51	2.54
20	MeI + 4-ClPhNMe_2	1.13	0.62	**59**	Dimroth E_T26	−3.54	2.07
21	MeI + 3-MePhNMe_2	1.10	0.53	**60**	$Brooker_{XR}$	0.62	1.08
22	MeI + 4-MeOPhNMe_2	1.09	0.51	**61**	Davis *A*	−1.97	0.55
23	EtI + Et_3N	1.06	0.52	**62**	Davis *B*	−1.31	0.24
24	EtO_2CCH_2Br + Et_3N	0.09	0.50	**63**	Davis E_{CT}	−0.69	0.11
25	EtO_2CCH_2I + Et_3N	0.15	0.46	**64**	$HCONMe_2$	−1.48	3.20
26	4-O_2NPhF + Et_4N + N_{3^-}	−1.52	0.66	**65**	$POCl_3$	2.01	2.25
27	$PhSO_2Cl$ + $PhNH_2$	−1.33	−0.11	**66**	Me_2CHCH_2Cl, trans	3.47	1.55
28	$ClSO_2NCO$ + hexene	1.26	0.37	**67**	Me_2HPO, band 1	−3.35	3.96
29	TCNE + 4-methoxystyrene	0.86	0.71	**68**	$(Me_3C)_2NO$, N	−0.18	0.11
30	Br_2 + 1-pentene	0.62	1.15	**69**	piperidyloxy, N	−0.16	0.10
31	Br_2 + Me_4Sn	0.17	1.14	**70**	pyrrolinyloxy, N	−0.16	0.11
32	2-PhSPhCO_3CMe_3	−0.23	0.33	**71**	4-AcC_5H_4NMe, 2-H	−0.14	0.15
33	Berson Ω	−0.04	0.04	**72**	4-AcC_5H_4NMe, 3-H	−0.19	0.13
34	sulfoxide rearrangement	−0.10	0.12	**73**	$4\text{-AcC}_5\ H_4NMe$, 5-H	−0.18	0.13
35	$PhCO_2H$	0.78	0.20	**74**	4-AcC_5H_4NMe, 6-H	−0.18	0.15
36	2-O_2NPhOH	0.63	0.13	**75**	4-AcC_5H_4NMe, Ac-H	−0.61	0.28
37	picramic acid	1.73	0.25	**76**	2-fluoropicoline, F	−1.14	0.36
38	*o*-vanillin	0.32	0.24	**77**	Et_3PO, P	−9.73	3.92
39	5-methylfurfural	0.37	0.19				

Appendix: A Full Description of the Conversion of A *and* B *to* N *and* Y

As already noted, the parameters are assumed to be separable into components as follows:

$$A_{hep}^{S1} = n_A N_{hep}^{S1} + y_A Y_{hep}^{S1} \tag{3}$$

$$B_{hep}^{S1} = n_B N_{hep}^{S1} + y_B Y_{hep}^{S1} \tag{4}$$

For a second solvent S2, equation 3 becomes

$$A_{hep}^{S2} = n_A N_{hep.}^{S2} + y_A Y_{hep}^{S2} \tag{10}$$

Calculating the *y*s and the Ratio of the *n*s. In our first calculation, we base the *Y* scale on *tert*-butyl chloride solvolysis rates as calculated from the expression given by Swain et al. (9): log *k* (*tert*-butyl chloride) = 7.37*A* + 5.65*B* − 6.10. We simply subtract the result of the calculation of log *k* (*tert*-butyl chloride) for the two solvents of interest to obtain the *Y* values to be used in equation 3. These *Y* values preserve the advantage of the Swain approach in that any unusually large error in one rate constant of a standard reaction is minimized, because the *A* and *B* values are optimized for a number of reactions. We again note that the Swain parameters are compatible with an infinite group of *N* scales. Any one of these scales may be obtained by assuming a value of the ratio, *R*, of nucleophilicity of two reference solvents designated as follows:

$$S1 = Sref1 \text{ and } S2 = Sref2$$

We replace N_{hep}^{S1} in equation 3 with RN_{hep}^{S2}. We then solve equation 3 for nN_{hep}^{S2}. We solve equation 10 for the same quantity, where S2 is the reference solvent Sref2. Two equations that are equal to *nN* of solvent ref2 result. Equating these equations and solving for y_A gives

$$y_A = \frac{RA_{hep}^{Sref2} - A_{hep}^{Sref1}}{RY_{hep}^{Sref2} - Y_{hep}^{Sref1}} \tag{5}$$

A similar expression having *B* in place of *A* and the same denominator is obtained for y_B. Dividing the equations gives

$$\frac{y_A}{y_B} = \frac{BA_{hep}^{Sref2} - A_{hep}^{Sref1}}{RB_{hep}^{Sref2} - Y_{hep}^{Sref1}} = f(R) \tag{6}$$

We may solve equations 3 and 4 for n_A and n_B in solvent S1 where S1 is any solvent. Dividing gives

$$\frac{n_A}{n_B} = \frac{BA_{\text{hep}}{}^{S2} - A_{\text{hep}}{}^{S1}}{RB_{\text{hep}}{}^{S2} - Y_{\text{hep}}{}^{S1}} = \text{constant} \qquad (7)$$

This equation says that the ratio of the proportions of nucleophilicity that are present in the *A* and *B* parameters may be obtained from the parameters for any solvent by subtracting the contribution of ionizing power from each (*A* and *B*) and dividing. Although the y values depend on the assigned value of the ratio, *R*, as do the y/y ratios (equation 6), the quotient in equation 7 is, remarkably, independent of the y values, within the error limits posed by significant figures. Sample calculations have confirmed this result. A proof that this statement is correct comes from an alternative way to get the ratio.

Calculating Sensitivities to Nucleophilicity and Ionizing Power. If the right-hand terms in equations 3 and 4 are put into equation 11, we can collect the terms

$$\Delta G = aA_{\text{hep}}{}^{S1} + bB_{\text{hep}}{}^{S1} \qquad (11)$$

that contain *N* and *Y*. The multipliers of *N* and *Y* are the sensitivities, usually called l and m (5) or s and m (7). As in our earlier conversion (7), sensitivities are

$$s_{\text{RX}} = a_{\text{RX}} n_A = b_{\text{RX}} n_B \qquad (8)$$

$$m_{\text{RX}} = a_{\text{RX}} y_A + b_{\text{RX}} y_B \qquad (9)$$

In the conversion of *A* and *B* to *N* and *Y*, the *Y* values represent the best fit of equation 1 to the rates of solvolysis of *tert*-butyl chloride. Clearly, none of the *N* solvent property should be added to *Y* to give a better fit, because the fit is already optimized. Therefore, $s = 0$ for the reactions of this compound. Equation 8 with $s = 0$ leads to

$$\frac{n_A}{n_B} = \frac{-b_{t\text{BuCl}}}{a_{t\text{BuCl}}} = \frac{-5.64}{7.37} = -0.76526 \qquad (12)$$

Within the limits imposed by the availability of two significant figures in *a*s and *b*s, this number agrees with the values obtained in equation 8 from *A*, *B*, and *Y* values in various solvents.

Finding the Nucleophilicities. Values of $n_A N$ or $n_B N$ for any solvent can be obtained from equations 3 or 4, because we now have the y values from equations 5 and 6. The nN values may be compared with the proposed sets of nucleophilicity values in the literature. Such a comparison will be appropriate only if we have chosen the ratio R for the nucleophilicities of acetic and formic acid to be 1, because this same assumption was made in setting up the scales in the literature. As has been noted, we have plotted the $n_A N$ values versus the N values of Schadt, Bentley and Schleyer (Figure 1).

The slope of the plot is the n_A value, which may be considered to be a scaling factor. Dividing nN by this n scale factor gives N values extracted from the A and B parameters. This N scale initially has no chosen zero point, although a few values exist that are not far from zero. The N value for water may be subtracted from the N value in each solvent to get values that are comparable to the published scales. All of these N values based on A and B are given in Table II. Y values obtained as described also appear in Table II. The s and m values (equations 8 and 9) are listed in Table III. In Table I, other numerical values and relationships are given.

Literature Cited

1. Cope, A. C.; Grisar, J. M.; Peterson, P. E. *J. Am. Chem. Soc.* **1959,** *81,* 1640–1642.
2. Peterson, P. E.; Kelly, R. E.; Belloli, R.; Sipp, K. A. *J. Am. Chem. Soc.* **1965,** *87,* 5169–5171.

3a. Peterson, P. E.; Bopp, R. J.; Chevli, D. M.; Curran, E. L.; Dillard, D. E.; Kamat, R. J. *J. Am. Chem. Soc.* **1967,** *89,* 5902–5910.

3b. Peterson, P. E . *Acc. Chem. Res.* **1971,** *4,* 407–413.

4. Peterson, P. E.; Waller, F. J. *J. Am. Chem. Soc.* **1972,** *91,* 991–992.
5. Schleyer, P. v. R.; Fry, J. L.; Lam, L. K.; Lancelot, C. J. *J. Am. Chem. Soc.* **1970,** *92,* 2542–2544.
6. Bentley, T. W.; Schadt, F. L.; Schleyer, P. v. R. *J. Am. Chem. Soc.* **1972,** *94,* 992–995.
7. Peterson, P. E.; Vidrine, D. W.; Waller, F. J.; Henrichs, P. M.; Magaha, S.; Stevens, B. *J. Am. Chem. Soc.* **1977,** *99,* 7968–7976.
8. Swain, C. G.; Mosely, R. B.; Bown, D. E. *J. Am. Chem. Soc.* **1955,** *77,* 3732–3734.
9. Swain, C. G.; Swain, M. S.; Powell, A. L.; Alunni, S. *J. Am. Chem. Soc.* **1983,** *105,* 503–513.
10. Taft, R. W.; Abbout, J. M.; Kamlet, M. J. *J. Org. Chem.* **1984,** *49,* 2001–2005.
11. Swain, C. G. *J. Org. Chem.* **1984,** *49,* 2005–2010.
12. Kevill, D. N. *J. Chem. Res.* **1984,** 86–87.

RECEIVED for review October 21, 1985. ACCEPTED June 30, 1986.

22

Solvolysis of Electron-Deficient 1-Arylethyl Tosylates

Kinetic and Stereochemical Tests for Nucleophilic Solvent Participation

Annette D. Allen, V. M. Kanagsabapathy, and Thomas T. Tidwell*

Department of Chemistry, University of Toronto, Scarborough Campus, Scarborough, Ontario M1C 1A4, Canada

The rates and stereochemistry of solvolysis of 1-arylethyl tosylates with electron-withdrawing substituents on the α-carbon or the aryl ring have been determined and suggest initial formation of ion pairs, which undergo competitive return to reactant or formation of products. The stereochemistry of substitution varies from major inversion to racemization to small net retention as a function of solvent and substrate. These electron-deficient carbocations show large magnitudes of ρ^+ *values but modest values of the solvent parameter* m. *These results indicate strongly nucleophilic solvents assist the ionization of the less reactive substrates, whereas in the weakly nucleophilic media, solvent attack on reversibly formed ion pairs is rate-limiting. The first intermediate interacts with a solvent shell but without specific attachment to a single solvent molecule.*

HOW NUCLEOPHILIC SUBSTITUTION OCCURS has been one of the preeminent problems in mechanistic organic chemistry for more than 50 years (*1–3*), and the importance of this process more than justifies this sustained interest. An area that has received particularly close scrutiny is the "borderline" or "combat zone" (*2*) region where potential competition between mechanisms of the S_N1 and S_N2 type exists. These processes involve formation of a nonspecifically solvated carbocation intermediate in the former case and a direct displacement of the leaving group by a solvent molecule in the latter and are designated k_c and k_s processes for solvolysis, respectively.

* To whom correspondence should be addressed.

0065-2393/87/0215-0315$06.00/0

The suggestion that a gradual transition occurs between these mechanisms so that in the intermediate region the two processes merge, and a five-coordinate intermediate exists on the reaction coordinate has received interest recently. The possibility of partial covalent character in the solvolysis transition state has been considered for a long time (*4*) and was given particular prominence by the report of Doering and Zeiss in 1953 (*5*) that methanolysis of 3,5-dimethyl-3-hexyl phthalate proceeded with net 54% inversion of configuration and 46% racemization. The structure of this intermediate was depicted as I (equation 1) (*5*), and the reaction coordinate for the merging of the mechanisms is depicted in Figure 1 (*6*).

$$\text{C–X} \longrightarrow \underset{\textbf{I}}{\text{S(H)O—C—X}} \longrightarrow \text{S(H)O—C—O(H)S} \qquad (1)$$

As the leaving group departs, the positive charge developing in the remaining organic moiety is not restricted to one atom but is delocalized throughout the structure where it interacts with many solvent molecules, and the electrophilic role of the solvent in assisting departure of the leaving group by hydrogen bonding has also been emphasized (*1–7*). The distinctive aspect of the "S_N2-intermediate" (*6*) or Doering-Zeiss (*5*) proposals would appear to be the suggestion of a particularly strong backside interaction by one specific solvent molecule that exists for a sufficiently long time that a discrete chemical intermediate exists.

As a test of the extent of nucleophilic solvent participation in transition states in the "borderline" or "combat zone" region, we studied 1-arylethyl

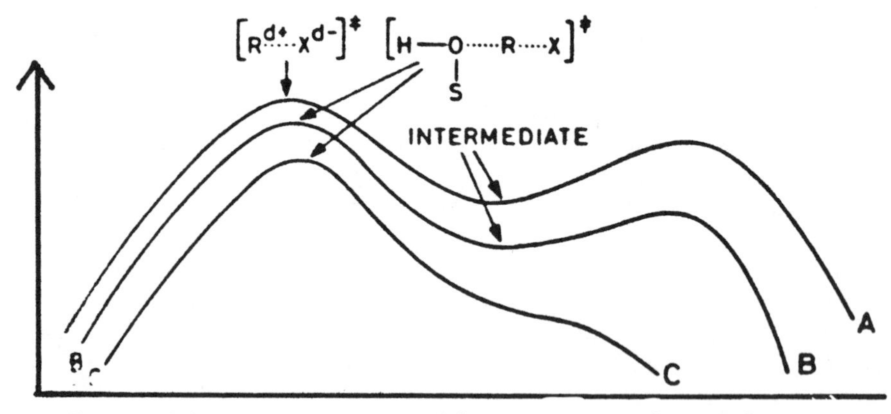

Figure 1. Schematic representation of the upper portion of potential energy surfaces for merging of substitution mechanisms (6). Reproduced from reference 6. Copyright 1981 American Chemical Society.

systems that are deactivated by electron-withdrawing fluorine atoms. These systems are chiral and thus permit determination of the stereochemistry of the substitution and furthermore have been the subject of intensive previous study (*1–3, 8*) and are known to have a reactivity in the region where S_N1 and S_N2 processes are becoming competitive and a merging of mechanisms might occur. The presence of electron-withdrawing fluorine atoms either on the carbon adjacent to the substitution center or as part of ring substituents would be expected to destabilize (*9*) the developing carbocations by increasing their electron deficiency (*10*) and should thereby favor nucleophilic participation by the solvent.

Results

The preparation of the 1-aryl-2,2,2-trifluoroethyl tosylates **I–III** and the mesylate **IV** was described (*11–16*) as well as the solvolytic rate constants of these derivatives in a range of solvents (*11–16*). Solvolytic and polarimetric rate constants were also obtained for optically active 1-phenyl-2,2,2-trifluoroethyl triflate **Va** and solvolytic rate constants for the α-deuterated analogue **Vb** (*11*). Solvolytic and polarimetric rate constants were also measured for optically active **VIa–c** (*17*).

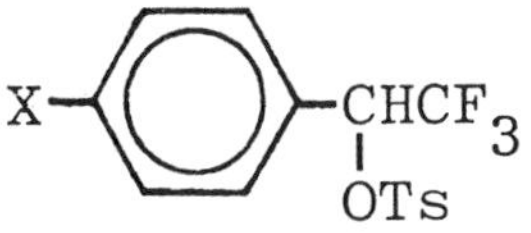

X–C₆H₄–CHCF₃(OTs)

Ia, X = CH_3O
Ib, X = CH_3
Ic, X = H
Id, X = Br

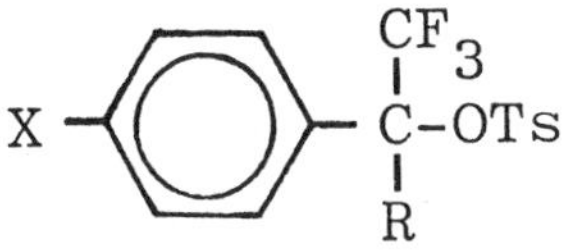

X–C₆H₄–C(CF₃)(R)–OTs

IIa, X = CH_3; R = CF_3
IIb, X = CH_3O; R = CF_3
IIc, X = H; R = CF_3
IId, X = H; R = CN
IIe, X = CH_3; R = CN
IIf, X = H; R = $^{13}CH_3$

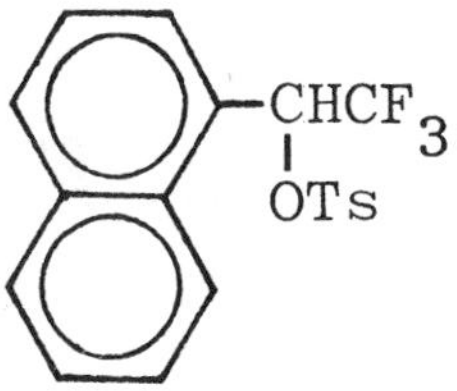

Naphthyl–CHCF₃(OTs)

III

Anthryl–CHCF₃(OMs)

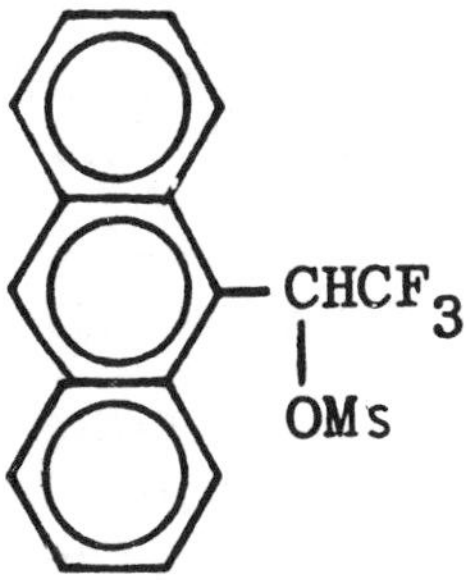

IV

For comparative purposes, the solvolytic rate constants for **I–VI** at 25 °C in hexafluoroisopropyl alcohol (HFIP) are given in Table I, together with the m values, which measure the dependence of the rates on the solvent ionizing power parameter Y_{OTs} by the equation $\log (k/k_0) = mY_{OTs}$ (*3, 6*).

Table I. Rate Dependence on Solvent Ionizing Power of Benzylic Sulfonates $ArCR(CF_3)OTs$ Relative to $C_6H_5CH_2OMs$ at 25 °C in HFIP

Ar	*R*	*No.*	m	k(25 °C)	k_{rel}	*Ref*
				$ArCR(CF_3)OTs$		
p-anis	H	**Ia**	0.76	0.583	63	11
p-tol	H	**Ib**	0.94	5.66×10^{-5}	6.1×10^{-3}	11
C_6H_5	H	**Ic**	0.69	4.03×10^{-8}	4.4×10^{-6}	11
C_6H_5	CH_3	**IIf**	1.01	2.95×10^{-3}	0.32	14
C_6H_5	CF_3	**IIc**			1.2×10^{-8}	12
C_6H_5	CN	**IId**			6.1×10^{-7}	12
p-tol	CN	**IIe**	0.66	1.94×10^{-4} [a]	2.1×10^{-2}	15
p-anis	CF_3	**IIb**	0.76	0.235a	25	12
1-naphthyl	H	**III**	0.94	1.50×10^{-3}	0.16	16
9-anthryl	H	**IV**	0.64	0.124	13	16
$C_6H_5CH_2OMs$				9.23×10^{-3}	1.0	32
				$ArCH(OTs)CH_3$		
m-BrC_6H_4		**VIa**	0.82[b]	0.267	29	17
p-$CF_3C_6H_4$		**VIb**	**0.78**[b]	2.19×10^{-2}	2.4	17
3,5-$(CF_3)_2C_6H_3$		**VIc**	-[c]	1.36×10^{-5}	1.5×10^{-3}	17

NOTE: $k_{rel} = k(25\ °C)/(9.23 \times 10^{-3})$.
[a] Interpolated from the mY correlation.
[b] Derived from the plot excluding CH_3CH_2OH solvents.
[c] Derived from rates for TFA and HFIP only: $\log k = 1.36Y_{OTs} - 9.74$.

The products of the reactions involve substitution at the central carbon with the exception of 1-(9-anthryl)-2,2,2-trifluoroethyl mesylate (**VI**), which undergoes extensive ring substitution at the 10-position (*16*), and **IIb** (*11, 12, 14–17*). The stereochemistry of substitution of **Va** and **VIa–c** was also determined (*11, 17*).

C_6H_5–$CRCF_3$ (OTf)

Va, R = H
Vb, R = D

$ArCHCH_3$ (OTs)

VIa, Ar = *m*-BrC_6H_4
VIb Ar = *p*-$CF_3C_6H_4$
VIc Ar = 3,5-$(CF_3)_2C_6H_3$

The reaction of **IIb** in TFE gave the expected substitution product **VIIa,** but solvolysis in CH_3CH_2OH gave a mixture of **VIIb** and **VIIc** in a ratio of 50:50 as determined by vapor pressure chromatographic separation and isolation (Chart I) (*15*). Solvolysis in CD_3OH gave **VIId** and **VIIe** in a ratio of

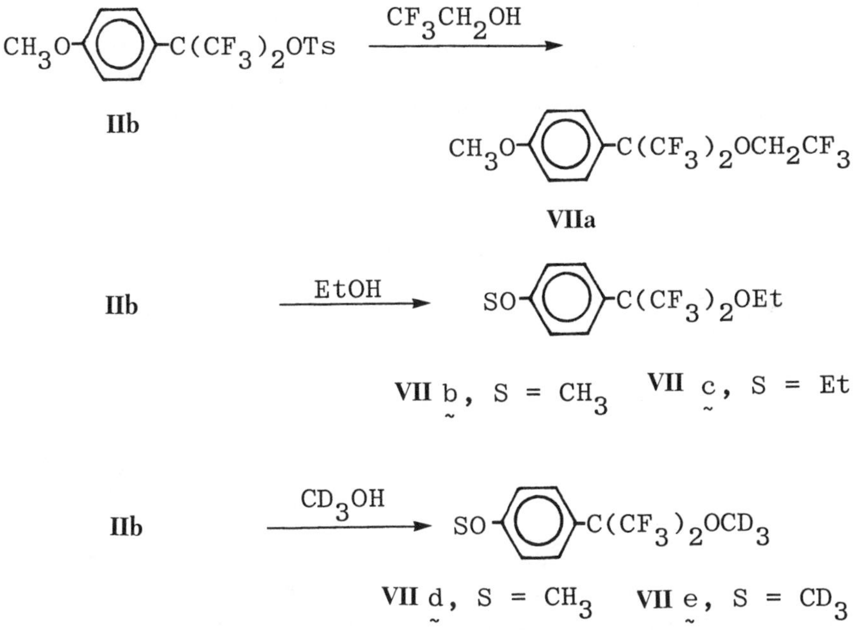

$CH_3O-C_6H_4-C(CF_3)_2OTs$ (**IIb**) $\xrightarrow{CF_3CH_2OH}$ $CH_3O-C_6H_4-C(CF_3)_2OCH_2CF_3$ (**VIIa**)

IIb $\xrightarrow{EtOH}$ $SO-C_6H_4-C(CF_3)_2OEt$

VII b, S = CH_3 **VII** c, S = Et

IIb $\xrightarrow{CD_3OH}$ $SO-C_6H_4-C(CF_3)_2OCD_3$

VII d, S = CH_3 **VII** e, S = CD_3

Chart I

69:31 as analyzed by the relative intensity of their respective M^+ peaks at 291 and 294.

Trifluoromethyl Substituent Effects

The destabilizing effect of the CF_3 group adjacent to a developing carbocation center is illustrated by the fact that CF_3-substituted alkenes are 10^5–10^7 times less reactive in protonation according to equation 2 (R = Ar or CH_3CHO) than the corresponding derivatives where hydrogen replaces CF_3 (*18, 19*). These rates are predicted by equation 3, where the electrophile substituent parameters σ_p^+ (*21*) for the group R and CF_3 are summed (*20*). The strong electron-withdrawing power of the CF_3 group is manifested by its σ_p^+ value of 0.61 (*21*), and use of this parameter permits quantitative prediction of the kinetic effect of this group by using equation 3.

$$CH_2{=}C(R)(CF_3) \xrightarrow{H^+} CH_3C^+(R)(CF_3) \quad (2)$$

$$\log k_{H^+} = -10.5\Sigma\sigma_p^+ - 8.92 \quad (3)$$

In solvolytic reactions, the presence of CF_3 groups in positions rather remote from the reaction center leads to significant rate depressions compared to hydrogen (*22–26*), as in examples **VIII–X** (*22–24*), where the rate accelerations for replacement of the CF_3 groups by hydrogen are shown in parentheses.

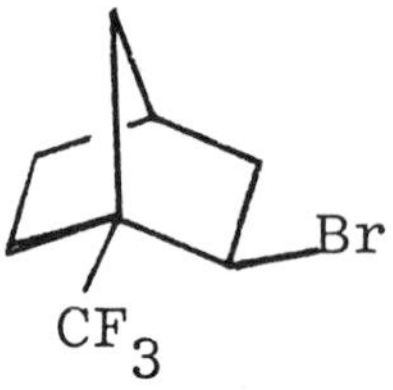

VIII (2×10^5) 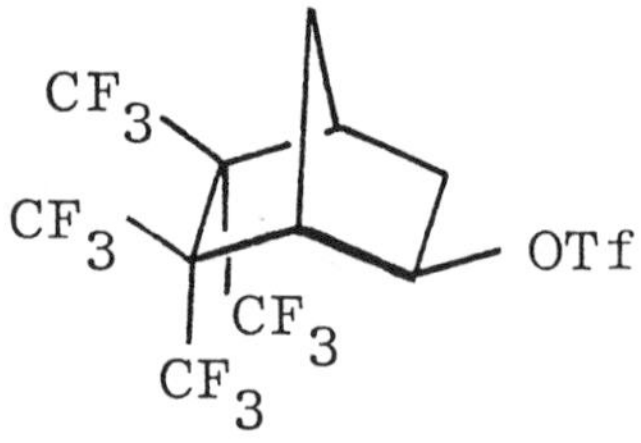 (*22*) **IX** (10^{11}) 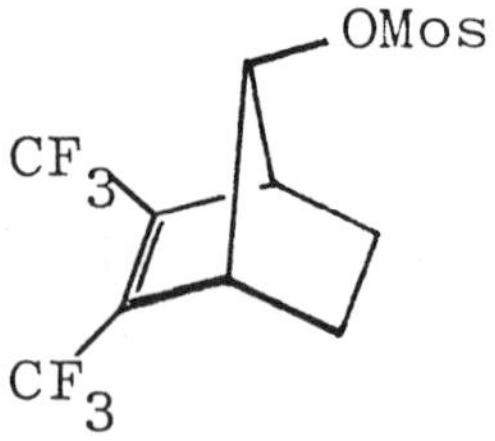 **X** (10^{12})

Mos = $p\text{-}CH_3OC_6H_4SO_2$

Trifluoromethyl substituents in the α position are also strongly destabilizing in the solvolysis of 2-(trifluoromethyl)propyl triflate (**XI**), which is less reactive than 2-propyl triflate by factors ranging from 1.5×10^4 in CH_3CH_2OH to 4×10^6 in TFA (*27*). Comparable factors were found for the tosylate corresponding to **XI** (*28*). These compounds may be interpreted as reacting through an ion-pair mechanism (equation 4) on the basis of the large rate deceleration due to the CF_3 group, the exclusive formation of 2-(trifluoromethyl)propene as the only product in all solvents, the large rate enhancements caused by basic salts, and the noncumulative isotope effect $k(d_0)/k(d_3) = 1.78$ and $k(d_0)/k(d_6) = 3.80$ for deuteration of one or both methyl groups (*27*).

$$\underset{\textbf{XI}}{CF_3C(CH_3)_2OTf} \rightleftarrows CF_3\overset{+}{C}(CH_3)_2\ OTf^- \longrightarrow CF_3C(CH_3){=}CH_2 \qquad (4)$$

The observed absence of a correlation of the rates of **XI** with the Y_{OTs} solvent ionizing power parameter (*3, 6*) or with the new Y_{OTf} parameters for triflates (*29, 30*) is also indicative of a significant kinetic effect of the solvent acting as a base in assisting the elimination according to equation 4. This process is an example of the $E2_C{}^+$ mechanism and has also been invoked to explain the results of solvolysis of some α-carbonyl- and α-phosphoryl-substituted mesylates (*31–34*).

Such solvent-assisted elimination could result by the "S_N2-intermediate" mechanism (*6*), but the high degree of carbocation character indicated by the large decelerating effect of CF_3 in **XI** and the noncumulative isotope effect, which was quantitatively accounted for on the basis of the ion-pair scheme in equation 4 (*27*), argues against the former interpretation.

Solvolyses of systems $ArAr'C(OTs)CF_3$ and the related bromides have also been studied and also show rate decelerations of approximately 10^5 for CF_3 relative to hydrogen (*35, 36*). The highly crowded nature of these substrates appears to preclude any kinetically important nucleophilic solvent participation. The constancy of the $k(H)/k(CF_3)$ rate ratios in comparison to less crowded systems suggests either that no steric effects involve the CF_3 groups on the k_c reactivity of these substrates or that any such effects tend to cancel.

Tertiary Benzylic Derivatives

The solvolyses of the tertiary α-trifluoromethyl-substituted benzylic tosylates **IIa–f** would be expected to be protected from any involvement of direct nucleophilic solvent participation during departure of the leaving group for steric reasons, and the experimental evidence is consistent with this interpretation. Thus, for the ring-substituted analogues of **IIc** and **IIf** ρ^+ values (*21*) of −10.7 (*15*) and −6.85 (*37*), respectively, indicate a very high degree of cationic character with a high demand for electron donation by the aryl substituent, particularly in the former doubly destabilized system. For **IIf,** the isotope effect $k(CD_3)/k(CH_3)$ of 1.3–1.6 (depending on solvent), the modest kinetic effects of added salts, and the calculated rate acceleration of 2×10^5 for replacement of the CF_3 by H are all indicative of a carbocation route, as indicated in equation 5 (*14*). The question of the reversibility of the initial ionization cannot be definitely answered in the absence of any data on optically active or specifically ^{18}O-labeled derivatives.

$$ArC(CF_3)(R)OTs \rightleftharpoons Ar\overset{+}{C}(CF_3)(R) + OTs^- \longrightarrow \text{products} \quad (5)$$

The dependence of the rates of solvolysis of the tertiary systems **IIb** and **IIf** on the solvent ionizing power by the single-parameter equation $\log (k/k_0) = mY_{OTs}$ (*6*) is listed in Table I along with a number of other values for destabilized systems. Interestingly, the m value of 0.76 for **IIb** is considerably less than that of 1.01 for **IIf,** and a plausible explanation of this result is that the former system, which is destabilized by two CF_3 groups, has greater charge delocalization into the ring and a consequent lower demand for charge dispersal by the solvent.

The product from **IIb** in CH_3CH_2OH or CD_3OH involves attack of the solvent on the ring, as evidenced by the formation of the ipso substitution products **VIIc** and **VIIe** (*15*). Evidently, the carbocation intermediate **XII** reacts with solvent to form the intermediate **XIII,** as shown in equations 6–8. Two processes may be envisaged for the conversion of **XIII** to the products, one involving reionization of **XIII** to **XII** and the other involving direct formation of **VII** from **XIII** by a 1,5 shift (equation 9).

$$CH_3O{-}C_6H_4{-}\overset{+}{C}(CF_3)_2 \underset{H^+(k_3)}{\overset{SOH(k_2)}{\rightleftharpoons}} (SO)(CH_3O)C_6H_4{=}C(CF_3)_2 \qquad (6)$$

XII **XIII** (a, S = Et; b, S = CD_3

$$\textbf{XII} \xrightarrow[SOH]{k_1} CH_3O{-}C_6H_4{-}C(CF_3)_2OS \qquad \textbf{VIIb}\ (S = Et),\ \textbf{VIId}\ (S = CD_3) \qquad (7)$$

$$\textbf{XIII} \xrightarrow[-CH_3OH]{SOH} SO{-}C_6H_4{-}C(CF_3)_2OS \qquad \textbf{VIIc}\ (S = Et),\ \textbf{VIIe}\ (S = CD_3) \qquad (8)$$

$$CH_3O,\ OS,\ CF_3,\ CF_3 \not\longrightarrow \textbf{VIIb, VIId} \qquad (9)$$

However, the mechanism of equations 6–8 is evidently favored, as the results are consistent with this process but not that of equation 9. Thus, no product was observed from reaction of **IIb** in either CH_3CH_2OH or CD_3OH that resulted from a shift of the CH_3O group to the benzylic carbon, and particularly for **XIIIb** such a shift should be equally probable as the shift of CD_3O.

Mass spectrometric analysis of the product in CD_3OH shows that the ratio of M^+ ions corresponding to **VIId** and **VIIe** is 69/31. A simple steady-state analysis of equations 6–8 predicts the ratio of products **VIIIb/VIIc** or **VIId/VIIe** will equal $2k_1/k_2$, so that for CD_3OH, $k_2 = 0.9k_1$. For CH_3CH_2OH, **VIIb** equals **VIIc,** so $k_2 = 2k_1$. The greater preference of CH_3CH_2OH for attack at the ipso carbon relative to CH_3OH may reflect a steric preference with the larger nucleophile attacking at the para position.

Secondary Derivatives

The solvolyses of the secondary 1-aryl-2,2,2-trifluoroethyl derivatives **I** and **III–V** would be more likely candidates for kinetically significant solvent participation than the tertiary derivatives because of the greater steric accessibility to the carbinyl center by solvent and because of the increased electron demand of the secondary derivatives compared to **IIf.** However, by a number of criteria no strong evidence exists for a major role for rate-limiting solvent participation concurrent with leaving-group departure.

Thus, the ρ^+ values for **I** range from -6.7 to -11.9 depending upon the solvent, with the smallest magnitude of ρ^+ occurring for TFA (*11*). The large magnitudes of these values provide strong evidence for carbocationic intermediates. The isotope effects $k(\mathrm{H})/k(\mathrm{D})$ for **V** range from 1.21 to 1.34 depending upon the solvent, and this finding is also consistent with rate-limiting formation of carbocations (*11*).

The evidence on the stereochemistry of the solvolysis was obtained from optically active $C_6H_5CH(OTf)CF_3$ (**Va**), which gave ratios of k_α, the polarimetric rate constant, and k_{uv}, the rate of product formation, of 7.3, 27, 1.2, 1.1, and 1.0 in TFA, HFIP, TFE, HOAc, and CH_3CH_2OH, respectively (*11*). The acetolysis product from (−)-**Va** was 41% inverted and 59% racemized (*11*).

These kinetic and stereochemical results for the tosylates **I** and triflates **V** may be consistently interpreted in terms of initial ionization to an ion pair that undergoes some return to optically inactive starting material and is converted to product by attack of solvent (Scheme I). The observation of substrate racemization during solvolysis appears to be incontrovertible evidence for an ion pair that has a rather long lifetime in the poorly nucleophilic TFA and HFIP and hence has considerable opportunity for racemization and ion-pair return. In the more nucleophilic solvents, the ion pair is captured by the solvent more rapidly so that racemization and ion-pair return are no longer competitive. The fact that acetolysis of (−)-**Va** gave partial net inversion indicates the triflate leaving group effectively shielded one side of the cation so that the solvent preferentially attacked the other side.

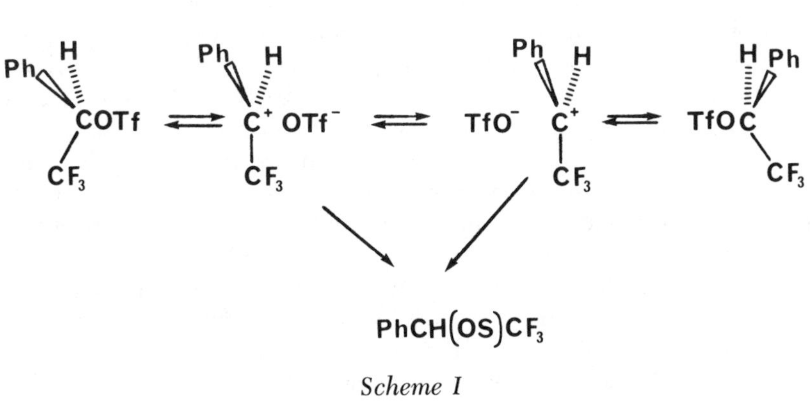

Ph(H)C(OTf)CF₃ ⇌ Ph(H)C⁺ OTf⁻ (CF₃) ⇌ TfO⁻ C⁺(Ph)(H)(CF₃) ⇌ TfOC(H)(Ph)CF₃

$PhCH(OS)CF_3$

Scheme I

The m values for **Ia–c, III,** and **IV** from the relationship lot $(k/k_0) = mY_{OTs}$ (*6, 38*) are all less than 1.0; thus, a lower response to solvent ionizing power results for these benzylic substrates than for 2-adamantyl tosylate. A plausible explanation of this result is that the developing charge in the former systems is largely delocalized onto the aryl ring and therefore less demand exists for solvent stabilization. Also, a number of studies of formation of carbocationic systems with adjacent carbonyl groups show similarly low m values (*31–34*). These latter systems may have charge delocalization onto the adjacent carbonyl group and in some cases adjacent aryl groups as well and, hence, a lower demand for solvent stabilization.

These correlations of reaction rate with solvent ionizing power Y_{OTs} are somewhat scattered, so some uncertainty exists in the resultant m values, but nevertheless a sufficient number of examples are available to suggest that a charge delocalization effect on the response to solvent ionizing power occurs. Such an effect might well also occur in systems reacting by k_Δ (neighboring group participation) routes in which the developing charge is partially dispersed onto a participating neighboring group. Studies of solvent effects on reactivity in such systems are just beginning to emerge and will provide an independent test of this theory of solvent effects. In one recent study, the solvolysis of mustard chlorohydrin ($HOCH_2CH_2SCH_2CH_2Cl$), a substrate known to react by a k_Δ path, showed abnormally low rates in mixed solvents containing CF_3CH_2OH, as compared to rates in aqueous CH_3CH_2OH and acetone (*39*). The cause of this effect has not been established, but the solvation requirements of the delocalized k_Δ transition state are clearly relevant to this phenomenon.

The naphthyl and anthryl derivatives **III** and **IV** are considerably more reactive than the corresponding phenyl or tolyl compounds **Ib** and **Ic** (Table I), a result expected for the greater electron-donating ability of the naphthyl and anthryl rings in carbocation stabilization (*16, 40, 41*). A value of 1.3 for k_α/k_{uv} (the ratio of the polarimetric rate constant and the rate of product formation determined spectrophotometrically) was found for **IV** in TFE, but the high reactivity of **IV** precluded polarimetric rate measurements in other fluorinated solvents (*16*). Because of the stabilizing character and large size of the aromatic rings in **III** and **IV**, solvent participation at the central carbon is quite unlikely in these compounds, and the reactivity data obtained are consistent with an ion-pair mechanism similar to Chart I or equation 5.

The formation of the ring-substituted products **XIV** during solvolysis of **IV** provides a unique probe of the reaction mechanisms in this case, as both the reactant and product are chiral. A possible mechanism for formation of the product could involve solvent attack at the relatively unhindered 10-position of **IV** or on a chiral ion pair with a stereoelectronic preference for addition either syn or anti to the leaving group, analogous to proposals for the S_N2' mechanism (*42*). However the products **XIV** from optically active **IV** were completely racemic, ruling out the operation of such a process.

CF_3
CHOMs

SOH →

H CF_3

H OS

(–)-**IV**

XIV, S = CH_2CH_3 or CH_2CF_3

1-Arylethyl Derivatives

The electron-withdrawing effect of the α-CF_3 group in **I–V** would be expected to reduce the possibility of carbocation formation, but the net effect of an α-CF_3 group is also known to inhibit nucleophilic attack, possibly due to the unfavorable interaction of the nucleophile with the lone pairs on fluorine (*43*). Therefore, the 1-arylethyl tosylates **VIa–c** bearing electron-withdrawing halogen groups on the ring were examined as substrates where the effects of nucleophilic solvent participation would be likely to be manifested (*17*).

The rates of **VIa** and **VIb** in different solvents gave somewhat different correlations by the equation $\log (k/k_0) = mY_{OTs}$ for aqueous ethanols and for other solvents, with m values of 0.82 and 0.78 for the latter, whereas **VIc** gave a very scattered correlation by this equation (*17*). Therefore, the rates were correlated by the equation $\log (k/k_0 = mY_{OTs} + lN$ (*38*), where N is a measure of solvent nucleophilicity and l expresses the dependence on N. For **VIa, VIb,** and **VIc,** values obtained in this way for m and l were 0.90, 0.91, and 0.93 and 0.26, 0.37, and 0.72, respectively, with correlation coefficients of 0.994, 0.994, and 0.977, repsectively. Whatever the merits of the dual correlation with Y_{OTs} and N, all three substrates showed enhanced rates in the ethanol solvents compared with the trends for the other solvents; this result suggests that nucleophilic interactions are important.

The ρ^+ values derived from the rates of reaction of **VIa–c** in different solvents are −4.9 (TFA), −5.1 (100% HFIP), −5.9 (97% HFIP), −6.3 (100% TFE), −5.1 (97% TFE), −3.9 (HOAc), −3.0 (80% EtOH), −3.3 (90% CH_3CH_2OH), and −3.4 (100% CH_3CH_2OH). These ρ^+ values may be compared to those of −5.0 to −6.3 for 1-arylethyl derivatives attributed to reactions involving carbocations by Richard and Jencks (*8*) and that of −2.9 for the S_N2 reaction of 1-arylethyl derivatives with azide (*8*).

The net stereochemistry of the solvolysis of **VIa–c** in different solvents and the ratios of polarimetric rate constants k_α and the rate constants for product formation k_{uv} are summarized in Table II (*17*).

The results in the less nucleophilic solvents, particularly the large magnitudes of ρ^+ and the high k_α/k_{uv} ratios, are consistent with reversible

Table II. k_{α}/k_{uv} and Net Stereochemistry (in Parentheses) for Solvolysis of $ArCH(OTs)CH_3$ (VI), 25 °C (*17*).

Solvent	*3-BrC$_6$H$_4$* (**VIa**)	*3-CF$_3$C$_6$H$_4$* (**VIb**)	*3,5-(CF$_3$)$_2$C$_6$H$_3$* (**VIc**)
TFA			1.4[a]
HFIP		9.2[b]	2.1[c]
TFE	1.9 (rac)	1.4[c]	
HOAc	1.1 (40% inv)	1.1[d]	—[e]
EtOH	1.1 (72% inv)	1.0[f]	—[g]

[a] 6% retention.
[b] Racemic.
[c] Product optically active, stereochemistry unknown.
[d] 46% inversion.
[e] 77% inversion.
[f] 74% inversion.
[g] 87% inversion.

formation of ion pairs that undergo rate-limiting attack by the solvent as shown in Scheme I. The occurrence of 6% net retention in the trifluoroacetolysis of **VIc** could occur through an interaction of the leaving group with the solvent directing attack to the front side and has been observed in other systems (*17, 34, 44,* and references cited therein).

The lower magnitude of the ρ^+ values for **VIa–c** in the more nucleophilic solvents, particularly those containing CH_3CH_2OH, is evidence for nucleophilic solvent participation in these solvents. A quantitative estimate of the acceleration due to this cause in 100% CH_3CH_2OH may be obtained by comparing the observed rates to those calculated from mY_{OTs} correlations for the remaining solvents in which such solvent participation is less important or absent. Thus, m values of 0.82, 0.78, and 1.35 are obtained for **VIa–c,** respectively, by using data for all the non-CH_3CH_2OH solvents in the former two cases and for TFA and HFIP in the latter. Rate constants calculated in this way in CH_3CH_2OH at 25 °C in the absence of nucleophilic solvent participation are 1.6×10^{-5}, 2.9×10^{-6}, and 8×10^{-13} s^{-1} for **VIa, VIb,** and **VIc,** respectively. The last value, based on extrapolation of a two-point correlation, is only approximate.

Comparison of the calculated nucleophilic accelerations of 9, 17, and 10^6 to the degree of inversion (72, 74, and 87%, respectively) for **VIa–c** shows that even though the kinetic effects of solvent assistance are significant, this fact does not lead to the complete inversion characteristic of the S_N2 reaction as found in acetolysis of 2-octyl tosylate (*45*). The best description of the first intermediate in the solvolysis of **VIa–c** would thus appear to be the solvated ion pair shown in Scheme II. In this species, the solvent is involved in nucleophilic solvation of the central carbon as well as the remainder of the carbocation and also participates in electrophilic solvation of the anion. Numerous solvent molecules are involved, and no strong interaction of a single nucleophilic solvent molecule at the central carbon leading to exclusive inversion occurs.

Thus, these studies of 1-arylethyl systems provide excellent insight into the behavior of reactive systems where mechanistic changeover occurs.

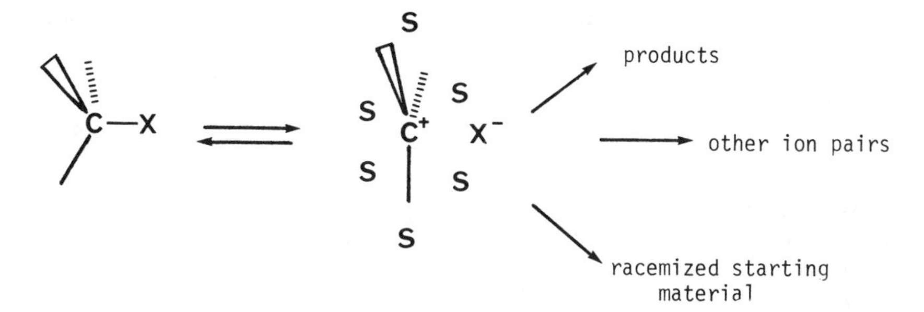

Scheme II. Solvated ion pairs in solvolysis.

Other quite different approaches to the same problems exist, and the continued pursuit of such studies will provide increasingly more accurate understanding of transition-state structure.

Acknowledgment

This work was made possible by the contributions of our co-workers listed in the references, the continued research support of the Natural Sciences and Engineering Research Council of Canada, and a NATO grant that permitted collaboration with M. Charpentier and her associates at the CNRS, Thiais, France.

Literature Cited

1. Ingold, C. K. *Structure and Mechanism of Organic Chemistry;* Cornell University: Ithaca, NY, 1953.
2. Jencks, W. P. *Chem. Soc. Rev.* **1981,** *10,* 345–375.
3. Bentley, T. W.; Schleyer, P. v. R. *Adv. Phys. Org. Chem.* **1977,** *14,* 1–67.
4. Winstein, S.; Grunwald, E.; Jones, H. W. *J. Am. Chem. Soc.* **1951,** *73,* 2700–2707.
5. Doering, W. v. E.; Zeiss, H. H. *J. Am. Chem. Soc.* **1953,** *75,* 4733–4738.
6. Bentley, T. W.; Bowen, C. T.; Morten, D. H.; Schleyer, P. v. R. *J. Am. Chem. Soc.* **1981,** *103,* 5466–5475.
7. Raber, D. J.; Harris, J. M.; Schleyer, P. v. R. In *Ions and Ion Pairs in Organic Reactions;* Szwarc, M., Ed.; Wiley: New York, Vol. 2, 1974; Chapter 3, pp 247–374.
8. Richard, J. P.; Jencks, W. P. *J. Am. Chem. Soc.* **1984,** *106,* 1383–1396.
9. Tidwell, T. T. *Angew. Chem., Int. Ed. Engl.* **1984,** *23,* 20–32.
10. Gassman, P. G.; Tidwell, T. T. *Acc. Chem. Res.* **1983,** *16,* 279–285.
11. Allen, A. D.; Ambidge, I. C.; Che, C.; Micheal, H.; Muir, R. J.; Tidwell, T. T. *J. Am. Chem. Soc.* **1983,** *105,* 2343–2350.
12. Allen, A. D.; Kanagasabapathy, V. M.; Tidwell, T. T. *J. Am. Chem. Soc.* **1983,** *105,* 5961–5962.
13. Kanagasabapathy, V. M.; Sawyer, J. F.; Tidwell, T. T. *J. Org. Chem.* **1985,** *50,* 503–509.
14. Allen, A. D; Jansen, M. P.; Koshy, K. M.; Mangru, N. N.; Tidwell, T. T *J. Am. Chem. Soc.* **1982,** *104,* 207–211.

15. Allen, A. D.; Kanagasabapathy, V. M.; Tidwell, T. T. *J. Am. Chem. Soc.* **1986,** *108*, 3470–3474.
16. Allen, A. D.; Girdhar, R.; Jansen, M. P.; Mayo, J. D.; Tidwell, T. T. *J. Org. Chem,* **1986,** *51*, 1324–1329.
17. Allen, A. D.; Kanagasabapathy, V. M.; Tidwell, T. T. *J. Am. Chem. Soc.* **1985,** *107*, 4513–4519.
18. Koshy, K. M.; Roy, D.; Tidwell, T. T. *J. Am. Chem. Soc.* **1979,** *101*, 357–363.
19. Allen, A. D.; Shahidi, F; Tidwell, T. T. *J. Am. Chem. Soc.* **1982,** *104*, 2516–2518.
20. Nowlan, V. J.; Tidwell, T. T. *Acc. Chem. Res.* **1977,** *10*, 252–258.
21. Brown, H. C.; Okamoto, Y. *J. Am. Chem. Soc.* **1958,** *80*, 4979–4987.
22. Lenoir, D. *Chem. Ber.* **1975,** *108*, 2055–2072.
23. Kirmse, W.; Mrotzeck, U.; Seigfried, R. *Angew. Chem., Int. Ed. Engl.* **1985,** *24*, 55–56.
24. Gassman, P. G.; Hall, J. B. *J. Am. Chem. Soc.* **1984,** *106*, 4267–4269.
25. Gassman, P. G.; Harrington, C. K. *J. Org. Chem.* **1984,** *49*, 2258–2273.
26. Poulter, C. D.; Rilling, H. C. *Acc. Chem. Res.* **1978,** *11*, 307–313.
27. Jansen, M. P.; Koshy, K. M.; Mangru, N. N.; Tidwell, T. T. *J. Am. Chem. Soc.* **1981,** *103*, 3863–3867.
28. Koshy, K. M.; Tidwell, T. T. *J. Am. Chem. Soc.* **1980,** *102*, 1216–1218.
29. Creary, X.; McDonald, S. R. *J. Org. Chem.* **1985,** *50*, 474–479.
30. Kevill, D. N.; Anderson, S. W. *J. Org. Chem.* **1985,** *50*, 3330–3333.
31. Creary, X. *J. Am. Chem. Soc.* **1984,** *106*, 5568–5577.
32. Creary, S.; Geiger, C. *J. Am. Chem. Soc.* **1982,** *104*, 4151–4162.
33. Creary, X.; Geiger, C. C.; Hilton, K. *J. Am. Chem. Soc.* **1983,** *105*, 2851–2858.
34. Creary, X.; McDonald, S. R.; Eggers, M. D. *Tetrahedron Lett.* **1985,** *26*, 811–814.
35. Liu, K.-T.; Kuo, M.-Y. *Tetrahedron Lett.* **1985,** *26*, 355–358.
36. Skrobul, A. P. Ph.D. Thesis, Purdue University, Lafayette, IN, 1981; *Diss. Abstr. Int. B* **1981,** *42* 1900.
37. Liu, K.-T.; Kuo, M.-Y.; Sheu, C.-F. *J. Am. Chem. Soc.* **1982,** *104*, 211–215.
38. Schadt, F. L.; Bentley, T. W.; Schleyer, P. v. R. *J. Am. Chem. Soc.* **1976,** *98*, 7667–7674.
39. McManus, S. P.; Neamati-Mazraeh, N.; Hovanes, B. A; Paley, M. S.; Harris, J. M. *J. Am. Chem. Soc.* **1985,** *107*, 3393–3395.
40. Berliner, E.; Shieh, N. *J. Am. Chem. Soc.* **1957,** *79*, 3849–3854.
41. Stock, L. M.; Brown, H. C. *Adv. Phys. Org. Chem.* **1963,** *1*, 35–154.
42. Bach, R. D.; Wolber, G. J. *J. Am. Chem. Soc.* **1985,** *107*, 1352–1357.
43. Smart, B. E. In *The Chemistry of Halides, Pseudo-Halides and Azides;* Patai, S.; Rappoport, Z., Eds.; Wiley: New York, 1983; Supplement D, Chapter 14.
44. McManus, S. P.; Zutaut, S. E. *Tetrahedron Lett.* **1984,** *25*, 2859–2862.
45. Streitwieser, A., Jr.; Walsh, T. D.; Wolfe, J. R., Jr. *J. Am. Chem. Soc.* **1965,** 87, 3682–3685.

RECEIVED for review October 21, 1985. ACCEPTED June 30, 1986.

NOVEL NUCLEOPHILIC REACTIONS

23

Relationship between Nucleophilic Reactions and Single-Electron Transfer

Application to Reactions of Radical Cations

Addy Pross

Department of Chemistry, Ben-Gurion University of the Negev, Beersheva, Israel,* and Department of Chemistry, Stanford University, Stanford, CA 94305

Nucleophilic reactions often compete with single-electron-transfer (SET) processes. The configuration mixing (CM) model, which builds up reaction profiles qualitatively, is utilized to provide simple experimental criteria for predicting the factors likely to encourage one pathway over the other pathway. The analysis suggests both *nucleophilic and SET processes involve a single-electron shift. Factors that favor a SET process include (1) strong donor–acceptor pair ability of nucleophile and electrophile, (2) steric interactions in the transition state, (3) delocalization of the odd electrons that make up the nucleophile–substrate bond, and (4) the nucleophile–substrate bond strength. The apparent reluctance of aromatic radical cations to undergo direct nucleophilic attack is explained on the basis of the single-electron shift model for nucleophilic reactions.*

THE CLASS OF SINGLE-ELECTRON-TRANSFER (SET) processes in organic chemistry (for a recent review on organic electron-transfer reactions, *see* reference 1) has expanded enormously over recent years so that organic reaction mechanisms may be broadly divided into two general classes: the polar reactions in which electrons seem "to move about in pairs" and the so-called one-electron processes in which electrons are transferred one at a time (2). Yet, strangely, the relationship between these two pathways is far from clear. For the specific case of nucleophilic reactivity, the question arises:

* Where correspondence should be sent.

0065-2393/87/0215-0331$06.00/0

Why do nucleophiles at times follow a polar pathway, for example, the S_N2 reaction or nucleophilic addition to the carbonyl group, whereas at other times these same nucleophiles might react via a SET process? What is the relationship between these two general pathways and what factors influence which reaction pathway will be followed in any particular case? This chapter analyzes this problem using the configuration mixing (CM) model (*3–5*) by comparing both reaction processes.

Finally, we will turn to the chemistry of odd electron species. The reaction of radical cations with nucleophiles has been extensively studied over recent years. In a detailed review, Parker (*6*) concluded that certain radical cations, which were previously thought to undergo direct nucleophilic attack, actually react via a series of electron-transfer steps. Using the CM analysis, this chapter provides a simple explanation for the puzzling reactivity properties of odd-electron species and shows how their behavior fits into the polar–SET mechanistic picture.

Discussion

An S_N2 reaction, depicted in equation 1, seems to come about by an electron pair on the nucleophile "displacing" a second electron pair—the R–X σ bond. Four valence electrons appear to be involved. The problem with this representation is that the electronic rearrangement seems radically different from that in a SET process, such as the initiation step of the $S_{RN}1$ process (*7*), equation 2. Clearly, just a single electron has been transferred from the nucleophile to RX. As a consequence of these quite different descriptions, the relationship between the two processes becomes obscure. What factors encourage one pathway over the other is not clear.

$$\text{N:}^- + \text{R—X} \longrightarrow \text{N—R} + \text{:X}^- \tag{1}$$

$$\text{N:}^- + \text{R—X} \longrightarrow \text{N}\cdot + (\text{R}\dot{-}\text{X})^- \tag{2}$$

Despite the fact that polar nucleophilic reactions are commonly termed "two-electron" processes, what the term really signifies must be understood. The term does *not* mean that during the course of the reaction, electron pairs relocate within the molecule two by two, in the way that the curly arrow convention implies. Two-electron processes merely indicate that all electrons that were spin-coupled in reactants remain spin-coupled in the products. Actually, the S_N2 process and indeed all other polar nucleophilic reactions are really just single-electron-shift processes (*2, 3*). The approximate wave function that describes the reactants of the S_N2 process (equation 1) is given by χ_R, equation 3, and differs from the corresponding wave function describing products, χ_P, equation 4, by just a single-electron shift. The single-electron-shift nature of the reaction becomes even more apparent when simple VB structures are used to represent χ_R and χ_P, as shown in **I** and **II** respectively.

$$\chi_R = \frac{1}{\sqrt{2}} [\phi_N(1)\bar{\phi}_N(2)\bar{\phi}_R(3)\phi_X(4) - \phi_N(1)\bar{\phi}_N(2)\phi_R(3)\bar{\phi}_X(4)] \quad (3)$$

$$\chi_P = \frac{1}{\sqrt{2}} [\phi_N(1)\bar{\phi}_R(2)\bar{\phi}_X(3)\phi_X(4) - \phi_N(1)\bar{\phi}_R(2)\bar{\phi}_X(3)\phi_X(4)] \quad (4)$$

$$\underset{\textbf{I}}{N\!:^{-} \quad R\cdot \quad \cdot X} \qquad\qquad \underset{\textbf{II}}{N\cdot \quad \cdot R \quad :X^{-}}$$

To convert **I** to **II** all that is needed is to shift a single electron from $N\!:^-$ to $\cdot X$. In other words, the barrier to an S_N2 reaction may be thought of as coming about through the avoided crossing of χ_R and χ_P (*3, 8–10*) or, using the D (donor)–A (acceptor) terminology, by a $DA–D^+A^-$-avoided crossing (*9, 11*). Consideration of the S_N2 process in these terms provided a means of assessing the factors governing reactivity in these systems (*8, 11, 12*).

In the context of this chapter, we demonstrate that only by considering polar nucleophilic processes as SET processes can the factors that govern the competition between polar and SET processes as well as nucleophilic reactions of radical cations be adequately understood.

Consider the reaction of a carbonyl compound with a nucleophile, the hydroxide ion:

$$HO\!:^- \; + \; >C{=}O \;\rightarrow\; HO{-}\overset{|}{\underset{|}{C}}{-}O\!:^- \quad (5)$$

For this reaction, reactant and product configurations are depicted by **III** and **IV**, respectively. Here again, we see that reaction comes about by a single-electron shift. The product configuration has two spin-paired electrons on O and C that can form a bond once that electron shift has occurred. So we see that the difference between a nucleophilic addition process and SET lies not in the number of electrons that are shifted but in whether two coupled electrons in close proximity are generated following the electron shift. This statement is the essence of the polar–SET competition. In a SET pathway D, A reacts to form D^+,A^-, while in a polar process such as that in equation 5,

$$\underset{\textbf{III}}{HO\!:^- \quad \dot{C}{-}\dot{O}} \qquad\qquad \underset{\textbf{IV}}{HO\cdot \quad \cdot C{-}\ddot{O}^-}$$

D,A react to form $D^+{-}A^-$. The difference is that in the polar pathway two odd electrons on $D\cdot^+$ and $A\cdot^-$, brought about by the single-electron shift, are paired into a single bond. In the case of the SET process, no such interaction occurs.

SET versus Polar Pathways

Any factor that inhibits or disallows coupling of the two odd electrons after the single-electron shift will encourage SET over a polar nucleophilic pathway. This statement forms the basis for understanding the competition between these two routes. The following factors will have a bearing on the polar–SET competition (*2*).

Donor–Acceptor Pair Ability. The better the donor–acceptor pair, the earlier the avoided crossing between DA and D^+A^- configurations so that the degree of coupling between the two odd electrons is reduced. This relationship will lead to an increase in the likelihood of a SET process, and numerous examples where this trend is observed exist (*13–18*).

Steric Interactions between D and A. If D or A is sterically hindered, then coupling between $D\cdot^+$ and $A\cdot^-$ is impeded and a SET pathway is encouraged (*13, 19, 20*).

D–A Bond Strength. The stronger the D–A bond, the more likely D–A coupling will occur. If the D–A bond that is to form is weak, the SET is encouraged. The tendency for iodide ion to act as an electron donor but fluoride ion as a nucleophile (*21–22*) may be explained in this manner.

Radical Delocalization. If the two radical centers on D and A are extensively delocalized, coupling is inhibited and a SET pathway is encouraged. This actually represents a special case of D–A bond strength, because the coupling of delocalized radicals leads to weak bonds. Numerous examples of all of these predictions exist (*2*); these examples make the foregoing analysis a most useful one. We see therefore that only by viewing the polar process as a single-electron shift does the relationship between the polar and SET pathways become clear.

Nucleophilic Attack on Radical Cations

What is the mechanism of attack of nucleophiles on radical cation species such as anthracene or thianthrene radical cations? Extensive studies by a number of groups have been conducted on the reaction of radical cations with nucleophiles (*6, 23, 24*). Two main mechanisms have been proposed: the disproportionation pathway, equations 6 and 7, in which the nucleophile,

Nu, attacks the dication formed by the disproportionation of the cation radical $A^{\cdot +}$; and the half-regeneration mechanism, equations 8 and 9, in which nucleophilic attack takes place directly on the radical cation. A third mechanism, termed the complexation mechanism, is closely related to the disproportionation mechanism but differs from it in that one of the reacting radical cation molecules of equation 6 is complexed to a molecule of the nucleophile in a donor–acceptor charge-transfer π complex. The role of the nucleophile donor is to facilitate electron transfer to the second radical cation group. The disproportionation step of the complexation mechanism is indicated in equation 10.

$$2A^{\cdot +} \longrightarrow A^{2+} + A \qquad (6)$$

$$A^{2+} + Nu \longrightarrow (A\text{–}Nu)^{2+} \qquad (7)$$

$$A^{\cdot +} + Nu \rightleftarrows (A\text{–}Nu)^{\cdot +} \qquad (8)$$

$$(A\text{–}Nu)^{\cdot +} + A^{\cdot +} \longrightarrow (A\text{–}Nu)^{2+} + A \qquad (9)$$

$$A^{\cdot +}/Nu + A^{\cdot +} \longrightarrow A^{2+}/Nu + A \qquad (10)$$

The actual mechanism that is followed in any given case has been a subject of considerable controversy. As Parker and co-workers noted (*6, 24*), these reactions are exceedingly complex because they involve multistep pathways through a large number of intermediates. Despite the complexity, evidence for each of the three pathways has been presented (*6, 24*). Most recently, however, in an excellent review of the subject, Parker (*6*) reassessed the existing data and concluded that in certain cases the half-regeneration pathway is not operative, as was initially thought. Thus, for example, diphenylanthracene radical cation in its reaction with pyridine is now thought to react via the complexation pathway (*6*), and not via the half-regeneration pathway (*25*) (for a selection of papers on the reaction of radical cations with nucleophiles, see references 26–37; for a more extensive list of references, see references 6 and 24). Indeed, on the basis of Parker's analysis, we believe that the reaction of radical cations with nucleophiles proceeds predominantly, if not exclusively, via the disproportionation mechanism (or the closely related complexation mechanism).

The question now arises: Why does a dication, whose formation is governed by an equilibrium constant of approximately 10^{-9}, appear to be the species that actually undergoes nucleophilic attack? What is the factor that inhibits direct attack of the nucleophile on the highly reactive radical cation itself? On the basis of the CM analysis, the likelihood of direct attack—the central feature of the half-regeneration pathway—is considered to be slight, as shown by comparing the reactions of normal cations, R^+, with their radical cation counterpart, $A^{\cdot +}$.

The direct attack of a nucleophile, N, on a cation, R^+, involves a single electron shift (*3, 38*). The reaction may be described by the avoided crossing of DA and D^+A^- curves as indicated in Figure 1a. A single-electron shift from N: to R^+ in the R^+ :N pair generates the $R\cdot \cdot N^+$ radical pair, which can collapse to form an R–N σ bond. The case of a radical cation is, however, quite different. An electron shift from Nu: to a radical cation $A^{\cdot +}$ merely regenerates A, the parent hydrocarbon; therefore, a simple nucleophilic

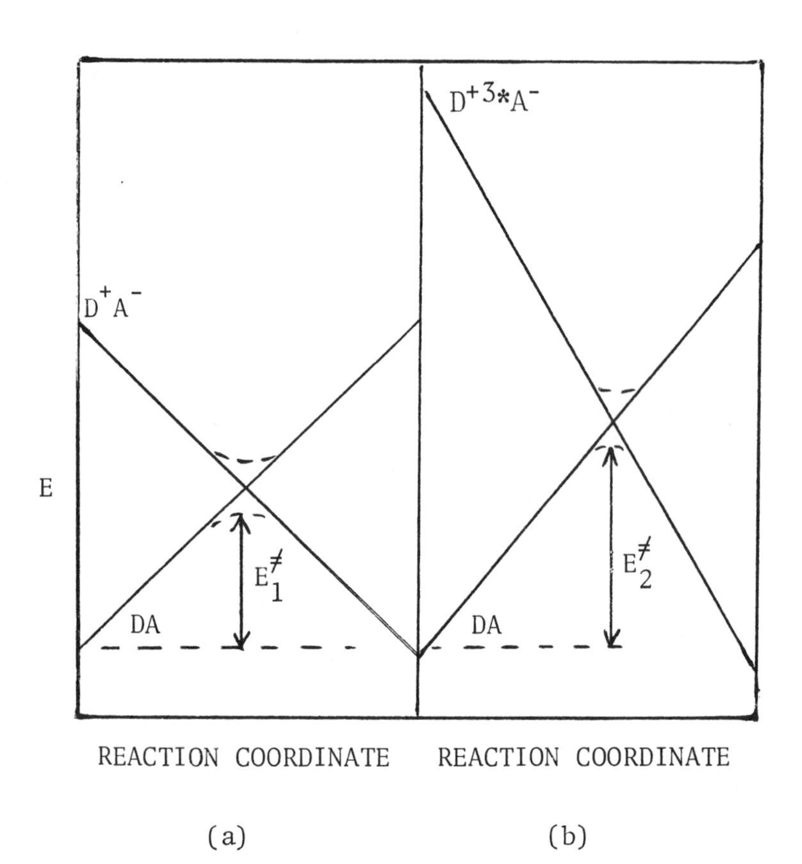

Figure 1. (a) Schematic energy diagram illustrating the way in which the reaction profile for nucleophilic attack on a normal cation may be built up from the avoided crossing of DA and D^+A^- configurations.
(b) Corresponding diagram for nucleophilic attack on a radical cation, in which the product configuration is now $D^+\,{}^{3*}A^-$. Because $D^+\,{}^{3*}A^-$ is doubly excited with respect to DA, while D^+A^- is just singly excited,
$E^{\ddagger}_2 > E^{\ddagger}_1$.

addition process is precluded. In simple terms, after the electron shift no odd electron exists in the hydrocarbon moiety with which the nucleophile radical, $Nu\cdot^{+}$, can couple. Thus, a $DA–D^{+}A^{-}$ crossing for a radical cation–nucleophile pair does *not* describe a nucleophilic addition process but describes an electron transfer reaction from the nucleophile to the radical cation. A nucleophilic addition process on a radical cation is brought about by exciting the DA configuration to $D^{+}\ {}^{3*}A^{-}$ (Figure 1b). Only this doubly excited configuration is electronically set up to bring about the nucleophilic addition reaction: the shift of an electron from the nucleophile to the radical cation prepares the nucleophile for covalent bonding, while excitation of the hydrocarbon moiety to the triplet state uncouples an electron pair; therefore, the electron pair is also prepared for covalent bonding. However, as seen in Figure 1b, a high-energy product configuration is likely to lead to a high-energy pathway, so that we believe direct nucleophilic attack on a radical cation is an unfavorable process. For this reason, competing pathways, which only involve singly excited $D^{+}A^{-}$ product configuration (i.e., electron-transfer reactions), are preferred for radical cations. This result is indeed what is observed (*6*). The disproportionation and complexation mechanisms involve the radical cation species in just electron-transfer steps. Only at the dication stage does direct nucleophilic attack occur. At this point, a $DA–D^{+}A^{-}$ crossing, that is, a single-electron shift, *will* lead to a nucleophilic addition. Formation of the radical cation species as a consequence of the electron shift from N to A^{2+} leads to an $A\cdot^{+}\ \cdot N^{+}$ radical pair, which *can* collapse to give $A^{+}–N^{+}$; nucleophilic attack has occurred. The inexplicably slow protonation reaction of radical anions may be understood in similar terms and will be discussed elsewhere (A. Pross, to be published).

Acknowledgment

I am indebted to Sason Shaik, whose collaboration in the development of the CM model led to many of the ideas discussed.

Literature Cited

1. Eberson, L. *Adv. Phys. Org. Chem.* **1982,** *18,* 79.
2. Pross, A. *Acc. Chem. Res.* **1985,** *18,* 212.
3. Pross, A.; Shaik, S. S. *Acc. Chem. Res.* **1983,** *16,* 363.
4. Pross, A. *Adv. Phys. Org. Chem.* **1985,** *21,* 99.
5. Pross, A.; Shaik, S. S. *J. Am. Chem. Soc.* **1982,** *104,* 187.
6. Parker, V. D. *Acc. Chem. Res.* **1984,** *17,* 243.
7. Bunnett, J. F. *Acc. Chem. Res.* **1978,** *11,* 413.
8. Shaik, S. S.; Pross, A. *J. Am. Chem. Soc.* **1982,** *104,* 2708.
9. Pross, A.; Shaik, S. S. *J. Am. Chem. Soc.* **1982,** *104,* 1129.
10. Pross, A.; Shaik, S. S. *Tetrahedron Lett.* **1982,** 5467.
11. Shaik, S. S. *J. Am. Chem. Soc.* **1983,** *105,* 4359.
12. Shaik, S. S. *Prog. Phys. Org. Chem.* **1985,** *15,* 197.

13. Rollick, K. L.; Kochi, J. K. *J. Am. Chem. Soc.* **1982,** *104,* 1319.
14. House, H. O. *Acc. Chem. Res.* **1976,** *9,* 59.
15. Evans, J. F.; Blount, H. N. *J. Am. Chem. Soc.* **1978,** *100,* 4191.
16. Troughton, E. B.; Molter, K. E.; Arnett, E. M. *J. Am. Chem. Soc.* **1984,** *106,* 6726.
17. Russell, G. A.; Jawdosiuk, M.; Makosa, M. *J. Am. Chem. Soc.* **1979,** *101,* 2355.
18. Zieger, H. E.; Angres, I.; Mathisen, D. *J. Am. Chem. Soc.* **1976,** *98,* 2580.
19. Ashby, E. C.; Goel, A. B. *J. Am. Chem. Soc.* **1981,** *103,* 4983.
20. Ashby, E. C.; Goel, A. B.; DePriest, R. N. *J. Org. Chem.* **1981,** *46,* 2329.
21. Evans, T. R.; Hurysz, L. F. *Tetrahedron Lett.* **1977,** 3103.
22. Rozhkov, I. N.; Gambaryan, R. P.; Galpern, E. G. *Tetrahedron Lett.* **1976,** 4819.
23. Bard, A. J.; Ledwith, A.; Shine, H. J. *Adv. Phys. Org. Chem.* **1976,** *13,* 156.
24. Hammerich, O.; Parker, V. D. *Adv. Phys. Org. Chem.* **1984,** *20,* 55.
25. Manning, G.; Parker, V. D.; Adams, R. N. *J. Am. Chem. Soc.* **1969,** *91,* 4584.
26. Shine, H. J.; Murata, Y. *J. Am. Chem. Soc.* **1969,** *91,* 1872.
27. Murata, Y.; Shine, H. J. *J. Org. Chem.* **1969,** *34,* 3368.
28. Parker, V. D.; Eberson, L. *J. Am. Chem. Soc.* **1970,** 92, 7488.
29. Marcoux, L. *J. Am. Chem. Soc.* **1971,** 93, 537.
30. Svanholm, U.; Hammerich, O.; Parker, V. D. *J. Am. Chem. Soc.* **1975,** 97, 101.
31. Svanholm, U.; Parker, V. D. *J. Am. Chem. Soc.* **1976,** *98,* 997, 2942.
32. Kim, K.; Hull, V. J.; Shine, H. J. *J. Org. Chem.* **1974,** *39,* 2534.
33. Evans, J. F.; Blount, H. N. *J. Org. Chem.* **1977,** *42,* 976.
34. Evans, J. F.; Blount, H. N. *J. Am. Chem. Soc.* **1978,** *100,* 4191.
35. Evans, J. F.; Blount, H. N. *J. Phys. Chem.* **1979,** *83,* 1970.
36. Hammerich, O.; Parker, V. D. *Acta Chem. Scand., Ser. B* **1981,** *35,* 341.
37. Cheng, H. Y.; Sackett, P. H.; McCreery, R. L. *J. Am. Chem. Soc.* **1978,** *100,* 962.
38. Hoz, S. *J. Org. Chem.* **1982,** *47,* 3545.

RECEIVED for review October 21, 1986. ACCEPTED January 31, 1986.

Reduction versus Substitution in the Reaction of Nitroaryl Halides with Alkoxide Ions

Cristina Paradisi and Gianfranco Scorrano *

Centro Meccanismi Reazioni Organiche del C.N.R., Dipartimento di Chimica Organica, Via Marzolo 1, 35131 Padova, Italy

The competition between reduction of the nitro group and nucleophilic substitution of chlorine in the reaction of 4-chloronitrobenzene with alkoxide ions in alcoholic solutions was studied. The substitution reaction is sluggish except for short-chain primary alkoxides but is greatly activated by the addition of suitable cation complexing agents or tetraalkylammonium ions. When oxygen is carefully excluded from the reaction mixture, the reduction path prevails. Rather surprisingly, however, reduction is strongly inhibited in the presence of crown ether complexed cations, to the point that, even in deoxygenated solutions, only the substitution reaction takes place. Mechanistic features are discussed on the basis of results of electron paramagnetic resonance and electrochemical investigations and of product and kinetic analysis.

THE NATURE OF THE INTERACTION between an electron-rich species, a nucleophile $Nu:^-$, and a substrate S capable of electrophilic reactivity has concerned chemists for a long time. A widely used mechanistic distinction is made between single-electron transfer (SET), leading to radical species (Scheme I, path a), and two-electron processes, involving direct coordination via a pair of unshared electrons (Scheme I, path b).

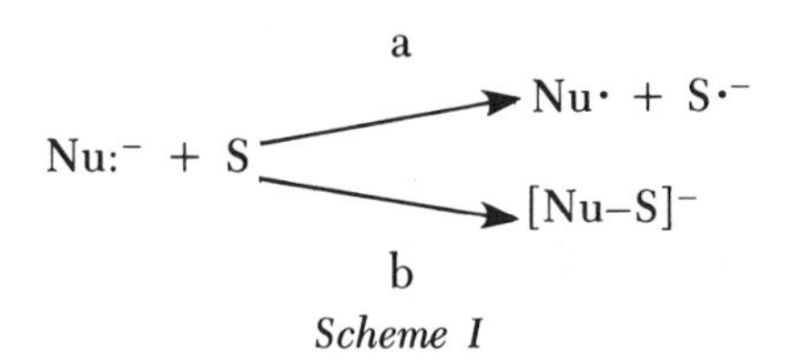

Scheme I

0065-2393/87/0215-0339$06.00/0

When S bears a nucleofugal substituent, Y, substitution of Y by Nu may occur. Examples are provided by $S_{RN}1$ (*1–3*), S_N2 (*4*), and S_NAr (*5, 6*) reactions.

Mechanistic assignment in terms of path a or b is not always straightforward. Recently, the proposal was advanced that also "apparent" two-electron processes, such as S_N2 substitution reactions, proceed via single-electron transfer with synchronous coupling of the electron pair of the ensuing radical pair (*7, 8*). The opposite situation has also been encountered. Evidence was recently reported that the reaction of nitrobenzene with *t*-BuOK in THF, a system in which radicals are observed, is not a straightforward electron-transfer process but proceeds via initial nucleophilic attack (*9*).

Perhaps of more general interest, also for the synthetic chemist, are systems in which channels a and b compete to give different products. In the course of our studies of the reactivity of nitroaromatic compounds in alkaline alcoholic solutions, we encountered one such system when we used 4-chloronitrobenzene (*10*). 4-Chloronitrobenzene falls into the category of those aryl halides, bearing electron-withdrawing substituents (W) in ortho–para arrangement, that are considered activated toward nucleophilic aromatic substitution according to the S_NAr mechanism. Alkoxide ions form an important class of nucleophiles in these reactions, leading to alkyl aryl ethers (*5, 6, 11*) (X is a halide):

$$\text{X–C}_6\text{H}_4\text{–W} + \text{RO}^- \rightleftarrows [\text{(X)(OR)C}_6\text{H}_4\text{W}]^- \longrightarrow \text{RO–C}_6\text{H}_4\text{–W} + \text{X}^- \quad (1)$$

Nitroarenes, on the other hand, are strong electron acceptors and easily undergo one-electron reduction (*12, 13*). Thus, nitrobenzene, to cite one example, has been customarily used as an effective "quencher" in chain reactions involving radical anion intermediates, such as in $S_{RN}1$ reactions (*3*). Under different conditions, nitroarene radical anions are reactive species. In particular, Zinin (*14*) reported that treatment of nitroarenes with hot alkaline alcoholic solutions results in products of reduction, mainly the azoxy derivative (equation 2). These complex multistep processes involve nitroarene radical anion intermediates and are quite effectively inhibited by oxygen (*10, 15, 16*). In 1964, Russell et al. (*17*) wrote that "apparently much of the chemistry of aromatic nitro, nitroso and azo compounds in basic solution involves electron-transfer processes".

$$\text{ArNO}_2 \xrightarrow[\Delta]{\text{ROH–RO}^-} \underset{\text{major product}}{\text{Ar}\overset{+}{\text{N}}(\text{O}^-)\text{=NAr}} + \text{other products} \quad (2)$$

We found that, under certain conditions, with 4-chloronitrobenzene products of substitution and of reduction, are formed independently and in competition along channels b and a (Scheme I). The relative para orientation of the two functionalities is necessary to have comparable kinetic reactivities, the ortho isomer being too reactive and the meta totally unreactive in the S_NAr mode.

In this chapter, we describe this competition, the factors that influence it, and strategies to obtain the desired product in good yield by properly adjusting the reaction conditions. Some remarkable ion-pairing effects disclosed recently will be described in particular detail. These studies have led to significant progress in our understanding of the mechanism of the reduction process.

Substitution Channel

4-Chloronitrobenzene is a rather poor substrate in S_NAr reactions with alkoxide ions in the parent alcohol. The percent yield of 4-nitroalkoxybenzene obtained in the reactions $4\text{-ClC}_6H_4NO_2 + RO^- \longrightarrow (ROH)\ 4\text{-ROC}_6H_4NO_2 + Cl^-$ is as follows: for MeO^- and EtO^-, >95; for $1\text{-C}_8H_{17}O^-$, <5; for 2-PrO^-, >50; and for $t\text{-BuO}^-$, <2. Except for short-chain primary alkoxides (*10*), the substitution product is obtained in very poor yield: a mere few percent in the case of 1-octanol (*18*, *19*) and *tert*-butyl alcohol (*10*, *20*) and less than 50% in 45 h even with the simplest secondary alcohol, 2-propanol (*10*). The data in 2-propanol refer to experiments run with oxygen bubbling through the reaction mixture in order to "quench" the reduction process. Without this expedient, the yield of ether product is usually lower because of the competing reduction process. In equation 3, the product distribution is reported for a typical reaction of 4-chloronitrobenzene with 2-PrOK in 2-PrOH at 75 °C, run without atmospheric control. The variety of products obtained and the highly irreproducible yields observed under these conditions reflect the complexity of these systems.

$$\text{Cl-C}_6H_4\text{-NO}_2 \xrightarrow[75\ ^\circ\text{C, air}]{\text{2-PrOK–2-PrOH}} \underset{9\%}{\text{2-PrO-C}_6H_4\text{-NO}_2} + \underset{5\%}{\text{HO-C}_6H_4\text{-NO}_2} +$$

$$\underset{33\%}{\text{Cl-C}_6H_4\text{-N=N-C}_6H_4\text{-Cl}} + \underset{23\%}{\text{Cl-C}_6H_4\text{-N=N-C}_6H_4\text{-Cl}} + \underset{28\%}{\text{Cl-C}_6H_4\text{-NH}_2} \qquad (3)$$

More reproducible behavior is observed under controlled atmosphere. Scheme II is a summary of the behavior of 4-chloronitrobenzene in 2-PrOK–2-PrOH solutions under oxygen and argon. The effect of oxygen as a powerful inhibitor of the reduction process is clearly evident. Accordingly, we found (*18, 19*) that reflux conditions, which lead to degassing of the solution, favor the reduction path. $t_{1/2}$ data reported in Scheme II allow for a crude comparison of the reactivities of 4-chloronitrobenzene in the S_NAr and reduction processes. Given a chance to occur, by excluding oxygen, reduction is significantly faster than substitution.

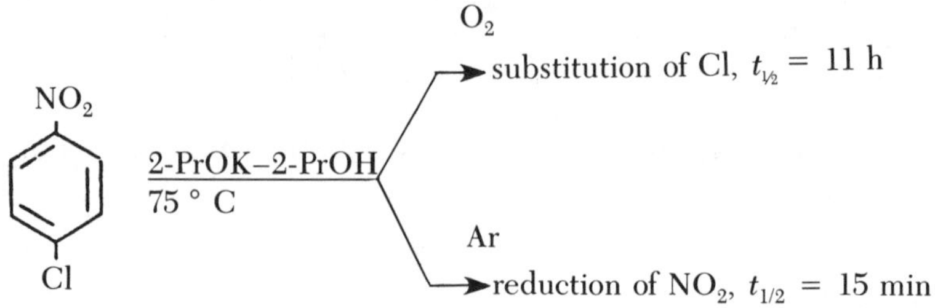

NO_2, Cl

2-PrOK–2-PrOH, 75 ° C

O_2 → substitution of Cl, $t_{1/2}$ = 11 h

Ar → reduction of NO_2, $t_{1/2}$ = 15 min

Scheme II

We found (*18, 19*) that ion pairing is one of the most important factors that influence the competition between reduction and substitution. Ion pairing usually depresses the rate of these S_NAr reactions because "free" alkoxide ions display a superior reactivity, although the opposite situation was found in some instances (*21*). Examples of rate enhancements observed upon addition of cation complexing agents or tetraalkylammonium ions are reported in Table I. The use of n-Bu_4NBr, for example, allows one to obtain substitution products in good yields and convenient times in reactions with otherwise poorly reactive alkoxides.

Table I. Data Relative to the Reaction

$$\underset{\mathbf{I}}{4\text{-}ClC_6H_4NO_2} \xrightarrow[\text{air}]{ROH\text{–}KOH} \underset{\mathbf{II}}{4\text{-}ROC_6H_4NO_2} + Cl^-$$

ROH	*Catalyst*	*Reaction Time (h)*	*Temperature (°C)*	*% Yield* *I*	*% Yield* *II*
2-propanol	—[a]	46	75	12	59
2-propanol	18-crown-6	3	75	2	80
2-propanol	n-Bu_4NBr	3	75	2	79
2-propanol	Carbowax 20M	5	75	3	79
2-propanol	Glyme 5	22	75	1	75
1-octanol	—[a]	0.75	100	>0.5	3
1-octanol	n-Bu_4NBr	0.28	100	1	96
2-butanol	—[a]	22.4	90	8	33
2-butanol	n-Bu_4NBr	1.7	90	1	81
PhOK in PhCl	—[a]	5	reflux	93	1
PhOK in PhCl	n-Bu_4NBr	5	reflux	>0.5	93

[a] None.

More quantitative indications of the impact of ion-pairing effects on these S_NAr processes emerged from kinetic studies of the reaction of 4-chloronitrobenzene with MeOK, EtOK, and 2-PrOK in their respective alcohols with and without 18-crown-6 (*19*). Complexation of K^+ by the crown compound results in increased S_NAr reactivity, the effect being maximum for the 2-PrOK–2-PrOH system. Rate enhancements for MeO^-, EtO^-, and 2-PrO^- are 1.6, 4.0, and 28.0, respectively, and follow the trend of the association constant for ROK in ROH (*22*). When ion association is eliminated or reduced as in the case of crown-complexed K^+, the observed relative reactivities, 1.0 for MeO^-, 3.6 for EtO^- and 11.4 for 2-PrO^-, follow the order of the alkoxides basicity [pK_a values for MeOH, EtOH, and 2-PrOH are 15.09, 15.93, and 17.1, respectively (*23*)].

The rate enhancements observed in these nucleophilic aromatic substitution reactions when using crown-complexed ions and tetraalkylammonium ions are indeed not surprising. Many examples are known of increased reactivity in nucleophilic substitutions due to complexation with crown compounds (*24*), also in S_NAr reactions (*25, 26*). In our system, however, this "normal" effect is accompanied by an inhibiting effect on the competing reduction path, which is discussed under Reduction Channel.

Reduction Channel

As anticipated in Scheme II, reaction of 4-chloronitrobenzene with 2-PrOK in 2-PrOH at 75 °C under argon leads exclusively to products of reduction of the substrate (equation 4) and of oxidation of the solvent to acetone (*27*). A

$$Cl{-}C_6H_4{-}NO_2 \xrightarrow[75\,°C,\ Ar]{2\text{-PrOK–2-PrOH}} \underset{57.2\%}{Cl{-}C_6H_4{-}\overset{+}{N}(O^-){=}N{-}C_6H_4{-}Cl} +$$

$$\underset{12.1\%}{Cl{-}C_6H_4{-}NH_2} + \underset{5.6\%}{Cl{-}C_6H_4{-}N{=}N{-}C_6H_4{-}Cl} + \underset{1.0\%}{Cl{-}C_6H_4{-}NHCOOCH(CH_3)_2} +$$

$$\underset{6.0\%}{Cl{-}C_6H_4{-}NHCHO} + \underset{1.0\%}{Cl{-}C_6H_4{-}NHCOCH(OH)CH_3} +$$

$$\underset{1.0\%}{Cl{-}C_6H_4{-}NHCH_2COOCH(CH_3)_2} \qquad (4)$$

common precursor of all the products of reduction is the nitroso derivative, ArNO. This intermediate reacts further according to two competing paths. The first and major path involves one-electron reduction followed by dimerization and leads to the azoxy derivative ArN(O)NAr (*10, 27, 28*). The second path involves condensation with acetone enolate ion to give the imino derivative $ArN{=}CHCOCH_3$. This compound, in turn, can undergo hydrolysis to $ArNH_2$ or engage in further redox processes leading eventually to the remaining minor products of equation 4 (*27, 29*). A detailed description of all the processes involved, now sufficiently well understood, is beyond the scope of this paper, and the interested reader is referred to the cited literature. The outline sketched in Scheme III suffices for our purposes. Worth mentioning is the fact that conditions have been optimized to obtain either anilines or azoxybenzenes cleanly and in good yields (*30*).

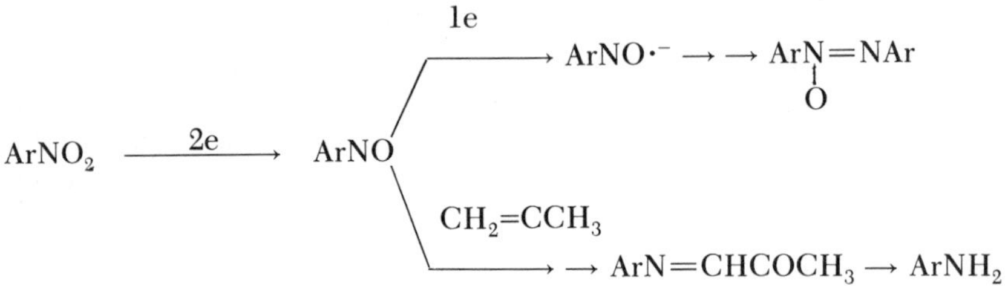

$$ArNO_2 \xrightarrow{2e} ArNO \begin{cases} \xrightarrow{1e} ArNO^{\cdot -} \rightarrow \rightarrow ArN(\rightarrow O){=}NAr \\ \xrightarrow{CH_2{=}CCH_3} \rightarrow ArN{=}CHCOCH_3 \rightarrow ArNH_2 \end{cases}$$

Scheme III

Under the reaction conditions employed, reduction of ArNO is much faster than of the corresponding $ArNO_2$. The rate-limiting step in the reduction of $ArNO_2$ is, therefore, in the first stage of the overall process (equation 5).

$$ArNO_2 \xrightarrow[-HO^-]{+2e,\ +H^+} ArNO \quad (5)$$

Kinetic studies (*31*) have disclosed the following about these reactions: (1) They are first order in $ArNO_2$ and approximately first order (1.2) in alkoxide. The deviation from unit slope of the log k vs. log alkoxide concentration is probably caused, at least in part, by the use of concentration rather than activity data. This value can be a rather severe approximation, particularly in the high concentration region of the range used, 0.05–0.54 M. (2) These reactions are activated by electron-withdrawing substituents, a ρ value of +2.35 being derived from a rate correlation with the σ_p constants. (3) The reactions have "normal" activation parameters (E_a = 21.5 kcal mol^{-1}).

A study (*32*) on the effects of changing the ion association state has been very informative. We found that ion pairing has a significant kinetic effect on the reduction path, although in striking contrast to that on the S_NAr reaction.

Reduction is indeed favored by ion association phenomena. We found, for example, that the rate of reduction of nitrobenzene in degassed solutions of 2-PrOK in 2-PrOH decreases as a function of the concentration of 18-crown-6, as shown in Figure 1. The leveling off corresponding to a rate reduction of a factor of approximately 200 is reached when approximately 2 equiv of crown compound/equiv of 2-PrOK is present; under these conditions, complexation of K^+ is presumably quantitative (*24*).

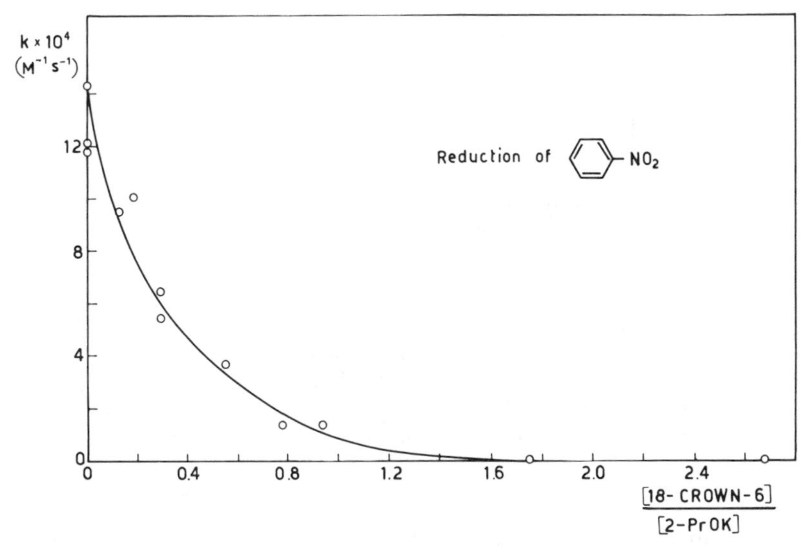

Figure 1. Influence of 18-crown-6 on the rate of reduction, k, *of nitrobenzene in 2-PrOK–2-PrOH at 75 °C under Ar.* k *was derived from the slope of plots of ln* [*nitrobenzene*] *vs. time.*

An analogous study involving 4-chloronitrobenzene (*32*) provided a nice picture of how subtly the balance between substitution and reduction is determined by the extent of ion association phenomena. In Figure 2, plots are shown of the dependence of the rate constants for the substitution (k_s) and the reduction (k_r) processes on the [18-crown-6]:[2-PrOK] ratio. As is evident, k_s and k_r follow opposite trends.

Tetraalkylammonium ions have an effect similar to that of crown-complexed K^+ in depressing the rate of reduction, much as they similarly enhance the S_NAr reactivity as seen in Table I. With K^+ as a reference, relative rates for reduction of 3-chloronitrobenzene with 2-PrO^-M^+ in 2-PrOH at 75 °C under argon are as follows: (M^+, k^{k+}/k^{m+}): K^+, 1.00·Na^+, 0.90; K^+–18-crown-6, 0.09; and n-Bu_4N^+, 0.10 ($k^{k+} = 7.0 \times 10^{-3}$ $m^{-1}s^{-1}$). (C. Paradisi and G. Scorrano, unpublished results).

Interestingly, significant counterion effects are also found in the electrochemical reduction of nitroarenes in 2-PrOH at a graphite electrode

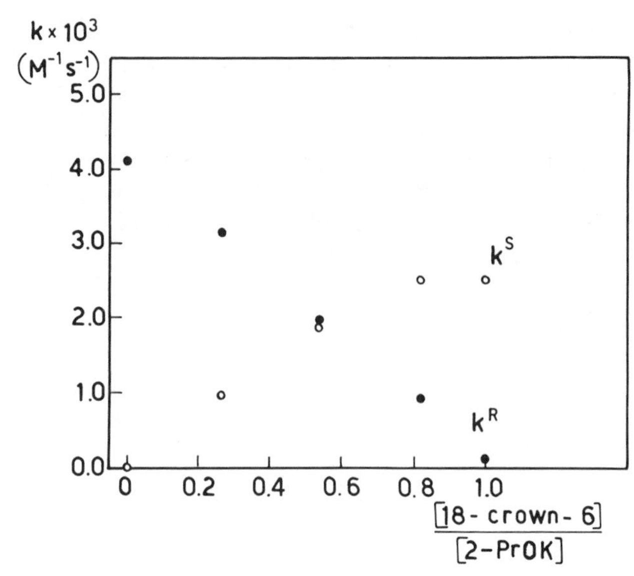

Figure 2. Influence of 18-crown-6 on the rates of competing substitution (k_s) *and reduction* (k_r) *for 4-chloronitrobenzene in 2-PrOK–2-PrOH at 75 °C under Ar.* k_s *was determined directly as the slope of plots 1n [4-(2-PrO)-* $C_6H_4NO_2\cdot$*] vs. time.* k_r *was calculated as the difference* k − k_s, *where* k *is the observed rate of disappearance of 4-chloronitrobenzene.*

studied by means of cyclic voltammetry (CV) (*33*). CV profiles for reduction of 4-chloronitrobenzene in 2-PrOH in the presence of various cations are shown in Figure 3. Marked differences are clearly evident. In the presence of n-Bu_4N^+ and of crown-complexed alkali ions, the CV reduction profile consists of two peaks, the first one due to the reversible one-electron transfer step $ArNO_2 \rightleftarrows ArNO_2\cdot^-$ and the second due to an irreversible overall three-electron process, comprising also chemical steps and leading to ArNHOH (*13*). In the presence of uncomplexed alkali ions, the second peak is progressively anticipated ($K^+ < Na^+ < Li^+$), to the point that it merges into the first. The peak potential of the first reduction wave is, on the other hand, unaffected by the nature of the positive ion. Nitrobenzene and 3-chloronitrobenzene behave similarly.

Observations of this sort led us to propose (*33*) that the mechanism of electrochemical reduction of nitroarenes in 2-propanol involves short-lived nitroarene dianion intermediates (Scheme IV). These intermediates form via one-electron reduction of $ArNO_2\cdot^-$ at the electrode and engage in a series of chemical and electrochemical steps, initiated by protonation. Ion pairing facilitates electrochemical reduction of nitroarenes in 2-propanol by lowering the reduction potential of $ArNO_2\cdot^-$; this reduction thus makes the second stage of the overall process more accessible. No effect is observed on the

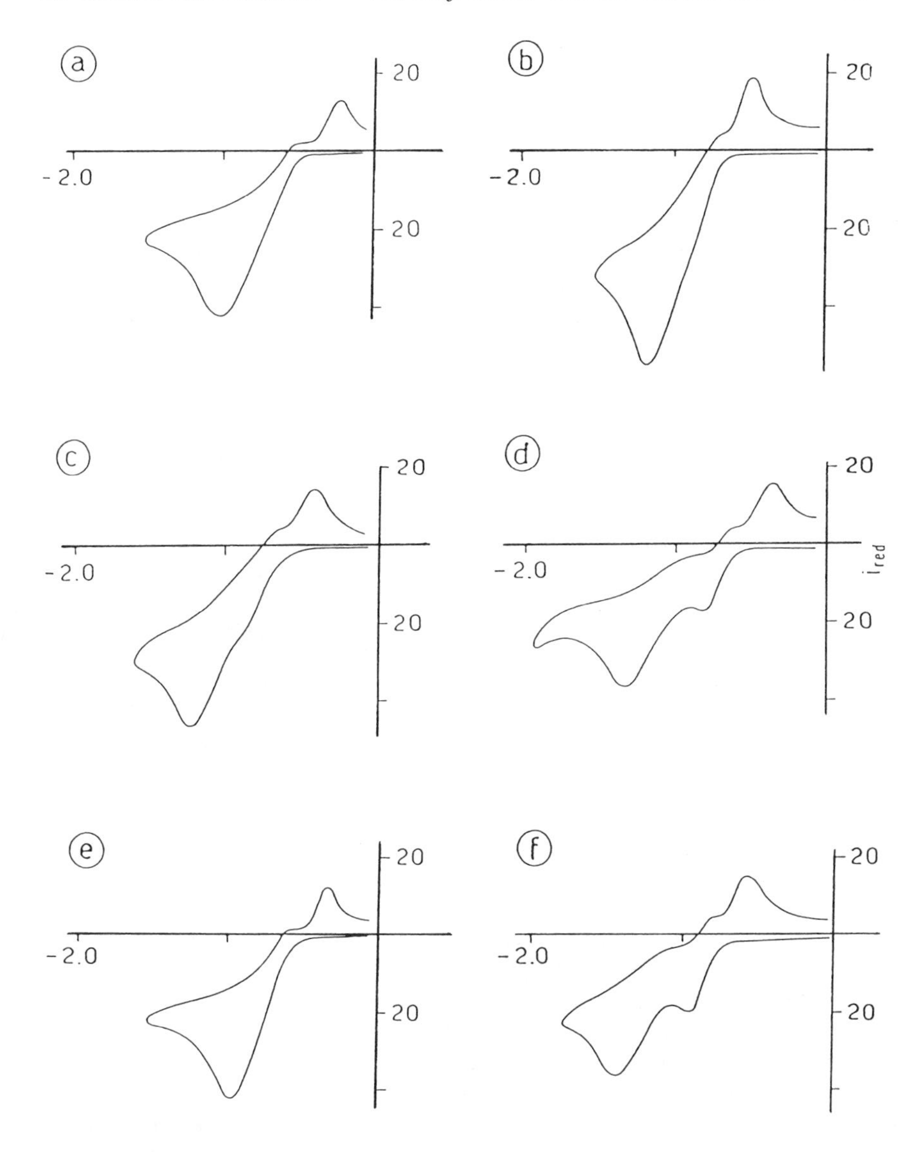

Figure 3. CV curves for the reduction of 4-chloronitrobenzene (1.3×10^{-3} M) in 2-propanol on a glassy carbon electrode in the presence of 0.08 M M^+SCN^-, and, for e–g, of 0.08 M 18-crown-6. M^+: Li^+ (a, e), Na^+ (b, f), K^+ (c, g), and n-Bu_4N^+ (d). Units are volts, V, for the abscissa (potentials are more negative toward the left) and μA for the ordinate. The broken line in plot g refers to an experiment in which the potential sweep was reversed at a value of −1.3 V. (Reprinted with permission from reference 31. Copyright 1981 S. Ventura.) *Continued on p. 348.*

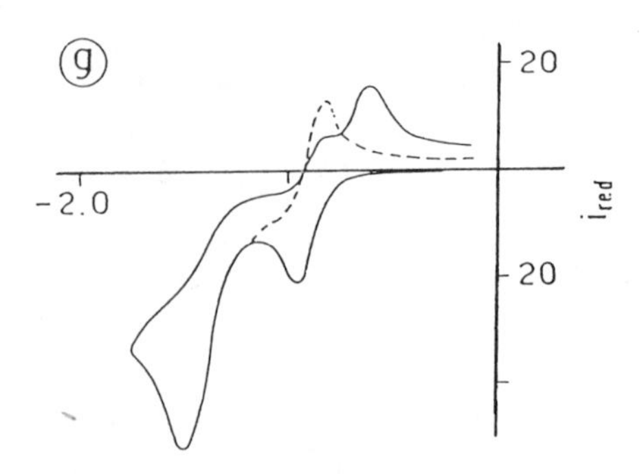

Figure 3 continued

potential of the first peak; thus, reduction of $ArNO_2$ to its radical anion is not facilitated by the possibility of ion pairing.

$$ArNO_2 \xrightarrow{e} ArNO_2^{\cdot -} \xrightarrow[7.3]{e} ArNO_2^{2-} \xrightarrow[11.8]{H^+}$$

$$ArN(OH)O^- \xrightarrow{-HO^-} ArNO \xrightarrow[2H^+]{2e} ArNHOH$$

Scheme IV

An analogous explanation can be invoked to account for the counterion effects observed in the base-promoted reduction of nitroarenes described earlier. Indications in this sense have been obtained by means of in situ electron paramagnetic resonance (EPR) analysis of reacting solutions (*32*). In Figure 4, for example, plots are shown of the concentration of $4\text{-}ClC_6H_4NO_2$ and of the intensity of the EPR signal due to $4\text{-}ClC_6H_4NO_2^{\cdot -}$ as a function of time. Experiments of this sort provide evidence that these alkoxy-promoted reductions involve nitroarene radical anion intermediates.

We also found (unpublished results) that the amount of $ArNO_2^{\cdot -}$ formed during these reactions, crudely estimated from the intensity of the EPR signal, depends on the specific substrate used. In particular, slow-reacting substrates give rise, during their reactions, to most intense EPR signals. Thus, the signal intensity due to $ArNO_2^{\cdot -}$ decreases in the order $C_6H_5NO_2 >$ $4\text{-}ClC_6H_4NO_2 > 3\text{-}ClC_6H_4NO_2$, which is exactly the opposite of the reactivity order (relative reactivities are 1.0, 3.2, and 5.4, respectively). We thus conclude that the overall reactivity is determined by the rate of decay of the first reaction intermediate, $ArNO_2^{\cdot -}$.

Reactions of nitrobenzene and of 3-chloronitrobenzene, substrates not subject to competing S_NAr, were also monitored by EPR in solutions of 2-PrOK in 2-PrOH containing 18-crown-6. Far from inhibiting the formation of the $ArNO_2^{\cdot -}$ intermediate, disruption of ion pairing by crown-complexation of K^+ caused minor changes in the fine structure of the signal but apparently

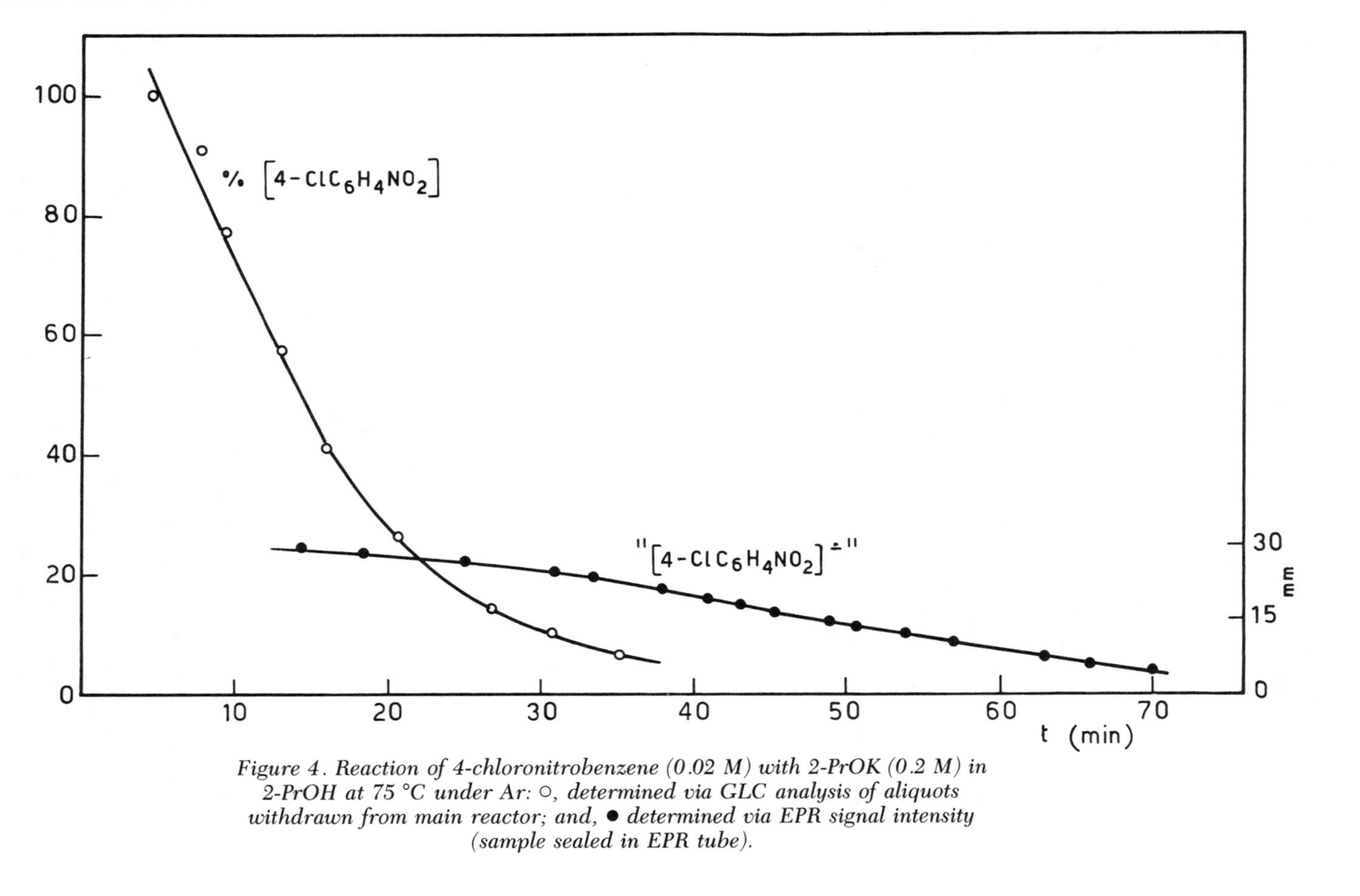

Figure 4. Reaction of 4-chloronitrobenzene (0.02 M) with 2-PrOK (0.2 M) in 2-PrOH at 75 °C under Ar: ○, determined via GLC analysis of aliquots withdrawn from main reactor; and, ● determined via EPR signal intensity (sample sealed in EPR tube).

no reduction in its intensity. In fact, an increase in the concentration of $ArNO_2^{\cdot-}$ in crown-containing solutions could be deduced from the observation of line broadening phenomena in the case of nitrobenzene reactions. Thus, apparently $ArNO_2^{\cdot-}$ intermediates accumulate in the presence of crown-complexed K^+ ion. The step in which $ArNO_2^{\cdot-}$ is consumed must benefit from the possibility of ion pairing.

In the light of these observations, we propose that reduction of $ArNO_2$ to ArNO proceeds via two successive reduction steps leading, respectively, to $ArNO_2^{\cdot-}$ and $ArNO_2^{2-}$, much as is the case for the electrochemical process (Scheme V, path a). The possibility that $ArNO_2^{\cdot-}$ intermediates, species with $pK_a < 4$ (*34*), undergo protonation prior to further reduction (Scheme V, path b) is unlikely under the strongly basic conditions employed. Protonation of $PhNO_2^{\cdot-}$ by 2-propanol is unfavorable in liquid ammonia (*35*). Moreover, protonation is not compatible with the counterion effects observed in these processes. Proton transfer involving oxygen and nitrogen anions should not be significantly influenced by ion pairing. Contrary to carbanions (*36*, *37*), protonation should actually be faster for the "free" anions (*38*).

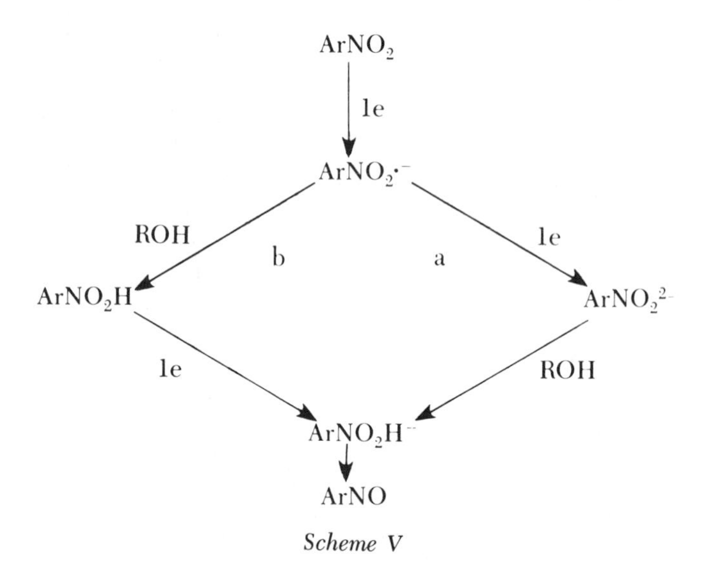

Scheme V

Where Does the Electron Come From?

Crucial questions concern the identity of the species that act as electron donors in forming the $ArNO_2^{\cdot-}$ and $ArNO_2^{2-}$ intermediates. The simplest possibility is that $ArNO_2^{\cdot-}$ is formed via direct SET from the alkoxide ion (Scheme I, path a). The resulting alkoxy radical should rapidly give an alkyl radical via H abstraction from the solvent. The α-hydroxy radical so produced is in acid–base equilibrium with its conjugate base, the acetone radical anion. The equilibrium constant for the dissociation of the 2-propanol radical, $K = [(CH_3)_2\,CO^{\cdot -}][H^+]/[(CH_3)_2COH^{\cdot}]$, is 6.3×10^{-13} in aqueous solution (*39*, *40*). These processes are described by equations 6–8. Alternatively, $(CH_3)_2CO^{\cdot -}$ can form directly from the alkoxy radical via H abstraction from an alkoxide ion (corresponding to the sum of equations 7 and 8). Both radicals, $(CH_3)_2COH^{\cdot}$ and its ionized form, $(CH_3)_2CO^{\cdot -}$, are known (*40–43*) to reduce effectively nitrobenzene to its radical anion in aqueous solution (equations 9 and 10); the rate constants are 1.6×10^9 and 3.0×10^9 $mol^{-1}ls^{-1}$, respectively (*42*).

$$ArNO_2 + (H_3C)_2CHO^- \rightarrow ArNO_2^{\cdot -} + (H_3C)_2CHO^{\cdot} \qquad (6)$$

$$(H_3C)_2CHO^{\cdot} + (H_3C)_2CHOH \rightarrow (H_3C)_2CHOH + (H_3C)_2\dot{C}OH \qquad (7)$$

$$(H_3C)_2{\cdot}COH + (H_3C)_2CHO^- \rightleftarrows (H_3C)_2{\cdot}CO^- + (H_3C)_2CHOH \qquad (8)$$

$$ArNO_2 + (H_3C)_2{\cdot}COH \rightarrow ArNO_2^{\cdot -} + (H_3C)_2C{=}O + H^+ \qquad (9)$$

$$ArNO_2 + (H_3C)_2{\cdot}CO^- \rightarrow ArNO_2^{\cdot -} + (H_3C)_2C{=}O \qquad (10)$$

The feasibility of electron transfer from alkoxides to acceptor species, attractive as it is for its simplicity, has been prospected as a real possibility only in a few instances (*44*, *45a*) and, in some cases, questioned (*9*, *45b*, *46a*). Buncel and Menon (*45b*) estimated that electron transfer from *t*-BuO$^-$ to 4-nitrotoluene in *t*-BuOH is energetically unfavorable to the extent of about 3 eV. The estimate is based on calculations involving gas-phase electron affinity data for *t*-BuO$^-$ (1.93 eV) instead of its oxidation potential in solution, which, as is the case for other alkoxides, is not known. The approximation is necessarily crude, because solvation effects could be of significant magnitude; nevertheless, the estimated value has met wide acceptance.

Carbanions of many different types, on the other hand, are known to undergo one-electron oxidation at the expense of nitroarenes (*17*). Examples range from enolates to the conjugate base of $(CH_3)_2SO$ (*11*) and include the interesting studies on fluorene derivatives of Guthrie (*37*). Because of the ascertained ability of carbanions to produce nitroarene radical anions, mechanistic proposals have been advanced for the alkoxide-promoted reduction based on the involvement of carbanions of some sort in the key electron-transfer step. These comprise Meisenheimer complexes or Janovski-type intermediates and nitroaryl anion (*17*), according to Schemes VI and VII, respectively, shown here for nitrobenzene.

Scheme VI is not consistent with our findings, as it requires the formation of alkoxy-substituted products in significant amount. Scheme VII is more attractive. Evidence has been reported that nitroarenes undergo aromatic proton abstraction in basic solutions (*46b*). Moreover, preliminary experiments performed with an ion cyclotron resonance (ICR) mass spectrometer indicate that proton abstraction is a major path in the reaction of t-BuO$^-$ with 4-chloronitrobenzene in the gas phase at low pressures (C.

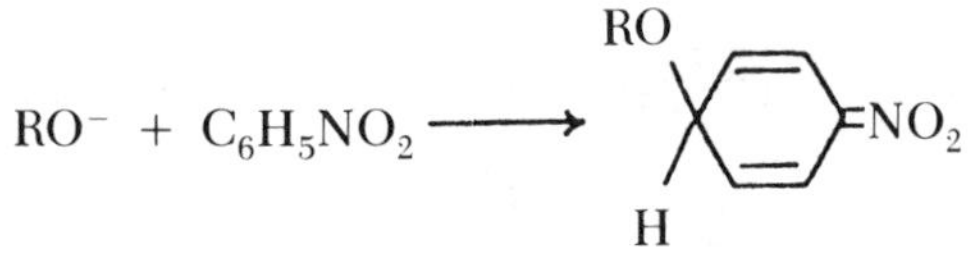

$$RO^- + C_6H_5NO_2 \longrightarrow \text{(RO)(H)C}_6\text{H}_5\text{=NO}_2^- \quad \textbf{I}$$

$$\text{I} + C_6H_5NO_2 \longrightarrow C_6H_5NO_2^{\cdot -} + \text{(RO)(H)C}_6\text{H}_5^{\cdot}\text{-NO}_2 \quad \textbf{II}$$

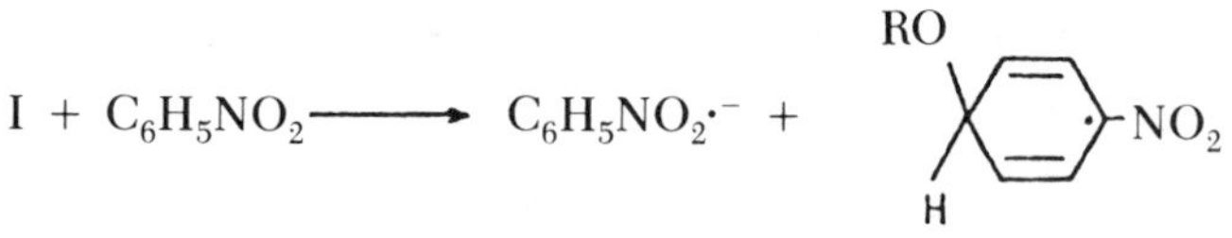

$$\textbf{II} + RO^- \rightarrow ROH + ROH_6H_4NO_2^{\cdot -}$$

$$ROC_6H_4NO_2^{\cdot -} + C_6H_5NO_2 \rightleftarrows ROC_6H_5NO_2 + C_6H_4NO_2^{\cdot -}$$

Scheme VI

Giancaspro, C. Paradisi, G. Scorrano, and M. Speranza, unpublished results). However, difficulties arise in deriving a kinetic expression with first-order dependence in alkoxide and nitroarene.

$$C_6H_5NO_2 + RO^- \rightleftarrows C_6H_4NO_2^- + ROH$$

$$C_6H_4NO_2^- + C_6H_4NO_2 \rightarrow C_6H_5NO_2^{\cdot} + C_6H_5NO_2^{\cdot -}$$

$$C_6H_4NO_2^{\cdot} + HCOH \rightarrow C_6H_5NO_2 + {\cdot}\!>\!COH$$

$${\cdot}\!>\!COH + C_6H_5NO_2 + RO^- \rightarrow C_6H_5NO_2^{\cdot -}\ >\!C{=}O\ ROH$$

Scheme VII

An alternative proposal involves H transfer from the α-C of the alkoxide to one of the oxygens of the nitro group followed by ionization of the neutral radical so obtained to its conjugate base, the nitroarene radical anion (*16*).

This mechanism, however, is certainly not of general validity, because nitroarenes are also reduced in t-BuO$^-$–t-BuOH solutions (*10, 15*).

A complete mechanistic description is not possible at present. Kinetic analysis needs to include, possibly, a description in terms of the radical anion intermediate. The data presently available seem to be best accommodated by a scheme comprising reversible electron transfer from the alkoxide ion, equation 6, or possibly the enolate ion of acetone adventitiously present in the solvent (*47*), followed by rate-limiting reduction of $ArNO_2{\cdot}^-$ to $ArNO_2^{2-}$. This step may occur via dismutation of $ArNO_2{\cdot}^-$ (equation 11), or via one-electron reduction of $ArNO_2{\cdot}^-$ by the strongly reducing radicals present in

$$2ArNO_2{\cdot}^- \longrightarrow ArNO_2^{2-} + ArNO_2 \qquad (11)$$

the system (equations 7 and 8). Dismutation of nitroarene radical anions is amply precendented. Behar and Neta (*43*) found that $ArNO_2{\cdot}^-$ species (including Ar = 4-ClC_6H_4) generated via electron transfer from e_{aq} or $(CH_3)_2COH$ in irradiated aqueous solutions decay by a second-order process. Moreover, Guthrie and Cho (*48*) reported a clean second-order decay of the EPR signal intensity of $PhNO_2{\cdot}^-$ generated photochemically in methanol–methoxide solution. We note that both systems differ from ours in that the decay of $ArNO_2{\cdot}^-$ is monitored in the absence of the $ArNO_2^-$ forming step. This procedure is done because the source of radicals is removed before analysis occurs. We also note that transfer of a second electron (e_{aq}) on $ArNO_2{\cdot}^-$ to give $ArNO_2^{2-}$ species is kinetically as fast as the first (*49*). Therefore, one-electron reduction of $ArNO_2{\cdot}^-$ by the strongly reducing acetone radical anion (or its conjugate acid) could successfully compete with dismutation under our conditions.

Literature Cited

1. Russell, G. A.; Danen, W. C. *J. Am. Chem. Soc.* **1966,** *88*, 5663.
2. Kornblum, N. *Angew. Chem., Int. Ed. Engl.* **1975,** *14*, 734.
3. Bunnett, J. F. *Acc. Chem. Res.* **1978,** *11*, 413.
4. Ingold, C. K. *Structure and Mechanism in Organic Chemistry,* 2nd ed.; Cornell University: Ithaca, NY, 1969; Chapter VII.
5. Miller, J. *Aromatic Nucleophilic Substitution;* American Elsevier: New York, 1968.
6. Buncel, E.; Crampton, M. R.; Strauss, M. J.; Terrier, F. *Electron Deficient Aromatic- and Heteroaromatic-Base Interactions;* Elsevier: Amsterdam, 1984.
7. Pross, A.; Shaik, S. S. *Acc. Chem. Res.* **1983,** *16*, 363.
8. Pross, A. *Acc. Chem. Res.* **1985,** *18*, 212.
9. Guthrie, R. D.; Nutter, D. E. *J. Am. Chem. Soc.* **1982,** *104*, 7478.
10. Bassani, A.; Prato, M.; Rampazzo, P.; Quintily, U.; Scorrano, G. *J. Org. Chem.* **1980,** *45*, 2263.
11. Feuer, H.; Hooz, J. In *The Chemistry of the Ether Linkage;* Patai, S., Ed.; Interscience: New York, 1967; Chapter 10.
12. Buck, P. *Angew. Chem., Int. Ed. Engl.* **1969,** *8*, 120.
13. Fry, A. J. In *The Chemistry of Amino, Nitroso and Nitro Compounds and Their Derivatives;* Patai, S., Ed.; Wiley: New York, 1982; Chapter 8.

14. Zinin, N. *J. Prakt. Chem.* **1845,** *36,* 93.
15. Russell, G. A.; Janzen, E. G. *J. Am. Chem. Soc.* **1962,** *84,* 4153.
16. Bellobono, I. R.; Gamba, A.; Sala, G.; Tampieri, M. *J. Am. Chem. Soc.* **1972,** *94,* 5781.
17. Russell, G. A.; Janzen, E. G.; Strom, E. T. *J. Am. Chem. Soc.* **1964,** *86,* 1807.
18. Paradisi, C.; Quintily, U.; Scorrano, G. *J. Org. Chem.* **1983,** *48,* 3022.
19. Prato, M.; Quintily, U.; Salvagno, S.; Scorrano, G. *Gazz. Chim. Ital.* **1984,** *114,* 413.
20. Pietra, F.; Vitali, D.; Del Cima, F.; Cardinali, G. *J. Chem. Soc. B* **1970,** 1659.
21. Buncel, E.; Dunn, E. J.; Bannard, R. A. B.; Purdon, J. G. *J. Chem. Soc., Chem. Commun.* **1984,** 162.
22. Barthel, J.; Wachter, R.; Knerr, M. *Electrochimo Acta* **1971,** *16,* 723.
23. Murto, J. *Acta Chem. Scand.* **1964,** *18,* 1043.
24. Hiraoka, M. *Crown Compounds. Their Characteristics and Applications;* Kodansha: Tokyo, and Elsevier: Amsterdam, 1982; Chapter 4.
25. Del Cima, F.; Biggi, G.; Pietra, F. *J. Chem. Soc., Perkin Trans. 2* **1973,** 55.
26. Mariani, C.; Modena, G.; Scorrano, G. *J. Chem. Res. (S)* **1978,** 392.
27. Prato, M.; Quintily, U.; Scorrano, G. *J. Chem. Soc., Perkin Trans. 2,* in press.
28. Russell, G. A.; Geels, E. *J. Am. Chem. Soc.* **1965,** *87,* 122.
29. Prato, M.; Quintily, U.; Scorrano, G. *Gazz. Chim. Ital.* **1984,** *114,* 405.
30. Prato, M.; Quintily, U.; Scorrano, G. *Bull. Soc. Chim. France,* in press.
31. Ventura, S. Thesis, University of Padova, Padova, 1981.
32. Maggini, M. Thesis, University of Padova, Padova, 1984.
33. Maggini, M.; Paradisi, C.; Scorrano, G.; Daniele, S.; Magno, F. *J. Chem. Soc., Perkin Trans. 2,* **1986,** 267.
34. Hayon, E.; Simic, M. *Acc. Chem. Res.* **1974,** *7,* 114.
35. Smith, W. H.; Bard, A. J. *J. Am. Chem. Soc.* **1975,** *97,* 5203.
36. Dorfman, L. M.; Sujdak, R. J.; Bockrath, B. *Acc. Chem. Res.* **1976,** *9,* 352.
37. Guthrie, R. D. In *Comprehensive Carbanion Chemistry. Part A. Structure and Reactivity;* Buncel, E.; Durst, T., Eds.; Elsevier: Amsterdam, 1980; Chapter 5.
38. Halliday, J. D.; Bindner P. E. *J. Phys. Chem.* **1979,** *83,* 1829.
39. Asmus, K. D.; Henglein, A.; Wingler, A.; Beck, G. *Ber. Bunsenges. Phys. Chem.* **1966,** *70,* 756.
40. Adams, G. E.; Willson, R. L. *J. Chem. Soc., Faraday Trans. 1* **1973,** *69,* 719.
41. Eiben, K.; Fessenden, R. W. *J. Phys. Chem.* **1971,** *75,* 1186.
42. Asmus, K. D.; Wigger, A.; Henglein, A. *Ber. Bunsenges. Phys. Chem.* **1966,** *70,* 862.
43. Behar, D.; Neta, P. *J. Phys. Chem.* **1981,** *85,* 690.
44a. Ciminale, F.; Bruno, G.; Testaferri, L.; Tiecco, M.; Martelli, G. *J. Org. Chem.* **1978,** *43,* 4509.
44b. Ashby, E. C.; Argyropoulos, J. N. *Tetrahedron Lett.* **1986,** *27,* 465.
45. Buncel, E.; Menon, B. C. *J. Am. Chem. Soc.* **1980,** *102,* 3499.
46a. Buncel, E. In *The Chemistry of Amino, Nitroso and Nitro Compounds and their Derivatives;* Patai, S., Ed.; Wiley: New York, 1982; Chapter 27.
46b. Eberson, L. *Adv. Phys. Org. Chem.* **1982,** *18,* 79.
47. Lempert, K.; Simig, G.; Tamas, J.; Toth, G. *J. Chem. Soc., Perkin Trans. 2* **1984,** 1927.
48. Guthrie, R. D.; Cho, N. S. *J. Am. Chem. Soc.* **1979,** *101,* 4698.
49. Gruenbein, W.; Henglein, A.; Steven, G.; Beck, G. *Ber. Bunsenges. Phys. Chem.* **1971,** *75,* 126.

RECEIVED for review October 21, 1985. ACCEPTED February 24, 1986.

25

Reactions of Alkyl Radicals with Nucleophiles

Glen A. Russell and Rajive K. Khanna

Department of Chemistry, Iowa State University, Ames, Iowa 50011

The relative reactivities of nucleophiles in the series $R^1R^2C{=}NO_2^-$, $R^1R^2C{=}C(O^-)Ph$ and $RC(CO_2Et)_2^-$ or toward $Me_3C\cdot$ and $PhCOCH_2\cdot$ reveal strong polar effects in the addition of a radical to a nucleophile ($R\cdot + N^- \rightarrow RN\cdot^-$). Toward the electron donor $Me_3C\cdot$, reactivity decreases (R^1, R^2 = Ph > H > Me) with an increase in basicity of the nucleophile, whereas toward the electrophilic $PhCOCH_2\cdot$, reactivity increases (R^1 and R^2 = H < Me) with basicity. Benzylic-type anions undergo electron transfer ($R\cdot + N^- \rightarrow R^- + N\cdot$) with $PhCOCH_2\cdot$ but not with $Me_3C\cdot$. The transition state for the addition of a radical to a nucleophile seems to be controlled by SO molecular orbital–lowest unoccupied molecular orbital interactions for nucleophilic radicals like $Me_3C\cdot$ but by SO molecular orbital–highest occupied molecular orbital interactions by electrophilic radicals such as $PhCOCH_2\cdot$.

THE REACTION OF A FREE RADICAL with an anion to form a radical anion (equation 1) appears to be a simple reaction whose rate might be expected to increase with the exoergicity of the reaction. Indeed, the reaction is recognized to occur only when the fragments R or N contain unsaturation such that $RN\cdot^-$ is a reasonably stable species. Among the functional groups that can lead to stability of a radical anion are aromatic rings and nitro, cyano, or carbonyl groups, that is, groups recognized to readily undergo facile one-electron reduction by suitable reducing agents such as alkali metals.

$$R\cdot + N^- \rightarrow R{-}N\cdot^- \quad (1)$$

Substituent effects in reaction 1 have received little attention because of the lack of suitable methods of generating specific free radicals under basic conditions. One general process involving reaction 1 is the $S_{RN}1$ substitution

0065-2393/87/0215-0355$06.00/0

$$
\begin{array}{l}
R\cdot + N^- \longrightarrow RN\cdot^- \\
RN\cdot^- + RX \longrightarrow RN + RX\cdot^- \\
\underline{RX\cdot^- \longrightarrow R\cdot + X^- \qquad\qquad} \\
RX + N^- \longrightarrow RN + X^-
\end{array}
$$

Scheme I

process (Scheme I) where R· can be either an alkyl or aryl radical (*1–3*). This reaction sequence was originally recognized in aliphatic substitution reactions in which R· was an α-nitroalkyl, α-(*p*-nitrophenyl)alkyl, or α-(*p*-nitrobenzoyl)alkyl radical derived from RX·⁻ where X = halogen, NO_2, PhS, PhSO, $PhSO_2$, SCN, Me_3N^+, or Me_2S^+ (*4, 5*). Aliphatic free radicals not containing a nitro group will participate in the $S_{RN}1$ scheme if a nitronate anion ($R^1R^2C{=}NO_2^-$) is used as the radicalphile (*6–14*). Some examples of such systems and the substrate (RX) leading to RX·⁻ and R· are given in Table I.

Table I. Aliphatic Radicals (R·) Not Containing a Nitro Group That Adds to $Me_2C{=}NO_2^-$

R·	*X of RX·⁻*	*Reference*
$(EtO_2C)_2CEt$	NO_2	6
Me_2CCN[a]	NO_2	6–9
$Me_2CC({=}O)R$, Me_2CCO_2R	NO_2	7, 8
$RCMe_2$ (R = alkyl, aryl)[b]	NO_2	10
n-C_6F_{13}, *n*-C_8F_{17}	I	11
$PhCOCMe_2$, *p*-$NCC_6H_4COCMe_2$	Cl, Br	12
p-NCC_6H_4CHBr, *p*-$MeO_2CC_6H_4CHBr$	Br	13
p-$NCC_6H_4CH_2$	$N(Me)_3^+$	14
p-$NCC_6H_4CMe_2$	Cl	14
o,p-$(NC)_2C_6H_3CMe_2$[a]	$PhSO_2$	14

[a]Radicals also trapped by $(EtO_2C)_2CR^-$ and PhS^-.
[b]Trapped by $H_2C{=}NO_2^-$.

Alkyl and substituted alkyl radicals can also be generated from the corresponding alkylmercury halides or carboxylates by the chain sequence of Scheme II (*15–17*). The reactivity of RHgCl in Scheme II decreases drastically from R = *t*-Bu to R = *i*-Pr to R = *n*-Bu as expected for a dissociative electron transfer from RN·⁻ to RHgCl (*18*). The following discussion will concentrate on results obtained by competitive reactions of pairs of nucleophiles (N_A^-, N_B^-) with R· = $Me_3C\cdot$ or $PhCOCH_2\cdot$ generated from the corresponding RHgCl in the photostimulated chain reaction of Scheme II.

$$R\cdot + N^- \longrightarrow RN\cdot^-$$
$$\underline{RN\cdot^- + RHgX \longrightarrow RN + R\cdot + Hg^\circ + X^-}$$
$$RHgX + N^- \longrightarrow RN + Hg^\circ + X^-$$

Scheme II

By use of large excess (~5–20-fold) of the radicalphiles, the relative reactivities (k_A/k_B) can be calculated from the relative yields of the products, $R{-}N_A$ and $R{-}N_B$ (reaction 2). The relative yields of $R{-}N_A$ and $R{-}N_B$ in a competitive chain process will be determined only by the relative values of k_A and k_B when all intermediate radical anions ($RN_A\cdot^-$, $RN_B\cdot^-$) are converted to products by electron transfer to RHgCl. Thus, at a kinetic chain length of > 20, the ratio of products formed will be essentially independent of the possibly different rate constants for the electron-transfer reaction of $RN_A\cdot^-$ and $RN_B\cdot^-$. The photostimulated reaction of *t*-BuHgCl with $Me_2C{=}NO_2^-$ gives an inhibition period with $(t\text{-Bu})_2NO\cdot$ from which, under the conditions employed (0.1 M *t*-BuHgCl, 0.1 M $Me_2C{=}NO_2^-/K^+$-18-crown-6, and Me_2SO, 35 °C), it is calculated that the rate of photostimulated initiation is 9.3×10^{-7} M s^{-1}. In the absence of inhibitor, or after the inhibitor is consumed, the rate of the free radical chain substitution leading to $Me_3CCMe_2NO_2$ is 4.5×10^{-5} M s^{-1}, that is, a kinetic chain length of ~50. Longer kinetic chain lengths are observed in the absence of 18-crown-6, which apparently can divert *t*-Bu· from the desired chain reaction.

$$R\cdot + \begin{matrix} N_A^- \xrightarrow{k_A} RN_A\cdot^- \\ \\ N_B^- \xrightarrow{k_B} RN_B\cdot^- \end{matrix} \xrightarrow{RHgX} \begin{matrix} R{-}N_A \\ \\ R{-}N_B \end{matrix} + R\cdot + Hg^\circ + X^- \qquad (2)$$

Evidence for Alkyl Radical Intermediates

Determination that a given reaction proceeds by a free-radical chain mechanism is usually experimentally simple because a chain mechanism involves initiation and termination reactions. Thus, reactions of $Me_2C(Cl)NO_2$, *p*-$O_2NC_6H_4CH_2Cl$, *p*-$O_2NC_6H_4C({=}O)CMe_2Cl$, or RHgCl (R = 1°, 2°, and 3° alkyl and benzyl) with $Me_2C{=}NO_2^-$ do not occur in the absence of photostimulation at ambient temperatures, and the irradiated reactions can be inhibited by radical traps such as O_2, $Ph_3C\cdot$, or $(t\text{-Bu})_2NO\cdot$. Powerful electron acceptors such as $(O_2N)_2C_6H_4$ are also excellent inhibitors for processes involving radical anions as intermediates (*19*).

Evidence for the intermediacy of R· in the free-radical chain reaction of RHgCl with $Me_2C{=}NO_2^-$ was obtained by a study of the reaction of the 5-hexenylmercurial with $Me_2C{=}NO_2^-$. The reaction yields the unrearranged and cyclized cyclopentylcarbinyl derivatives ($RCMe_2NO_2$) in a ratio linearly dependent upon $[Me_2C{=}NO_2^-]$. By use of the known cyclization rate constant of the 5-hexenyl radical as a standard, the bimolecular rate constant for trapping of the 5-hexenyl radical by $Me_2C{=}NO_2^-$ is calculated to be 2.4 $\times$ 10^5 M^{-1} s^{-1} at 40 °C in Me_2SO or HMPA with Me_4N^+ as the counterion (*20*). This rate constant decreases by a factor of 2–3 when Li^+ is used as the counterion.

In Scheme II, the reaction between $RN\cdot^-$ = $Me_3CC(Me)_2(NO_2)\cdot^-$ and Me_3CHgCl must occur quite rapidly and irreversibly because not only does $Me_3CC(Me)_2NO_2$ fail to retard the reaction but also better electron acceptors such as $PhNO_2$ have a negligible effect on the rate of the substitution process and $C_6H_4(NO_2)_2$ is not an efficient inhibitor for the substitution process whereas $(Me_3C)_2NO\cdot$ is effective.

If a free alkyl radical is being trapped by anions in a competitive reaction sequence (e.g., reaction 2), the relative reactivity (k_A/k_B) should be independent of the group X in the radical anion ($RX\cdot^-$) that is the precursor of $R\cdot$. For example, the reaction of $XCMe_2NO_2$ with pairs of anions from the group $(EtO_2C)_2CMe^-$, $Me_2C{=}NO_2^-$, $(EtO)_2PO^-$, and $(EtO)_2PS^-$ gives a relative reactivity independent of the nature of X for X = Cl, NO_2, and SO_2Ph (*21*). This observation excludes substitution processes involving direct reaction between the nucleophiles and $XCMe_2NO_2\cdot^-$ and requires the intermediacy of $Me_2C(NO_2)\cdot$ in the chain reactions.

Important effects of ionic association are found when the relative reactivity of anions toward $Me_2C(NO_2)\cdot$ are thus measured (*21*). Toward the free anions observed in Me_2SO–K^+-[2.2.2]-cryptand, the relative reactivities of $(EtO_2C)_2CMe^-$:$Me_2C{=}NO_2^-$:$(EtO)_2PS^-$:$(EtO)_2PO^-$ toward $Me_2C(NO_2)\cdot$ are 10:1.0:0.90:0.54. However, in the presence of Li^+ where selective ion pairing can occur, the observed relative reactivities were 0.24:1.0:1.2:<0.05 ($[N^-]_0$ = $[Li^+]$ = 0.2 M). The reactivities of $MeC(CO_2Et)_2^-$ and $(EtO)_2PO^-$ are drastically reduced relative to the reactivity of $Me_2C{=}NO_2^-$ or $(EtO)_2PS^-$ because of preferential ionic association, which decreases the reactivity of the ion-paired nucleophiles in reaction 1. Solvents such as Me_2SO or HMPA and counterions such as R_4N^+ or K^+-18-crown-6 are often the preferred choices for aliphatic $S_{RN}1$ substitution reactions. In a less polar solvent such as THF with Li^+ as the gegenion, extensive ionic association occurs with both $Me_2C{=}NO_2^-$ and $MeC(CO_2Et)_2^-$. Now the relative reactivity toward $Me_2C(NO)_2$ greatly favors the chelated ion pair of the malonate anion by at least a factor of 70. If structural effects upon reaction 1 are analyzed, the effects of ion pairing between the gegenion and N^- must be taken into consideration.

Reactivity of a Series of Anions Toward the tert-*Butyl Radical*

Table II summarizes the yields of the substitution products obtained in the photostimulated reaction of *t*-BuHgCl with a variety of anions. In the dark, the reactions occurred at a significant rate only above 60 °C where the

Table II. Reaction of Nucleophiles with *t*-BuHgCl

Nucleophile[a]	*Time (h)*	*Product (% Yield)*[b]
$Me_2C{=}NO_2^-$	2	$Me_3CCMe_2NO_2$ (69)
$MeCH{=}NO_2^-$	2	$Me_3CCH(Me)NO_2$ (74)
$MeC(Ph){=}NO_2^-$	2	$Me_3CC(Me)(Ph)NO_2$ (67)
$H_2C{=}NO_2^-$	2	$Me_3CCH_2NO_2$ (68)
$PhCH{=}NO_2^-$	2	$Me_3CCH(Ph)NO_2$ (71)
NO_2^-	2	Me_3CNO_2 (71)
(Phthalimide)$^-$[c]	5	*N-tert*-butylphthalimide (72)
N_3^-[c]	8	Me_3CN_3 (34)
$PhCHCN^-$	5	$Me_3CCH(Ph)CN$ (4); $PhCH_2CMe_3$ (11)
Ph_2CCN^-	2	Me_3CCPh_2CN (48); $Ph_2C{=}C{=}NCMe_3$ (26)
Ph_3C^-	2[d]	Ph_3CCMe_3 (39); 6-*tert*-butyl-3-benzhydrylidene-1,4-cyclohexadiene (21); *p*-$Me_3CC_6H_4CPh_2CMe_3$ (5)
Ph_2CH^-	2[d]	Ph_2CHCMe_3 (36)
(Fluorenyl)$^-$	2[d]	9-*tert*-butylfluorene (44)
$PhC(CO_2Et)_2^-$	7	$PhC(CO_2Et)_2CMe_3$ (43)
$Me_3C(O^-){=}CH_2$	8	$Me_3CCOCH_2CMe_3$ (7)
$Me_3C(O^-){=}CPh_2$	6	$Me_3CCOCPh_2CMe_3$ (6)
$PhC(O^-){=}CH_2$	6	$PhCOCH_2CMe_3$ (54)
$PhC(O^-){=}CHMe$	4	$PhCOCH(Me)CMe_3$ (34)
$PhC(O^-){=}CMe_2$	5	$PhCOCMe_2CMe_3$ (21)
$PhC(O^-){=}CHPh$	2	$PhCOCH(Ph)CMe_3$ (63)
$PhC(O^-){=}CPh_2$	2	$PhCOCPh_2CMe_3$ (57)
$PhC(O^-){=}$Fluorene	8	9-*tert*-butyl-9-benzoylfluorene (8)

NOTE: Reactions were performed in nitrogen-purged Me_2SO in the presence of equimolar amounts of 18-crown-6, with irradiation from a 275-W sunlamp positioned approximately 15 cm from the Pyrex reaction flask.
[a] Generated by the action of potassium *tert*-butoxide on the conjugate acid.
[b] Yields determined by ^{1}H-NMR spectroscopy and GLC on a 1-mmol scale for reactions 0.1 M in *t*-BuHgCl and N^-.
[c] Commercially available potassium salts were used.
[d] Hexamethylphosphoric triamide solvent.

thermolysis of *t*-BuHgCl to form $Me_3C\cdot$ becomes important. This dark reaction was completely inhibited by 10 mol % of $(t\text{-}Bu)_2NO\cdot$. In addition to nitronate anions, the anions derived from phenones [$PhC(O^-){=}C(R^1)(R^2)$] and phenylacetonitriles [$PhC(R){=}C{=}N^-$] were capable of trapping *t*-Bu radicals. Benzylic anions such as $(Ph)_3C{:}^-$, $(Ph)_2CH^-$, or (fluorenyl)$^-$ also yielded tert-butylation products. In the case of Ph_2CCN^-, products of both C and N alkylation were observed while $Ph_3C:^-$ gave substitution at both the α and para positions:

$$Me_3C\cdot + \begin{array}{l} Ph_2\ddot{C}{=}C{=}N^- \\ \\ Ph_2C{=}C{=}N:^- \end{array} \longrightarrow \begin{array}{l} Ph_2C(CMe_3)CN\cdot^- \\ \\ Ph_2C{=}C{=}NCMe_3\cdot^- \end{array} \xrightarrow{-e} \begin{array}{c} Ph_2C(CMe_3)CN \\ + \\ PH_2C{=}C{=}NCMe_3 \end{array} \quad (3)$$

$$Me_3C\cdot + Ph_3C:^- \longrightarrow \begin{array}{l} Ph_3CCMe_3\cdot^- \\ \\ Ph_2C{=}C_6H_4(CMe_3)(H)\cdot^- \end{array} \xrightarrow{-e} \begin{array}{c} Ph_3CCMe_3 \\ + \\ Ph_2C{=}C_6H_4{-}CMe_3 \end{array} \quad (4)$$

Among the anions that failed to give at least 3% of the substitution products in 8 h were $HC(NO_2)_2^-$, $C(NO_2)_3^-$, $EtO_2CCPh_2^-$, $Me_3CC(O^-){=}CPh_2$, $MeC(CO_2Et)_2^-$, $PhC(O^-){=}CHCOPh$, $PhC(O^-){=}C(Ph)COPh$, $PhC(O^-)CHCN$, $PhC(O^-)CHCO_2Et$, and (9-nitrofluorene)$^-$.

The relative reactivities of a series of carbanions determined by competitive reactions using $Me_2C{=}NO_2^-$ as a standard nucleophile are summarized in Table III. The absolute rate constant for the trapping of 5-hexenyl radical by $Me_2C{=}NO_2^-$ in Me_2SO at 40 °C is $\sim 2.5 \times 10^5$ $M^{-1}s^{-1}$ (*20*). Presumably, the rate constant for trapping of $Me_3C\cdot$ will be less than that for a primary alkyl radical although aliphatic $S_{RN}1$ processes are remarkably insensitive to steric effects. Those anions that were unreactive in the absence of $Me_2C{=}NO_2^-$ not only failed to react in competitive experiments in the presence of $Me_2C{=}NO_2^-$ but also failed to interfere with the trapping of $Me_3C\cdot$ by $Me_2C{=}NO_2^-$. A maximum reactivity of not greater than 0.005 that of $Me_2C{=}NO_2^-$ was observed for $O_2NCH{=}NO_2^-$, $(O_2N)_2C{=}NO_2^-$, $Fl{=}NO_2^-$, $Me_3CC(O^-){=}CPh_2$, $HC(CO_2Et)_2^-$, $MeC(CO_2Et)_2^-$, $PhC(O^-){=}CHCOPh$, $PhC(O^-){=}CPhCOPh$, $PhC(O^-){=}CHCN$, and $PhC(O^-){=}CHCO_2Et$.

As illustrated in Scheme III, the basicity of an anion plays an important role in determining the change in standard free energy ($\Delta G°$) for reaction 1 [$\Delta G°(A^-)$] (*22*). When comparing a series of nucleophiles with the same functional group, literature data indicates that the reduction potential ($E°$) remains essentially constant. If $\Delta G°(AH)$ remains constant, variation in the pK_a of the conjugate base of the nucleophile results in a change of $\Delta G°$ for reaction 1 equal to $-1.4\ \Delta pK_a$: that is, an increase in basicity of the nucleophile by 1 pK_a unit increases the exoergicity of reaction 1 by 1.4 kcal/mol if $E°$ and $\Delta G°(AH)$ remain constant. Thus, if the rate of reaction 1 is

$$\begin{array}{ccc} A^- + H^+ + Me_3C\cdot & \xrightarrow{\Delta G^\circ(A^-)} & Me_3C{=}A\cdot^- + H^+ \\ \updownarrow (k_a) & & \updownarrow (E^\circ) \\ AH + Me_3C\cdot & \xrightarrow{\Delta G^\circ(AH)} & Me_3C{=}A + H\cdot \end{array}$$

Scheme III

controlled by the change in free energy of the reaction, the rate should increase with the basicity of the anion if the stability of the radical anion is held constant. This situation is obviously not the case for many of the entries in Table III. In Table IV, the pertinent pK_a data for the various acetophenone derivatives and calculated $\Delta G^\circ(A^-)$ values are listed. The calculated values are based upon literature E° values and values of $\Delta G^\circ(AH)$ calculated by Benson's group additivity rules (*18*).

Table III. Relative Reactivity of Anions toward $Me_3C\cdot$ at 35 °C in Dimethyl Sulfoxide (K^+-18-Crown-6)

Anions	*Relative Reactivities*	*Anions*	*Relative Reactivities*
$Me_2C{=}NO_2^-$	1.00	$PhC(O^-){=}CHMe$	0.20
$MeCH{=}NO_2^-$	6.1	$PhC(O^-){=}CMe_2$	0.03
$H_2C{=}NO_2^-$	35	$PhC(O^-){=}CHPh$	1.1
$PhCH{=}NO_2^-$	1.0	$PhC(O^-){=}CPh_2$	2.2
$PhC(Me){=}NO_2^-$	7.4	$PhC(O^-){=}$fluorene	0.01
NO_2^-	0.4	$PhC(CO_2Et)_2^-$	0.02[a]
Ph_2CCN^-	6.5	$EtC(CO_2Et)_2^-$, $HC(CO_2Et)_2^-$	<0.005
$PhCHCN^-$	<0.1		

[a]Limiting value at high $[Me_3CHgCl]$ (>0.3 M) with $[N^-] = [Me_2C{=}NO_2^-] = 0.1$ M.

Table IV. Reaction of $Me_3C\cdot$ with $PhC(O^-){=}C(R^1)(R^2)$ to Yield $PhC(O^-)C(R^1)(R^2)CMe_3$

R^1	*R^2*	pK_a	ΔG°	$\Sigma\sigma_{R^1,R^2}$	*Relative Reactivity*
H or Ph	PhCO	~9	~16	2-3	<0.005
o,o'-biphenylenyl(Fl)		10.1	14.4	~1.4	0.01
Ph	Ph	—[a]	—[a]	1.2	2.2
H	Ph	21.5	−1.6	1.1	1.1
H	Me	24.4	−6.4	0.5	0.2
Me	Me	26.3	−8.6	0.0	0.03

[a]Not determined.

Table IV reveals an inverted reactivity order as a function of pK_a, $\Delta G°$, or $\Sigma\sigma^*$ for R^1 and R^2. For the enolate anions of low basicity (R^1 and R^2 = H and PhCO; Ph and PhCO; H and CN; and H and CO_2Et), process 1 is highly endoergic and no reaction is observed. The reactivity increases sharply from R^1 and R^2 = *o,o'*-biphenylenyl (pK_a = 10.1) to R^1 = R^2 = Ph (pK_a $\simeq$ 20), but then the reactivity decreases as the enolate anion is made more basic by changing R^1 and R^2 from phenyl to methyl or hydrogen (*17*). This decrease in reactivity parallels the σ^* values for the substitutents R^1 and R^2 (ρ^* = 1.5).

The *tert*-butyl radical is a nucleophilic species and prefers to react via a transition state in which an electron has been transferred to the substrate. Apparently, the nucleophilicity of the *tert*-butyl radical becomes the dominant factor for the exoergic reactions of Table IV where an early transition state is involved. The reactions of nitronate anions seem to follow a similar rate profile. Very weakly basic anions such as $CH(NO_2)_2^-$, $C(NO)_2)_3^-$, or fluorene=NO_2^- fail to trap $Me_3C\cdot$. With R^1 and R^2 = H and C_6H_5 ($\Sigma\sigma^*$ = 1.1), a lower reactivity is observed than for the more basic CH_2=NO_2^- ($\Sigma\sigma^*$ = 0.98). However, as the electron-donating ability of the substituents increases further, the reactivity now decreases in a linear fashion with $\Sigma\sigma^*$ giving ρ^* = 1.6 (*r* = 0.997) for H_2C=NO_2^-, MeCH=NO_2^-, MeC(Ph) =NO_2^-, and Me_2C=NO_2^-. The same effect seems to occur with acetonitrile derivatives where Ph_2CCN^- is 65 times more reactive than the more basic $PhCHCN^-$.

Reactivity of a Series of Anions Toward the Phenacyl Radical

The phenacyl radical ($PhCOCH_2\cdot$) can be generated by electron transfer to $PhCOCH_2HgCl$. Because of the stability of $PhC(O^-)$=CH_2, the phenacyl radical should be an electrophilic species in contrast to the nucleophilic character of $Me_3C\cdot$ and other unsubstituted alkyl radicals. Reaction of PhC-OCH_2HgCl with N^- = Me_2C=NO_2^- leads to the alkylation product PhC-$OCH_2CMe_2NO_2$, which readily undergoes an E_2 elimination of HNO_2 to yield PhCOCH=CMe_2 as the major product.

When the photostimulated reaction of $PhCOCH_2HgCl$ with a series of anions was surveyed (Table V), several obvious differences between $PhCOCH_2HgCl$ or $PhCOCH_2\cdot$ and *t*-BuHgCl or *t*-Bu· are apparent. The mercurial now will undergo alkyl exchange with CH_2=NO_2^-, leading to $PhCOCH_3$ and isolable O_2NCH_2HgCl. Phenylated anions that had given substitution products with *t*-BuHgCl [e.g., Ph_2C=C=N^-, Ph_2C=$C(O^-)$Ph, PhCH= $C(O^-)$Ph, $PhC(CH_3)$=NO_2^-, PhCH=NO_2^-, and $PhC(CO_2Et)_2^-$] now give products of oxidative dimerization [e.g., $Ph_2C(CN)C(CN)Ph_2$] with $PhCOCH_2HgCl$, presumably from electron transfer with $PhCOCH_2\cdot$ (Scheme IV). In reactions with anions such as Ph_2C=C=N^-, $PhC(O^-)$=CPh_2, or $PhC(O^-)$=CHPh, the phenacyl radical is similar to the electrophilic 2-nitro-2-propyl radical, which reacts with

Table V. Photostimulated Reaction Products between $PhCOCH_2HgCl$ and Various Nucleophiles

Nucleophile[a]	*Time (h)*	*Products (% Yield)*[b]
$Me_2C{=}NO_2^-$	2	$PhCOCH_2CMe_2NO_2$ (22); $PhCOCH{=}CMe_2$ (65)
$MeCH{=}NO_2^-$	4	$PhCOCH_2CHMeNO_2$ (32); $PhCOCH{=}C(Me)NO_2$ (4)
$H_2C{=}NO_2^-$	5	$PhCOCH_2CH_2NO_2$ (3); O_2NCH_2HgCl; $PhCOCH_3$
$PhCH{=}NO_2^-$	3	$PhCOCH_2CH(Ph)NO_2$ (7); $PhCH{=}C(Ph)NO_2$ (48); $PhCH(NO_2)CH(NO_2)Ph$ (11)
$PhC(Me){=}NO_2^-$	3	$PhC(Me)(NO_2)C(Me)(NO_2)Ph$ (56); $PhC(Me){=}C(Me)Ph$ (13)
$EtC(CO_2Et)_2^-$	2	$PhCOCH_2C(Et)(CO_2Et)_2$ (70)
$MeC(CO_2Et)_2^-$	2	$PhCOCH_2C(Me)(CO_2Et)_2$ (61)
$HC(CO_2Et)_2^-$	6	$PhCOCH_2CH(CO_2Et)_2$ (27)
$PhC(CO_2Et)_2^-$	2	$PhCOCH_2CH(Ph)(CO_2Et)_2$ (21); $PhC(CO_2Et)_2C(CO_2Et)_2Ph$ (64)
$Me_3C(O^-){=}CH_2$	4	$PhCOCH_2CH_2COCMe_3$ (37)
$PhC(O^-){=}CMe_2$	2	$PhCOCH_2CMe_2COPh$ (52)
$PhC(O^-){=}CHMe$	4	$PhCOCH_2CH(Me)COPh$ (24)
$PhC(O^-){=}CHPh$	3	$PhCOCH(Ph)CH_2COPh$ (4); $PhCOCH(Ph)CH(Ph)COPh$ (63)
$PhC(O^-){=}CPh_2$	2	$PhCOC(Ph)_2C(Ph)COPh$ (71)
$Ph_2C{=}C{=}N^-$	2	$Ph_2C(CN)C(CN)Ph_2$ (69)

NOTE: Reactions were performed in hexamethylphosphoric triamide under N_2 with photostimulation from a 275-W sunlamp approximately 15 cm from the Pyrex reaction flask. Reactions were performed on a 1-mmol scale with $[PhCOCH_2HgCl] = [N^-] = 0.1$ M.
[a]Generated by the reaction of the conjugate acid with Me_3COLi.
[b]Yields were determined by ^1H-NMR and GLPC analysis.

these anions by essentially complete electron transfer to form the oxidative dimerization products, that is, Scheme IV with O_2NCMe_2Cl and $Me_2C(NO_2)\cdot$ in place of $PhCOCH_2HgCl$ and $PhCOCH_2\cdot$ (23). The phenacyl radical adds to some anions [e.g., $N^- = HC(CO_2Et)_2^-$, $MeC(CO_2Et)_2^-$, or $Me_3CC(O^-){=}CH_2$] that are unreactive toward *tert*-butyl radicals. In these cases, apparently the stability of $Ph\dot{C}(O^-)CH_2N$ relative to $Me_3CN\cdot^-$ causes reaction 1 to be more exoergic with $R\cdot = PhCOCH_2\cdot$ than with $R\cdot = Me_3C\cdot$.

$$PhCOCH_2\cdot + Ph_2C{=}C{=}N^- \rightarrow PhCOCH_2^- + Ph_2\dot{C}CN$$
$$Ph_2\dot{C}CN + Ph_2C{=}C{=}N^- \rightarrow Ph_2C(CN)C(CN)Ph_2\cdot^-$$
$$Ph_2C(CN)C(CN)Ph_2\cdot^- + PhCOCH_2HgCl \rightarrow Ph_2C(CN)C(CN)Ph_2 + PhCOCH_2\cdot^- + Hg^0 + Cl^-$$

$$2\,Ph_2C{=}C{=}N^- + PhCOCH_2HgCl \rightarrow Ph_2C(CN)C(CN)Ph_2 + PhCOCH_2^- + Hg^0 + Cl^-$$

Scheme IV

The reactions of $PhCOCH_2HgCl$ with the nucleophiles of Table V were retarded by the presence of 18-crown-6. Higher yields were observed in $(Me_2N)_3PO$ (HMPA) than in Me_2SO. In HMPA and in the absence of 18-crown-6, essentially the same yields were observed with K^+ and Li^+ as the counterions. As in the case of $Me_3C\cdot$, no reaction was observed between $PhCOCH_2\cdot$ and very stable anions such as $O_2NCH{=}NO_2^-$, $(O_2N)_2C{=}NO_2^-$, $PhC(O^-){=}CHCOPh$, and $PhC(O^-){=}CHSO_2Ph$.

Competitive reactions with $Me_2C{=}NO_2^-$ as the standard were performed with three series of anions, $R^1R^2C{=}NO_2^-$, $RC(CO_2Et)_2^-$, and $R^1R^2C{=}C(O^-)Ph$ in the HMPA–Li^+ system. Ion pairing may have some effect on the reactivity observed, but for a given series of closely related anions, this effect should be reasonably constant. The results of these competitive reactions are given in Table VI where the relative reactivity is based on the yield of the coupling product ($PhCOCH_2N$) or its elimination product formed by loss of HNO_2. Phenylated anions, such as $Ph_2C{=}C(O^-)Ph$, have a much higher total relative reactivity than the values in Table VI because of the formation of oxidative dimerization products. For example, in the competition of equal molar amounts of $Me_2C{=}NO_2^-$ and $Ph_2C{=}C(O^-)Ph$ for $PhCOCH_2\cdot$, only 4.3% of the $PhCOCH_2\cdot$ was trapped to yield $PhCOCH_2CMe_2NO_2\cdot^-$, although approximately 96% of the $PhCOCH_2\cdot$ reacted to give $Ph_2\dot{C}COPh$, and coupling products from $PhCOCH_2CPh_2COPh\cdot^-$ were not detected. In terms of total reactivity toward $PhCOCH_2\cdot$, $Ph_2C{=}$

Table VI. Relative Reaction of Nucleophiles in Reaction 1 toward $PhCOCH_2\cdot$

Nucleophile	*Relative Reactivity*[a]
$Me_2C{=}NO_2^-$	1.00
$MeCH{=}NO_2^-$	0.65; 0.56[b]
$CH_2{=}NO_2^-$	0.17
$PhCH{=}NO_2^-$	0.10
$EtC(CO_2Et)_2^-$	1.98
$MeC(CO_2Et)_2^-$	1.87
$HC(CO_2Et)_2^-$	0.22
$PhC(CO_2Et)_2^-$	0.02
$PhC(O^-){=}CMe_2$	0.69
$PhC(O^-){=}CHMe$	0.27
$PhC(O^-){=}CHPh$	0.09
$PhC(O^-){=}CPh_2$	<0.01

[a]Based upon a series of experiments with different N^- and $Me_2C{=}NO_2^-$ concentrations (0.05–0.20 M) with $[PhCOCH_2HgCl]$ = 0.01 M at 35 °C in hexamethylphosphoric triamide–Li^+. Yields of substitution derived products were in the range of 65–95% except for $CH_2{=}NO_2^-$ or the benzylic-type carbanions where yields were ~25%.
[b]Dimethyl sulfoxide–K^+-18-crown-6.

$C(O^-)Ph$ is about 25 times as reactive as $Me_2C{=}NO_2^-$, whereas in terms of formation of the addition products from reaction 1 ($RN\cdot^-$), $Me_2C{=}NO_2^-$ is more reactive than $Ph_2C{=}C(O^-)Ph$.

Comparison of an Electrophilic and a Nucleophilic Radical in Reaction 5

The data of Tables IV and VI form an interesting comparison. Within the series $R^1R^2C{=}NO_2^-$, $R^1R^2C{=}C(O^-)Ph$, or $RC(CO_2Et)_2^-$, the relative reactivities in reaction 5 vary in completely opposite directions for $Me_3C\cdot$ and $PhCOCH_2\cdot$: that is, for $Me_3C\cdot$, $H_2C{=}NO_2^- > MeCH{=}NO_2^- > Me_2C{=}NO_2^-$; $PhC(O^-){=}CPh_2 > PhC(O^-){=}CHPh > PhC(O^-){=}CHMe > PhC(O^-){=}CMe_2$; and $PhC(CO_2Et)_2^- > CH_3C(CO_2Et)_2^-$. For $PhCOCH_2\cdot$, $H_2C{=}NO_2^- < MeCH{=}NO_2^- < Me_2C{=}NO_2^-$; $PhC(O^-){=}CPh_2 < PhC(O^-){=}CHPh < PhC(O^-){=}CHMe < PhC(O)^-){=}CMe_2$; and $PhC(CO_2Et)_2^- < HC(CO_2Et)_2^- < MeC(CO_2Et)_2^-$. Toward $PhCOCH_2\cdot$, reactivity increases with the basicity of the nucleophile, whereas toward $Me_3C\cdot$, reactivity decreases with basicity. This decrease in reactivity with increasing exoergicity of reaction 1 with $Me_3C\cdot$ seems to be mainly an effect on the value of $\Delta H^\ddagger$, at least as judged from the temperature effect on the competition between $MeCH{=}NO_2^-$ and $Me_2C{=}NO_2^-$. When competitive thermally initiated reactions of *t*-BuHgCl with $Me_2C{=}NO_2^-$ and $MeCH{=}NO_2^-$ were performed at 55–85 °C in the dark, the relative reactivity of $MeCH{=}NO_2^-$ decreased from 5.0 at 55 °C to 4.4 at 65 °C to 4.05 at 75 °C and 3.8 at 85 °C. These data yielded $\Delta H^\ddagger(Me_2C{=}NO_2^-) - \Delta H^\ddagger(MeCH{=}NO_2^-) = 2.2$ kcal/mol and $\Delta S^\ddagger(Me_2C{=}NO_2^-) - \Delta S^\ddagger(MeCH{=}NO_2^-) = 3.6$ eu. Extrapolation to 35 °C yields $\Delta G^\ddagger(Me_2C{=}NO_2^-) - \delta G^\ddagger(MeCH{=}NO_2^-) = 1.1$ kcal/mol and a calculated relative reactivity of $MeCH{=}NO_2^-$ of 6.2 versus the photostimulated value of 6.1.

$$PhCOCH_2\cdot + R_2\bar{C}\text{-}Q \rightarrow PhCOCH_2CR_2Q\cdot^- \tag{5}$$

The energy of activation in the addition of a radical to a nucleophile is a function of the oxidation–reduction potentials of both the anion and the radical. An easily oxidized radical such as $Me_3C\cdot$ has a lower energy of activation for attack upon an anion if the anion can accept an electron in its lowest unoccupied molecular orbital. Phenylated anions are particularly reactive toward electron-donor radicals such as $Me_3C\cdot$ because of this SO molecular orbital (SOMO)–LUMO interaction in the transition state, (**I**). Substitution of methyl for hydrogen at the reacting center of the carbanion leads to a decrease in reactivity because of a higher energy of the transition state **I** despite an increase in the exoergicity of the overall reaction. On the

$$Me_3C\cdot + R_2CQ^{-1} \rightarrow \underset{\textbf{I}}{[Me_3C^+ \;\; R_2CQ^{-2}]} \rightarrow Me_3CCR_2Q\cdot^- \tag{6}$$

other hand, an electrophilic radical such as $PhCOCH_2\cdot$ prefers to add to the more highly methylated anion (ie., the more basic anion) for the pairs, $Me_2C{=}NO_2^- > H_2C{=}NO_2^-$; $MeC(CO_2Et)_2^- > HC(CO_2Et)_2^-$; $Me_2C{=}C(O^-)Ph > MeCH{=}C(O^-)Ph$. This finding is consistent with a transition state resembling **II** in which the SOMO–HOMO interaction is dominant. However, if the transition state resembles **II**, low reactivity of the phenylated nucleophiles in this reaction is now surprising. Although the phenylated anions are admittedly less basic than the hydrogen analogues, the stability of **II** should be greater with the phenyl than with the methyl substituent. Indeed, the phenylated anions react to give nearly exclusively the oxidative dimerization products derived from radicals such as $Ph_2\dot{C}COPh$. If reaction 5 with transition state **II** and reaction 8 were occurring in competition, the rates of both reactions should increase in the order R = Ph > Me > H. Reaction 8 does occur much more readily for R = Ph than for R = Me or H when the attacking radical is electrophilic in character, whereas with the nucleophilic $Me_3C\cdot$, no evidence for reaction 8 is observed even with R = Ph. Toward $PhCOCH_2\cdot$, reactions 5 (proceeding via transition state **II**) and 8 cannot be independent reactions occurring in competition with each other. Instead, if reaction 5 proceeds by transition state **II**, it must merge into reaction 8 as the stability of the species $R_2\dot{C}Q$ increases. Of course, **II** may be a transition state for R = H or Me but may become a reaction intermediate for R = Ph. Alternately, **II** may be a real intermediate for R = H, Me, or Ph, but cage collapse to $PhCOCH_2CR_2Q\cdot^-$ occurs for R = H or Me although escape from the cage is the preferred route when R = Ph and $Ph_2\dot{C}Q$ is less reactive and more persistent. A third possibility is that the products of reaction 8 result from the instability of the product from reaction 5, for example, $PhCOCH_2\cdot + R_2\bar{C}Q \rightarrow PhCOCH_2CR_2Q\cdot^- \rightarrow PhCOCH_2^- + R_2\dot{C}Q$. Attempts to identify the adduct radical anion as an intermediate in oxidative dimerization processes by using an excess of an electron acceptor, such as $PhCOCH_2HgCl$ or O_2NCMe_2Cl, have been unsuccessful in converting an oxidative dimerization sequence (Scheme IV) into a substitution process (Scheme I). The conclusion is that the two processes do not involve $PhCOCH_2CR_2Q\cdot^-$ as a common intermediate, although an intermediate such as **II** may be common to both processes.

$$PHCOCH_2\cdot + R_2CQ^{-1} \rightarrow [PhCOCH_2{:}^- \; R_2\dot{C}\text{-}Q] \rightarrow PhCOCH_2CR_2Q\cdot^- \quad (7)$$

II

$$PhCOCH_2\cdot + R_2\bar{C}\text{-}Q \rightarrow PhC(O^-){=}CH_2 + R_2\dot{C}Q \quad (8)$$

Although the reactions of $PhCOCH_2\cdot$ with $PhCOCR_2^-$ and $R_2C{=}NO_2^-$ can perhaps be rationalized by a transition state resembling **II**, the reactivities in the series $RC(CO_2Et)_2^-$ is difficult to explain by this transition state or intermediate. The relative reactivity decreases from R = Me = 1.0 to R = H = 0.12 to R = Ph = 0.01 (addition, reaction 5) and 0.03 (electron transfer, reaction 8). Here the total reactivity of $PhC(CO_2Et)_2^-$ is less than that of the methyl or hydrogen analogue with the total reactivity increasing with the exoergicity of the reaction. For malonate, and perhaps for $PhCOCR_2^-$ and $R_2C{=}NO_2^-$ nucleophiles, the relative reactivities in reaction 5 seem to be best explained by a highest occupied molecular orbital (HOMO)–SOMO interaction with a transition resembling the product $Ph\dot{C}(O^-)C(R)(CO_2Et)_2$ (i.e., a transition state with extensive carbon–carbon bond formation). Apparently, the relative orbital energies are such that the SOMO–HOMO interaction is more stabilizing as the HOMO orbital energy of the nucleophile increases from R = Ph to H to Me. Reaction 7 would thus be a separate and competing process whose rate increases from R = H to Me to Ph.

A given radical can be considered to be either nucleophilic or electrophilic in character depending upon the nature of the nucleophile being attacked in reaction 5. Thus, toward nitronate anions in H_2O, methyl radical gives a nucleophilic relative reactivity series comparable to $Me_3C\cdot$ in Me_2SO ($H_2C{=}NO_2^- > MeCH{=}NO_2^- > Me_2C{=}NO_2^-$), that is, faster reaction with the least basic anion (*24*). On the other hand, toward the more easily oxidized Ph_3C^- and $p\text{-}PhC_6H_4CPh_2^-$, $Me\cdot$ adds preferentially to the strongest base (Ph_3C^-) even though the exoergicity of this process is less than for addition to $p\text{-}PhC_6H_4CPh_2^-$ (*22, 25*).

Substitution reactions occurring by reaction 5 are of considerable value in organic chemistry, particularly in view of the tolerance of this reaction to appreciable steric repulsion (e.g., tertiary radicals will add to tertiary carbanions). The scope of the reaction is limited by the fact that highly nucleophilic radicals will add only to anions of low basicity irrespective of the overall exoergicity of reaction 5. However, with electrophilic radicals, the rate of reaction 5 increases with the basicity of the carbanion and with the exoergicity of the reaction, but now oxidative dimerization can become an important competing reaction.

Acknowledgment

Acknowledgment is made to the donors of the Petroleum Research Fund, administered by the American Chemical Society, for support of this work. This work was also supported by a grant from the National Science Foundation (CHE–8119343).

Literature Cited

1. Kornblum, N.; Michel, R. E.; Kerber, R. C. *J. Am. Chem. Soc.* **1966,** *88*, 5660, 5662.
2. Russell, G. A.; Danen, W. C. *J. Am. Chem. Soc.* **1966,** *88*, 5663.

3. Kim, J. K.; Bunnett, J. F. *J. Am. Chem. Soc.* **1970,** 92, 7463, 7464.
4. Russell, G. A.; Norris, R. K. *Rev. React. Species in Chem. React.* **1973,** *1*, 65.
5. Kornblum, N. *Angew. Chem., Int. Ed. Engl.* **1975,** *14*, 734.
6. Russell, G. A.; Norris, R. K.; Panek, E. J. *J. Am. Chem. Soc.* **1971,** *93*, 5839.
7. Kornblum, N.; Boyd, S. D.; Stuchal, F. W. *J. Am. Chem. Soc.* **1970,** 92, 5783.
8. Kornblum, N.; Boyd, S. D. *J. Am. Chem. Soc.* **1970,** 92, 5784.
9. Kornblum, N.; Singh, H. K.; Boyd, S. D. *J. Org. Chem.* **1984,** *49*, 358.
10. Kornblum, N.; Erickson, A. S. *J. Org. Chem.* **1981,** *46*, 1037.
11. Feiring, A. *J. Org. Chem.* **1983,** *48*, 347.
12. Russell, G. A.; Ros, F. *J. Am. Chem. Soc.* **1985,** *107*, 2506.
13. Freeman, D. J.; Norris, R. K.; Woolfenden, S. K. *Aust. J. Chem.* **1978,** *31*, 2477.
14. Kornblum, N.; Fifolt, M. J. *J. Org. Chem.* **1980,** *45*, 360.
15. Russell, G. A.; Hershberger, J.; Owens, K. *J. Am. Chem. Soc.* **1979,** *101*, 1312.
16. Russell, G. A.; Hershberger, J.; Owens, K. *J. Organomet. Chem.* **1982,** *225*, 43.
17. Russell, G. A.; Khanna, R. K. *J. Am. Chem. Soc.* **1985,** *107*, 1450.
18. Russell, G. A.; Khanna, R. K. *Tetrahedron* **1985,** *41*, 4133.
19. Kerber, R. C.; Urry, G. W.; Kornblum, N. *J. Am. Chem. Soc.* **1965,** *87*, 4520.
20. Russell, G. A.; Guo, D. *Tetrahedron Lett.* **1984,** 25, 5239.
21. Russell, G. A.; Ros, F.; Mudryk, B. *J. Am. Chem. Soc.* **1980,** *102*, 7601.
22. Tolbert, L. M. *J. Am. Chem. Soc.* **1980,** *102*, 3531.
23. Russell, G. A.; Jawdosiuk, M.; Makosza, M. *J. Am. Chem. Soc.* **1979,** *101*, 2355.
24. Veltwisch, D.; Asmus, K.-D. *J. Chem. Soc., Perkin Trans. 2* **1982,** 1143.
25. Tolbert, L. M. *J. Am. Chem. Soc.* **1980,** *102*, 6808.

RECEIVED for review October 21, 1985. ACCEPTED FEBRUARY 10, 1986.

26

Reactions of Ambident Nucleophiles with Nitroaromatic Electrophiles and Superelectrophiles

E. Buncel[1], J. M. Dust[1], K. T. Park[1], R. A. Renfrow[2], and M. J. Strauss[2]

[1] Department of Chemistry, Queen's University, Kingston, Ontario K7L 3N6, Canada
[2] Department of Chemistry, University of Vermont, Burlington, VT 05405

This study reports on the reactions of ambident nucleophiles with electron-deficient nitroaromatic and heteroaromatic substrates; anionic σ complex formation or nucleophilic substitution result. Ambident behavior is observed in the case of phenoxide ion (O versus C attack) and aniline (N versus C attack). O or N attack is generally kinetically preferred, but C attack gives rise to stable σ complexes through thermodynamic control. "Normal" electrophiles such as 1,3,5-trinitrobenzene or picryl chloride are contrasted with superelectrophiles such as 4,6-dinitrobenzofuroxan or 4,6-dinitro-2-(2,4,6-trinitrophenyl)benzotriazole 1-oxide (PiDNBT), which give rise to exceptionally stable σ complexes. Further interesting information was derived from the presence in PiDNBT of two electrophilic centers (C-7 and C-1′) susceptible to attack by the ambident nucleophilic reagent. The superelectrophiles are found to exhibit lesser selectivity toward different nucleophilic centers of ambident nucleophiles compared with normal electrophiles.

THE ANIONIC σ COMPLEXES FORMED between polynitroaromatic compounds and bases (*1, 2*), commonly known as Meisenheimer complexes, are used as models of the reaction intermediates that are considered to be formed in activated nucleophilic aromatic substitution reactions (*3–6*), as well as being of intrinsic interest. Thus, numerous studies describe the formation and transformation of such σ complexes (*7–14*). As a result, a variety of structural types of these species have been characterized and subjected to detailed investigation. A number of theoretical studies relating to these species have also been reported (*15*).

0065-2393/87/0215-0369$06.00/0

Attention focused recently on the interaction of nitroaromatic and heteroaromatic compounds with potentially ambidentate nucleophiles. These studies revealed novel structure–reactivity relationships.

Acetonate σ Complexes of Nitroaromatics

The first observations in this area were made 100 years ago, when in 1886 Janovsky and Erb (*16*) reported the formation of an intensely colored purple substance in the reaction of acetone with 1,3-dinitrobenzene (DNB) in alkaline solution. After a period of controversy as to whether the structure of the species formed was **1** or **2** (*17–19*), the **2** was unambiguously proven by ^{1}H-NMR spectroscopy (*20*). The corresponding adduct of 1,3,5-trinitrobenzene (TNB), which is also obtained readily on solvolysis of the TNB·CH_3O^- complex in acetone, was similarly shown by ^{1}H-NMR spectroscopy to have the structure **3** (*21*). More recent $^{13}C^-$ and ^{15}N-NMR spectroscopic studies confirmed these formulations (*22*). The structures of σ complexes such as **3** are now more correctly depicted as the delocalized species, in accord with experimental results (e.g., X-ray crystal structure) and theoretical considerations (*23*)

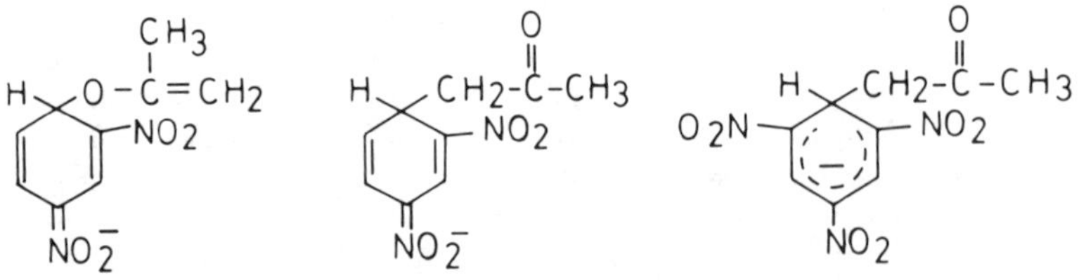

Thus, the α-carbon of enolate anion has a much greater basicity toward the electron-deficient carbon of DNB or TNB than does the enolate oxygen. A comparison of the thermodynamic stabilities of acetonate adducts of nitroaromatics with those formed by oxygen nucleophiles confirms this conclusion (*24*). Moreoever, kinetic studies have shown that the unusual stability of enolate complexes derives primarily from the low rate of their uncatalyzed decomposition (*25*). The poor leaving-group ability of enolate ions can be explained by the requirements of rehybridization and solvent reorganization that must accompany the expulsion of this carbanion nucleofuge, as compared with an oxygen nucleofuge such as methoxide ion.

Phenoxide Ion as Ambident Nucleophile

The first instance of the ambident reactivity of phenoxide ion toward nitroaromatics was reported from this laboratory. Thus, phenoxide ion reacted with TNB to yield the carbon-bonded adduct **7** (*26*); up to that time only the oxygen-bonded adduct **8** had been reported (*27*):

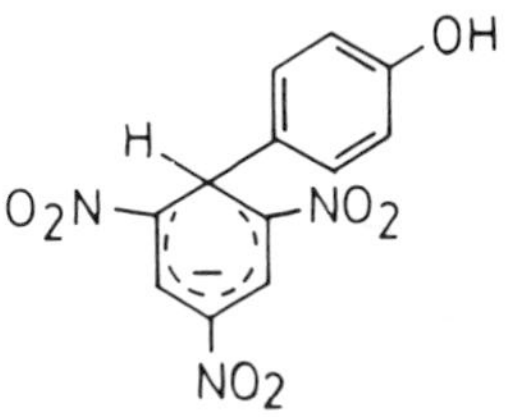

7 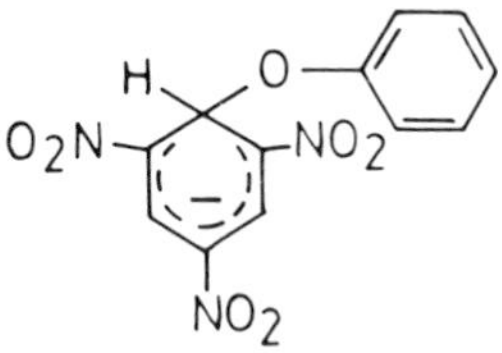 8

These results can be explained by the reactions in Scheme I and the potential energy profile in Figure 1. Attack by $C_6H_5O^-$ via oxygen is kinetically preferred; the energy barrier for this process should be lower than that for attack via carbon, because in formation of the C adduct **6** aromaticity would be disrupted. However, whereas the O-bonded adduct **8** can revert back to the reactants, the C-bonded adduct **6** initially formed will rapidly rearomatize by proton loss to give the final product **7**; this pathway is effectively irreversible. The C-bonded adduct is therefore obtained as the product of thermodynamic control.

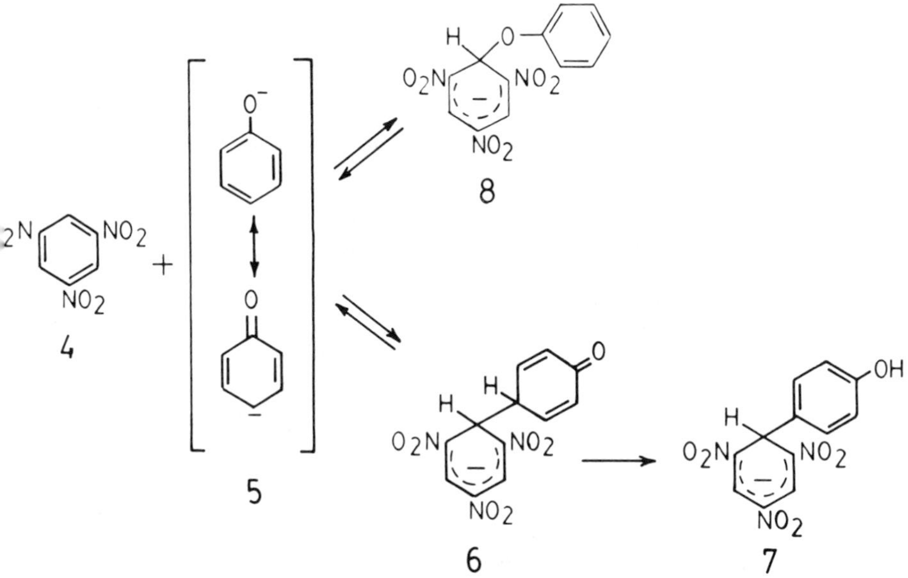

Scheme I

A variety of other C-bonded phenoxide complexes of nitroaromatics have since been reported; in most cases, the oxygen-bonded adduct is not observed. For example, in the reaction of TNB with 1-napthoxide, only the 1:1 and 2:1 C-bonded adducts **9** and **10** were formed (*28*):

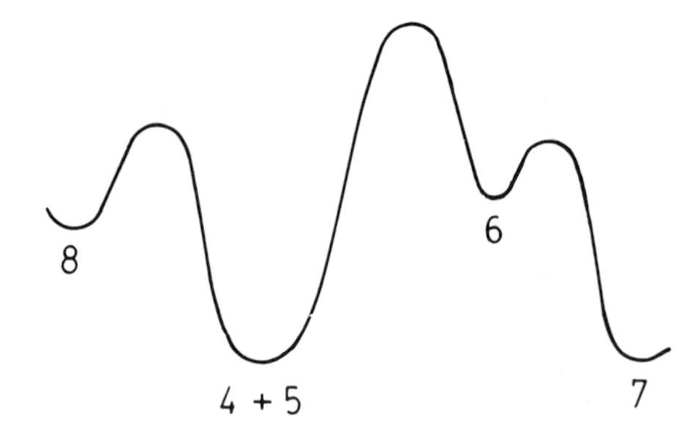

Figure 1. Free energy profile for reaction of TNB with phenoxide ion showing kinetic and thermodynamic control in σ-complex formation.

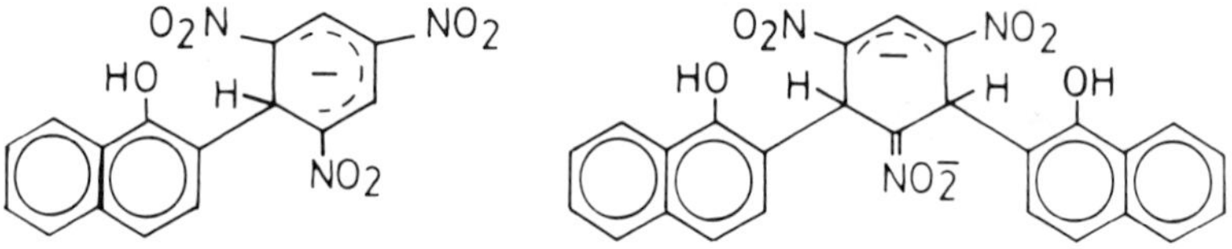

9 10

Interestingly, the reaction of 1-phenyl-2,4,6-trinitrobenzene gives only the C-bonded 1,3 adduct **11** formed by attack at C-3 of the substrate:

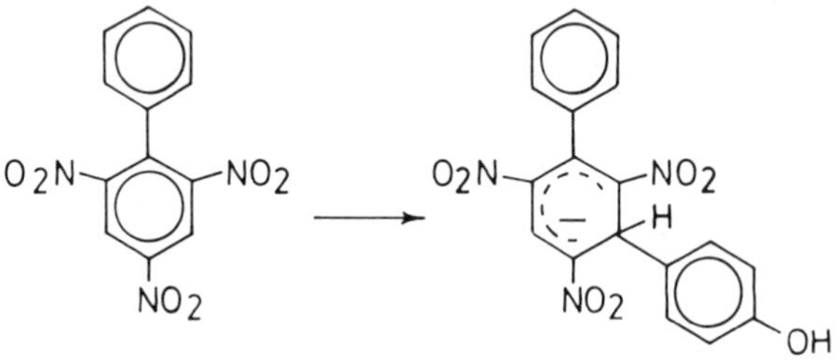

11 (1)

No evidence was found for the formation of a 1,1 C adduct or for any O adducts. Buncel et al. (*29*) concluded that F strain, relief of steric compression, and delocalizability considerations play dominant roles in accounting for the observed reaction course in this system.

When the ortho and para positions on the phenoxide are all substituted by alkyl groups, attack by oxygen of the phenoxide moiety is realized. Thus, for 2,4,6-trimethylphenoxide, the O-bonded adduct **12** was identified by ^{1}H-NMR spectroscopy as well as by its characteristic electronic absorption spectrum (*30, 31*). However, a competitive displacement of NO_2 by phenoxide takes place in this system to give the diphenyl ether **13** as the final product (Scheme II).

A direct comparison of the rates and equilibria associated with O and C attack by phenoxide on TNB is not currently available. However, in the case of 2,4,6-trinitroanisole (TNA), rate data for O attack by phenoxide can be compared with that by methoxide (*32*). Using fast reaction techniques (stopped flow and T jump), Bernasconi and Muller (*32*) found that the reaction of TNA with $C_6H_5O^-$ in $(CH_3)_2SO$–water media gives rise in a rapid process to the O-bonded 1,1 phenoxide adduct as a transient species, which is then converted to the 1,3 hydroxide adduct of TNA in a slower process. The data showed that $C_6H_5O^-$ attack (H_2O) is faster than CH_3O^- attack (CH_3OH) by a factor of 2.9, but $C_6H_5O^-$ expulsion is faster by 4.5×10^6, with the result that the equilibrium constant for 1,1 phenoxide adduct formation is smaller than for 1,1 methoxide adduct formation by 1.5×10^6.

Picryl Chloride–Phenoxide and TNB–Phenoxide Contrast

Whereas in the reaction of TNB with phenoxide ion the adduct from C attack is obtained, picryl chloride and other nitroaryl halides react with phenoxide ion to give only **14**, the product of O attack. This is in fact the normal route for the formation of picryl phenyl ethers, and no instance of a picryl halide

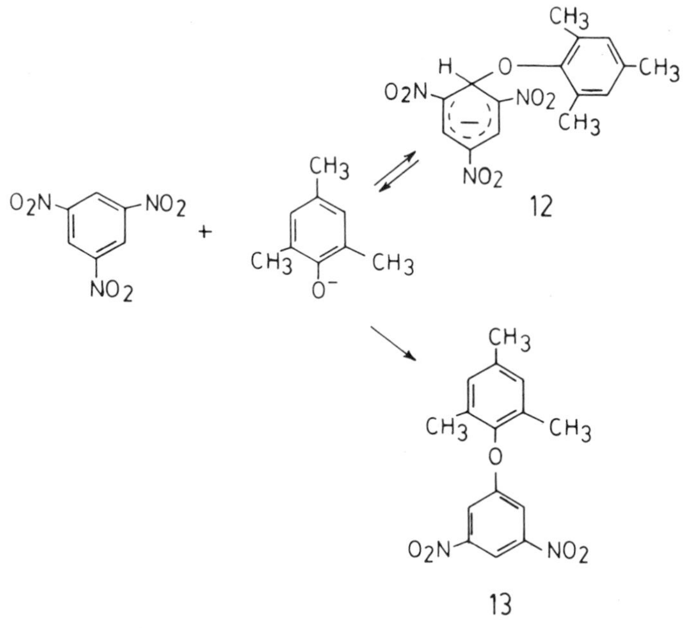

Scheme II

reacting with phenoxide ion to give the product of C attack, the biphenyl derivative **15,** has been reported.

A rationalization of this seeming anomaly follows on consideration of Scheme III and the accompanying reaction profile in Figure 2. The scheme depicts the possible pathways in the S_NAr reaction of a picryl halide (PiCl) with phenoxide ion. The boxed-in portion of this scheme is viewed as an extension of the TNB–$C_6H_5O^-$ system in which case X is H, as compared with X = Cl for the PiCl–$C_6H_5O^-$ system.

The potential energy profile for the PiCl–$C_6H_5O^-$ system (Figure 2) can similarly be viewed as an extension of that for the TNB–$C_6H_5O^-$ system (Figure 1). The energy barriers to formation of the initial O and C adducts in the two cases, that is, when X is H or Cl, should be quite similar, and the

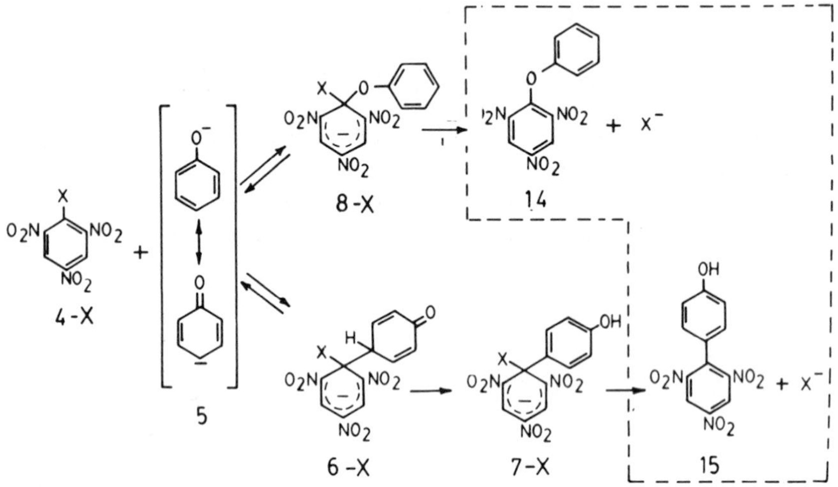

Scheme III

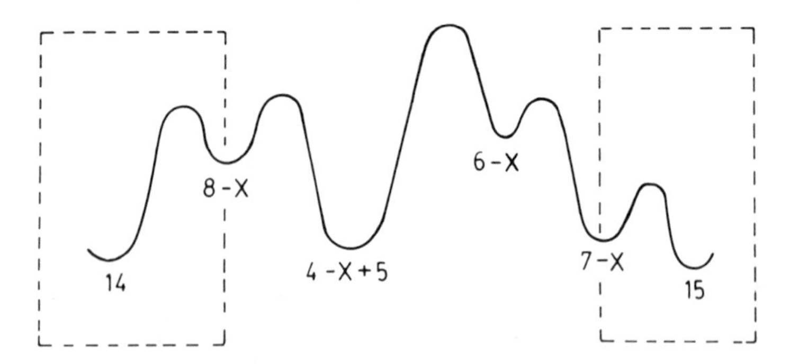

Figure 2. Free energy profile for reaction of picryl halides with phenoxide ion shown as an extension of the TNB–$C_6H_5O^-$ system.

stabilities of the initially formed adducts **8** and **6** will also be nearly independent of whether X is Cl or H. However, the O-bonded adduct **8**–X can now lose Cl^- as a good leaving group to form the displacement product **14** so that this reaction will occur preferentially. Therefore, in the PiCl system nucleophilic displacement by the oxygen of phenoxide is the kinetically preferred pathway over displacement by the carbon of phenoxide.

Aniline as an Ambident Nucleophile: Normal versus Superelectrophiles

The demonstration of ambident nucleophilic reactivity of aryl oxides in their reactions with nitroaromatics raises the question whether aromatic amines might function similarly. However, TNB does not react with aniline alone to form a stable σ complex; only a charge transfer or π complex is formed (*33*). This result is perhaps surprising because TNB is well-known to form stable σ complexes with primary or secondary aliphatic amines (*7, 8*).

We showed, however, that in the presence of a tertiary amine catalyst [$(CH_3CH_2)_3N$ or diazabicyclo[2,2,2]octane (DABCO)] TNB readily reacts with aniline or with *N*-methylaniline to form spectrally observable anilide σ complexes, that is, via N bonding (*34, 35*). σ-Complex formation via C bonding was not detected.

4,6-Dinitrobenzofuroxan (DNBF) as a Superelectrophile

Possibly a stronger electrophile than TNB might react with aniline via its carbon center. Recent work (*36–38*) has shown that DNBF (**16**) is a much stronger electrophile than TNB. For example, DNBF reacts with neutral H_2O or MeOH to give the respective σ complexes **17** (equation 2), whereas TNB requires HO^- or CH_3O^- for reaction to occur. DNBF has therefore been called a superelectrophile. (In equation 2, R is H or CH_3.)

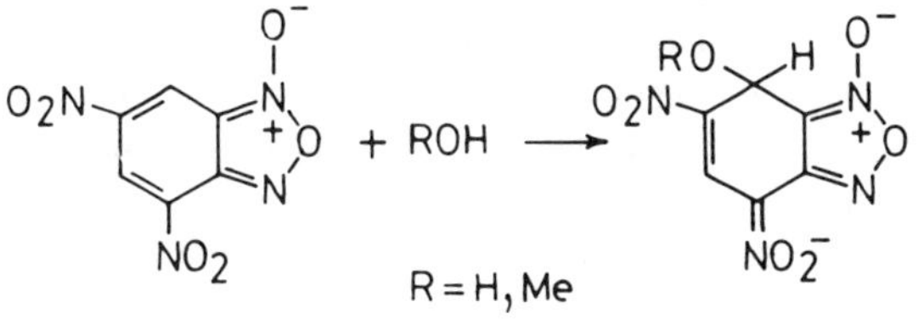

16 + ROH ⟶ **17** (2)

R = H, Me

We found in fact that DNBF reacts with aniline in $(CH_3)_2SO$ or $(CH_3)_2SO$–methanol to give initially the N-bonded adduct **19** but that this reverts to the C-bonded adducts **21** and **22**, which are formed as the thermodynamically preferred species (*39, 40*). The reaction sequence is shown in Scheme IV, which is the analogue of Scheme I for the TNB–$C_6H_5O^-$ system.

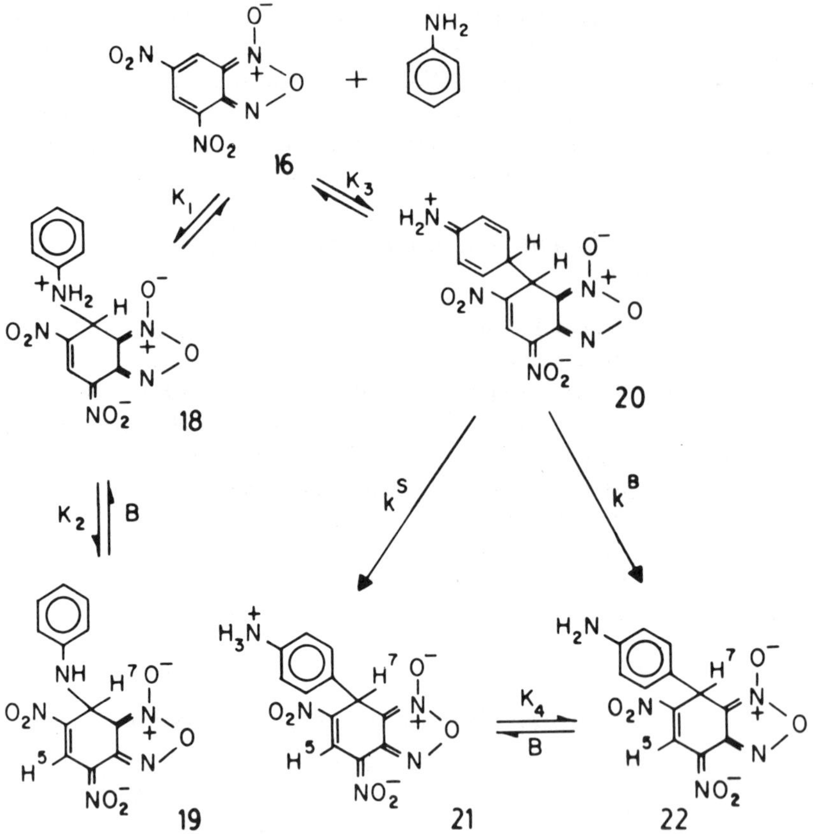

Scheme IV

In the TNB–aniline system, a detailed kinetic analysis showed that the lack of formation of a stable σ complex in the absence of a tertiary amine catalyst was the result of a thermodynamic, rather than a kinetic, factor (*34*). That is, the equilibrium constant for formation of the initial zwitterionic adduct $(TNB \cdot NH_2C_6H_5)^{\pm}$ is very small, such that in the absence of tertiary amine the overall equilibrium for complex formation is unfavorable. Addition of a tertiary amine, however, causes the equilibrium to be shifted to the right, due to the much more favorable deprotonation of the zwitterionic adduct to the anionic adduct, corresponding to the increased basicity of $(CH_3CH_2)_3N$ or DABCO as compared to $C_6H_5NH_2$.

The contrasting behavior in the DNBF–$C_6H_5NH_2$ system, where tertiary amine catalysis is not required, reflects the greater stability of σ complexes formed by this highly electron deficient substrate. These results and Scheme IV indicate that *the more reactive DNBF electrophile can differenti-*

ate less between the two nucleophilic centers of aniline, compared with TNB, which would be expected on the basis of reactivity–selectivity considerations.

Also, the σ complex **23** could be prepared, in which one DNBF moiety is bonded to aniline via the para carbon center and another DNBF moiety is bonded via the nitrogen center. This diadduct was readily obtained by addition of DNBF to a solution of **21** in $(CH_3)_2SO$ followed by 2 equiv of triethylamine (*39*).

Phenoxide ion was found to react with DNBF to yield the carbon-bonded adduct **24**. The ^{1}H-NMR chemical shifts for **24** are close to those of the corresponding protons in **7**, the carbon-bonded phenoxide adduct derived from TNB. Also, the adduct **24** is stable to acidification, as was found to be the case with the TNB–phenoxide adduct **7**. No evidence for the formation of the oxygen-bonded adduct **25** was found.

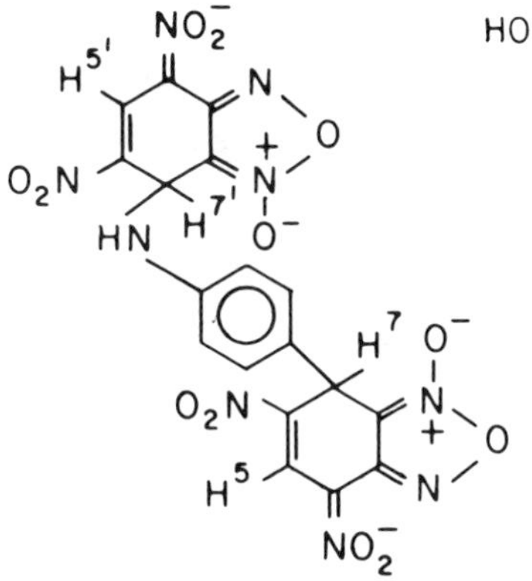

23

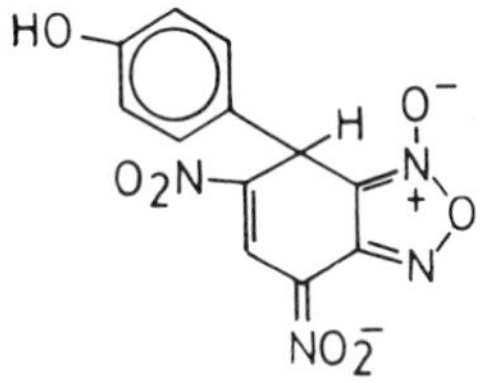

24

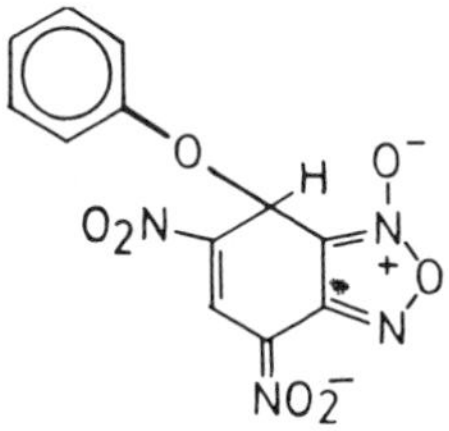

25

Contrasting with the reaction of phenoxide ion, thiophenoxide ion reacts with DNBF to give only the S-bonded adduct. 4-Aminothiophenoxide ion reacts in a similar fashion; C-bonded complexes could not be detected in these systems.

The π-excessive five-membered ring heterocycles furan, pyrrole, and *N*-methylpyrrole add readily to C-7 of DNBF to yield the C(α) adducts **27** (*41*):

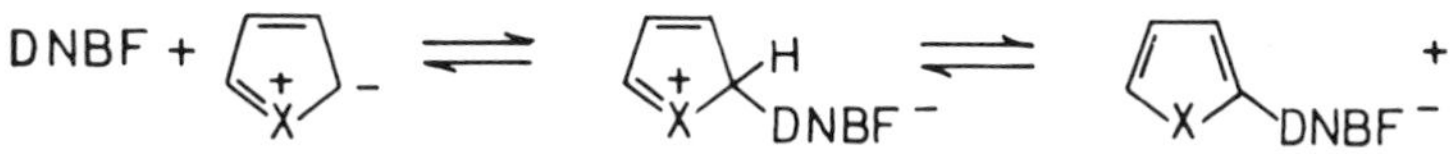

DNBF + (heterocycle) ⇌ **26** (H, DNBF⁻) ⇌ **27** (DNBF⁻) + H^+ (3)

X = O, NH, NMe

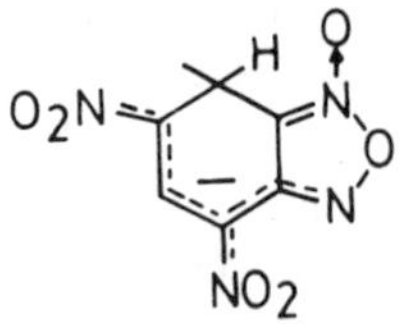

O_2N H O N O N NO_2 DNBF⁻ ⁻FBND X DNBF⁻ DNBF⁻ N R

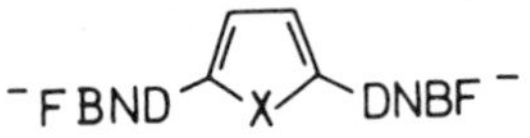

28 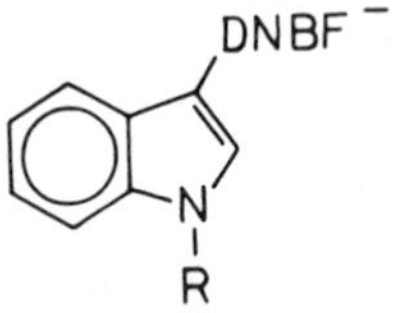 29

The intermediate zwitterionic species **26** could not be detected, which is in accord with a rapid rearomatization of the five-membered ring as a driving force in this process. Further reaction of **27** with DNBF yields the C(α),C(α′) diadducts **28**, which are formed as a 1:1 mixture of two diastereomers. DNBF also reacts at the 3-position of indole and *N*-methylindole to give the corresponding adducts **29** (R is H and CH_3) (*42*).

Interestingly, pyrrolide, indolide, and imidazolide ions react with TNB to give initially the N-bonded adducts, which subsequently undergo conversion to the C adducts (*43*). The fact that anionic nucleophiles are required for reaction with TNB, as compared with neutral bases in the DNBF case, is in accord with the less electron deficient character of TNB relative to DNBF. Further, for the anionic nucleophiles, a charge-controlled process takes place to give initially the kinetically preferred N adducts. The neutral bases, however, react with the softer carbon nucleophilic center in an orbital-controlled process. Further work will be required to establish the generality of these observations.

4,6-Dinitro-2-(2,4,6-trinitrophenyl)benzotriazole 1-Oxide (PiDNBT) as an Ambident Superelectrophile

A new superelectrophile was recently prepared in our laboratories, namely, the picryl dinitrobenzotriazole derivative **30** (*44*, *45*). This interesting molecule has two likely electrophilic sites available for attack, that is, via path a at C-7 of the benzotriazole moiety or via path b at C-1′ of the picryl moiety (*see* Scheme V). Path a gives rise to a spectrally observable adduct **31**, however, the adduct **32** formed via path b loses the benzotriazole moiety as the nucleofuge and the products of displacement **33** and **34** are obtained (Scheme V).

Phenoxide ion reacts with PiDNBT via path b as an oxygen nucleophile. Also, aniline reacts via path b as a nitrogen nucleophile. However, with the more sterically hindered 2,6-dimethylaniline, or *N*,*N*-dimethylaniline, route a is followed with the amines acting as carbon-nucleophiles to yield the adducts **35** and **36**.

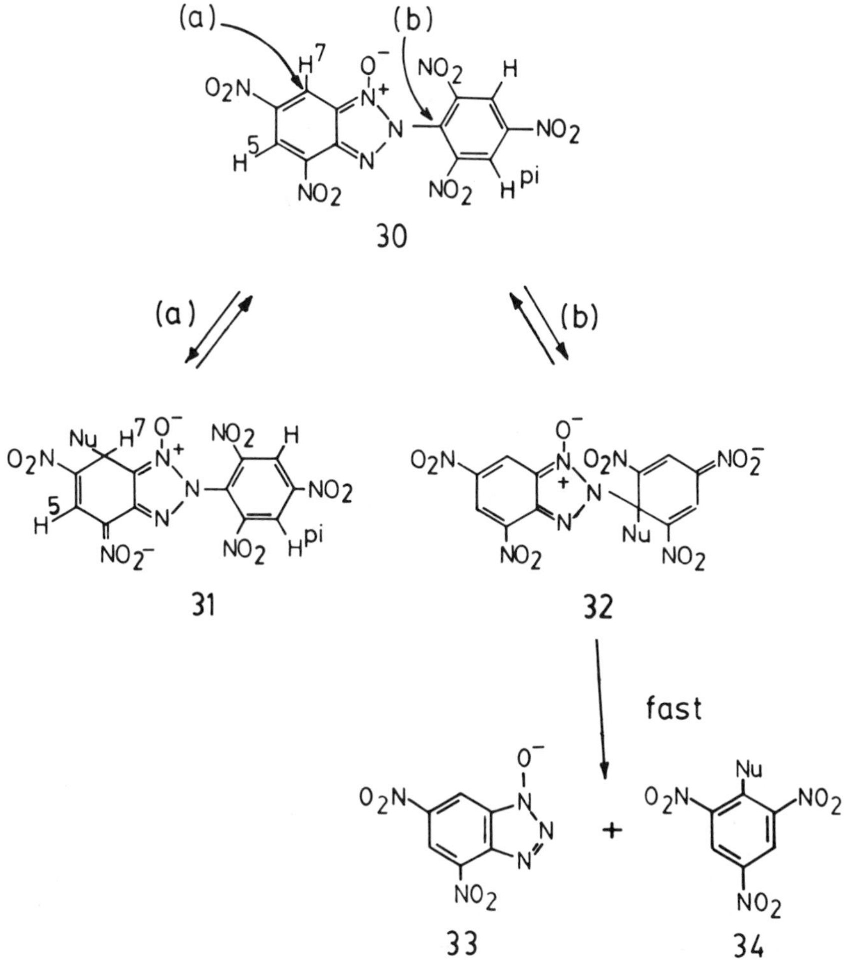

(a) (b)
O_2N H^7 O^- N^+ N N NO_2 H NO_2 H^5 NO_2 NO_2 H^{pi}
30

(a) (b)

O_2N Nu H^7 O^- N^+ N N NO_2 H NO_2 H^5 NO_2^- NO_2 H^{pi}
31

O^- O_2N N + N N O_2N NO_2^- NO_2 Nu NO_2
32

fast

O^- O_2N N N N NO_2
33

+

Nu O_2N NO_2 NO_2
34

Scheme V

These results thus reveal further contrasts in the reactivity of PiDNBT compared with that of TNB as the standard electrophile, because TNB does not react with dimethylaniline to give a stable adduct. Formation of the carbon-bonded complexes **35** and **36** substantiates the powerful electrophilic properties of this novel heterocyclic system.

The potential energy profile for the reaction of PiDNBT with ambident nucleophiles is unusual, because two reaction pathways must be represented corresponding to two electrophilic sites in the substrate and each pathway should accommodate the possibility of attack by two nucleophilic centers in the ambident reagent. However, a simplified view can be given in the form of

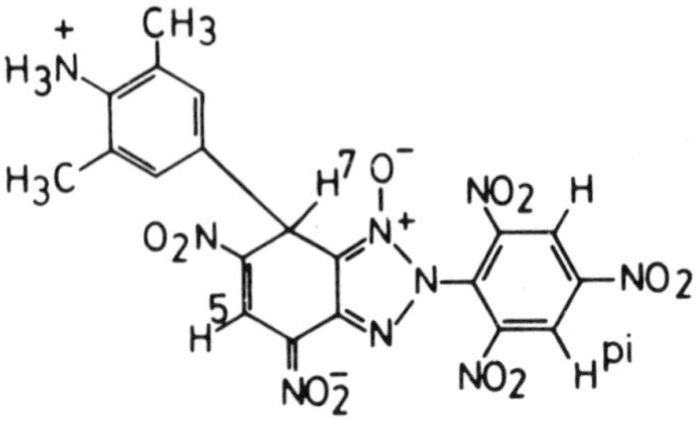

35

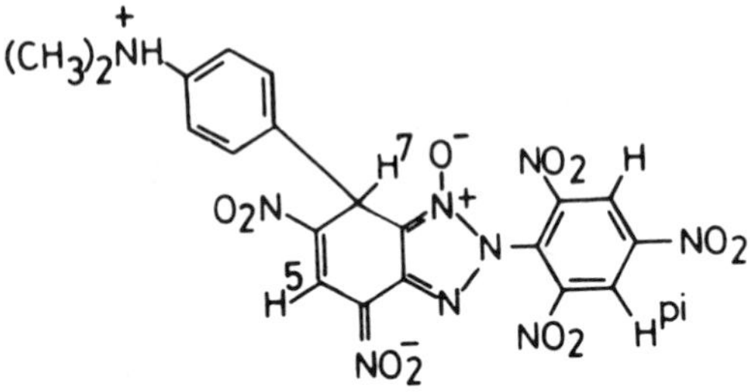

36

a dual potential energy–reaction coordinate diagram (not shown). Thus, attack via path a at C-7 of the benzotriazole moiety would be represented by the analogue of Figure 1 for the attack of an ambident nucleophile on TNB giving rise to products of kinetic and thermodynamic control. On the other hand, path b would be represented by the analogue of Figure 2 for the attack of an ambident nucleophile on picryl chloride, with the benzotriazole moiety taking the place of the chloride nucleofuge in PiCl. (The profiles for the ambident aniline–nitroaromatic system would include in each case an additional energy minimum corresponding to the zwitterionic species $ArX \cdot NH_2C_6H_5^{\pm}$ where ArX represents the nitroaromatic moiety, that is, TNB, PiCl, or PiDNBT, but the overall argument and conclusions would be analogous to the phenoxide system.)

Therefore, the reactions of phenoxide ion or aniline with PiDNBT follow path b and are analogues of the PiCl–$C_6H_5O^-$ system as depicted in Figure 2, while 2,6-dimethylaniline and *N*,*N*-dimethylaniline follow path a, being analogues of the TNB–$C_6H_5O^-$ system shown in Figure 1. Though in this case the change in reaction course, from path b to path a, could be influenced by inhibiting the reactivity of the aniline through ortho or *N*-methyl substitution, in general which pathway, a or b, will be favored in a given system cannot readily be predicted. Kinetic studies currently under way (*46*) should shed further light on this problem. To what extent reactivity–selectivity correlations apply in these systems then can be determined in a more quantitative manner.

Experimental Section

Preparation. Dimethyl sulfoxide was dried by stirring with calcium hydride and distilled under nitrogen. $(CD_3)_2SO$ was stored over molecular sieves. Methanol was distilled from barium oxide.

Anhydrous potassium phenoxide was prepared by the method of Kornblum and Lurie (*47*). Picryl chloride was prepared from picric acid by published procedures (*48*). DNBF (**16**) was prepared by nitration of benzofuroxan (*49*) and recrystallized from glacial acetic acid. The potassium salt of the methoxy adduct **17** was obtained from the reaction of methanolic DNBF with potassium hydroxide (*50*). PiDNBT (**30**) was prepared by cyclization of 2,2′,4,4′,6,6′-hexanitrohydrazobenzene (*51*) with concentrated sulfuric acid and recrystallized from glacial acetic acid: yellow plates; mp 291–294 °C (decomposition); NMR [100 MHz, $(CH_3)_2SO$-d_6] 9.09 (d, 1 H, J = 1.9 Hz), 9.48 (s, 2 H), 9.45 (d, 1 H, J = 1.9 Hz).

Caution: PiDNBT (**30**) is a sensitive high explosive and should be handled with caution. Impact sensitivity studies show **30** to be more sensitive than dry picric acid and approximately as sensitive as 1,3,5-trinitrohexahydro-1,3,5-triazine.

NMR Experiments. For monitoring by NMR spectroscopy, reactions of the nitroaromatics with nucleophiles were generally carried out in situ by addition of the appropriate reagent to a $(CD_3)_2SO$ solution of the given nitroaromatic (DNBF, 0.4 M; PiDNBT, 0.1 M). The reagents were applied to the sides of the NMR tube, which was then capped and shaken vigorously. Spectra were recorded at given time intervals on a JEOL MH-100 spectrometer with tetramethylsilane as the internal standard.

The σ complexes showed the following NMR parameters.

DNBF Complexes. **19:** δ 8.74 (s, 1 H, H-5), 6.08 (s, 1 H, H-7), 6.9 (m, 5 H, ArH). **21:** δ 8.79 (s, 1 H, H-5), 5.40 (s, 1 H, H-7), 7.41, 7.26 (A_2B_2, J = 8 Hz, 4 H, ArH), 9.72 (br, s, 3 H, NH_3^+). **22:** δ 8.78 (s, 1 H, H-5), 5.18 (s, 1 H, H-7), 6.96, 6.61 (A_2B_2, J = 9 Hz, 4H, ArH). **23:** δ 8.72 (s, 1 H, H-5), 6.05 (d, J = 9 Hz, 1 H, H-7), 8.79 (s, 1 H, H-5′), 5.17 (s, 1 H, H-7′), 6.94, 6.64 (A_2B_2, J = 9 Hz, 4 H, ArH), 6.35 (d, J = 9 Hz, 1 H, NH). **24:** δ 8.72 (s, 1 H, H-5), 5.45 (s, 1 H, H-7), 6.97 (m, 4 H, ArH), 9.63 (br s, 1 H, OH).

PiDNBT Complexes. **35:** δ 9.27 (s, 2 H, H^π), 8.79 (s, 1 H, H-5), 5.46 (s, 1 H, H-7), 7.06 (s, 2 H, ArH), 2.29 (s, 6 H, CH_3). **36:** δ 9.26 (s, 2 H, H^π), 8.79 (s, 1 H, H-5), 5.57 (s, 1 H, H-7), 7.43 (s, 4 H, ArH), 3.12 (s, 6 H, CH_3).

Acknowledgments

Support of this research by grants from the Natural Sciences and Engineering Research Council of Canada (E. Buncel) and the Naval Surface Weapons Center at Silver Spring, MD (M. J. Strauss), is gratefully acknowledged.

Literature Cited

1. Jackson, C. L.; Boos, W. F. *Am. Chem. J.* **1898,** *20*, 444.
2. Meisenheimer, J. *Justus Liebigs Ann. Chem.* **1902,** *323*, 205.

3. Buncel, E.; Crampton, M. R.; Strauss, M. J.; Terrier, F. *Electron Deficient Aromatic- and Heteroaromatic-Base Interactions: The Chemistry of Anionic Sigma Complexes;* Elsevier: Amsterdam, 1984.
4. Miller, J. *Nucleophilic Aromatic Substitution;* Elsevier: Amsterdam, 1968.
5. Bernasconi, C. F. *MTP Int. Rev. Sci.: Org. Chem., Ser. One* **1973,** *3,* 33.
6. Bunnett, J. F.; Cartano, A. V. *J. Am. Chem. Soc.* **1981,** *103,* 4861.
7. Terrier, F. *Chem. Rev.* **1982,** *82,* 77.
8. Artamkina, G. A.; Egorov, M. P.; Beletskaya, I. P. *Chem. Rev.* **1982,** *82,* 427.
9. Buncel, E. *The Chemistry of Amino, Nitro and Nitroso Compounds;* Patai, S., Ed.; Wiley: London, 1982; Supplement F.
10. Bernasconi, C. F. *Chimia* **1980,** *34,* 1.
11. Fendler, J. H.; Hinze, W. L.; Liu, L. J. *J. Chem. Soc., Perkin Trans. 2* **1975,** 1768.
12. Fyfe, C. A. *The Chemistry of the Hydroxy Group;* Patai, S., Ed., Wiley: London, 1971; Part 1.
13. Strauss, M. J. *Chem. Rev.* **1970,** *70,* 667.
14. Crampton, M. R. *Adv. Phys. Org. Chem.* **1969,** *7,* 211.
15. Strauss, M. J.; Fleischman, S.; Buncel, E. *J. Mol. Struct. Theochem.* **1985,** *121,* 37, and references cited therein.
16. Janovsky, J. V.; Erb, L. *Chem. Ber.* **1886,** *19,* 2155.
17. Gitis, S. S. *J. Gen. Chem. USSR (Engl. Transl.)* **1957,** *27,* 1956.
18. Gitis, S. S.; Kaminskii, A. Ya. *Usp. Khim.* **1978,** *47,* 1970.
19. Canback, T. *Sven. Farm. Tidskr.* **1949,** *53,* 151.
20. Foster, R.; Fyfe, C. A. *Chem. Commun.* **1967,** 1219.
21. Foster, R.; Fyfe, C. A. *Tetrahedron* **1965,** *21,* 3363.
22. Mahacek, V.; Sterba, V.; Lycka, A.; Snobl, D. *J. Chem. Soc., Perkin Trans. 2* **1982,** 355.
23. Destro, R.; Gramaccioli, C. M.; Simonetta, M. *Acta Crystallogr., Sect. B* **1979,** *35,* 733.
24. Murphy, R. M.; Wulff, C. A.; Strauss, M. J. *J. Am. Chem. Soc.* **1974,** *96,* 2678.
25. Kolb, I.; Mahacek, V.; Sterba, V. *Collect. Czech. Chem. Commun.* **1976,** *41,* 1914.
26. Buncel, E.; Webb, J. G. K. *J. Am. Chem. Soc.* **1973,** *95,* 8470.
27. Shein, S. M.; Byval'kevich, O. G. *Zh. Org. Khim.* **1972,** *12,* 328.
28. Mahacek, V.; Sterba, V.; Sterbova, A. *Collect. Czech. Chem. Commun.* **1976,** *41,* 2556.
29. Buncel, E.; Murarka, S. K.; Norris, A. R. *Can. J. Chem.* **1984,** *62,* 534.
30. Buncel, E.; Moir, R. Y.; Norris, A. R.; Chatrousse, A. P. *Can. J. Chem.* **1981,** *59,* 2470.
31. Buncel, E.; Jonczyk, A.; Webb, J. G. K. *Can. J. Chem.* **1975,** *53,* 3761.
32. Bernasconi, C. F.; Muller, M. C. *J. Am. Chem. Soc.* **1978,** *100,* 5530.
33. Ross, S. D.; Labes, M. M. *J. Am. Chem. Soc.* **1957,** *79,* 76.
34. Buncel, E.; Eggimann, W.; *J. Am. Chem. Soc.* **1977,** *99,* 5958.
35. Buncel, E.; Jarrell, H.; Leung, H. W.; Webb, J. G. K. *J. Org. Chem.* **1974,** *39,* 272.
36. Terrier, F.; Millot, F.; Norris, W. P. *J. Am. Chem. Soc.* **1976,** *98,* 5883.
37. Buncel, E.; Chuaqui-Offermanns, N.; Moir, R. Y.; Norris, A. R. *Can. J. Chem.* **1979,** *57,* 494.
38. Terrier, F.; Halle, J. C.; Simonnin, M. P.; Pouet, M. J. *J. Org. Chem.* **1984,** *49,* 4363.
39. Strauss, M. J.; Renfrow, R. A.; Buncel, E. *J. Am. Chem. Soc.* **1983,** *105,* 2473.
40. Read, R. W.; Spear, R. J.; Norris, W. P. *Aust. J. Chem.* **1984,** *37,* 985.

41. Halle, J. C.; Simonnin, M. P.; Pouet, M. J.; Terrier, F. *Tetrahedron Lett.* **1983,** *24,* 2255.
42. Terrier, F.; Debleds, F.; Halle, J. C.; Simonnin, M. P. *Tetrahedron Lett.* **1982,** *23,* 4079.
43. Halle, J. C.; Terrier, F.; Pouet, M. J.; Simonnin, M. P. *J. Chem. Res., Synop.* **1980,** 360.
44. Renfrow, R. A.; Strauss, M. J.; Cohen, S.; Buncel, E. *Aust. J. Chem.* **1983,** *36,* 1843.
45. Buncel, E.; Renfrow, R. A.; Strauss, M. J. *Can. J. Chem.* **1983,** *61,* 1690.
46. Buncel, E.; Dust, J.; Park, K. T. *Abstracts,* 68th Canadian Chemical Conference, Kingston, Ontario, Canada, June 1985; OR-C3-7.
47. Kornblum, M.; Lurie, A. P. *J. Am. Chem. Soc.* **1959,** *81,* 2705.
48. Boyer, R.; Spencer, E. V.; Wright, G. F. *Can. J. Res., Sect. B.* **1946,** *24,* 200.
49. Drost, P. *Justus Liebigs Ann. Chem.***1899,** *49,* 307.
50. Boulton, A. J.; Clifford, D. P. *J. Chem. Soc.* **1965,** 5414.
51. Leeman, H.; Grandmougin, E. *Chem. Ber.* **1908,** *41,* 1295.

RECEIVED for review October 21, 1985. ACCEPTED JANUARY 24, 1986.

27

Sulfonyl Transfer to Nucleophiles

J. F. King, S. Skonieczny, K. C. Khemani, and J. D. Lock

University of Western Ontario, London, Ontario N6A 5B7, Canada

Sulfonyl-transfer mechanisms are briefly reviewed, and two recent studies described in greater detail. (1) From comparison of the labeled $CH_3SO_3^-$ from the reaction of CH_3SO_2Cl with $(CH_3)_3N$ in a buffered D_2O–organic medium with that from the products of the reactions of $CH_3SO_2N^+(CH_3)_3FSO_3^-$ and $BrCH_2SO_2^-$ under similar conditions, part of the reaction with $(CH_3)_3N$ (~40%) was found to occur by net direct displacement of Cl^- by $(CH_3)_3N$, with most of the remainder by elimination of HCl to give sulfene. (2) Hydrolysis of 2-hydroxyethanesulfonyl chloride is shown to proceed largely by way of β-sultone, which reacts with water or added nucleophiles (Nu^-) to give $HOCH_2CH_2SO_3^-$ and $NuCH_2CH_2SO_3^-$, respectively; the possibility of sulfur–oxygen cleavage of β-sultone by Cl^- is discussed.

SULFONYL TRANSFER TO NUCLEOPHILES may be illustrated by such reactions as (1) sulfonyl chloride hydrolysis ($RSO_2Cl + H_2O \longrightarrow RSO_3H + HCl$), (2) esterification of an alcohol or phenol ($RSO_2X + R'OH \longrightarrow RSO_2OR' + HX$), (3) synthesis of sulfonamides ($RSO_2X + 2R'_2NH \longrightarrow RSO_2NR'_2 + R'_2NH_2 + X^-$), (4) desulfonation of an episulfone (the final stage of the Ramberg–Bäcklund reaction) (episulfone + $OH^- \longrightarrow CH_2{=}CH_2 + HSO_3^-$), (5) the sulfonamide–aminosulfone rearrangement [$C_6H_5NHRSO_2Ar \longrightarrow$ (R'Li) o-$RNHC_6H_4SO_2Ar$], and (6) arene transsulfonylation [$ArSO_2CH_3 + Ar'H \longrightarrow (H^+)\ ArH + Ar'SO_2CH_3$]. Most of these reactions have been well-known for many years, but efforts to elucidate the mechanism of these processes are, however, much more recent and by no means complete. In this chapter, we shall briefly outline current knowledge of the mechanisms of some of these reactions and then present a progress report on two specific topics currently under study in our laboratory.

The older work was summarized by Vizgert (*1*), and aspects of the more recent studies up to about 1978 were critically reviewed by Kice (*2*). Some topics in sulfonyl transfer are not discussed here: intramolecular sulfonyl transfers with sulfonic esters (*3, 4*), episulfone cleavage (*5*), the sul-

0065-2393/87/0215-0385$06.00/0

fonamide–aminosulfone rearrangement (*6–8*), and arene transsulfonylation (*9*).

The earliest mechanistic studies of sulfonyl transfer were concerned primarily with solvolysis in alcoholic or aqueous media and were invariably interpreted in terms of a direct displacement process (*1, 2*):

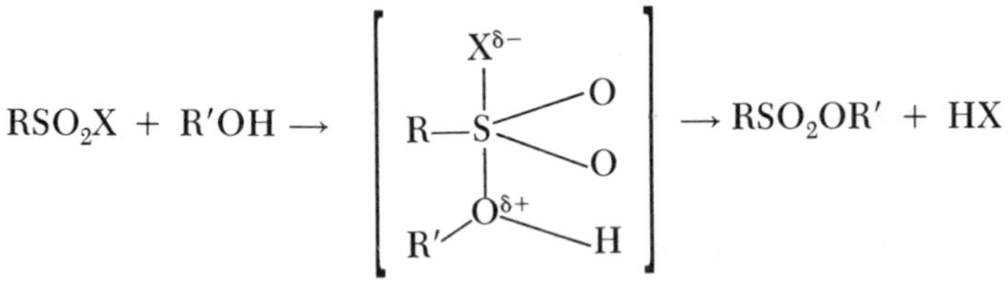

$$RSO_2X + R'OH \rightarrow \left[\begin{array}{c} X^{\delta-} \\ | \\ R{-}S(O)(O) \\ | \\ R'{-}O^{\delta+}{-}H \end{array} \right] \rightarrow RSO_2OR' + HX$$

I

A much discussed question arising in the direct displacement process concerns the timing of bond formation and breakage, that is, whether **I** (or a similar species) corresponds to a transition state or intermediate. A good case for a two-step process proceeding via a pentacoordinated intermediate has been put by Kice (*2*) and supported by the isolation of hypervalent analogues by Perkins and Martin (*10*). Williams and co-workers (*11, 12*), however, argue that in the particular case of displacement of aryloxide from arenesulfonic ester with hydroxide, their evidence is in better agreement with a one-step mechanism.

Such uncatalyzed solvolytic processes are synthetically unimportant; as a procedure for making esters, for example, the reaction is usually highly inefficient because of the tendency of these esters to alkylate an alcohol about as fast as the sulfonyl chloride sulfonylates an alcohol, and so by the time the starting material is consumed, extensive further reaction of the product will have occurred. On the other hand, the promotion of esterification by tertiary amines is a routine synthetic procedure with both arenesulfonyl and alkanesulfonyl species.

For the reaction of arenesulfonyl chlorides with pyridine bases, Rogne (*13, 14*) concluded that the amines function as nucleophilic catalysts, for example, **II**.

$$ArSO_2X + C_5H_5N \longrightarrow [ArSO_2N^+C_5H_5\ X^-] \xrightarrow{ROH} ArSO_2OR + HX$$

II

When the aryl group is replaced by a 1-alkenyl (e.g., vinyl) grouping, nucleophilic catalysis of a different kind is observed (*15*):

$$CH_2{=}CHSO_2Cl + C_5H_5N \rightarrow [C_5H_5N^+CH_2CH{=}SO_2] + Cl^-$$

III

$$[C_5H_5N^+CH_2CH{=}SO_2] + ROH \rightarrow C_5H_5N^+CH_2CH_2SO_2OR$$

III **IV**

$$\rightarrow CH_2{=}CHSO_2OR + C_5H_5NH^+$$

V

In this case, the nucleophile attacks the vinylogous carbon to form the cationic sulfene **III**, which on subsequent reaction gives either the betylate (**IV**, R = alkyl or aryl group) or the simple overall substitution product (**V**).

With alkanesulfonyl halides bearing an α-hydrogen, however, the normal attack of a tertiary amine is evidently at the α-hydrogen with elimination to form the sulfene (**VI**). The key piece of evidence was the observation, in our laboratory (*16*) and that of Truce (*17*), of the monodeuterated product (**VII**) in the presence of the deuterated reagent (e.g., ROD).

$$RCH_2SO_2X + (CH_3CH_2)_3N \rightarrow [RCH{=}SO_2] + (CH_3CH_2)_3NH^+ X^-$$

VI

$$[RCH{=}SO_2] + R'OD \rightarrow RCHDSO_2OR'$$

VII

These experiments, however, were interpreted by the two groups somewhat differently (*18*, *19*). Truce and Campbell (*19*) noted that under their conditions [CH_3SO_2Cl in benzene at room temperature with 1.3 equiv of $(CH_3CH_2)_3N$ and 1.2 equiv of CH_3OD] typically about half of the product was monodeuterated (with the rest undeuterated); these researchers suggested that the monodeuterated material arose from the sulfene (**VI**) and the undeuterated product from the sulfonylammonium salt, $CH_3SO_2N^+(CH_2CH_3)_3\ Cl^-$ (**VIII**), which was formed both (1) by direct displacement of chloride ion by triethylamine and (2) from the sulfene (**VI**), which "collapses with the triethylammonium chloride produced" (*19*). Our conditions (using a large excess of ROD, often as the solvent) led to a large proportion (usually >90%) of the monodeuterated product (**VII**), and our view was that most of the product (>90%) was formed directly from the sulfene (**VI**) with much of the undeuterated material being formed by trapping of the sulfene with return of the protium originally on the sulfonyl chloride (and which had been removed in forming the sulfene). Truce and Campbell regarded this latter process as unlikely, citing control experiments in which no exchange was observed between CH_3OD and preformed crystalline triethylammonium chloride in benzene. In our opinion, however, the possibility of protonation in the sulfene trapping process or of H–D exchange with ROD *prior* to precipitation is not ruled out by the control studies. Truce and Campbell explained their results by invoking participation of triethylammonium ion in an unprecedented reaction (with sulfene); our simple notion that the same species is involved in proton transfer or exchange seems much more attractive to us.

This chapter describes subsequent experiments to clarify the mechanism of reactions of alkanesulfonyl chlorides and related compounds with water, alcohols, and aromatic amines. In the course of these studies, we also noted a third-order term in the rate law and further work on this led to the observation of hydrogen–deuterium multiexchange with small tertiary amines. These results prompted further study, which has proceeded gradually over a number of years. The current state of this work is described.

Mechanism of Reaction of Methanesulfonyl Chloride and Anhydride with Trialkylamines and Water or Alcohols: Multiexchange of Hydrogen with Small Amines

Reaction of methanesulfonyl chloride with excess D_2O and trimethylamine in 1,2-dimethoxyethane (DME) gave a mixture of isotopically labeled methanesulfonate salts of the following composition (*20*, *21*): $CH_3SO_3^-$, 1.8%; $CH_2DSO_3^-$, 25.6%; $CHD_2SO_3^-$, 24.7%; and $CD_3SO_3^-$, 48.0%. Rather more CD_3 product was obtained with quinuclidine or 1,4-diaza[2.2.2]bicyclooctane (DABCO), although successive replacement in the trimethylamine of the methyl groups by ethyl reduced the proportion of CD_3 and CHD_2 products until with triethylamine the product consisted of 9.6% CH_3, 89.8% CH_2D, and 0.5% CHD_2 materials. A reasonable route to these products is shown in Scheme I. In accord with this, we found a number of years ago that sulfonylammonium salts, for example, $CH_3SO_2N^+(CH_3)_3$ FSO_3^-, give a larger proportion of the CD_3 product than the sulfonyl chloride under the same conditions (22). To make a quantitative assignment of the relative importance of the competing pathways in Scheme I, especially of the ratio k_2/k_1, we devised a buffered aqueous–organic medium [K_2CO_3–$KDCO_3$ in D_2O–DME–MeCN (62:7:10) adjusted to the specific pH with DCl] (1) in which reagents and products were soluble and (2) that maintained pseudo-first-order reaction conditions. The reactions of a series of tertiary amines with methanesulfonyl chloride and with the corresponding sulfonyl-ammonium fluorosulfate were carried out, and though in principle the rate constant ratios of all of the competing reactions in Scheme I can be derived from the combined product compositions from the sulfonyl chloride and the sulfonylammonium salt, in practice the complexity of the mechanism and the limited accuracy of the data lead to considerable uncertainty in values so obtained (23).

To generate sulfene without any possibility of simultaneous formation of the sulfonylammonium salt, we turned to the "abnormal route" to sulfene (*17*), specifically that from bromomethanesulfinate anion, $BrCH_2SO_2^-$. This reaction, though not ideal because of complication by a competing displace-

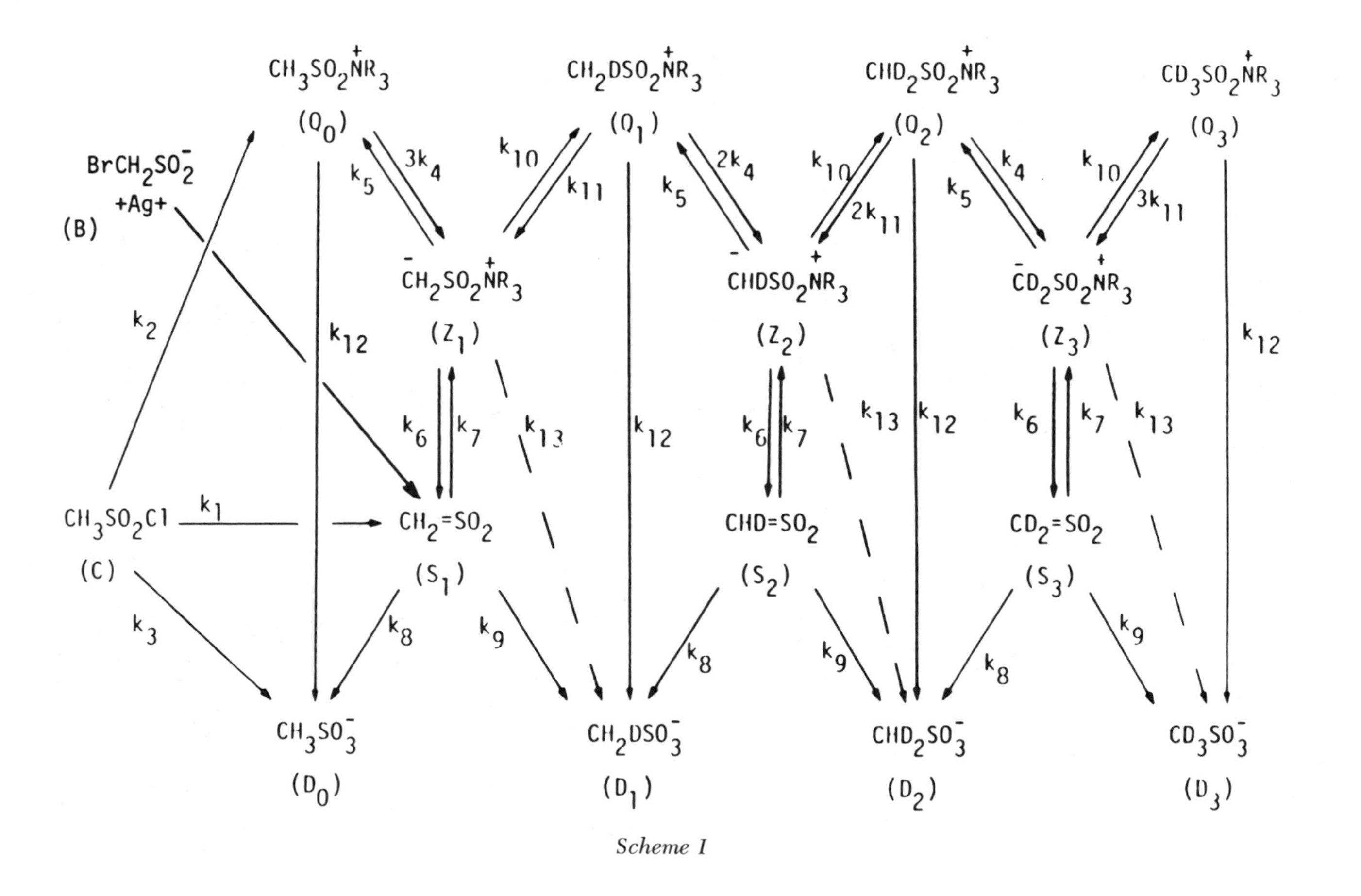

CH3SO2NR3+ (O0)
CH2DSO2NR3+ (O1)
CHD2SO2NR3+ (O2)
CD3SO2NR3+ (O3)
BrCH2SO2- +Ag+ (B)
-CH2SO2NR3+ (Z1)
-CHDSO2NR3+ (Z2)
-CD2SO2NR3+ (Z3)
CH3SO2Cl (C)
CH2=SO2 (S1)
CHD=SO2 (S2)
CD2=SO2 (S3)
CH3SO3- (D0)
CH2DSO3- (D1)
CHD2SO3- (D2)
CD3SO3- (D3)
k1
k2
k3
3k4
2k4
k4
k5
k6
k7
k8
k9
k10
k11
2k11
3k11
k12
k13

Scheme I

ment of Br by $(CH_3)_3N$ to give $(CH_3)_3NCH_2SO_2^-$ (25), is believed, nonetheless, to form methanesulfonate anion only by way of sulfene; the $CH_3SO_3^-$ so obtained shows a small but clearly observable proportion of the CD_3 product (*see* Table I). Simple comparison of the products from the sulfonyl chloride with those of the sulfonylammonium salt the bromomethanesulfinate salts would lead to the conclusion that $k_2 \sim k_1$ in the reaction of CH_3SO_2Cl. This result was placed on a more quantitative basis by deriving analytical expressions relating products to rate constant ratios for a simplified scheme in which proton uptake from the medium was neglected. The parameters so derived were then used in a computer-simulated version of the full scheme, and these parameters were allowed to vary systematically so as to generate a minimum value for the squares of the differences between the found and simulated percentages of the labeled products. In this way, a set of parameters (listed in Table II) was obtained which provided the simulated values shown in Table I. The agreement of the found and simulated numbers is well

Table I. Experimental and Simulated Values for the Percentages of the Labeled Methanesulfonate Anions

Substrate	*CH_3*		*CH_2D*		*CHD_2*		*CD_3*	
	Found[a]	*Simulated*[b]	*Found*[a]	*Simulated*[b]	*Found*[a]	*Simulated*[b]	*Found*[a]	*Simulated*[b]
$CH_3SO_2N^+(CH_3)_3\ FSO_3^-$	1	1	4	4	8	8	88	88
$BrCH_2SO_2^-$	8	8	79	79	2	2	11	11
CH_3SO_2Cl	8	8	47	46	3	5	42	41
$CH_3SO_2OSO_2CH_3$	21	20	12	12	6	8	62	60

NOTE: The reaction was carried out by addition of the substrate (4–5 mmol) in CH_3CN (10 mL) to a solution of trimethylamine (80 mmol) in D_2O (~62 mL) plus DME (7 mL) with K_2CO_3–$KDCO_3$ buffer (1 M) adjusted to pH 10.2 with DCl, at 25 °C.

[a]Experimental value from mass or ^{13}C-NMR spectra.

[b]Obtained by computer simulation using parameters initially calculated from analytical expressions derived from a simplified scheme (ignoring k_5 and k_8 terms) and then modified fitting (least squares) to a computer simulation of the full reaction scheme. Values for k_{12}/k_4, k_9/k_7, k_{11}/k_4, and k_6/k_{10} were obtained in this way from the sulfonylammonium salt and bromomethanesulfinate results; these parameters were held constant while k_2/k_1 and k_3/k_1 were optimized for the reactions of the sulfonyl chloride and anhydride.

within experimental uncertainty and is therefore consistent with the mechanism in Scheme I. Though high accuracy is not claimed for the parameters so obtained, the values in Table II (which include also those for experiments with diethylmethylamine) show a satisfying consistency. For example, as expected from simple considerations of steric accessibility, for the trimethylamine reaction k_2/k_1 is larger and both k_9/k_7 and k_6/k_{10} are smaller than those for diethylmethylamine. Interestingly, direct attack on sulfur is the *major* initial reaction of the anhydride (~70% of the total) with direct sulfene formation contributing only a minor part (~11%). Very recent experiments at

Table II. Optimized Parameters for Hydrolyses of Methanesulfonyl Chloride and Methanesulfonic Anhydride

Reaction[a]	k_2/k_1	k_3/k_1	k_{12}/k_4	k_9/k_7	$k_6/k_{10}0$
$CH_3SO_2Cl + (CH_3)_3N$	0.77	0.1	0.013	5.9	0.029
$CH_3SO_2Cl + (CH_3CH_2)_2NCH_3$	0.46	0.004	0.02	28.	0.33
$CH_3SO_2OSO_2CH_3 + (CH_3)_3N$	6.5	1.8	0.013	5.9	0.029

NOTE: The parameters were determined as described in footnote *b* of Table I. All *k* values except k_6 are pseudo-first-order rate constants and therefore all ratios refer to the specific concentrations used.

[a]Reaction conditions as described in the note of Table I except for the experiment with $(CH_3CH_2)_2NCH_3$, which was carried out at pH 10.7.

lower pH raise the interesting possibility of significant reversal of the formation of the sulfonylammonium chloride. In this event, the various k_2/k_1 ratios as found in this work may not reflect simply the relative rates of reaction of the base at sulfur versus hydrogen but may well require more complex interpretation; further work is in progress.

We have not prepared $CH_3SO_2N^+(CH_3CH_2)_3X^-$ and therefore cannot determine k_2/k_1 for triethylamine in the same way. Extrapolating from the diethylmethylamine results indicates that the nucleophilic catalysis route would only be a minor, but not necessarily negligible, pathway. Because we saw no sign of $CD_3SO_3^-$ with triethylamine, we may conclude that the sulfonylammonium salt, if formed, goes to the sulfene, that is, k_6/k_{10} is fairly large.

We also have carried out a parallel series of experiments with CH_3OD in organic solvents such as benzene and methylene chloride and find a very similar pattern, but limitations of space preclude further description here.

Returning to the Truce–Campbell scheme, our more recent work, including that using pure sulfonylammonium salts as starting materials, indicates that some reactions ascribed by these authors to CH_3SO_2Cl and $CH_3SO_2N^+(CH_3CH_2)_3$ do not occur. We suggest that the most likely mechanism for the reaction of methanesulfonyl chloride with triethylamine and D_2O or CH_3OD is one proceeding almost entirely via sulfene that has been formed largely (at least 80%) by a direct elimination. In the light of our previous observation that ethanesulfonyl chloride gave only 13% $CH_3CD_2SO_3^-$ with DABCO–D_2O, apparently the direct elimination process is the normal route for the reaction between a typical tertiary amine and most alkanesulfonyl chlorides bearing an α-hydrogen.

One final note concerning Truce and Campbell's mechanism. Their proposal apparently arose from a belief that the hydrogen removed from the sulfonyl chloride (CH_3SO_2Cl in this case) could not become incorporated in the CH_3SO_2 group of the $CH_3SO_2OCH_3$ product. The following experiment shows, however, that such an exchange can take place. Reaction of an equimolar mixture of CH_3SO_2Cl and $C_6H_5CD_2SO_2Cl$ (the latter compound being much more reactive) with triethylamine and excess methanol in benzene gave methyl methanesulfonate containing ~12% of $CH_2DSO_2OCH_3$; clearly, some of the hydrogen from one sulfonyl chloride is capable of appearing in the product derived from the other sulfonyl chloride.

Hydrolysis of 2-Hydroxyethanesulfonyl Chloride

We recently reported (26) on the synthesis and some aspects of the reactivity of 2-hydroxyethanesulfonyl chloride (9) (Scheme II). Our early experiments gave not only the expected 2-hydroxyethanesulfonate anion (**XI**) but also other materials as well: the reaction in water yielded (except at the lowest initial concentrations) a clearly detectable amount of 2-chloroethanesulfonate (**XII**), which increased with addition of chloride ion to the reaction mixture. When the reaction was promoted by a tertiary amine, the major product was the betaine (**XIII**). These products are conveniently accounted for by assuming that β-sultone (**X**) is formed in the major path from **IX** as in Scheme II. β-Sultone itself (1,2-oxathietane 2,2-dioxide, X) is unknown though simple substituted β-sultones were first shown to be formed as rather fragile species on careful sulfonation of olefins by Bordwell and co-workers (27–29). Those β-sultones with electron-withdrawing groups on position 4 of the ring may be made by the formal cyclization of sulfene with the appropriate carbonyl precursor (e.g., chloral) and are stable, readily isolable materials (for a summary of the literature, *see* reference 30). The reactivity of β-sultones lacking electron-withdrawing groups, however, would appear to increase with diminishing substitution, and **X** would be expected to be very reactive indeed. Hence, more insight into its chemistry might well be obtained from a study of the reactions of **IX** than by possibly fruitless attempts at isolation.

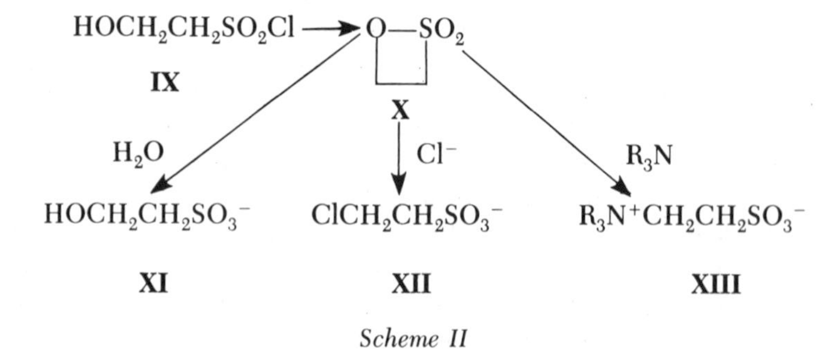

Scheme II

The rates of hydrolysis were conveniently measured in water maintained at constant ionic strength (0.3 M, with $NaClO_4$) at 25.0 °C by the pH-stat technique as described elsewhere (*15*). These data gave the pH–rate profile shown in Figure 1 and which is described by the rate law $k_{obsd} = 2.40 \times 10^{-4} + (5.38 \times 10^6)[OH^-]$. The product of the reaction under kinetic conditions ($c_0 = 10^{-3}$–10^{-5} M) was almost entirely **XI** with small amounts (~2–5%) of **XII**. Addition of sodium chloride to the reaction mixture led to more **XII**. In the presence of sodium thiocyanate **IX** gave a mixture of **XI** and $NSCCH_2CH_2SO_3^-$ (**XIV**) with the concentration dependence shown (for reactions at two different pH values) in Figure 2. In addition, k_{obsd} was found to be independent of the thiocyanate concentration; that finding is, of course, the classic rate–product criterion for the presence of an intermediate formed in a rate-determining reaction and consumed in a fast step. From the fact that each of the lines in Figure 2 tends to approach a constant limiting value, we conclude that in both the uncatalyzed and hydroxide-induced processes another route, in addition to the reaction proceeding by way of the intermediate that reacts with thiocyanate, exists that is unaffected by thiocyanate.

The reaction in the presence of added chloride ion showed two further features of interest: (1) a small but unmistakable rate suppression (*see* Figure 3 and the dashed curve in Figure 1) and (b) an informative pH–product ratio profile (Figure 4). These observations led to the mechanism shown in Scheme III (A and B indicate acid and base, respectively.)

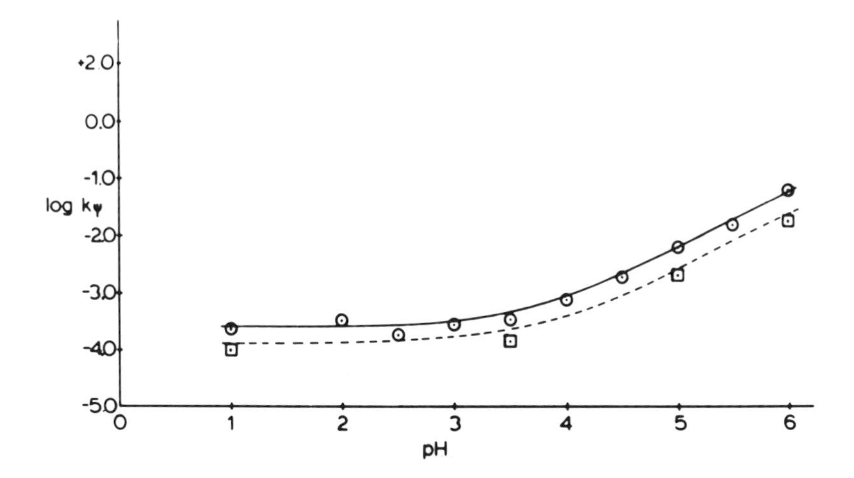

Figure 1. pH–Rate profile for hydrolysis of 2-hydroxyethanesulfonyl chloride (**IX**) *in water at 25.0 °C; circles (⊙) refer to reaction in a medium maintained at ionic strength 0.3 M with* $NaClO_4$*; squares (⊡) refer to the reaction in 1 M NaCl (no* $NaClO_4$*).*

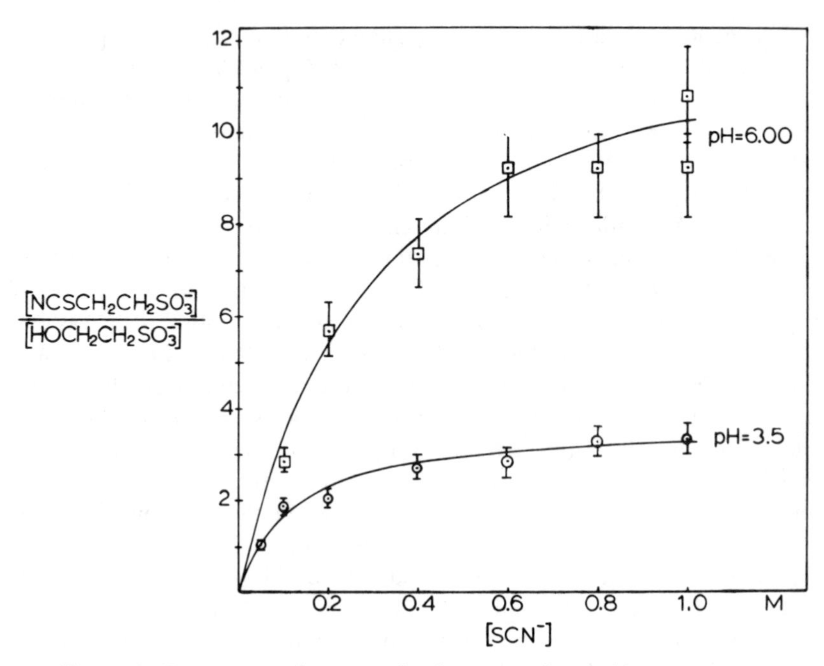

*Figure 2. Variation in the ratio of 2-thiocyanoethanesulfonate to 2-hydroxyethanesulfonate (**XIV:XI**) with concentration of thiocyanate anion in the hydrolysis of 2-hydroxyethanesulfonyl chloride; circles (○) refer to the reaction at pH 3.50, squares (□) to the reaction at pH 6.00.*

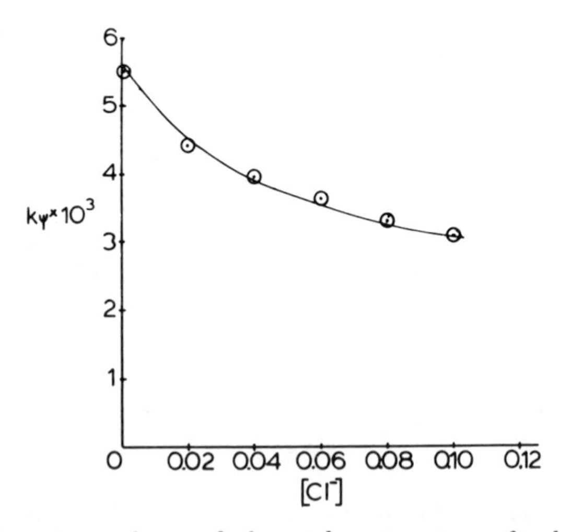

*Figure 3. Variation in the pseudo-first-order rate constant for the reaction of 2-hydroxyethanesulfonyl chloride (**IX**) with aqueous sodium chloride solution (pH 5.00 and ionic strength was 0.3 M with $NaCl0_4$, 25.0 °C).*

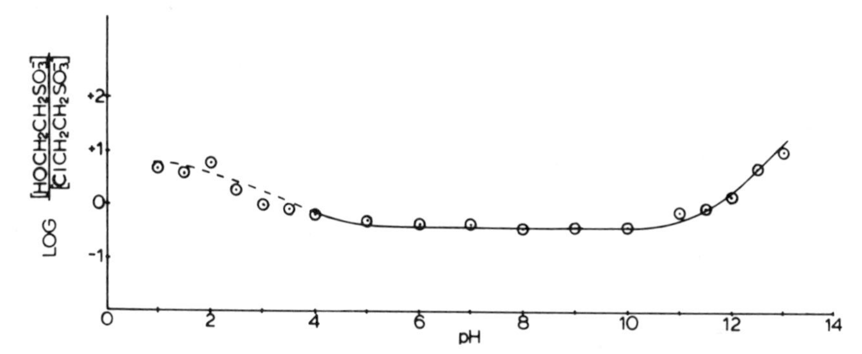

*Figure 4. pH–product ratio (14:11) profile for the hydrolysis of 2-hydroxyethanesulfonyl chloride (**IX**) in the presence of NaCl (1 M) at 25.0 °C.*

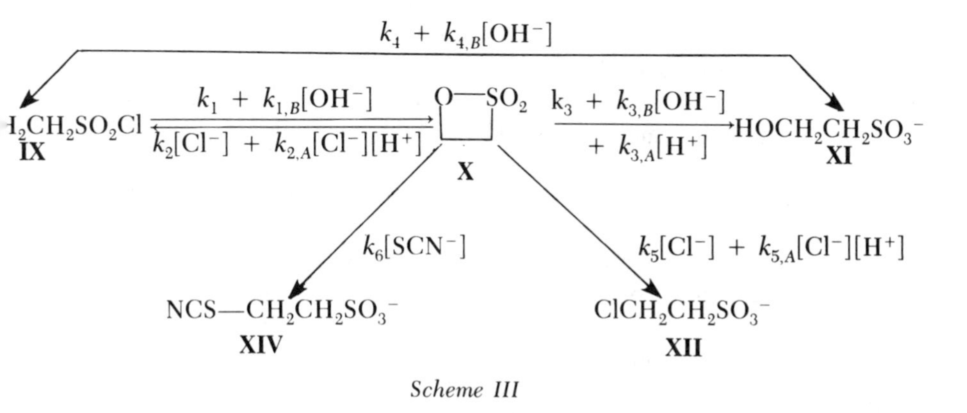

Scheme III

Observation of the reaction course by ^{1}H-NMR spectrometry showed no sign of any intermediates; we therefore applied the steady-state assumption and obtained the following expressions:

$$k_{obsd} = \frac{a(k_1 + k_{1,B}[OH^-]) + (a + k_2[Cl^-]/k_3 + k_{2,A}[H^+][Cl^-]/k_3)(k_4 + k_{4,B}[OH^-])}{a + k_2[Cl^-]/k_3 + k_{2,A}[H^+][Cl^-]/k_3}$$

where $a = 1 + k_{3,A}[H^+]/k_3 + k_{3,B}[OH^-]/k_3 + k_5[Cl^-]/k_3 + k_{5,A}[H^+][Cl^-]/k_3$ and

$$\frac{[\mathbf{XII}]}{[\mathbf{XI}]} = \frac{b(k_5[Cl^-]k_3 + k_{5,A}[Cl^-][H^+]/k_3}{bc + (k_4 + k_{4,B}[OH^-])\{(k_{2,A} + k_{5,A})[Cl^-][H^+]/k_3 + (k_2 + k_5)[Cl^-]/k_3 + c\}}$$

where $b = k_1 + k_{1,B}[OH^-]$ and $c = k_{3,B}[OH^-]/k_3 + k_{3,A}[H^+]/k_3 + 1$. From these equations with the data shown in Figures 1–4, we obtained the

2-Hydroxyethanesulfonyl Chloride Hydrolysis: Estimated Rate Constants and Ratios

$k_1 = 1.62 \times 10^{-4}\ s^{-1}$
$k_{1,B} = 5.0 \times 10^{6}\ M^{-1}\ s^{-1}$
$k_4 = 7.8 \times 10^{-5}\ s^{-1}$
$k_{4,B} = 3.8 \times 10^{5}\ M^{-1}\ s^{-1}$
$k_2/k_3 = 20$
$k_5/k_3 = 8$
$k_6/k_3 = 50$
$k_{3,B}/k_3 = 1.1 \times 10^{3}\ M^{-1}$
$(k_{2,A}/k_3 = 6 \times 10^{3}\ M^{-1})$
$(k_{3,A}/k_3 = 6 \times 10^{2}\ M^{-1})$
$(k_{5,A}/k_3 = 6 \times 10^{2}\ M^{-1})$

NOTE: The ratios for the acid-catalyzed terms are approximate only and are given in parentheses.

approximate (and preliminary) rate constants and ratios given in the Box; these values in turn were used to obtain the actual lines shown in Figures 1–4. Except perhaps for the low-pH region (note the dashed line in Figure 4 and the footnote to the Box), the agreement between calculation and experiment is acceptable, and we conclude that the mechanism in Scheme III is consistent with observation.

Of the features that emerge from an inspection of this mechanism and the contents of the Box, perhaps the most remarkable is that chloride would appear to react with β-sultone faster at the *sulfur* than at the carbon atom. This finding is noteworthy for two reasons. (1) The usual reaction of nucleophiles with alkyl esters of sulfonic acids—even those of hindered alkyl groups—is *alkyl* not sulfonyl transfer. (2) With β-propiolactone, the carbonyl analogue of **XII**, the usual reactivity pattern is also altered, but in precisely the opposite way, that is, from the usual carbonyl–oxygen cleavage to alkyl–oxygen breaking of the β-lactone (*31*, *32*). To be sure, reaction of the trichloromethyl-substituted β-sultone **XV** reportedly gives sulfonamides (*33*), but the S_N2 reactivity of a carbon bearing a trichloromethyl group is well-known to be dramatically reduced and no special effect specifically favoring attack at sulfur need be invoked for this reaction. In the present study, however, alkyl transfer would be expected to be, at most, only slightly inhibited by the β-sulfonyl group (*34*). If indeed the sulfur–oxygen cleavage by Cl^- is real (and the rate suppression not merely due to a special salt effect, for example), we must look for a factor that accelerates sulfonyl relative to alkyl transfer. Inspection of molecular models and looking at the internuclear distances reported for the isolable analogue of the supposed intermediate or

O—SO_2
Cl_3C—⌊__⌋

XV

transition state for substitution at the sulfonyl group (*19*) suggest that the energy of the transition state is probably significantly raised by nucleophile–O_s and nucleofuge–O_s nonbonding interactions (where O_s refers to a sulfonal oxygen). Incorporation of the sulfur atom in a four-membered ring would be expected to lead to a distortion of the usual 180° angle of the Nu–S–Lg system such that the Nu · · · O_s and O_r · · · O_s distances (where O_r refers to the ring oxygen) would be increased and the nonbonding interactions thereby diminished. Reasonably, smaller nonbonding interactions can be expected in the attack at an ordinary methylene group and the analogous effect for attack on carbon in the four-membered ring would be smaller. We must caution, however, that sulfur–oxygen cleavage of **X** by Cl^- (or other nucleophiles) is only a hypothesis at this stage, and further experiments are being carried out.

Acknowledgments

We thank S. K. Wong and Doug Hairsine for the mass spectra and J. B. Stothers and Valerie Richardson for the ^{13}C-NMR spectra. This work was supported by grants from the Natural Sciences and Engineering Research Council of Canada.

Literature Cited

1. Vizgert, R. V. *Russ. Chem. Rev. (Engl. Transl.)* **1963,** *32,* 1–20.
2. Kice, J. L. *Adv. Phys. Org. Chem.* **1980,** *17,* 65–181 (especially pp 156–172).
3. Andersen, K. K.; Gowda, G.; Jewell, L.; McGraw, P.; Phillips, B. T. *J. Org. Chem.* **1982,** *47,* 1884–1889.
4. Thea, S.; Guanti, G.; Hopkins, A. R.; Williams, A. *J. Org. Chem.* **1985,** *50,* 3336–3341.
5. King, J. F.; Hillhouse, J. H.; Khemani. K. C. *Can. J. Chem.* **1985,** *63,* 1–5.
6. Hellwinkel, D.; Supp, M. *Angew. Chem., Int. Ed. Engl.* **1974,** *13,* 270.
7. Shafer, S. J.; Closson, W. D. *J. Org. Chem.* **1975,** *40,* 889–892.
8. Hellwinkel, D.; Lenz, R. *Chem. Ber.* **1985,** *118,* 66–85.
9. El-Khawaga, A. M.; Roberts, R. M. *J. Org. Chem.* **1985,** *50,* 3334–3336.
10. Perkins, C. W.; Martin, J. C. *J. Am. Chem. Soc.* **1983,** *105,* 1377–1379.
11. Deacon, T.; Farrar, C. R.; Sikkel, B. J.; Williams, A. *J. Am. Chem. Soc.* **1978,** *100,* 2525–2534.
12. Thea, S.; Williams, A. *J. Chem. Soc., Perkin Trans. 2* **1981,** 72–77.
13. Rogne, O. *J. Chem. Soc. B* **1970,** 727–730.
14. Rogne, O. *J. Chem. Soc. B* **1971,** 1334–1337.
15. King, J. F.; Hillhouse, J. H.; Skonieczny, S. *Can. J. Chem.* **1984,** *62,* 1977–1995.
16. King, J. F.; Durst, T. *J. Am. Chem. Soc.* **1964,** *86,* 287–288.
17. Truce, W. E.; Campbell, R. W.; Norell, J. R. *J. Am. Chem. Soc.* **1964,** *86,* 288.
18. King, J. F.; Durst, T. *J. Am. Chem. Soc.* **1965,** *87,* 5684–5692.
19. Truce, W. E.; Campbell, R. W. *J. Am. Chem. Soc.* **1966,** *88,* 3599–3604.
20. King, J. F.; Luinstra, E. A.; Harding, D. R. K. *J. Chem. Soc., Chem. Commun.* **1972,** 1313–1315.

21. Luinstra, E. A. Ph.D. Thesis, University of Western Ontario, London, Ontario, Canada, 1971.
22. King, J. F.; du Manoir, J. R. *J. Am. Chem. Soc.* **1975,** *97,* 2566–2567.
23. Lock, J. D. Ph.D. Thesis, University of Western Ontario, London, Ontario, Canada, 1984.
24. King, J. F.; Beatson, R. P.; Buchshriber, J. M. *Can. J. Chem.* **1977,** *55,* 2323–2330.
25. King, J. F.; Skonieczny, S. *Phosphorus Sulfur* **1985,** *25,* 11–20.
26. King, J. F.; Hillhouse, J. H. *Can. J. Chem.* **1983,** *61,* 1583–1593.
27. Bordwell, F. G.; Peterson, M. L.; Rondestvedt, C. S., Jr. *J. Am. Chem. Soc.* **1954,** *76,* 3945–3950.
28. Bordwell, F. G.; Peterson, M. L. *J. Am. Chem. Soc.* **1954,** *76,* 3952–3956.
29. Bordwell, F. G.; Peterson, M. L. *J. Am. Chem. Soc.* **1954,** *76,* 3957–3961.
30. Timberlake, J. W.; Elder, E. S. In *Comprehensive Heterocyclic Chemistry;* Lwowski, W., Ed.; Pergamon: Oxford and New York, 1984; Vol. 7, Part 5, pp 449–489.
31. Cowdrey, W. A.; Hughes, E. D.; Ingold, C. K.; Masterman, S.; Scott, A. D. *J. Chem. Soc.* **1937,** 1252–1271.
32. Bartlett, P. D.; Rylander, P. N. *J. Am. Chem. Soc.* **1951,** *73,* 4275–4280, and references cited therein.
33. Hanefeld, W.; Kluck, D. *Arch. Pharm. (Weinheim, Ger.)* **1978,** *311,* 698–703.
34. Bordwell, F. G.; Brannen, W. T., Jr. *J. Am. Chem. Soc.* **1964,** *86,* 4645–4650.

RECEIVED for review October 21, 1985. ACCEPTED January 28, 1986.

28

Nucleophilicity in Reactions at a Vinylic Carbon

Zvi Rappoport

Department of Organic Chemistry, The Hebrew University, Jerusalem 91904, Israel

The probes and methods for determining the relative nucleophilicities of nucleophiles toward electrophilic olefins and toward vinyl cations were examined. Literature data were used in an attempt to construct a substrate-independent nucleophilicity scale toward vinylic carbon. The nucleophilicities are found to be dependent on electronic, steric, and symbiotic effects, and limited series obeyed a "constant selectivity", a "reactivity–selectivity", or a dual-parameter linear free-energy relationship. The conclusion made was that because of different blends of the effects, the construction of a substrate-independent nucleophilicity scale was impossible at present, but an approximate scale was presented. In nucleophilic reactions on relatively long lived vinyl cations, the steric effects predominate, but at constant steric effects, reactivity–selectivity relationships were found for very short series of substrates. Additional data are required for constructing more reliable nucleophilicity scales toward neutral and positively charged vinylic carbons.

NUCLEOPHILICITY TOWARD VINYLIC CARBON is discussed in this review. This review is not intended to be comprehensive and it avoids overlap with reviews of Bernasconi and Hoz in the present volume, which are related to this topic. We divide our discussion into two parts. In the first, we will discuss nucleophilicity in processes where bond formation between the nucleophile (Nu) and the vinylic carbon of a neutral electrophilic olefin is rate-determining. In the second, we will present the limited data related to nucleophilicity toward vinyl cations.

Nucleophilicity in Reactions of Electrophilic Olefins

Scheme I presents several nucleophilic reactions on electrophilic olefins **1** that lead to different processes and types of products (*1–3*). Y or Y′ are

0065-2393/87/0215-0399$06.50/0

carbanion-stabilizing groups. R^1 is a substituent, X may be another substituent or a leaving group, and the nucleophile can be anionic as shown in Scheme I, can be neutral (e.g., RNH_2), or can carry a leaving group (e.g., ClO^-). The basic assumption is that the nucleophilic attack is rate-determining and it generates the carbanion **2** when the nucleophile is negatively charged or the corresponding zwitterion when the nucleophile is neutral. The fate of **2** depends on the system (*1*). Addition of proton can lead to an adduct in a Michael-type reaction. Intramolecular rotation around the central C–C bond followed by expulsion of the nucleophile can lead to an $E \leftrightarrows Z$ isomerization. When X is a leaving group, its expulsion will give substitution (*2–3*), and if the nucleophile carries a leaving group X′, it may be expelled with the formation of a (usually small) ring.

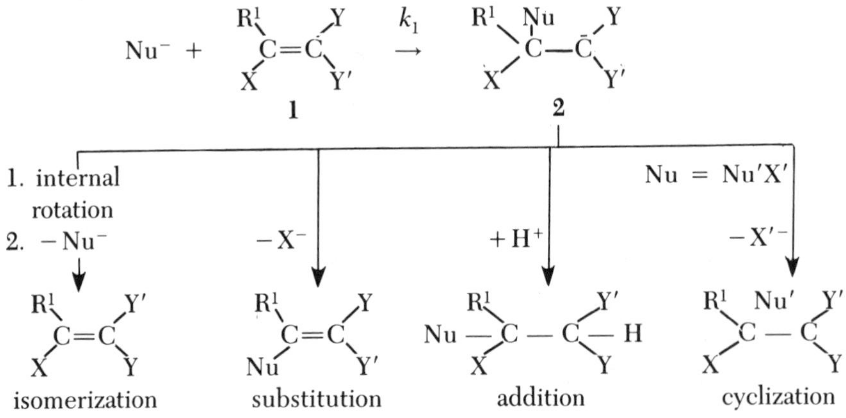

$$Nu^- + \underset{\mathbf{1}}{(R^1)(X)C{=}C(Y)(Y')} \xrightarrow{k_1} \underset{\mathbf{2}}{(R^1)(X)(Nu)C{-}\bar{C}(Y)(Y')}$$

From **2**:

1. internal rotation 2. $-Nu^-$ → $(R^1)(X)C{=}C(Y')(Y)$ — isomerization

$-X^-$ → $(R^1)(Nu)C{=}C(Y)(Y')$ — substitution

$+H^+$ → $Nu{-}C(R^1)(X){-}C(Y')(Y){-}H$ — addition

Nu = Nu′X′; $-X'^-$ → $(R^1)(X)C(Nu')C(Y')(Y)$ (ring) — cyclization

Scheme I

A priori, each of these processes, as well as others (*1*), can serve as a probe for evaluating the nucleophilicity order of a series of nucleophiles. However, several practical and mechanistic problems should be recognized in trying to chose a probe process or in assembling literature data in an attempt to construct a nucleophilicity scale.

First, the process may not involve a nucleophilic attack on the vinylic carbon. This fact is well recognized in vinylic substitution (*2, 3*), and a typical example is the substitution of (*E*)- and (*Z*)-β-halovinyl sulfones (**3** and **4**; X = Cl or Br) by PhS^- and MeO^- in MeOH (*4–6*). Both reactions of both substrates are of a second order and give retention of configuration, and the "element effects" k_{Br}/k_{Cl} are 2.3 (*E*) and 2.2 (*Z*) with PhS^- and 0.84 (*E*) with MeO^-, values that are consistent with rate-determining nucleophilic attack on the vinylic carbon (*2*). However, for the Z isomer, k_{Br}/k_{Cl} with MeO^- is 185, and because α-hydrogen exchange is rapid under the substitution conditions, the reaction of the Z-bromide probably proceeds via elimination–addition. This finding would not be recognized if the only mechanistic information is the reaction order and the stereochemistry and could lead to a distortion in a nucleophilicity scale based on **4**, X = Br.

$$\underset{\mathbf{3}}{TolSO_2(H)C{=}C(H)X} \qquad \underset{\mathbf{4}}{TolSO_2(H)C{=}C(X)H}$$

A different problem exists if k_1 is not the rate-determining step. This situation is sometimes easily recognized by the kinetics. A large number of nucleophilic additions (*7, 8*), isomerizations (*9, 10*) and substitutions (*11, 12*) by amines are of a kinetic order higher than that in the amine, and proton transfer in a step following the initial nucleophilic attack is probably the rate-determining step. However, this result is not always revealed by the kinetics.

An example is the use of $E \leftrightarrows Z$ isomerization as a nucleophilicity probe. This process has practical advantages because it does not require the characterization of different products for each nucleophile, and the thermodynamics of the process is independent of the nucleophile. Table I gives the kinetic parameters for the amine-catalyzed (*Z*)-**5** to (*E*)-**5** isomerization (equation 1) (*13, 14*). The k_2 values are easily interpreted in terms of electronic and steric effects on a rate-determining nucleophilic attack. However, the very low $\Delta H^{\ddagger}$ values suggest that the observed rate constant is not k_1 but a more complex expression, and the internal rotation step k_{rot} may be rate-determining. In view of this, this process should not be used as a probe. Very low activation enthalpies in several vinylic substitutions [e.g., $\Delta H^{\ddagger}$ = 0.8–2.0 kcal mol^{-1} for the reaction of the para position of *N,N*-dialkylanilines with tricyanovinyl chloride in chloroform (*15*)] may also indicate a composite rate constant.

$$R_2NH + \underset{(Z)\text{-}\mathbf{5}}{o\text{-}MeOC_6H_4(H)C{=}C(CO_2Et)CN} \underset{k_{-1}}{\overset{k_1}{\rightleftharpoons}} o\text{-}MeOC_6H_4(H)(R_2\overset{+}{N}H)C{-}\overset{-}{C}(CO_2Et)CN \overset{k_{rot}}{\rightleftharpoons}$$

$$o\text{-}MeOC_6H_4(H)(R_2\overset{+}{N}H)C{-}\overset{-}{C}(CN)CO_2Et \underset{k'_1}{\overset{k'_1}{\rightleftharpoons}} \underset{(E)\text{-}\mathbf{5}}{o\text{-}MeOC_6H_4(H)C{=}C(CN)CO_2Et} + R_2NH \qquad (1)$$

Equation 1

Finally, whereas most vinylic substitutions proceed via Scheme I, the nucleophilic attack and leaving-group expulsion may be concerted in rela-

Table I. Kinetic Parameters for the (*Z*)-5 ⇌ (*E*)–5 Isomerization in Benzene at 40 °C (*14*)

Parameter	Bu_2NH	i-Bu_2NH	Bu_3N	i-Pr_2NH	C_5H_5N	*2,6-Lutidine*
10^2k_2 (L mol^{-1} s^{-1})	750	149	10.6	5.8	1.1	0.09
$\Delta H^{\ddagger}$ (kcal mol^{-1})	4.4	2.2	2.1	4.0	2.5	2.6
$-\Delta S^{\ddagger}$ (eu)	43	53	58	54	66	63

tively unreactive systems (*16*). The nucleophilicity scales for processes that involve and that do not involve a C–X bond cleavage in the rate-determining step should not necessarily be identical, and data for processes of Scheme I and for a suspected concerted vinylic substitution should not be combined.

Taking into account these reservations (as much as possible), we searched the literature to answer the following questions: (1) Is there a substrate-independent nucleophilic scale toward vinylic carbon? (2) If not, are different scales applicable for addition and substitution? (3) What is the role of electronic and steric effects and hard–soft interactions on the nucleophilic order? Because of the lack of space, we will mention only briefly the solvent dependence of the nucleophilicity.

In a search for a nucleophilicity scale, correlations of nucleophilic (*17*, *18*) reactivities in terms of the Ritchie's "constant selectivity" N_+ scale, the Swain–Scott scale (*19*), or other scales should be attempted. A remarkable correlation with the N_+ scale was reported recently by Hoz and Spiezman (*20*, *21*). A plot of log *k* for the reaction of the fluorenylidene derivative **6** versus N_+, which includes eight nucleophiles covering 11 orders of magnitude in reactivity (slower Nu = MeOH; faster Nu = N_3^- in Me_2SO), was linear. Fewer nucleophiles gave a similar plot for **7**, whereas **8** gave substitution of the nitro group by three nucleophiles and addition to C-9 by two nucleophiles. MeO^- in MeOH gave reactions at both positions. The slopes for **6** and **7** were 1.23 and 1.29, and the conclusion is that a selectivity parameter should be incorporated in Ritchie's equation. In spite of the position-dependent selectivity and the nonunity slope, the obedience to the N_+ scale for **6** and **7** is remarkable. An explanation in terms of an extensive electron transfer from the nucleophiles to the low lowest unoccupied molecular orbital (LUMO) substrates (e.g., **6**) in the transition state was offered (*20*, *21*).

Few other attempts to correlate data with a constant selectivity scale were less successful. Ritchie studied the substitution of *trans*-3-methoxy- (**9**) and *trans*-3-(methylthio)acrylophenones (**10**) with nucleophiles (*22*). Ratios of nucleophilicities for two nucleophiles toward **9** and **10** were not constant, for example, k_{OH^-}/k_{MeONH_2} = *22 and 0.3, respectively, whereas a ratio of 7.4 is predicted from the* ΔN^+. Correlations with N_+ were not discussed, and we found them to be nonlinear.

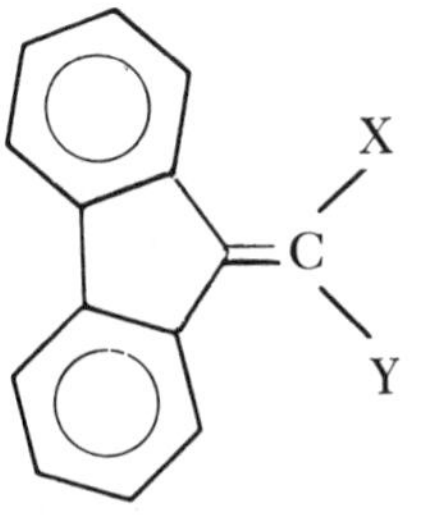

X
C
Y

6, X = Y = NO_2
7, X = Y = CN
8, X = H; Y = NO_2

MeOCH=CHCOPh (**9**) MeSCH=CHCOPh (**10**) MeOCH=CHCOC$_6$H$_4$OMe–*p* (**11**)

The log k values were correlated with two other sets of data. When plotted against the log k for reaction of the [(*p*-dimethylamino)-phenyl]tropylium ion with nucleophiles, two approximately parallel lines, one for reactions in MeOH and one for reactions with water, were obtained. A plot of log k for the reaction of **9** versus log k for the reaction of 2,4-dinitrophenyl acetate with several nucleophiles was linear; thus, similarities existed in the transition states for nucleophilic attack on activated vinylic and aromatic carbon.

The substitution of the analogue **11** with imidazole, OH^-, *n*-BuNH_2; and ethyl glycinate gave a linear log k versus N_+ plot with a slope of 1.02 ± 0.23 (*23*). However, other nucleophiles, morpholine, for example, deviated from this plot. The possibility was raised that the N_+ treatment is invalid for these systems.

If the scale applies to another highly activated system, benzylidene-Meldrum acid, **12**, then the slope calculated for the pair of nucleophiles PhO^- and OH^- is higher than unity, because $\log (k_{PhO^-}/k_{OH^-}) = 1.39$ whereas $\Delta N^+ = 0.85$, but the scale is lower than unity when calculated for the OH^-–N_2O and PhO^-–H_2O pairs (*24*).

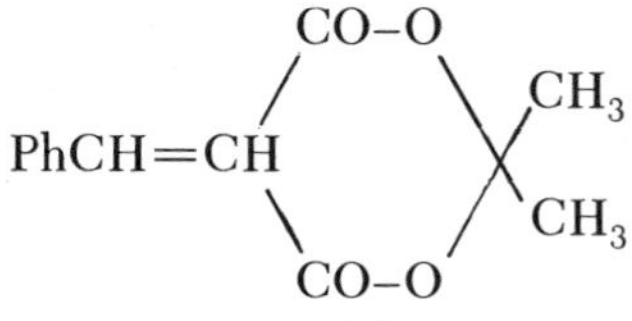

CO–O
PhCH=CH C(CH$_3$)$_2$
CO–O
12

An extensive work by Friedman and co-workers (*25–27*) on the addition of amino acids and peptides to singly activated electrophilic olefins serves as an additional example for the operation of a constant selectivity relationship for closely related nucleophiles and substrates. The active nucleophile is the amino acid anion (A^-) (equation 2; R = H or Me; Y = CN, CO_2Me, or $CONH_2$) and plots of log k_{A-} versus the pK_a of the acid are linear and parallel for addition to several RCH=CHY systems (β_{Nu} = 0.43) when the steric environments of the nucleophilic site are similar (*26*). For a few thio amino acids, β_{Nu} = 0.45 (*27*). Parallel lines were obtained in the addition of nucleophiles to the same substrate for amino groups attached to primary, secondary, and tertiary carbons (*25*). Three conclusions emerge from the data. (1) A nearly constant selectivity of pairs of amino nucleophiles or amino and thio nucleophiles of the same steric environment exist in their addition to different, but sterically similar, olefins. (2) Brønsted plots with $\beta_{Nu} < 0.5$ reflect the response to basicity at a constant steric environment. (3) Steric effects decrease the reactivity of crowded nucleophiles.

$$^{-}OOCCH_2CH_2NH_2 + RCH{=}CHY \longrightarrow {}^{-}OOCCH_2CH_2\overset{+}{N}H_2CH_2\bar{C}HY \underset{}{\overset{-H^+}{\rightleftharpoons}} {}^{-}OOCCH_2CH_2NHCH_2CH_2Y \quad (2)$$

In the majority of reactions studied, a constant selectivity relationship does not apply. In the addition of morpholine, glycine, and MeO^- to CH_2=CHX systems, X=$PO(OEt)_3$, $CONH_2$, CN, CO_2Me, SO_2Me, COMe, COPh, and CHO plots of log k versus pK_a (MeX) are linear but not parallel (*28*); thus, the reactivity ratios for each pair of nucleophiles are not constant.

The reactivity ratios of several pairs of nucleophiles in their reactions with different substrates are compared in Tables II–VIII. A constant ratio should indicate a Ritchie-type behavior. Each table reflects a different aspect of the nucleophilicity. Table II gives selected reactivity ratios for piperidine and morpholine, two amines of the same bulk that differ by 2.8 pK_a units in water. The ratios are not constant, and values between 1.5 and 22.3 were found. A characteristic feature is the decrease of the ratios from MeCN to tetrahydrofuran to EtOH (*31*). The lower values for substitution compared with addition in protic solvents may have mechanistic significance, but generalizations are unwarranted. In each case, β_{Nu} for these amines is low; this finding indicates an early transition state for the addition.

Table III gives $k_{N_3^-}/k_{piperidine}$ ratios. The values for the three β-chlorovinyl ketones (*35*) demonstrate two important features. First, they increase in the dipolar aprotic solvent, by relative increase in the reactivity of the anionic nucleophile. Second, and less expected, the order of the ratios changes from N_3^- < piperidine for *E*-PhCOC(Ph)=CHCl to N_3^- > piperidine for the cyclic β-haloketone. Consequently, difficulties are expected in attempts to construct even a qualitative nucleophilicity scale toward vinylic carbon.

Table II. Several $k_{piperidine}/k_{morpholine}$ Ratios

Substrate	*Solvent*	T (°C)	k_{pip}/k_{mor}	*Reference*
$CH_2{=}CHCN$	H_2O	25	10.1	29
$CH_2{=}CHCO_2Me$	MeOH	45	17.4	30
$p\text{-}O_2NC_6H_4C(OTs){=}C(CO_2Et)_2$	MeCN	30	15.2	31
	THF	30	9.1	31
	EtOH	30	5.5	31
$p\text{-}O_2NC_6H_4C(OMs){=}C(CO_2Et)_2$	EtOH	30	22.3	31
(*E*)-PhC(Cl)=C(NO_2)Ph	EtOH	30	3.2	32
(*E*)-PhC(I)=C(NO_2)Ph	EtOH	30	4.8	32
(*E*)-PhC(NO_2)=C(NO_2)Ph	EtOH	30	3.4	32
PhCH=C(COO)(COO)C(CH_3)$_2$ (cyclic)	50% Me_2SO-50% H_2O	20	1.5	33
PhCH=C$(CN)_2$	50% Me_2SO-50% H_2O	20	3.4	34

Table III. $\dfrac{k_{N_3^-}(98\%\ EtOH)}{k_{piperidine}(EtOH)}$ **(*A*) and** $\dfrac{k_{N_3^-}(98\%\ DMF)}{k_{piperidine}(DMF)}$ **(*B*) Ratios**

Compound	A	B
(NC)(Ph)C=C(H)(Cl)[a]	0.04 (EtOH)	—[b]
(PhCO)(Ph)C=C(H)(Cl)[c]	0.03	0.39
(PhCO)(Me)C=C(H)(Cl)[c]	0.14	2.92
1-COMe-2-Cl-cyclopentene[c]	2.67	38.40

[a]Reference 36.
[b]Not determined
[c]Reference 35.

Table IV. Several k_{PhS^-}/k_{MeO^-} Reactivity Ratios in MeOH at 0° C

Substrate	E	Z	*Reference*
$2,4\text{-}(O_2N)_2C_6H_3CH{=}CHBr$	13,445	—[a]	37
$p\text{-}O_2NC_6H_4CH{=}CHBr$	4,060	—	37
$p\text{-}O_2NC_6H_4CH{=}CHCl$	1,069	—	38
$p\text{-}O_2NC_6H_4CH{=}CHF$	28	—	38
$p\text{-}MeOC_6H_4COCH{=}CHBr$	540	—	39
$p\text{-}O_2NC_6H_4SO_2CMe{=}CHBr$	43.3	128.8	40
$p\text{-}TolSO_2CH{=}CHBr$	16.2	0.14	4
$p\text{-}TolSO_2CH{=}CHCl$	6.3	24.2	5, 6
$p\text{-}O_2NC_6H_4SO_2CH{=}CHCl$	13.7	17.1	5

[a]Not determined.

Table IV compares the reactivity ratios of a soft (PhS^-) to a hard (MeO^-) nucleophile in vinylic substitution. PhS^- is always more reactive, and ratios lower than unity, as for **4**, X = Br (*4*), are certainly due to elimination–addition with MeO^-. The ratios change by >2000-fold and are sensitive to the geometry of the substrate. An important feature is that for β-halo-*p*-nitrostyrenes the ratio decreases strongly with the increased "hardness" of the β-halogen (*38*). The lowest ratios are for the β-fluoro derivative, whereas the differences between the chloro and bromo compounds are not so large. This behavior is similar to that in S_NAr reactions. This behavior can be rationalized by symbiotic effects, which favor the soft–soft PhS^-–Br interaction and the hard–hard MeO^-F interaction. A reactivity–selectivity relationship for vinyl bromides of different electrophilicities does not exist.

Table V compares EtS^- and EtO^- and shows similar features. The k_{EtS^-}/k_{EtO^-} ratios are relatively low for the vinyl fluorides (*41*). Although values at the same temperature are not available for the heavier halogens, k_{EtS^-}(18 °C)/k_{EtO^-}(0 °C) values for $XCH{=}CHCO_2Et$ are 344 (Cl), 329 (Br), and 1991 (I) (*42*). The operation of steric effects is shown by the $k_{EtS^-}k/_{t\text{-}BuS^-}$ ratios of >20 in spite of the expected higher basicity of $t\text{-}BuS^-$. A combina-

Table V. Reactivity Ratios for Substitutions in EtOH at 0 °C (*41*)

Compound	k_{EtS^-}/k_{EtO^-}	$k_{EtS^-}/k_{t\text{-}BuS^-}$	$k_{EtO^-}/k_{t\text{-}BuS^-}$
(*E*)-$MeCF{=}CHCN$	15.7	31.6	2.0
(Z)-$MeCF{=}CHCN$	5.0	21.0	4.2
(*E*)-$MeCF{=}CHCO_2Et$	9.4	37.8	4.0
(Z)-$MeCF{=}CHCP_2Et$	28.1	39.1	1.4

Table VI. Reactivity Ratios of RO^- in Vinylic Reactions

Substrate	*Process*	*T* (°C)	$k_{EtO^-}(EtOH)/k_{MeO^-}(MeOH)$	$k_{i\text{-}PrO^-}(i\text{-}PrOH)/k_{MeO^-}(MeOH)$	Reference
$CH_2{=}CHCOMe$	addition	25	4.0	8.1	43
$CH_2{=}C(Me)COMe$	addition	24	—[a]	3.3	44
$CH_2{=}CHCN$	addition	20	9.1	49.6	45
$PhSO_2CH{=}CH_2$	addition	25	6.1	—	46
$PhSO_2CH{=}CHBr$	substitution	0	3.4 (*E*); 19.2 (*Z*)	—	4
$TolSO_2CH{=}CHOPh$	substitution	25	5.4 (*E*); 9.0 (*Z*)	—	47
$p\text{-}O_2NC_6H_4CH{=}C(Cl)Me$	substitution	30	15.4	—	48

[a]Not determined.

tion of the symbiotic and steric effects makes EtO^- a better nucleophile than t-BuS^-. Again, the order of thio and oxygen nucleophiles seems to be substrate-dependent.

Table VI shows that although the nucleophilicity of alkoxy nucleophiles follows their basicity, the ratios are substrate-dependent.

Comparison of relative reactivities at constant steric effects in terms of a few selected Hammett's ρ values is shown in Table VII for ArS^- and in Table VIII for anilines. Although some of the data in Table VII are based only on two points, clearly the ρ values are structure-dependent. Moreover, these values show no clear trend. The reactivity of β,β-dihalovinyl sulfones are similar to those of the β-halovinyl sulfones, but their ρ values are much lower. The addition reactions with ArS^- show higher ρ values than for most of the substitutions, but for the structurally similar vinyl sulfones, the ρ for the substitution is higher. In general, the ρ values for the anilines are significantly higher, but this fact does not necessarily mean an earlier transition state for the anionic nucleophiles because the ρ values for the equilibrium acidities of the anilinium ions are higher than those for the thiols.

Table VII. Several ρ Values for Reactions of ArS^- with Electrophilic Olefins

Substrate	*Process*[a]	T (°C)	*Solvent*	N[b]	ρ	*Reference*
p-$MeOC_6H_4SO_2CH{=}CBr_2$	S	0	MeOH	6	−0.84[c]	49
p-$ClC_6H_4SO_2CH{=}CCl_2$	S	0	MeOH	5	−0.72[c]	50
p-$ClC_6H_4SO_2CH{=}CCl_2$	S	30	MeOH	5	−0.36[c]	50
(E)-$TolSO_2CH{=}CHCl$	S	0	MeOH	2	−1.55	5
$PhSO_2CH{=}CH_2$	A	25	50% EtOH	6	−1.22	51
(E)-ClCH=C(CN)Ph	S	30	EtOH	2	−0.51	36
p-$O_2NC_6H_4C(OTs){=}C(CO_2Et)_2$	S	30	MeOH	2	−0.39	31
N-Ethylmaleimide	A	25	95% EtOH	3	~−2	52

[a]A represents addition; S represents substitution.
[b]Number of substituents studied.
[c]The best correlation is with σ^+.

Table VIII. ρ Values for Substitution of Vinyl Halides by $ArNH_2$

Substrate	*Solvent*	T (°C)	N[a]	ρ	*Reference*
PhCOCH=CHBr	i-PrOH	25	8	−2.35	53
$ClCH{=}C(CO_2Me)_2$	MeCN	30	5	−2.40	54
(E)-$PhC(Cl){=}CHNO_2$	MeCN	30	7	−2.99	55
p-$O_2NC_6H_4C(Br){=}C(CN)_2$	MeCN	30	2	−3.5	31
p-$O_2NC_6H_4C(Br){=}C(CN)_2$	THF	30	2	−3.5	31

[a]Number of substituents studied.

The contribution of increased steric effects to a lower nucleophilicity was mentioned earlier. Russian workers (*29*, *56*) who studied addition and substitution of a large number of amines with electrophilic olefins analyzed the reactions in terms of dual-parameter equations involving polar and steric effects.

For example, substitution of the vinyl sulfone **13** by primary, secondary, and teriary amines (equation 3) obeys a single equation (equation 4) (*56*), where $\Sigma\sigma^*$ is the sum of the inductive effects of the amine substituents, and E_N is an isosteric parameter where the steric effect of an amine R^1R^2NH is taken as similar to E_s of CHR^1R^2. Twenty-two amines obeyed the equation with $\rho^* = -4.8$ and $\delta = 1.70$ where ρ^* and δ are the sensitivity parameters to the inductive and steric effects, respectively. As a demonstration of the electronic effects, $k_{cyclohexylamine}/k_{aniline} \sim 10^4$ in spite of their similar size, whereas steric effects are reflected in the k_{Et_2NH}/k_{i-Pr_2NH} of approximately 3000 in spite of the similar basicities. Similar LFERs were observed in other reactions (Table IX) (*29*).

$$(Z)\text{-}p\text{-}O_2NC_6H_4SO_2CH{=}CHCl \xrightarrow{R^1R^2NH} p\text{-}O_2NC_6H_4SO_2CH{=}CHNR^1R^2$$

$$\mathbf{13} \xrightarrow{R^1R^2R^3N} p\text{-}O_2NC_6H_4\dot{S}O_2CH{=}CH\overset{+}{N}R^1R^2R^3 \quad (3)$$

$$\log k = \log k_0 + \rho^*\Sigma\sigma^* + \delta E_N \quad (4)$$

Nevertheless, the less bulky primary amines are often less reactive than secondary amines. In addition to tolyl vinyl sulfone, $k_{Et_2N}/k_{BuNH_2} = 2.9$ in EtOH (*57*). A preferred solvation of the primary amine was held responsible for this result, and indeed, in the bulky, less solvating *t*-BuOH, the ratio decreases to 0.09.

An interesting contribution of steric effects to solvation is the addition of benzenethiolate ions to *N*-ethylmaleimide in 95% EtOH (*52*). The reactivity ratio of the 4-H:4-Me:2-Me:2,6-Me_2 derivatives is 1:2.16:1.6:1.6; thus, a small steric retardation is superimposed on a small electronic acceleration. However, the 2-*t*-Bu derivative is 8.7 times faster than its 4-*t*-Bu isomer. This finding was ascribed to a rate-enhancing steric inhibition to solvation of the anion, which raises its reactivity by more than the reactivity decrease due to crowding in the transition state.

Both electronic and steric effects also operate in addition reactions of carbon nucleophiles. For addition of $R\bar{C}(NO_2)_2$ to methyl acrylate, the reactivity order for R of 1 (Me) < 1.2 (Et) < 1.6 (Cl) < 3688 (F) was observed (*58*) and ascribed to the destablization of the ground state of the fluorocarbanion, for example, by a $p(\ddot{F})$–$p(C^-)$ electron repulsion. The electronic effects are less pronounced in the addition of $XC_6H_4\bar{C}(NO_2)_2$ to methyl vinyl ketone,

Table IX. Parameters of Equation 4 for Reactions of Amines with Electrophilic Olefins at 25 °C

Substrate	*Process*[a]	*Solvent*	$-\rho^*$	δ	$\log k_0$	N[b]	*Reference*
(*E*)-*p*-$O_2NC_6H_4SO_2CH{=}CHCl$	S	PhCl	4.8	1.70	4.00	22	56
$CH_2{=}CHCN$	A	H_2O	3.9	0.84	1.67	15	29
p-$MeOC_6H_4COCH{=}CH_2$	A	10% EtOH	3.4	0.63	3.58	11	29
$TolSO_2CH{=}CH_2$	A	EtOH	4.0	1.36	2.07	11	29, 57

[a]A represents addition; S represents substitution.
[b]Number of amines used in the correlation.

Table X. Nucleophilicities in Substitutions of Chloroolefins

Nucleophile	(Z)-*$TolSO_2CH{=}CHCl$, MeOH, 0 °C (5)*	(Z)-*$ClCH{=}C(CN)Ph$, EtOH, 30 °C (36)*	(E)-*$PhC(Cl){=}(NO_2)Ph$, EtOH, 30 °C (32)*
$PhCH_2S^-$	17.4	—[a]	—
p-$TolS^-$	1.1	191	13,625
Piperidine	1.0	1.0	1.0
p-$XC_6H_4S^-$	0.67[b]	110;[c] 0.19[b]	5,875[c]
p-$XC_6H_4O^-$	—		6.3;[d] 2.0[c]
RO^-	0.11[e]	0.98[f]	—
Morpholine	—	0.16	0.31
Bu_2NH	0.048	—	—
c-$C_6H_{11}NH_2$	0.038	—	—
SCN^-	—	—	0.43
N_3^-	0.024	0.041	—

NOTE: The nucleophilicities are relative to piperidine for each substrate.
[a]Not determined.
[b]X = H.
[c]X = Cl.
[d]X = Me.
[e]R = Me.
[f]R = Et.

where $\rho = -0.33$ to -0.38 at 0–50 °C and $\beta_{Nu} = 0.21–0.25$ (*59*). However, ortho substituents deviated from the linear plot, but a linear correlation was found between log k for them and the difference in the van der Waals radii of H and X.

From this discussion, clearly, a quantitative nucleophilicity scale toward vinylic carbon cannot be constructed. Neither Ritchie's nor Swain–Scott's correlations are applicable. Different blends of contribution of polar, steric, and symbiotic effect can change the reactivity order. Whether a qualitative order prevails could be inferred by comparing the three substitution reactions of chloro olefins, which are the only processes for which a relatively extensive change in the type of nucleophile was conducted (Table X).

Piperidine, which is common to the three series, serves as the standard. Comparison of the $k_{TolS^-}/k_{piperidine}$ ratios show a very large variation, which is not reflected to the same extent in reactions of other nucleophiles. For most other systems studied (*31, 36*), these ratios are very appreciable. Moreover, PhO^- appears in a different relative position in the second and the third series.

Some short reaction series can supplement the data of Table X. The halide ions Cl^-, Br^-, and I^- should be added at the bottom of any nucleophilicity list as inferred from the very slow rates of substitution of β-halo-*p*-nitrostyrenes (*60*), the exchange of labeled chloride with the 1,1-diaryl-2-haloethylene system (*61*), and the very slow halide promoted $E \rightleftharpoons Z$ isomerization of activated haloolefins (*62; 63;* A. Gazit and Z. Rappoport, unpublished results).

The $EtS^- > t\text{-}BuS^- > EtO^-$ order can be taken from Table V. In the substitution of *trans*-3-methoxy-4′-(dimethylamino)acrylophenone, $k_{OH^-}/k_{morpholine} = 0.27$, that is, morpholine $> OH^-$ (*64*). In addition to substituted **12**, the order $H_2O \ll OH^- < p\text{-}BrC_6H_4O^- < PhO^-$ was established (*24*). The order $MeO^- > CN^-$ in MeOH for **6** and **7** can also be added (*20, 21*).

In the basic epoxidation of *o*-chlorobenzylidenemalononitrile, where the nucleophilic species is XO^- (X = Cl, OH), the reactivity order is $HOO^- > ClO^-$ (*65*). In addition to acrylonitrile, the relative order is piperidine (*51*) $> SO_3^{2-}$ (18) $> Bu_2NH$ (0.76) (*29, 66*), whereas in the Michael addition to acrylonitrile, the reactivity order is $MeO^- > \bar{C}H(COMe)_2 > \bar{C}H(COMe)$-$CO_2Et$ (*67*).

We tried to construct a qualitative nucleophilicity scale by combining the data for the three main systems of Table X and addition of the other data from the short reaction series. In most cases, the relationship between a nucleophile from the main series and only one of the nucleophiles in the short series is known. Hence, our combined nucleophilic scale is given in Figure 1 in the form of two lines. The main one (in boldface type) is a single reactivity scale, whereas short series are introduced above in the appropriate places (other orders are $HOO^- > ClO^-$ and $FC^-(NO_2)_2 > RC^-(NO_2)_2$ where

$EtS^- > t\text{–}BuS^-$ pyrrolidine > $> SO_3^{2-}$,

$i\text{–}PrO^- >$ $> OH^- \geqslant NH_2OH; > CN^- > CH(COMe)_2 > MeCOCHCO_2Et$ $> i\text{–}Pr_2NH$

$\mathbf{PhCH_2S^-}$ τ $\mathbf{ArS^-} > \mathbf{ArO^-} \approx$ **piperidine** > $\mathbf{EtO^-} > \mathbf{SCN^-} >$ morpholine $\geqslant \mathbf{MeO^-} > \sim PhO^- > Bu_2NH > c\text{–}C_6H_{11}NH_2 > N_3^- >> \mathbf{H_2O} > \mathbf{Br^-}$, $\mathbf{Cl^-}$, $\mathbf{I^-}$

Figure 1. A qualitative nucleophilic scale for reactions at vinylic carbon.

Table XI. α and α′ Values for Capture of α-Anisylvinyl Cations Formed in the Solvolysis of AnC(Br)=CRR′ (*69*)

Ion	$\alpha = k_{Br^-}/k_{AcO^-}$ *in AcOH*	$\alpha' = k_{Br^-}/k_{SOH}$ *in* *TFE*	*80% EtOH*
$An{-}\overset{+}{C}{=}CH_2$[a]	0.0	—[b]	0
$An{-}\overset{+}{C}{=}CMe_2$	4.3	394	0
$An{-}\overset{+}{C}{=}CAn_2$	21.7	78	3
$An{-}\overset{+}{C}{=}$(anthrone-10-ylidene)$=0$	75.0	1200	158

[a] An = *p*-Anisyl
[b] Not determined.

R = Me, Et, or Cl). We are well aware of the limitation of this series, such as the possibility of reversal of the order (hence the question mark before N_3^- and the ≈ sign for ArO^- and piperidine) as a function of the substrate and the reaction conditions. These drawbacks were mentioned previously. Solvent effects that should be important were also neglected. We believe that a further study of a few selected series with a large series of nucleophiles would be very helpful in solving some of the discrepancies and in filling the gaps.

An interesting conclusion is that in spite of the fact that the chloro-substituted vinylic carbon is relatively soft, several of the reactive nucleophiles of the Swain–Scott scale such as N_3^-, Br^-, SCN^-, and SO_3^{2-} show relatively low nucleophilicity in Table XI. When a more quantitative scale will be available, comparing it with the nucleophilicity order toward activated aromatic carbon would be constructive.

Nucleophilicity toward Vinyl Cations

A complementary part to the reaction with neutral, although polarized, vinylic carbon is the reaction of nucleophiles with vinyl cations **14** (equation 5). The data in this case are much more limited, for two reasons: (1) very few vinyl cations had been prepared with sufficiently long lifetime that allows their direct reaction with nucleophiles to be followed and (2) a very few capture experiments of a solvolytically generated vinyl cation by several nucleophiles were conducted.

$$R^2R^3C{=}C(R^1)X \xrightarrow{-X^-} \underset{\mathbf{14}}{R^2R^3C{=}\overset{+}{C}{-}R^1} \xrightarrow{Nu^-} R^2R^3C{=}C(R^1)Nu + R^2R^3C{=}C(Nu)R^1 \quad (5)$$

Four different probes gave short reactivity orders toward vinyl cations: (1) common ion rate depression in solvolysis; (2) competitive capture of solvolytically generated ions; (3) direct reaction of a vinyl cation with nucleophiles; and (4) competition between intra- and intermolecular nucleophilic capture. A short reactivity order is obtained in each case, but because of the different solvents and conditions the orders cannot be combined to a single series. However, a selectivity rule that governs the relative reactivities toward different vinyl cations in terms of a constant selectivity or a reactivity–selectivity relationship can be determined.

According to the Winstein–Ingold solvolysis scheme (*68*), the observation of a rate decrease in the solvolysis of RX by either the formed or an added X^- ("common ion rate depression", CIRD) serves as evidence for product formation from an intermediate free vinyl cation R^+. The simplified solvolysis scheme, when ion pairs are neglected and both the solvent SOH and its conjugate base SO^- may be present, is shown in Scheme II.

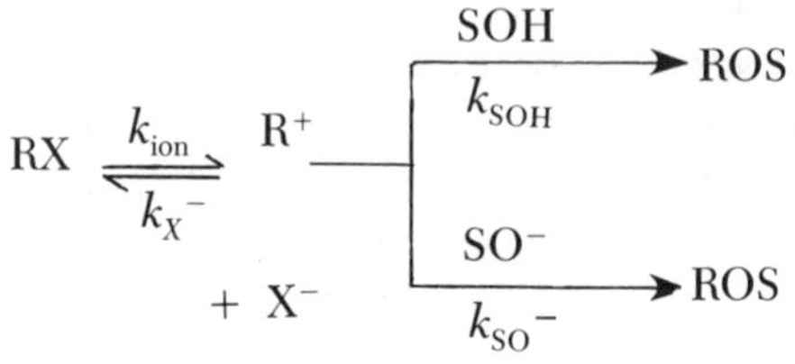

$$RX \underset{k_{X^-}}{\overset{k_{ion}}{\rightleftharpoons}} R^+ + X^- \quad \begin{cases} \xrightarrow[k_{SOH}]{SOH} ROS \\ \xrightarrow[k_{SO^-}]{SO^-} ROS \end{cases}$$

Scheme II

The rate equations for Scheme II are

$$k_t = k_{ion}/(1 + \alpha'[X^-]/[SOH]);\ \alpha' = k_{X^-}/k_{SOH} \tag{6}$$

$$k_t = k_{ion}/(1 + \alpha[X^-]/[SO^-]);\ \alpha = k_{X^-}/k_{SO^-} \tag{7}$$

The appearance of CIRD is a relatively rare phenomenon because it is observed only when R^+ is sufficiently selective to react competitively with the more nucleophilic X^- that is present in low concentrations and with the less nucleophilic SOH that is present in much higher concentrations. The selectivity constants α and α' are the ratios of the rate constant for the reverse reaction with X^- (k_{-X}) to the product-forming rate constant (k_{SOH} or k_{SO^-}). Consequently, α or α' could be used to evaluate the selectivity behavior. The experimental difficulties associated with the measure of α in vinylic systems, the question whether the capturing nucleophile in a buffered solvent is SOH or SO^-, and the reasons for the high selectivity are discussed extensively in reviews (*69, 70*) and will not be repeated here.

The largest number of α and α' values are available for the solvolysis of α-anisyl-β,β-disubstituted vinyl bromides. Evidence exists from CIRD that product formation is mainly or exclusively from the free ion (*69*), and α and α'

values for the derived ions (An = *p*-methoxyphenyl = anisyl) are given in Table XI. Clearly, Ritchie's constant selectivity (which requires that α and α' will be constant for different substrates) does not hold. The ratios change strongly with the nature of the ion. Comparison of the solvolytic reactivities of the precursor bromides shows that a reactivity–selectivity relationship, which calls for a higher k_{ion} for the more selective ion (i.e., higher α or α' accompanies higher k_{ion}) does not hold. Instead, the selectivities are governed mainly by steric effects: the bulkier is the β-substituent, the higher is the α's in AcOH–NaOAc or the α' in 80% EtOH and trifluoroethanol (TFE) (with one exception). This finding was ascribed to the severe steric hindrance to the in-plane approach of the nucleophiles to the vacant orbital. Because of this hindrance, a polarizable nucleophile such as Br^-, which can form a bond from a longer distance than a less-polarizable harder nucleophile such as AcO^-, becomes more reactive. The difference between the selectivities, that is, α (or α'), should therefore increase with the increased bulk of the β-substituents as was indeed observed.

What happens when the bulk of the β-substituents remains constant? Difficulties in obtaining the order of α values if their difference is small results, but for the anthronylidene derivatives **15** (Ar = An, Tol, *o*-An, or Ph), Rappoport et al. (*71*) showed that the log α values in 1:1 $AcOH–Ac_2O:AcO^-$ and the log α' values in TFE–2,6-lutidine increase linearly with log k_t (equation 8). The solvolysis rate constant k_t should be close to log k_{ion} (*72*), and for these structurally related systems, strong evidence exists that the products are formed from the free ion (*71*, *72*). This apparent reactivity–selectivity behavior finds strong support in the study of the solvolysis and $^{82}Br^-$ exchange of **15,** Ar = An or Tol, and of three triarylvinyl bromides in $AcOH–AcO^-–Et_4N^{81}Br$ (*73*). This technique measures independently k_t and α and is therefore superior to the CIRD method where k_t and α are obtained from a single experiment. The $k_{ex}{}^{cor}$ (exchange rate constants corrected for the natural decay of ^{82}Br), k_t, and α values are given in Table XII. Clearly, the α values for both series increase with the increase in k_t. Evidently, at constant steric effects, reactivity–selectivity behavior is obtained.

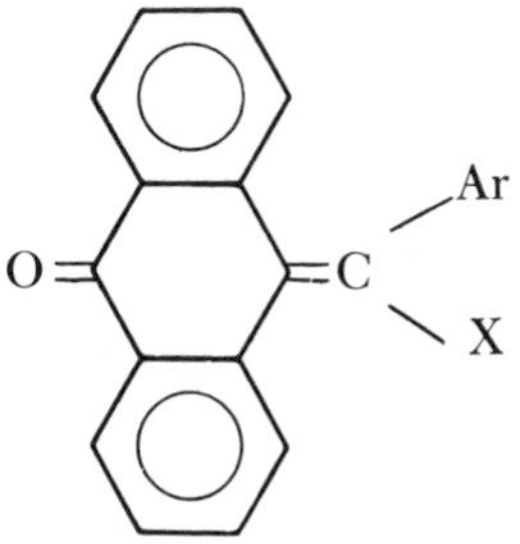

Ar

O= =C

X

15

$$\log k_t^\circ = -25.5 + 8.12 \log \alpha' \quad (8)$$

An interesting observation is that in the presence of a large excess of Br^-, the reaction of AcO^- with **15,** Ar = An, is of a second order (*71*). This result is a rare example of an $S_N2(C^+)$ route where the cation–anion recombination is rate-determining in a solvolysis reaction, and in principle, such a process could be used for obtaining directly rate constants for capture of vinyl cations.

Common ion rate depression for solvolysis of RX with the same R but with different leaving groups X was followed only for a single system. From the α values obtained in the solvolysis of (*E*)- and (*Z*)-1,2-dianisyl-2-phenylvinyl-X, **16,** in AcOH–AcO^- (equation 9), relative reactivities toward the derived ion **17** were measured (*74*) (nucleophile, relative reactivity): OMs^-, 0.16; OAc, 1.0; Cl^-, 15.2; Br^-; 45.5; and AcOH, 0.0024. From other data, I^- will probably be at the top of a similar order (*75*) and 2,4,6-trinitrobenzenesulfonate at the bottom (*76*).

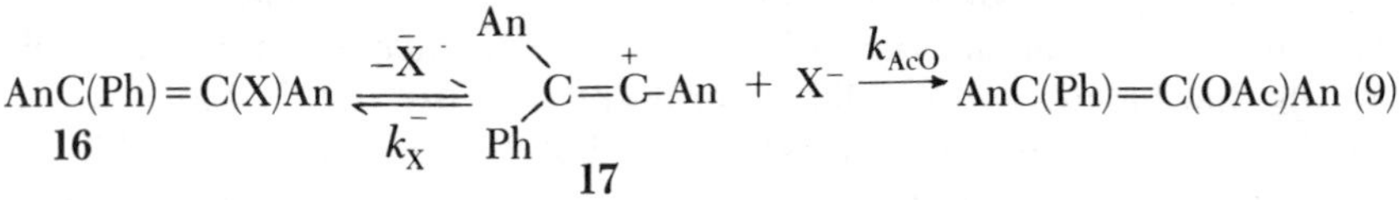

$$\underset{\mathbf{16}}{\mathrm{AnC(Ph)=C(X)An}} \underset{k_X}{\overset{-X^-}{\rightleftharpoons}} \underset{\mathbf{17}}{(\mathrm{An})(\mathrm{Ph})\mathrm{C}=\overset{+}{\mathrm{C}}\text{-}\mathrm{An}} + X^- \xrightarrow{k_{AcO}} \mathrm{AnC(Ph)=C(OAc)An} \quad (9)$$

The order of these data is similar to the Swain–Scott order in spite of the different solvents and the degree of hardness of the electrophilic center.

Table XII. Reactivity–Selectivity Relationships from Exchange–Solvolysis Experiments in AcOH–NaOAc–$Et_4N^{82}Br$ at 120 °C (73)

Compound[a]	$10^5 k_{ex}^{cor}$	$10^7 k_t$	α
$An_2C{=}C(Br)An$	22	114	21.7 ± 0.7
(*E*)-PhC(An)=C(Br)An	13	75	19.5 ± 1.6
$Ph_2C{=}C(Br)An$	8.8	53	18.6 ± 0.8
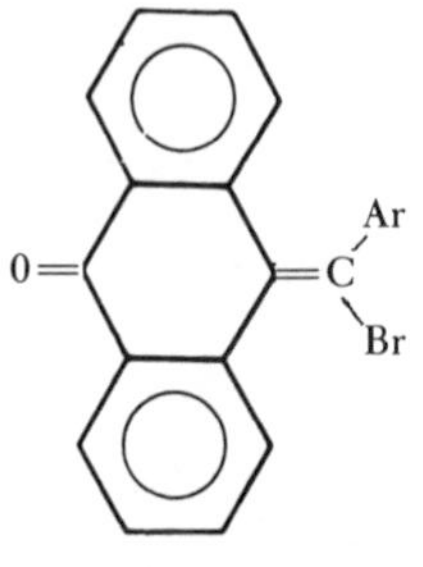 O=(anthraquinone-type ring)=C(Ar)Br : Ar = An	64	96	75.0 ± 1.4
Ar = Tol	0.084	0.77	12.2 ± 0.8

[a]An = *p*-anisyl; Tol = *p*–tolyl.

Table XIII. Capture Ratios in the Solvolysis of 18a in Binary Solvent Mixtures at 35 ° C (77, 78)

% TFE in TFE–H_2O	k_{H_2O}/k_{TFE}[a]	% TFE in TFE–EtOH	k_{EtOH}/k_{TFE}
97	1.50 (0.93)	90	11.0
94	1.43 (0.85)	80	9.4
90	1.25 (0.96)[b]	70	11.5
80	1.45 (0.76)	60	12.2

[a]The value for **18c** is in parentheses.
[b]1.37 for **18b**.

Interestingly, the order of the fractions of ion pairs returned from the **17**·X^- ion pairs (0.47, 0.38, and 0.24 for $X^- = Br^-$, Cl^-, and OMs^-, respectively), which reflects the intramolecular nucleophilicity in the ion pair, is parallel to the α values (*74*). The order of return from ion-paired X^- or from external X^- seems therefore to be governed by similar factors.

Solvolysis reactions of vinylic substrates in binary protic solvents are numerous (*69*), but in the majority of cases, the distribution of the two products is not useful for evaluating the relative reactivities of the two solvents. The main reason is that the vinyl ethers formed in H_2O–ROH mixtures, the vinyl formates, and sometimes the vinyl acetates formed in HCOOH–AcOH or AcOH–H_2O mixtures are frequently unstable under the solvolytic conditions. Their hydrolysis to the ketones will give erroneous capture ratios. In addition, in most cases the nature of the product-forming intermediate is not clear and it is frequently the ion pair.

The only values that seem reliable to us are given in Table XIII. Solvolysis of **18a** in buffered TFE–H_2O (*77*) and buffered TFE–EtOH (*78*) mixtures and of **18c** in buffered TFE–H_2O mixtures (*77*) was conducted over an extensive solvent composition range. The ethers are stable under the reaction conditions and **18b** forms the products from the free ion (*77*). The similarity of products from **18b** and **18a** suggested that products are formed in all cases from the free ion. The k_{H_2O}/k_{TFE} and k_{EtOH}/k_{TFE} ratios were calculated from the product ratios (cf. equation 10) and found to be reasonably constant in 97–80% TFE–H_2O and in 90–60% TFE–EtOH. A large change was observed in more aqueous solvents.

ArC(X)=CMe_2
18a, Ar = An; X = OTs
18b, Ar = An; X = Br
18c, Ar = *o*–An; X = OTs

$$\frac{k_{TFE}}{k_{H_2O}} \qquad \frac{[H_2O][ROCH_2CF_3]}{[TFE][ROH]} \qquad (10)$$

The ratios of Table XIII give the approximate reactivity order EtOH (11) > H_2O (1.4) > TFE (1.0) toward $An–\overset{+}{C}=CMe_2$ and H_2O (0.9) TFE (1.0) for its *o*-methoxy analogue. The former order is reasonable in terms of nucleophilicities of both species. The nearly similar reactivities of TFE and H_2O are somewhat surprising, but they do not differ much from the k_{TFE}/k_{H_2O} values calculated for the solvolysis of 1-adamantyl bromide and *t*-BuCl (*78*).

In a single unpublished study (M. Oka, H. Taniguchi, and S. Kobayashi), Taniguchi's group studied the competitive capture of the vinyl cation formed in the solvolysis **19** in 0.1 N aqueous NaOH by several nucleophiles. CIRD studies showed that >90% of the products are formed from the free ion. The data that are given in Table XIV in terms of log (k_{Nu^-}/k_{OH^-}) are compared with Richie's ΔN_+ values of the two nucleophiles. The agreement between only half of the values argues against correlation with N_+.

$$Me_2C=C(Br)C_6H_4OCH_2COO^- \text{-} p$$

19

A "direct" determination of the relative nucleophilicities of neutral nucleophiles toward vinyl cations was recently reported (*79*). The vinyl cations **20a–c** were generated by flash photolysis and their decay was followed simultaneously by UV spectroscopy and photocurrent measurements. The similar rates, the absence of effect of oxygen, and the identical spectra of **20b** formed from the precursor chloride and bromide argue strongly that the species studied are indeed the ions **20.**

$$R_2C=\overset{+}{C}–An$$

20a, R = An
20b, R = Ph
20c, R = Me

The decay rates in MeCN increased in the presence of nucleophiles such as alcohols, water, and THF. The order of decay rates MeOH > EtOH > *i*-PrOH > H_2O > *t*-BuOH was established for both **20a** and **20c** and log *k* for **20c** was linear with that for **20a** with a slope of 0.83. The reactivity order of the alcohols probably reflects the increased steric hindrance to capture with the increased bulk of the alcohol (the deviation of H_2O was ascribed to cluster formation). Because **20a** is more reactive than **20c,** the reactivity–selectivity principle applies for this limited series.

A similar reactivity order was found recently for competition between intramolecular cyclization and intermolecular capture. The $AgBF_4$-assisted solvolysis of trimesitylvinyl chloride 21 in alcohols gives both the ether **23** and 2,3-dimesityl-4,6-dimethylindene, **24** (equation 11; Mes = 2,4,6-$Me_3C_6H_2$) (*80*). These compounds are derived from the trimesitylvinyl cation

Table XIV. Reactivity Ratios [log (k_{Nu^-}/k_{OH^-})] for Capture of $Me_2C{=}\overset{+}{C}{-}C_6H_4OCH_2COO^-{-}p$ by Nucleophiles in 0.1 N NaOH at Room Temperature

Nu^-	log (k_{Nu^-}/k_{OH^-})	$N^+(Nu^-) - N_+(OH^-)$[a]
SCN^-	0.14	—[b]
N_3^-	−0.57	2.85
CN^-	−1.19	−1.08
F^-	−1.97	−1.30
NO_2^-	−1.97	−1.71

NOTE: Data are from Oka, M.; Taniguchi, H.; Kobayashi, S., unpublished results.
[a]From reference 17.
[b]Not determined.

22 by capture and by cyclization on a β-*o*-methyl group, respectively. The **23:24** ratios given in equation 11 strongly decrease with the increased bulk of the alcohol. Because the cyclization rate was assumed to be relatively solvent-insensitive, the main effect is on the capture rate. A log (k_{ROH}/k_{cyc}) plot for R = Me, Et, and *i*-Pr is linear with log k for the reaction of **20a** with the three alcohols. Consequently, steric effects on the nucleophilicity play a similar role in both cases.

$$Mes_2C{=}C(Mes)Cl\ (\mathbf{21}) \xrightarrow[ROH]{AgBF_4} Mes_2C{=}\overset{+}{C}{-}Mes\ (\mathbf{22}) \rightarrow Mes_2C{=}C(Mes)OR\ (\mathbf{23}) + \mathbf{24}$$

24: indene bearing Me, Me, Mes, Mes substituents

ROH	23:24
MeOH	89:11
EtOH	82:18
i-PrOH	24:76
t-BuOH	<5:>95

(11)

Other aspects of nucleophilicity toward vinyl cations are the site of capture of ambident ions and the easy intramolecular cyclization by *o*-methoxy and *o*-thiomethyl substituents on a β ring. The extensively studied β-aryl rearrangement across the double bond could be regarded as intramolecular substitution by the aryl ring, and data are available on the relative rate of rearrangement and capture by the solvent (*69, 70*). These topics are not discussed here for lack of space but should be addressed in a more complete discussion of the nucleophilicity.

Acknowledgments

I am indebted to my students who contributed to our studies on this topic. Thanks are due to H. Taniguchi for his permission to use the unpublished data of Table XIV.

Literature Cited

1. Patai, S.; Rappoport, Z. In *The Chemistry of Alkenes;* Patai, S., Ed.; Wiley-Interscience: New York, 1964; p. 469.
2. Rappoport, Z. *Adv. Phys. Org. Chem.* **1969,** *7,* 1.
3. Rappoport, Z. *Recl. Trav. Chim. Pays-Bas,* **1985,** *104,* 309.
4. Campagni, A.; Modena, G.; Todesco, P. E. *Gazz. Chim. Ital.* **1960,** *90,* 694.
5. Modena, G.; Todesco, P. E. *Gazz. Chim. Ital.* **1959,** *89,* 866.
6. DiNunno, L.; Modena, G.; Scorrano, G. *J. Chem. Soc. B* **1966,** 1186.
7. Menger, F. M.; Smith, J. H. *J. Am. Chem. Soc.* **1969,** *91,* 4211.
8. Lough, C. E.; Currie, D. J. *Can. J. Chem.* **1966,** *44,* 1563.
9. Nozaki, K. *J. Am. Chem. Soc.* **1941,** *63,* 2681.
10. Davies, M.; Evans, F. P. *Trans. Faraday Soc.* **1955,** *51,* 1506.
11. Rappoport, Z.; Ta-Shma, R. *J. Chem. Soc. B* **1971,** *871,* 1461.
12. Rappoport, Z.; Peled, P. *J. Chem. Soc., Perkin Trans. 2* **1973,** 616.
13. Patai, S.; Rappoport, Z. *J. Chem. Soc.* **1962,** 396.
14. Rappoport, Z.; Degani, C.; Patai, S. *J. Chem. Soc.* **1963,** 4513.
15. Rappoport, Z.; Greenzaid, P.; Horowitz, A. *J. Chem. Soc.* **1964,** 1334.
16. Rappoport, Z. *Acc. Chem. Res.* **1981,** *14,* 7.
17. Ritchie, C. D. *Acc. Chem. Res.* **1972,** *5,* 348.
18. Ritchie, C. D. *J. Am. Chem. Soc.* **1975,** *97,* 1170.
19. Swain, C. G.; Scott, C. B. *J. Am. Chem. Soc.* **1953,** *75,* 141.
20. Hoz, S.; Spiezman, D. *Tetrahedron Lett.* **1978,** 1775.
21. Hoz, S.; Spiezman, D. *J. Org. Chem.* **1983,** *48,* 2904.
22. Ritchie, C. D.; Kawasaki, A. *J. Org. Chem.* **1981,** *46,* 4706.
23. Lartey, P. A.; Fedor, L. *J. Am. Chem. Soc.* **1975,** *101,* 7385.
24. Bernasconi, C. F.; Leonarduzzi, G. D. *J. Am. Chem. Soc.* **1980,** *102,* 1361.
25. Friedman, M.; Wall, J. S. *J. Am. Chem. Soc.* **1964,** *86,* 3735.
26. Friedman, M.; Wall, J. S. *J. Org. Chem.* **1966,** *31,* 2888.
27. Friedman, M.; Cavins, J. F.; Wall, J. S. *J. Am. Chem. Soc.* **1965,** *87,* 3672.
28. Rappoport, Z.; Shenhav, C.; Patai, S. *J. Chem. Soc. B* **1970,** 469.
29. Dienys, G. J.; Kunskaite, L. J. J.; Vaitkevicius, A. K.; Klimavicius, A. V. *Org. React., Engl. Ed.* **1975,** *12,* 275.
30. Mallik, K. L.; Das, M. N. *Zeit. Phys. Chem.* **1960,** *25,* 205.
31. Rappoport, Z.; Topol, A. *J. Chem. Soc., Perkin Trans. 2* **1975,** 863.
32. Topol, A. Ph.D. Thesis, The Hebrew University, 1975.
33. Bernasconi, C. F.; Fornarini, S. *J. Am. Chem. Soc.* **1980,** *102,* 5329.
34. Bernasconi, C. F.; Fox, J. P.; Fornarini, S. *J. Am. Chem. Soc.* **1980,** *102,* 2810.
35. Beltrame, P.; Favini, G.; Cattania, M. G.; Guella, F. *Gazz. Chim. Ital.* **1968,** *98,* 380.
36. Rappoport, Z.; Topol, A. *J. Chem. Soc., Perkin Trans. 2* **1975,** 982.
37. Marchese, G.; Modena, G.; Naso, F. *Tetrahedron* **1968,** *24,* 663.
38. Marchese, G.; Naso, F.; Modena, G. *J. Chem. Soc. B* **1969,** 290.
39. Landini, D.; Montanari, F.; Modena, G.; Naso, F. *J. Chem. Soc. B* **1969,** 243.
40. Maioli, L.; Modena, G.; Todesco, P. E. *Bull. Sci. Fac. Chim. Ind. Bologne* **1960,** *18,* 66.

41. Chalchat, J.–C., Théron, F.; Vessiere, R. *Bull. Soc. Chim. Fr.* **1973,** 2501.
42. Biougne, J.; Théron, F.; Vessiere, R. *Bull. Soc. Chim. Fr.* **1972,** 2882.
43. Ferry, N.; McQuillin, F. J. *J. Chem. Soc.* **1962,** 103.
44. Ring, R. N.; Tesoro, G. C.; Moore, D. R. *J. Org. Chem.* **1967,** *32,* 1091.
45. Feit, B.–A.; Zilkha, A. *J. Org. Chem.* **1963,** *28,* 406.
46. Davies, W. G.; Hardisty, E. W.; Nevell, T. P.; Peters, R. H. *J. Chem. Soc. B* **1970,** 998.
47. Van Der Sluijs, M. J.; Stirling, C. J. M. *J. Chem. Soc., Perkin Trans. 2* **1974,** 1268.
48. Iskander, Y.; Nassar, A. M. G.; Tewfik, R. *J. Chem. Soc. B* **1970,** 412.
49. Shainyan, B. A.; Mirskova, A. N. *Zh. Org. Khim.* **1984,** *20,* 972.
50. Shainyan, B. A.; Mirskova, A. N. *Zh. Org. Khim.* **1980,** *16,* 2569.
51. De Maria, P.; Fini, A. *J. Chem. Soc. B* **1971,** 2335.
52. Semenow–Garwood, D. *J. Org. Chem.* **1972,** *37,* 3797.
53. Popov, A. F.; Litvinenko, L. M.; Kostenko, L. I. *Zh. Org. Khim.* **1973,** 982.
54. Rappoport, Z.; Topol, A. *J. Chem. Soc., Perkin Trans. 2* **1972,** 1823.
55. Rappoport, Z.; Hoz, S. *J. Chem. Soc., Perkin Trans. 2* **1975,** 272.
56. Popov, A.; Kravchenko, V.; Piskunova, Z.; Kostenko, A. *Org. React. (N.Y.)* **1980,** *17,* 325.
57. McDowell, S. J.; Stirling, C. J. M. *J. Chem. Soc. B* **1967,** 343.
58. Kaplan, L. A.; Pichard, H. B. *Chem. Commun.* **1968,** 1500.
59. Tselinsky, I. V.; Kolesetakaya, G. I. *Org. React. (N.Y.)* **1971,** *8,* 79.
60. Miller, S. I.; Yonan, P. K. *J. Am. Chem. Soc.* **1957,** *79,* 5931.
61. Beltrame, P.; Bellobono, I. R.; Fére, A. *J. Chem. Soc. B* **1966,** 1165.
62. Grünbaum, Z.; Patai, S.; Rappoport, Z. *J. Chem. Soc. B* **1966,** 1133.
63. Rappoport, Z.; Avramovitch, B. *J. Org. Chem.* **1982,** *47,* 1397.
64. Chu, J.–Y. H.; Murty, B. S. R.; Fedor, C. *J. Am. Chem. Soc.* **1976,** *98,* 3632.
65. Rosenblatt, D. H.; Broome, G. H. *J. Org. Chem.* **1963,** *28,* 1290.
66. Morton, M.; Lanfield, H. *J. Am. Chem. Soc.* **1952,** *74,* 3523.
67. Markisz, J. A.; Gettler, J. D. *Can. J. Chem.* **1969,** *47,* 1965.
68. Winstein, S.; Clippinger, E.; Feinberg, A. H.; Heck, R.; Robinson, G. C. *J. Am. Chem. Soc.* **1956,** *78,* 328.
69. Stang, P. J.; Rappoport, Z.; Hanack, M.; Subramanian, L. R. *Vinyl Cations;* Academic: New York, 1969; Chapter 6.
70. Rappoport, Z. In *Reactive Intermediates;* Abramovitch, R. A., Ed.; Plenum: 1983; Vol. 3, p 427.
71. Rappoport, Z.; Apeloig, Y.; Greenblatt, J. *J. Am. Chem. Soc.* **1980,** *102,* 3837.
72. Rappoport, Z.; Greenblatt, J. *J. Am. Chem. Soc.* **1979,** *101,* 1343.
73. van Ginkel, F. I. M.; Hartman, E. R.; Lodder, G.; Greenblatt, J.; Rappoport, Z. *J. Am. Chem. Soc.* **1980,** *102,* 7514.
74. Rappoport, Z.; Apeloig, Y. *J. Am. Chem. Soc.* **1975,** *97,* 821.
75. Miller, L. L.; Kaufman, D. A. *J. Am. Chem. Soc.* **1968,** *90,* 7282.
76. Modena, G.; Tonellato, U. *J. Chem. Soc. B* **1971,** 374.
77. Rappoport, Z.; Kaspi, J. *J. Am. Chem. Soc.* **1974,** *96,* 4518.
78. Rappoport, Z.; Kaspi, J. *J. Am. Chem. Soc.* **1980,** *102,* 3829.
79. Kobayashi, S.; Kitamura, T.; Taniguchi, H.; Schnabel, W. *Chem. Lett.* **1983,** 1117.
80. Biali, S. E.; Rappoport, Z. *J. Org. Chem.,* **1986,** *51,* 964.

RECEIVED for review October 21, 1985. ACCEPTED January 24, 1986.

ENHANCEMENT OF NUCLEOPHILICITY

29

Micellar Effects on Nucleophilicity

Clifford A. Bunton

Department of Chemistry, University of California, Santa Barbara, CA 93106

Aqueous cationic micelles speed and anionic micelles inhibit bimolecular reactions of anionic nucleophiles. Both cationic and anionic micelles speed reactions of nonionic nucleophiles. Second-order rate constants in the micelles can be calculated by estimating the concentration of each reactant in the micelles, which are treated as a distinct reaction medium, that is, as a pseudophase. These second-order rate constants are similar to those in water, except for aromatic nucleophilic substitution by azide ion, which is much faster than predicted. Ionic micelles generally inhibit spontaneous hydrolyses. But a charge effect also occurs, and for hydrolyses of anhydrides, diaryl carbonates, chloroformates, and acyl and sulfonyl chlorides and S_N hydrolyses, reactions are faster in cationic than in anionic micelles if bond making is dominant. This behavior is also observed in water addition to carbocations. If bond breaking is dominant, the reaction is faster in anionic micelles. Zwitterionic sulfobetaine and cationic micelles behave similarly.

SUBMICROSCOPIC, COLLOIDAL AGGREGATES can influence chemical reactivity. Aqueous micelles are the most widely studied of these aggregates, and these micelles form spontaneously when the concentration of a surfactant (sometimes known as a detergent) exceeds the critical micelle concentration, cmc (*1–3*). Surfactants have apolar residues and ionic or polar head groups, and in water at surfactant concentrations not much greater than the cmc, micelles are approximately spherical and the polar or ionic head groups are at the surface in contact with water. The head groups may be cationic, (e.g., trimethylammonium), anionic, (e.g., sulfate), zwitterionic (as in carboxylate or sulfonate betaines), or nonionic. The present discussion covers the behavior of ionic and zwitterionic micelles and their effects on chemical reactivity.

Micelles can incorporate hydrophobic solutes, and by virtue of their charge, ionic micelles attract counterions and repel co-ions. Micelles influence reaction rates and products in various ways. They provide a reaction medium apparently distinct from the bulk solvent, and rate constants may be

0065-2393/87/0215-0425$06.00/0

different in aqueous and micellar pseudophases. In addition, micelles can speed bimolecular, nonsolvolytic reactions by bringing reactants together or inhibit reactions by keeping reactants apart. For example, cationic micelles speed attack of nucleophilic anions, and anionic micelles inhibit these reactions, but depending upon the reaction type, spontaneous reactions may be speeded or retarded.

Quantitative analysis of these rate effects requires estimation of the contributions of the reactions in the bulk, aqueous medium and in the micellar pseudophase. This separation can be made provided that the reactant concentrations in each pseudophase can be estimated by direct measurement or by calculation (*4–16*).

The bulk of the experiments described were carried out by using surfactants of four different types: cationic [cetyltrimethylammonium salts, $C_{16}H_{33}N(CH_3)_3X$ (CTAX), X = halide, mesylate, azide, or hydroxide]; anionic [sodium dodecyl sulfate, $C_{12}H_{25}OSO_3Na$ (SDS)]; zwitterionic [*N*,*N*-dimethyl-*N*-dodecylglycine, $C_{12}H_{25}N^+(CH_3)_2CH_2CO_2^-$ (B1–12), or sulfobetaine, $C_{16}H_{33}N^+(CH_3)_2(CH_2)_3SO_3^-$ (SB3–16)]; and nonionic.

The following discussion covers the effects of these surfactants upon spontaneous reactions, but for bimolecular, nonsolvolytic reactions, only cationic and anionic micelles are discussed.

Water is the bulk solvent in all the experiments described here, although normal micelles form in a variety of three-dimensional associated solvents including 1,2-diols, formamide, and 100% sulfuric acid (*17–19*), and some kinetic work has been done on micelles in aqueous 1,2-diols (*20*).

Quantitative Treatment

The relation between rate constant and surfactant concentration is simple for spontaneous, or micellar-inhibited, nonsolvolytic reactions for which the distribution of only one reagent has to be considered (*11, 12*).

Distribution of substrate, S, between aqueous and micellar pseudophases (denoted by subscripts *W* and *M*, respectively) is written in terms of equation 1 where [D] − cmc is the concentration of micellized surfactant and K_S is the binding constant of S to micellized surfactant (*12*):

$$[S_M] = K_S[S_W]([D] - \text{cmc}) \qquad (1)$$

The first-order rate constant for the overall reaction, k_ψ, is given by

$$k_\psi = [k'_W + k'_M K_S([D] - \text{cmc})]/[1 + K_S([D] - \text{cmc})] \qquad (2)$$

where k'_W and k'_M are respectively the first-order rate constants in the aqueous and micellar pseudophases.

Equation 2 does not fit rate–surfactant profiles for micellar-assisted, bimolecular, nonsolvolytic reactions, because the distribution of both reactants must be considered (*4–10*). In addition, the question of the appropriate measure of concentration in the micelles must be answered.

For reaction of a nucleophile, Y, the first-order rate constants in equation 2 can be written as (*5, 9*)

$$k'_W = k_W[Y_W] \tag{3}$$

$$k'_M = k_M m_Y^S = k_M[Y_M]/([D] - cmc) \tag{4}$$

and k_M has the dimensions of reciprocal time, because mole ratios are dimensionless, so that

$$k_\psi = (k_W[Y_W] + k_M K_S[Y_M])/[1 + K_S([D] - cmc)] \tag{5}$$

Application of equation 5 requires estimation of $[Y_M]$. For some organic nucleophiles, the distribution of Y between water and micelles can be determined experimentally, and this approach has been used for reactions of imidazoles, oximes, amines, and phenoxide and thiophenoxide ions (*6–8, 13, 14*) (Table I).

Table I. Micellar Effects on Reactivity of Organic Nucleophiles

Reaction	k_{rel}	k_2^m/k_W
$C_6H_{13}CO_2C_6H_4$-4-NO_2 + imidazoles	—[a]	$\sim 10^{-2}$ [b]
$C_6H_{13}CO_2C_6H_4$-4-NO_2 + imidazolide ions	—[a]	~10 [c]
$CH_3CO_2C_6H_4$-4-NO_2 + $C_6H_5S^-$	69	0.42 [c]
$(C_6H_5O)_2PO_2C_6H_4$-4-NO_2 + $C_6H_5O^-$	3000	0.53 [d]
1-fluoro-2,4-dinitrobenzene + $C_6H_5NH_2$	3–8	0.17 [e]
1-fluoro-2,4-dinitrobenzene + $C_6H_5NH_2$	~3	0.12 [e,f]

NOTE: In CTABr unless specified. k_2^m is the second-order rate constant in the micelles.
[a] Not determined.
[b] Reference 6.
[c] Reference 8.
[d] Reference 14.
[e] Reference 31.
[f] In SDS.

The problem is more complex for bimolecular reactions of inorganic anions unless the ionic concentration in micelles (or water) can be measured directly (*15, 16*). Generally, the reaction solution contains both the reactive ion and the inert counterion of the surfactant, which compete for the micellar surface; Romsted showed (*4, 5*) how this competition can be described by equations similar to those applied to binding to ion-exchange resins, and his treatment has been applied to many micellar-assisted reactions. Table II gives some examples of reactions of nucleophilic anions in CTABr.

Table II. Micellar Effects on Reactions of Hydrophilic Anions

Reaction	K_{BrY}	k_2^m/k_W
$CH_3CO_2C_6H_4NO_2(4) + OH^-$	13	0.14[a]
$C_7H_{15}CO_2C_6H_4NO_2(4) + OH^-$	13	0.11[a]
NC–⌬N$^+C_{10}H_{21} + OH^-$	13	3.4[b]
H_3C–⌬–$SO_2CH_2OSO_2$–⌬–Cl + OH^-	25	0.14[c]
⌬N$^+C_{16}H_{33} + CN^-$	1	2.5[d]
$(C_6H_5CO)_2O + HCO_2^-$	10	0.21[e]
⌬⌬ (Cl, NO_2, NO_2) + N_3^-	2	400[f]
$C_6H_5CO_2C_6H_3$-2,4-$(NO_2)_2 + N_3^-$	2	1[f]

NOTE: In CTABr.
[a] Reference 73.
[b] Reference 5.
[c] Reference 74.
[d] Reference 22.
[e] Reference 75.
[f] Reference 37.

An alternative approach is to prepare a reactive-ion surfactant for which the counterion is itself the reactant and inert counterions and interionic competition are absent (*21–23*). In principle, this method simplifies estimation of the concentration of an ionic nucleophile, for example, in the micellar pseudophase. Both these treatments of ionic reactions involve assumptions and approximations that seem to be satisfactory, provided that ionic concentrations are low, e.g., <0.1 M. These assumptions and approximations fail when electrolyte concentrations are high (*24–25*). A more rigorous treatment is based on application of the Poisson–Boltzmann equation in spherical symmetry (*26–28*), and this treatment accounts for some of the failures of the simpler models (*29*, *30*).

Equations 2 and 5 assume that reactions are sufficiently slow for equilibrium to be maintained between aqueous and micellar pseudophases, but this condition is easily satisfied for most thermal reactions. The form of equation 5 is such that rate–surfactant profiles can be fit relatively easily, but the second-order rate constants, k_W and k_M, have different dimensions and cannot be compared directly. Comparison can be made by converting concentration in the micellar pseudophase from a mole ratio (equation 4) into a molarity. This conversion involves estimation of the molar volume element of reaction, V_M, in the micellar pseudophase (*4–10, 31*), and most workers use values of 0.14–0.37 L. If the lower value is taken, we obtain

$$k_2^m = 0.14k_M \quad (6)$$

and k_W and k_2^m (the second-order rate constant in the micelles) have the same dimensions and their values can be compared directly.

Thus, rate constants can be extracted for both spontaneous and bimolecular, nonsolvolytic reactions in micelles, and when these rate constants are compared with rate constants in water, the factors that control micellar rate enhancements can be identified. Micelles can exert a medium effect on reaction rate because the polarities of their surfaces appear to be lower than that of water (*32, 33*), and micelles could also, in principle, reduce the nucleophilicity or basicity of water. They could also affect the reactivity of nucleophilic anions, and this aspect of the problem will be considered first.

Nucleophilicities at Micellar Surfaces

Second-order rate constants at micellar surfaces can be calculated by using the pseudophase model, and a considerable amount of evidence exists for the generalization that for most nucleophiles second-order rate constants are similar in aqueous and micellar pseudophases (*5–10*). The evidence comes from the work of a number of groups and the quantitative conclusions depend on assumptions regarding the volume element of reaction and, for ionic reactions, the value of the ion-exchange constant (*5*). Various values of these parameters are used, which complicates precise comparisons of the results, but differences are generally within a factor of 2, so that qualitative comparisons are valid. An additional point is that a parameter such as the volume element of reaction probably depends upon the hydrophobicities of the reactants, because hydrophilic reactants will stay at the water-rich micellar surface whereas more hydrophobic reactants may penetrate more deeply into the micelle (*34, 35*). In addition, this volume may be sensitive to ionic concentrations (*36*). This problem is probably less serious than appears at first sight because most nucleophilic reactions involve polar substrates that are located at the surface and close to the micellar head groups. Detailed analysis of these questions is outside the scope of this chapter but is given elsewhere (*5, 7, 9*).

The second-order rate constants, k_2^m (Tables I and II), are calculated with volume elements of reaction varying between 0.14 and 0.37 depending upon the assumptions made by the various authors, but generally values of k_2^m and k_W are not very different. The reagents vary considerably in their affinities for micelles and a wide range of rate enhancements has been observed, depending largely upon the hydrophobicities of the reagents and the surfactant concentration. Only a very small sampling of the data is shown here, and much more extensive compilations are available (*5*). For reactions of nucleophilic anions, values of k_2^m calculated by using the ion-exchange model differ slightly from those based on the Poisson–Boltzmann equation (*30*), but these differences do not invalidate the general conclusions.

The data in Tables I and II, together with extensive additional evidence, allow several generalizations to be made about micellar effects upon bimolecular reactions (*5*). First, overall rate constants follow the distribution of both reactants between water and micelles. Second, second-order rate constants for reactions of nonionic nucleophiles are lower in micelles than in water. Third, second-order rate constants for reactions of anionic nucleophiles are similar in water and micelles except for some reactions of azide ion (*37*).

Second-order rate constants in the micelles (k_2^m) do not depend in any obvious way upon the hydrophobicities of the reactants or, for anionic nucleophiles, upon the surfactant counterion. The relatively small inhibitory micellar medium effect on reactions of nonionic nucleophiles is readily explained by the lower polarity of the micellar surface relative to water (*32*, *33*). The generally small effects of the medium upon the ionic reactions are also understandable because water activity and ionic hydration are similar at the micellar surface and in water (*34*, *35*, *38*).

Small rate enhancements occur for aromatic nucleophilic substitution as compared to deacylation or dephosphorylation. These effects may be due to favorable interactions between cationic micellar head groups and the transition states whose structures are akin to those of Meisenheimer complexes.

Nonionic micelles often have little effect on reactions of anionic nucleophiles, although if the substrate is very hydrophobic, it may penetrate the micelle and therefore be shielded from attack by hydrophilic anions (*39*).

Anionic micelles exclude reactive anions from their surfaces and therefore inhibit reaction. However, addition of high concentrations of electrolyte reduces the potential at the micellar surface so that anions may approach that surface sufficiently closely to react with micellar-bound substrate. However, these (residual) reactions are generally very much slower than reaction in water and can be neglected except in special cases (*40*).

Reactions of Azide Ion

Aromatic nucleophilic substitution, S_NAr, by azide ion is a significant exception to the generalization that second-order rate constants are similar in water and micelles (*37*). Values of k_2^m and k_W in cationic micelles are similar for deacylation, an S_N2 reaction and addition to (2,2′,4,4′,4″-pentamethoxy)-triphenylmethyl cation (Table II and references *37* and *41*). These results suggest that the nucleophilicity of N_3^- is not increased by interaction with cationic micelles. But k_2^m is considerably larger than k_W for aromatic nucleophilic substitution in reactions of N_3^- with 2,4-dinitrochlorobenzene,

2,4-dinitrochloronaphthalene, and hydrophobic *N*-alkyl-2-bromopyridinium ions (*37*). Therefore, the anomalous rate effect does not depend in any obvious way upon substrate hydrophobicity or charge.

In water, N_3^- is much less reactive in aromatic nucleophilic substitution than expected from its reactivity toward carbocations, that is, its N_+ value. Ritchie (*43*) initially developed his N_+ scale from nucleophilicities toward preformed carbocations and the scale fits the data for nucleophilicities toward many electrophiles, regardless of their charge. However, in water, and similar hydroxylic solvents, the nucleophilicity of azide ion, relative to that of other anions, seems to be related to the carbocation-like character of the electrophile. An acyl derivative with its sp^2 carbonyl group is somewhat akin to a carbocation stabilized by an alkoxide group, $>C{=}O \longleftrightarrow >C^+{-}O^-$, just as a triarylmethyl carbocation is stabilized by electron delocalization into the aryl groups and azide ion is a good nucleophile toward these electrophiles. As compared with anions such as OH^- or CN^-, azide ion, in water, is very reactive toward carbocations and in deacylation but is relatively unreactive toward dinitrohaloarenes (*44*).

The marked difference between k_2^m and k_W for reaction of azide ion and aromatic substrates may be due to unusually low reactivity in water, rather than high reactivity in micelles. Nucleophilicity of azide ion in micelles toward the various substrates is consistent with its N_+ value. Azide ion is unusual in that it is generally an effective nucleophile but also a reasonably good leaving group, and the different behaviors in water and micelles may also be related to partitioning of covalent intermediates in S_NAr reactions (*45*).

Generalizations on Nucleophilicity

The overall conclusion is that cationic micelles do not increase the nucleophilicity of anions (Tables I and II), except as noted for some reactions of azide ion. Nonionic nucleophilicity is slightly lower in micelles than in water simply because of their lower polarity. The overall rate enhancements are generally due to increased concentration of reagents in the micelles, and similar conclusions can be drawn for reactions in microemulsions (*46*) vesicles (*47*, *48*), and inverse micelles in apolar solvents (*49*).

Micelles of functional surfactants have the reactive group, typically an oximate, a hydroxamate, a thiolate, or an alkoxide, covalently bound in the surfactant head group. These micelles often give very large rate enhancements, which, as with nonfunctional micelles, are due largely to a high concentration of the reactive group at the micellar surface (*5*, *9*, *50*, *51*).

Any conclusion about solvent effects upon bimolecular, nonsolvolytic, reactions depends upon a definition of concentration. Conventionally, for solution reactions, concentration is written as molarity and generally, for nonmicellar reactions, in terms of total solution volume. The same approach is used for micellar reactions, although other measures, for example, mole fraction, might be more appropriate. The calculation of second-order rate constants in terms of an estimated molarity at the micellar surface is open to criticism, but the situation is no different when comparison is made between reactivities in different homogeneous solvents.

Spontaneous Uni- and Bimolecular Reactions

Spontaneous hydrolyses of alkyl halides, sulfonic esters, and acid chlorides and deacylations are typically micellar-inhibited. These reactions are also inhibited by a decrease in the water content of aqueous–organic solvents, although the extent of inhibition depends upon the reaction mechanism. Inhibition by micelles can be ascribed to their having lower polarities (*32, 33*) than bulk water, and comparisons of reaction rates in micelles and aqueous–organic solvents have been used to estimate effective dielectric constants or polarities at micellar surfaces (*52*). These comparisons require that the rate constant, k'_M (equation 2) can be estimated directly or indirectly by analysis of rate–surfactant profiles generally by use of equation 2, or equations similar to it (*12, 52–55*). k'_M is difficult to estimate for reactions that are strongly micellar inhibited, because even when most of the substrate is bound, the contribution of the residual reaction in water is important. Under these conditions, values of k'_M/k'_W, that is, k_{rel}, contain uncertainty, but this finding is not sufficient to invalidate the qualitative conclusions regarding micellar rate effects. Most of the original work with cationic micelles was done by using CTABr, whose limited solubility restricted the concentration that could be used (*54*). Recently, CTACl was used and its relatively high solubility allows use of higher [surfactant] and better estimation of k'_M.

In addition to a generalized medium effect, micelles also have a charge effect that seems to be related to the reaction mechanism. Most of the experiments were made by using CTAX as the cationic and SDS as the anionic surfactant, and the rate constants for reaction of fully micellar bound substrates are designated k^+ and k^-; that is, k^+ and k^- are values of k'_M in cationic and anionic micelles, respectively (*54, 55*).

All the results to date are covered by a simple generalization: if bond making is dominant in the transition state, $k^+/k^- > 1$, but if bond breaking is dominant, $k^+/k^- < 1$. Some examples of these S_N reactions are given in Table III, and compounds that react by mechanisms at the extremes of the S_N1–S_N2 spectrum fit the generalization. For example, hydrolyses of adamantyl substrates, for which $k^+/k^- < 1$, involve rate-limiting formation of a carbocation (or its ion pair):

$$\text{R–X} \longrightarrow \text{R}^+ + \text{X}^- \xrightarrow[\text{fast}]{\text{H}_2\text{O}} \text{ROH} + \text{H}^+ \qquad (7)$$

Table III. Spontaneous S_N Reactions

Substrate	$10^3k'_W$ *(s^{-1})*	10^3k_{rel}			k^+/k^-
		SDS	*CTABr*	*SB*	
$CH_3O_3SC_6H_5$	0.0	390	700[a]		1.8
2-Ad$O_3SC_6H_4$-4-Br	1.6	~4	1		0.2
2-Ad$O_3SC_6H_4$-4-NO^2	9.5	4	1	2	0.3
Pin-$O_3SC_6H_4$-4-Br	7.1	15	8		0.5
Pin-$O_3SC_6H_4$-4-NO_2	70.0	10	5	5	0.5
$(CH_3)_3CCHC_6H_5Cl$	0.5	7	1		0.1
$C_6H_5CH_2Br$	0.2	130	90		0.7

SOURCE: References 54 and 55.
[a] In cetyltrimethylammonium mesylate.

Reaction of methyl benzenesulfonate, for which $k^+/k^- > 1$, involves bimolecular nucleophilic attack by water in the transition state:

$$H_2O + CH_3O_3SC_6H_5 \rightarrow [H_2O\text{--}CH_3\text{--}O_3SC_6H_5] \rightarrow CH_3OH + C_6H_5SO_3H \quad (8)$$

The data also suggest that bond breaking plays a major role in most S_N hydrolyses in aqueous micelles except for a methyl substrate.

Bond making is clearly dominant in spontaneous hydolyses of carboxylic anhydrides and diaryl carbonates and here $k^+/k^- > 1$ (Table IV). These values of k^+/k^- are not related in any obvious way to the reactivity or hydrophobicities of the substrates, although hydrophobicity seems to affect the overall micellar inhibition, probably because the more hydrophobic substrates penetrate the micelles and are shielded from water molecules.

Table IV. Hydrolyses of Anhydrides and Diaryl Carbonates

Substrate	$10^3k'_W$ *(s^{-1})*	k_{rel}		k^+/k^-
$(4\text{-}O_2NC_6H_4CO)_2O$	26.60	0.06	0.20	3.4
$(C_6H_5CO)_2O$	0.33	0.02	0.06	2.2
$[4\text{-}t\text{-}(CH_3)_3CC_6H_4CO]_2O$	0.21	0.01	0.02	1.8
$(4\text{-}O_2NC_6H_4CO)_2CO$	0.45	0.10	0.54[a]	5.0

NOTE: Reference 54.
[a] In the sulfobetaine (SB3–16), $k_{rel} = 0.62$.

Acyl triazoles and carboxylic anhydrides have similar mechanisms of hydrolysis, and consistently ionic micelles inhibit hydrolysis of triazoles with $k^+/k^- > 1$ (52).

Hydrolysis of Acid Chlorides

Hydrolysis rates show a striking dependence upon micellar charge. Micellar inhibition is generally observed, but for a series of benzoyl chlorides there is a relationship between values of k^+/k^- and electronic effects of para substituents. Strongly electron-withdrawing substituents, for example, NO_2, Cl, or Br, lead to values of $k^+/k^- > 1$, whereas a strongly electron-donating substituent, for example, OCH_3, leads to $k^+/k^- < 1$ (Table V). These changes in k^+/k^- can be related to changes in the relative importance of bond making and breaking.

Table V. Hydrolysis of Acyl Chlorides and Chloroformates

Substrate	$10^3k'_W$ *(s^{-1})*	k_{rel} *SDS*	*CTAX*	*SB3–16*	k^+/k^-
$3,5\text{-}(O_2N)_2C_6H_3COCl$	200	0.30	>2	>7	
$4\text{-}O_2NC_6H_4COCl$	53	0.12	~2	~2.70	~16
$4\text{-}BrC_6H_4COCl$	190	0.01	0.50	0.07	5.0
$4\text{-}ClC_6H_4COCl$	214	0.01	0.05		5.0
C_6H_5COCl	1410	0.01	0.008[a]	0.02	0.6
$4\text{-}CH_3C_6H_4COCl$	~3 × 10^3	0.00	0.00		0.1
$4\text{-}CH_3OC_6H_4COCl$	~7 × 10^3	0.01	0.00	0.00	0.1
C_6H_5OCOCl	14	0.18	0.3[a]	0.48	1.7
$4\text{-}O_2NC_6H_4COCl$	76	0.32	1.7[a]	2.20	5.5

SOURCE: Reference 54 and unpublished data of M. M. Mhala and J. R. Moffatt.
NOTE: X = Cl unless specified.
[a] X = Br.

Hydrolyses of acyl halides are sometimes described in terms of the S_N1–S_N2 duality of the mechanism, or variants of it (*56, 57*), but these descriptions are unsatisfactory because they neglect the possibility of rehybridization of the carbonyl group in the course of reaction. Strongly electron withdrawing substituents favor nucleophilic addition by water to acyl centers, with assistance by a second water molecule acting as a general base (*58–60*), and good evidence for this mechanism exists in hydrolyses of carboxylic anhydrides and diaryl carbonates. This addition step should be followed by very rapid conversion of an anionic covalent intermediate into products, and the intermediate should have only a transient existence, at most, in polar, nucleophilic solvents.

At the other mechanistic extreme, hydrolysis of 4-methoxybenzoyl chloride, breaking of the C–Cl bond has probably made considerable progress in the transition state, and nucleophilic participation by water has made correspondingly less progress. However, little evidence exists for acyl cations as intermediates in hydrolyses of this, and similar, acyl chlorides, in water-rich solvents (*61*).

Scheme I illustrates these mechanistic pathways at the simplest level of description.

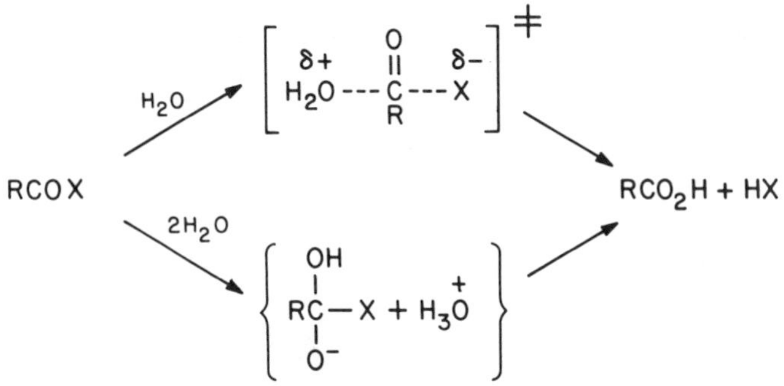

$$RCOX \xrightarrow{H_2O} \left[\overset{\delta+}{H_2O} \text{---} \underset{R}{\overset{O}{\overset{\|}{C}}} \text{---} \overset{\delta-}{X} \right]^{\ddagger} \longrightarrow RCO_2H + HX$$

$$RCOX \xrightarrow{2H_2O} \left\{ R\underset{O^-}{\overset{OH}{C}}\text{—}X + H_3\overset{+}{O} \right\} \longrightarrow RCO_2H + HX$$

Scheme I

Values of k^+/k^- (Table V) give an indication of the relative importance of bond making and breaking in hydrolyses of acyl chlorides, and the electronic effects upon these steps can be rationalized in terms of three-dimensional free-energy diagrams that consider reactions involving either prior addition or prior ionization as the mechanistic extremes (*62–64*). Aryl chloroformates should be similar to nitrobenzoyl chlorides in their hydrolytic mechanisms because aryloxy groups are strongly electron withdrawing and as expected $k^+/k^- > 1$ (Table V).

Most spontaneous deacylations are inhibited by aqueous ionic micelles, and the only exceptions to date are hydrolyses of 4-nitrophenyl chloroformate and nitrobenzoyl chlorides. These hydrolyses are faster in cationic micelles than in water, although anionic micelles of SDS inhibit reaction, and nucleophilic addition should be most dominant for these reactions.

Micellar effects upon these hydrolyses can be compared with those upon water addition to a preformed carbocation. The choice of carbocation is critical because, if it is too hydrophilic, it will not bind to a cationic micelle and, if it is too stable, reaction does not go to completion.

Addition to Carbocations

The 2,2′,4,4′,4″-pentamethoxytriphenyl methyl cation (**I**) is a suitable substrate for this work. Anionic micelles of SDS have little effect on the rate of water addition, but cationic micelles of CTACl and CTABr speed the reaction (*41*). This micellar effect of $k^+/k^- \sim 5$ is as predicted for a reaction in which water addition is dominant (Scheme II).

$$\left(\text{MeO}-\bigcirc\!\!\!\!\!\!\bigcirc-\text{OMe}\right)_2\text{C}-\bigcirc\!\!\!\!\!\!\bigcirc-\text{OMe} \xrightarrow{H_2O} \left(\text{MeO}-\bigcirc\!\!\!\!\!\!\bigcirc-\text{OMe}\right)_2\text{C(OH)}-\bigcirc\!\!\!\!\!\!\bigcirc-\text{OMe}$$

Scheme II

Hydrolysis of Benzenesulfonyl Chlorides

Most of the experiments have involved nucleophilic reactions at a carbon center, but the general principles can be applied to reactions at heteroatoms although only hydrolysis of arenesulfonyl chlorides has been examined to date (*65*).

Spontaneous hydrolyses of sulfonyl chlorides are believed to involve nucleophilic attack in the rate-limiting step although questions arise as to the timing of the bond-making and -breaking steps, because attack of water and loss of Cl^- could be concerted or stepwise (*66–68*). The rate sequence for reactions of para-substituted benzenesulfonyl chlorides in water is $H_3CO > CH_3 > H > Br < NO_2$.

Rate extremes with systematic variation of substituents are often considered to be evidence of changes in the molecularity of a reaction, but for hydrolyses of arenesulfonyl chlorides, rate extremes are more reasonably ascribed to variations in the extents of S–O bond making and S–Cl bond breaking. Variations in k^+/k^- support this hypothesis (Table VI), and as for hydrolyses of acid chlorides (Table V), bond making seems to be important but introduction of electron-donating groups increases the importance of bond breaking in the transition state.

Table VI. Hydrolysis of Benzenesulfonyl Chlorides

		k_{rel}			
Substituent	$10^3 k'_W$ (s^{-1})	*SDS*	*CTACl*	*SB3–16*	k^+/k^-
4-NO_2	2.45	0.04	0.86	0.89	21
4-Br	1.96	0.04	0.21	0.17	5
4-H	3.07	0.01	0.05	0.05	3
4-CH_3	3.86	0.01	0.03	0.03	3
4-CH_3O	6.10	0.01	0.01	0.01	1.1

NOTE: At 25.0 °C.
SOURCE: Reference 65.

Betaine Surfactants

Thus far, only ionic micelles have been discussed, but some results were obtained for reactions in zwitterionic micelles of the betaine (B1–12) and the sulfobetaine (SB3–16). These surfactants differ in that the carboxylate moiety of B1–12 could react nucleophilically, and this behavior is observed in deacylation and in hydrolyses of acid chlorides (*65*). In these reactions, the overall effect is due to nucleophilic attack, which opposes an inhibition due to the medium effect of the micelle. The balance between these effects depends largely upon the susceptibility of the substrate to nucleophilic attack.

The sulfonate moiety is only weakly nucleophilic in water, and the effects of micellized SB3–16 upon hydrolyses are generally very similar to those of cationic micelles (Tables IV–VI). This behavior suggests that in both systems the substrates bind close to the quaternary ammonium centers and that the sulfonate moiety, like a micellar-bound counteranion, is in the water and therefore not interacting strongly with the substrate (*65*). The only exceptions to this generalization were observed with hydrolyses of some acid chlorides (Table V). In these reactions, the balance between bond making and breaking seems to be very sensitive to the reaction medium (*58*).

Duality of Mechanism in Spontaneous Reactions

The molecularity of nonsolvolytic substitutions is given by the kinetic order, although second-order, bimolecular reactions may be concerted or stepwise. Other tests have to be used for spontaneous reactions, and many tests have been developed especially for reactions at saturated carbon (*69–71*). Compelling evidence suggests that ionization occurs in solvolyses of alkyl halides or arenesulfonates where a carbocation or its ion pair can be trapped in the course of reaction, and nucleophilic attack on methyl groups is all important, so that in these cases a clear distinction exists between the so-called S_N1 and S_N2 mechanisms. The concept of the duality of mechanism that developed from these reactions has been applied to a wide variety of organic and inorganic reactions and has tended to dominate thinking about reaction mechanisms in solution (*69*). Even here problems arise in the mechanistic description because nucleophilic protic solvents can not only participate at the electrophilic center nucleophilically (*70*) or by dipole–dipole interactions but can also solvate a leaving group (*72*).

Despite earlier controversy, the weight of evidence suggests that the S_N1 and S_N2 models are extremes in a mechanistic spectrum. Many solvolytic reactions at saturated carbon seem to involve linked bond making and breaking, as well as stabilization of the transition state by hydrogen bonding to a leaving anion, and the relative importance of these steps depends on substrate structure and on nucleophilicity of the solvent and its interaction with leaving groups.

Water is not an especially nucleophilic solvent, but water effectively solvates anionic leaving groups so that in water considerable ionization in the transition state should occur (*72*). The surfaces of micelles are water-rich (*34, 35*) and values of k^+/k^- for hydrolyses of alkyl halides and arenesulfonates suggest that nucleophilic participation is dominant in reactions at methyl groups but that increased alkyl and especially aryl substitution leads to bond breaking becoming dominant.

The situation is more complex for spontaneous hydrolyses of acyl derivatives, and values of k^+/k^- depend upon the ease of departure of the leaving group and attack of water and, for a given leaving group, upon electronic substituent effects.

For relatively poor leaving groups, for example, RCO_2^- or ArO^- where bond making is all important, a carbonyl addition mechanism can be written that is equivalent to the $B_{Ac}2$ mechanism of ester hydrolysis. The situation is more complicated when a good leaving group is present, as with acid chlorides or chloroformates. If strongly electron withdrawing groups are present, nucleophilic addition is dominant, and for hydrolysis of aryl chloroformates and nitrobenzoyl chlorides, the mechanism is similar to that of anhydride hydrolysis (Tables IV and V). But electron-donating groups favor bond breaking, and the mechanism of hydrolysis of benzoyl chloride and its 4-CH_3 and 4-CH_3O derivatives has considerable S_N2-like character, with extensive bond breaking in the transition state, which will have an open structure with extended bonding to the entering and leaving groups (*54*, *61*).

The relative importance of bond making and breaking in the transition state depends not only upon the substrate structure but also upon the reaction medium. Electronic effects upon solvolyses of benzoyl chlorides are very sensitive to changes in the solvent (58). In formic acid, a weakly nucleophilic solvent, but one that should effectively solvate leaving anions, a plot of log k against Hammett's ρ parameter has a slope of $\rho = -4.4$, but in a poorly ionizing solvent (95% acetone–5% H_2O), $\rho \sim 2$, but a shallow minimum occurs at $\sim \rho -0.2$ for strongly electron denoting groups (e.g., CH_3 and OCH_3). In 50% acetone–50% H_2O, the corresponding plot has a very well defined minimum at ~ 0.2. For reaction in water, $\rho = -3.4$, based on hydrolyses of benzoyl chloride and its 4-Cl, $-CH_3$, and $-OCH_3$ derivatives, but a minimum occurs at ~ 0.5. This behavior is understandable because water is a much better nucleophile than formic acid and a much better ionizing medium than aqueous acetone (*71*).

Whether changes in the relative importance of bond making and breaking, as in solvolyses of acyl chlorides, are to be regarded as changes in reaction mechanism is a matter of opinion, but clearly micelles, like any other reaction medium, can influence transition-state structure. Therefore, although values of k^+/k^- can be considered as indicative of "mechanism", the conclusions apply only to reactions taking place at micellar surfaces. However, these surfaces are water-rich, so the transition-state structures are expected to be similar to those in water.

All the evidence to date fits the hypothesis that micellar charge effects are related to mechanism, but the results are not so easy to explain. In a reaction dominated by bond breaking, positive charge developing at the reaction center should interact favorably with an anionic head group and the leaving anion will interact with water molecules adjacent to the micellar surface. For a spontaneous reaction, dominated by bond making, positive charge develops on the attacking water molecule but is distributed into nearby water molecules by hydrogen bonding. At the same time, negative charge will develop at, or adjacent to, the reaction center and will interact favorably with a micellar cationic head group. A preformed carbocation interacts unfavorably with a cationic head group, and this interaction will decrease as water adds to the carbocation, and for this reaction and those of strongly electrophilic acid chlorides, cationic micelles speed reaction, despite their general inhibitory effect upon spontaneous hydrolyses.

Acknowledgments

Support of this work by the National Science Foundation (Chemical Dynamics Program) and the Army Office of Research is gratefully acknowledged. The efforts of my co-workers who are cited here are very much appreciated.

List of Symbols

k_ψ Overall first-order rate constant, s^{-1}
k'_W First-order rate constant in water, s^{-1}
k_W Second-order rate constant in water, $M^{-1}s^{-1}$
k'_M First-order rate constant in the micelles, s^{-1}
k_M Second-order rate constant in the micelles, s^{-1}, with concentration written as a mole ratio
k_2^m Second-order rate constant in the micelles, $M^{-1}\ s^{-1}$; with concentration written as a molarity
k_{rel} Ratio of the overall rate constants in the presence and absence of micelles
K_S Binding constant of solute to the micelles
K_X^Y Ion-exchange constant for ions Y and X, given by $[Y_W][X_M]/[Y_M][X_W]$
m_Y^s Mole ratio of micellar-bound Y to micellized surfactant
cmc Critical micelle concentration
[D] Stoichiometric concentration of surfactant (detergent)
$[D_n]$ Concentration of micellized surfactant
V_M Volume element of reaction, sometimes identified with partial molar volume of a micelle
W Aqueous pseudophase when used as a subscript
M Micellar pseudophase when used as a subscript

Literature Cited

1. Fendler, J. H.; Fendler, E. J. *Catalysis in Micellar and Macromolecular Systems;* Academic: New York, 1975.
2. Fendler, J. H. *Membrane Mimetic Chemistry;* Wiley-Interscience: New York, 1982.

3. Cordes, E. H.; Gitler, C. *Prog. Bioorg. Chem.* **1973,** *2,* 1.
4. Romsted, L. S. In *Micellization, Solubilization and Microemulsions;* Mittal, K. L., Ed.; Plenum: New York, 1977; Vol. 2, p 509.
5. Romsted, L. S. In *Surfactants in Solution;* Mittal, K. L.; Lindman, B., Eds. Plenum: New York, 1984; Vol. 2, p 1015.
6. Martinek, K.; Yatsimirski, A. K.; Levashov, A. V.; Berezin, I. V. In *Micellization, Solubilization and Microemulsions;* Mittal, K. L., Ed.; Plenum: New York, 1977; Vol. 2, p 489.
7. Sudholter, E. J. R.; van der Langkruis, G. B.; Engberts, J. B. F. N. *Recl. Trav. Chim. Pays-Bas,* **1980,** *99,* 73.
8. Cuccovia, I. M.; Schroter, E. M., Monteiro, P. M.; Chaimovich, H. *J. Org. Chem.* **1978,** *43,* 2248.
9. Bunton, C. A. *Catal. Rev.—Sci. Eng.* **1979,** *20,* 1.
10. Chaimovich, H.; Bonilha, J. B. S.; Politi, M. J.; Quina, F. H. *J. Phys. Chem.* **1979,** *83,* 1851.
11. Bunton, C. A.; Fendler, E. J.; Sepulveda, L.; Yang, K.-U. *J. Am. Chem. Soc.* **1968,** *90,* 5512.
12. Menger, F. M.; Portnoy, C. E. *J. Am. Chem. Soc.* **1967,** *89,* 4698.
13. Epstein, J.; Kaminski, J. J.; Bodor, N.; Enever, R.; Sowa, J.; Higuchi, T. *J. Org. Chem.* **1978,** *43,* 2816.
14. Bunton, C. A.; Cerichelli, G.; Ihara, Y.; Sepulveda, L. *J. Am. Chem. Soc.* **1979,** *101,* 2429.
15. Larsen, J. W.; Tepley, L. B. *J. Colloid Interface Sci.* **1974,** *49,* 113.
16. Bunton, C. A.; Ohmenzetter, K.; Sepulveda, L. *J. Phys. Chem.* **1977,** *81,* 2000.
17. Ray, A.; Nemethy, G. *J. Phys. Chem.* **1971,** *75,* 809.
18. Ionescu, L. G.; Fung, D. S. *J. Chem. Soc., Faraday Trans. 1,* **1981,** *77,* 2907.
19. Menger, F. M.; Jerkunica, J. M. *J. Am. Chem. Soc.* **1979,** *101,* 1896.
20. Bunton, C. A.; Gan, L.-H.; Hamed, F. H.; Moffatt, J. R. *J. Phys. Chem.* **1983,** *87,* 336.
21. Bunton, C. A.; Romsted, L. S.; Savelli, G. *J. Am. Chem. Soc.* **1979,** *101,* 1253.
22. Bunton, C. A.; Romsted, L. S.; Thamavit, C. *J. Am. Chem. Soc.* **1980,** *102,* 3900.
23. Bunton, C. A.; Gan, L.-H.; Moffatt, J. R.; Romsted, L. S.; Savelli, G. *J. Phys. Chem.* **1981,** *85,* 4118.
24. Nome, F.; Rubira, A. F.; Franco, C.; Ionescu, L. G. *J. Phys. Chem.* **1982,** *86,* 1881.
25. Stadler, E.; Zanette, D.; Rezende, M. C.; Nome, F. *J. Phys. Chem.* **1984,** *88,* 1892.
26. Bell, G. M.; Dunning, A. J. *Trans. Faraday Soc.* **1970,** *66,* 500.
27. Mille, M.; Vanderkooi, G. *J. Colloid Interface Sci.* **1977,** *59,* 211.
28. Gunnarsson, G.; Jonsson, B.; Wennerstrom, H. *J. Phys. Chem.* **1980,** *84,* 3114.
29. Bunton, C. A.; Moffatt, J. R. *J. Phys. Chem.* **1985,** *89,* 4166.
30. Bunton, C. A.; Moffatt, J. R. *J. Phys. Chem.* **1986,** *90,* 538.
31. Bunton, C. A.; Carrasco, N.; Huang, S. K.; Paik, C.; Romsted, L. S. *J. Am. Chem. Soc.* **1978,** *100,* 5420.
32. Mukerjee, P. In *Solution Chemistry of Surfactants;* Mittal, K. L., Ed.; Plenum: New York, 1979; Vol. 1, p 153.
33. Ramachandran, C.; Pyter, R. A.; Mukerjee, P. *J. Phys. Chem.* **1982,** *86,* 3198.
34. Menger, F. M. *Acc. Chem. Res.* **1979,** *12,* 111.
35. Whitten, D. G., Bonilha, J. B. S.; Schanze, K. S.; Winkle, J. R. In *Surfactants in Solution·* Mittal, K. L.; Lindman, B.; Eds. Plenum: New York, 1984; Vol. 1, p 585.
36. Hicks, J. R.; Reinsborough, V. C. *Aust. J. Chem.* **1982,** *35,* 15.
37. Bunton, C. A.; Moffatt, J. R.; Rodenas, E. *J. Am. Chem. Soc.* **1982,** *104,* 2653.

38. Stigter, D. *J. Phys. Chem.* **1975,** *79,* 1008.
39. Bunton, C. A.; Robinson, L. *J. Org. Chem.* **1969,** *34,* 773.
40. Quina, F. H.; Politi, M. J.; Cuccovia, I. M.; Martins-Franchetti, S. M.; Chaimovich, H. In *Solution Behavior of Surfactants;* Mittal, K. L.; Fendler, E. J. Eds; Plenum: New York, 1982; Vol. 2, p 1125.
41. Bunton, C. A.; Cuenca, A. *Can. J. Chem.* **1986,** *64,* 1179.
42. Cuenca, A. Ph.D. Thesis, University of California, Santa Barbara, 1985.
43. Ritchie, C. D. *J. Am. Chem. Soc.* **1975,** *97,* 1170.
44. Ritchie, C. D.; Sawada, M. *J. Am. Chem. Soc.* **1977,** *99,* 3754.
45. March, J. *Advanced Organic Chemistry,* 3rd ed.; Wiley-Interscience: New York, 1985; Chapter 13.
46. Mackay, R. A. *Adv. Colloid Interface Sci.* **1981,** *15,* 131.
47. Fendler, J. H.; Hinze, W. *J. Am. Chem. Soc.* **1981,** *103,* 5439.
48. Cuccovia, I. M.; Quina, F. H.; Chaimovich, H. *Tetrahedron* **1982,** *38,* 917.
49. O'Connor, C. J.; Lomax, T. D.; Ramage, R. E. *Adv. Colloid Interface Sci.* **1984,** *20,* 21.
50. Fornasier, R.; Tonellato, U. *J. Chem. Soc., Faraday Trans. 1* **1980,** *76,* 1301.
51. Bunton, C. A.; Hamed, F. H.; Romsted, L. S. *J. Phys. Chem.* **1982,** 2103.
52. Fadnavis, N.; Engberts, J. B. F. N. *J. Org. Chem.* **1982,** *47,* 415.
53. Menger, F. M.; Yoshinaga, H.; Venkatasubban, K. S.; Das, A. R. *J. Org. Chem.* **1981,** *46,* 415.
54. Al-Lohedan, H.; Bunton, C. A.; Mhala, M. M. *J. Am. Chem. Soc.* **1982,** *104,* 6654.
55. Bunton, C. A.; Ljunggren. *J. Chem. Soc., Perkin Trans. 2* **1984,** *355.*
56. Williams, A.; Douglas, K. T. *Chem. Rev.* **1975,** *75,* 627.
57. Sneen, R. A. *Acc. Chem. Res.* **1973,** *6,* 46.
58. Johnson, S. L. *Adv. Phys. Org. Chem.* **1967,** *5,* 237.
59. Schowen, R. L. *Prog. Phys. Org. Chem.* **1972,** *9,* 275.
60. Menger, F. M.; Venkatasubban, K. S. *J. Org. Chem.* **1976,** *41,* 1868.
61. Bentley, T. W.; Carter, G. E.; Harris, H. C. *J. Chem. Soc., Chem. Commun.* **1984,** 398.
62. More O'Ferrall, R. A. *J. Chem. Soc. B.* **1970,** 274.
63. Jencks, W. P. *Chem. Rev.* **1972,** *72,* 705.
64. Lowry, T. H.; Richardson, K. S. *Mechanism and Theory in Organic Chemistry,* 2nd ed.; Harper and Row: New York, 1981; Chapter 4.
65. Bunton, C. A.; Mhala, M. M.; Moffatt, J. R. *J. Org. Chem.* **1985,** *50,* 4921.
66. Kice, J. L. *Adv. Phys. Org. Chem.* **1980,** *17,* 65.
67. Strangeland, L. J.; Senatore, L.; Ciuffarin, E. *J. Chem. Soc., Perkin Trans. 2* **1972,** 852.
68. Rogne, O. *J. Chem. Soc., Perkin Trans. 2* **1975,** 1486.
69. Ingold, C. K. *Structure and Mechanism in Organic Chemistry,* 2nd ed.; Cornell University: Ithaca, NY, 1969; Chapter 7.
70. Bentley, T. W.; Carter, G. E. *J. Am. Chem. Soc.* **1982,** *104,* 5741.
71. Bentley, T. W.; Schleyer, P. v. R. *Adv. Phys. Org. Chem.* **1982,** 4635.
72. Kevill, D. N.; Kamil, W. A.; Anderson, S. W. *Tetrahedron Lett.* **1982,** 4635.
73. Quina, F. H.; Politi, M. J.; Cuccovia, I. M.; Baumgarten, E.; Martins-Franchetti, S. M.; Chaimovich, H. *J. Phys. Chem.* **1980,** *84,* 361.
74. van der Langkruis, G. B.; Engberts, J. B. F. N. *J. Org. Chem.* **1984,** *49,* 4152.
75. Al-Lohedan, H.; Bunton, C. A. *J. Org. Chem.* **1982,** *47,* 1160.

RECEIVED for review October 21, 1985. ACCEPTED January 27, 1986.

30

Cation Effects on Solvents, Ligands, and Nucleophiles

Effect of Side Arms on Cation Binding by Macrocycles

G. W. Gokel, L. Echegoyen, K. A. Arnold, T. P. Cleary, V. J. Gatto, D. A. Gustowski, C. Hanlon, A. Kaifer, M. Kim, S. R. Miller, C. Minganti, M. Ouchi, C. R. Morgan, I. Posey, R. A. Schultz, T. Takahashi, A. M. Viscariello, B. D. White, and H. Yoo

Department of Chemistry, University of Miami, Coral Gables, FL 33124

Macrocyclic polyether compounds having one (lariat ethers) or two (bibracchial lariat ethers, BiBLEs) donor-group-bearing side arms exhibit Na^+-, K^+-, NH_4^+-, and Ca^{2+}-binding affinities and selectivities different from those of the unsubstituted macrocycles. Macrocycles utilizing a nitrogen atom as the point of attachment (pivot atom) show generally higher flexibility and binding strength than compounds having the sidearm(s) attached at a carbon (carbon pivot). The more flexible and less polar compounds favor K^+ over the more charge-dense cations irrespective of hole size. The cation binding involves both the macroring and the side arms. This fact is demonstrated for solutions as well as the solid state. N-pivot BiBLEs can be prepared by a very convenient single-step cyclization or by a two-step approach that is more conventional but that affords high yields of product. Both lariat ethers and BiBLEs can be electrochemically "switched" to alter the cation-binding affinities and strengths.

CROWN ETHERS AND THEIR RELATIVES have proved so interesting to the chemical community in part because of their ability to complex and stabilize various cations. In the process, these compounds may alter the properties of associated anions. Nowhere has this anion effect been more exploited than in phase-transfer catalysis (*1–3*).

The two classes of macrocyclic polydentate cation binders that have been known the longest and have been most studied are the crown ethers developed by Pedersen (*4*) and the cryptands invented by Lehn (*5, 6*). Not

0065-2393/87/0215-0443$07.00/0

long after their introduction, efforts were made to incorporate these macrocycles into polymeric matrices of one sort or another. This work had several goals. First, incorporation in the polymer matrix possibly would enhance catalyst stability. Second, unusual selectivities might be realized from the polymer that were not observed in the monomers. Third, improvement in recoverability and thus the economics of using these relatively expensive compounds was anticipated.

One method for attaching macrocycles to polymers is to use a molecular tether such as a hydrocarbon or polyethylene glycol chain. A hydrocarbon chain attached to the macrocycle should function largely in a mechanical sense to connect backbone and ring. A polyethylene glycol chain could serve as a mechanical link but also "cooperate" with the macroring in cation binding. Also, possibly a side arm containing donor groups could dominate the cation binding and render useless the crown ether. Two types of molecular tethers and two possible methods for attaching them to crown ethers are as follows:

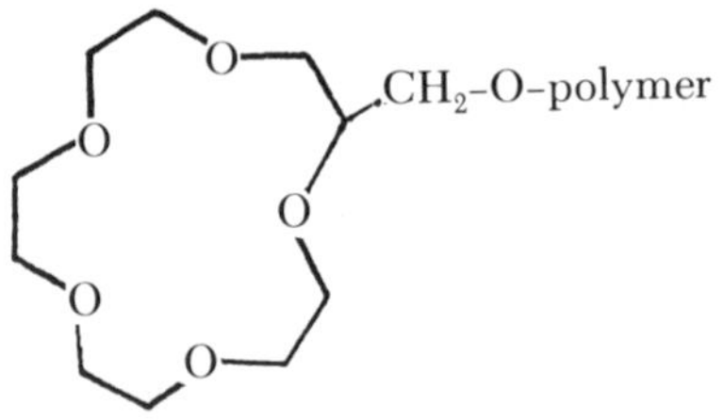

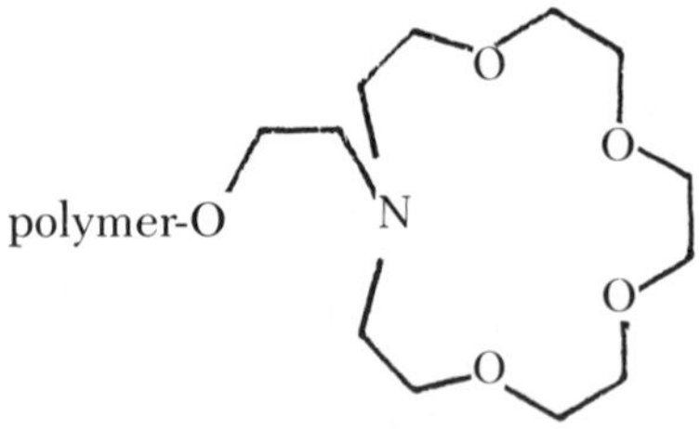

When the tether is attached to the macroring at carbon, the molecule is said to have a carbon pivot atom. When the side arm is attached as illustrated at the right, the pivot atom is nitrogen. Because the molecular models of these compounds resemble lassos and because the combination of side arm and macroring donors can permit a cation to be "roped and tied", we have called such macrocycles "lariat ethers".

Classes of Lariat Ethers

Because crown ethers and cryptands are composed most commonly of repeating $-CH_2CH_2O-$ or related units that differ by the number of carbons or the identity of heteroatoms, we can predict that side arms are likely to be attached either at carbon or at nitrogen. Sulfur and oxygen are usually divalent and therefore unavailable to serve as pivot atoms. To be effective, the side arms themselves must have embedded Lewis basic donor groups that can augment the macroring's cation-binding ability. Our own studies have thus far focused only on carbon- and nitrogen-pivot molecules and on compounds having only one or two arms. This focus was done so that detailed and systematic information about the cation-binding strengths and selectivities could be developed. Such a systematic survey should permit an understanding of cation binding in general, and this understanding, in turn, should permit an understanding of anion chemistry.

Single-Armed, Carbon-Pivot Lariat Ethers

Our survey of sidearm-bearing materials began with the readily accessible 15-membered ring, carbon-pivot systems. Glycerol, $HOCH_2–CHOH–CH_2OH$, was chosen as the basic pivot unit because it is both inexpensive and versatile. The glycerol unit was incorporated into the macrorings by using one of the primary and the secondary hydroxyl groups. The remaining hydroxyl group was used to attach the tether. Practically, attachment was accomplished by the reaction of epichlorohydrin with either an alcohol or a phenol as the nucleophile. The glycidyl ethers were then hydrolyzed to yield the $R–O–CH_2–CHOH–CH_2OH$ derivatives required for cyclization. The diols were cyclized in the standard way [NaH, tetrahydrofuran (THF)] with tetraethylene glycol ditosylate or tetraethylene glycol dimesylate (7).

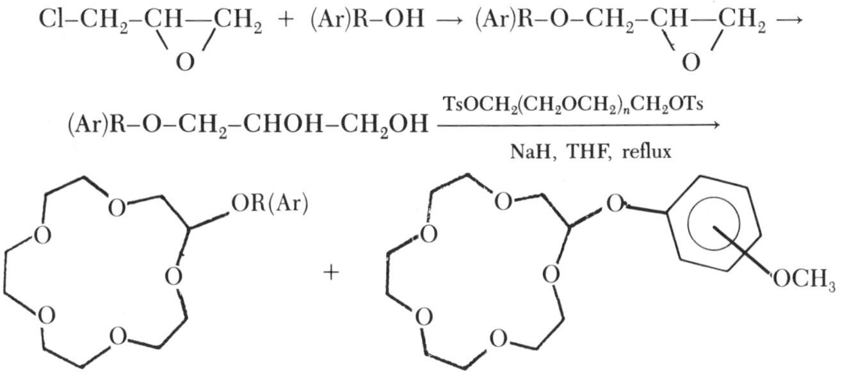

$$Cl–CH_2–\underset{\backslash\ O\ /}{CH—CH_2} + (Ar)R–OH \longrightarrow (Ar)R–O–CH_2–\underset{\backslash\ O\ /}{CH—CH_2} \longrightarrow$$

$$(Ar)R–O–CH_2–CHOH–CH_2OH \xrightarrow[\text{NaH, THF, reflux}]{TsOCH_2(CH_2OCH_2)_nCH_2OTs}$$

OR(Ar) + OCH₃

I, ortho
II, para

Early results with these systems were encouraging. A comparison of 2- (**I**) and 4-(methoxyphenoxy)methyl (**II**) ($CH_3–O–C_6H_4–O–CH_2–$) derivatives of 15-crown-5 (**III**) with the latter compound proved instructive. Liotta (8) reported that the cyclization yield for 15-crown-5 was 29%. We found that when the methoxy group was para and too remote to interact with a ring-bound cation, the cyclization yield was 30%. When the methoxy group was ortho, the cyclization yield more than doubled (ca. 70%). Two-phase extraction constant studies (9) also proved encouraging. The method involves extraction of sodium picrate from water into a nonpolar solvent like chloroform or dichloromethane. The picrate anion is highly colored and its crown-cation-facilitated extraction into the nonpolar solvent from water can be measured with considerable accuracy. 15-Crown-5 extracts about 7% of the available sodium picrate into chloroform. Para isomer **II** extracts about 6% and ortho isomer **I** extracts about 18% of the available Na^+.

At an early stage in the work, we felt two reservations about using the extraction technique for determining cation-binding information. First, the extraction "constant" values depend on a variety of factors such as ionic strength in the salt-containing aqueous solution, temperature, the particular solvent pair chosen, the solvent volumes, the ratio of metal cation to picrate anions, and so on. Although useful information can be obtained by using this technique, all of the many variables must remain constant for data to be comparable. Our second reservation about this method was our feeling that the best binding information available was equilibrium stability constant (K_S or log K_S) values determined by NMR, calorimetry, conductivity, and other methods. Tables of these values in a variety of solvents are now readily available (*10*). When we compared the homogeneous sodium cation binding strengths of **I–III** to the values obtained by the two-phase method, quite different results were obtained. In 90% aqueous methanol solution, the Na^+-binding constants (log K_S) were, for 15-crown-5 (**III**), 2.97; for para (**II**), 2.56; and for ortho (**I**), 2.97. In other words, the sidearm did not enhance binding at all and diminished it when the donor group was inappropriately placed on the sidearm.

Although we prepared a variety of carbon-pivot macrocycles, the cation-binding strengths of these molecules was never impressive. At first, we attributed this finding to "sidedness", that is, the sidearm must always be over the same face of the macroring. In collaboration with Okahara (*11, 12*), we prepared a number of macrocycles having a geminal methyl group at the pivot atom. The binding strengths for these quaternary-methyl lariats were substantially higher than when the methyl group was absent. We believe that this is due to a conformational effect that disfavors the best binding conformation of the ring. We therefore turned our attention to the nitrogen-pivot species, which have the advantage of rapid nitrogen inversion to enhance overall molecular flexibility.

Single-Armed, Nitrogen-Pivot Lariat Ethers

The monoaza crown ethers can be conveniently prepared from *N,N*-diethanolamine. Sidearms are incorporated as electrophiles rather than as nucleophiles in the carbon-pivot series. Alkylation is generally accomplished in acetonitrile solution using Na_2CO_3 as the base. The sidearm-bearing precursor fragment, $R{-}N(CH_2CH_2OH)_2$ is then cyclized. In this case, two ethyleneoxy units are incorporated from the precursor. Reaction of $R{-}N(CH_2CH_2OH)_2$ with triethylene glycol ditosylate affords the 15-membered ring system, and reaction with tetraethylene glycol ditosylate affords monoaza-18-crown-6 derivatives.

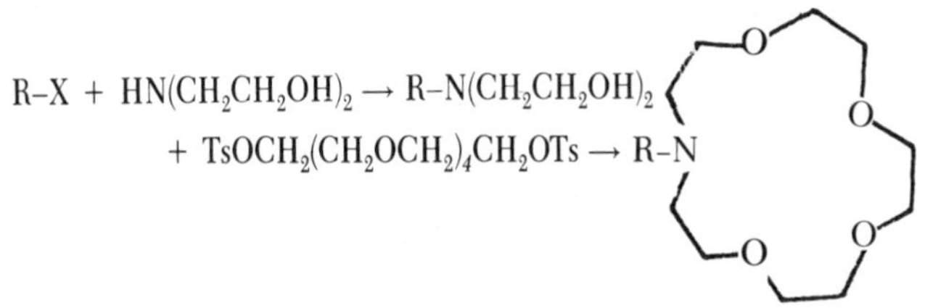

$$\text{R–X} + \text{HN(CH}_2\text{CH}_2\text{OH)}_2 \rightarrow \text{R–N(CH}_2\text{CH}_2\text{OH)}_2$$

$$+ \text{TsOCH}_2\text{(CH}_2\text{OCH}_2\text{)}_4\text{CH}_2\text{OTs} \rightarrow \text{R–N}$$

O O O O

Monoaza-12-crown-4 Derivatives

When certain long polyethyleneoxy side arms were present, formation of monoaza-15-crown-5 and 18-crown-6 derivatives by cyclization sometimes proved difficult. In such cases, the parent compounds were N-alkylated to give the desired lariat ethers. The parent compounds were obtained from the *N*-benzyl derivatives, prepared as described previously, which were hydrogenolyzed to the parent macrocycles. Such a procedure was used to prepare a number of monoaza-12-crown-4 derivatives as well (*13*).

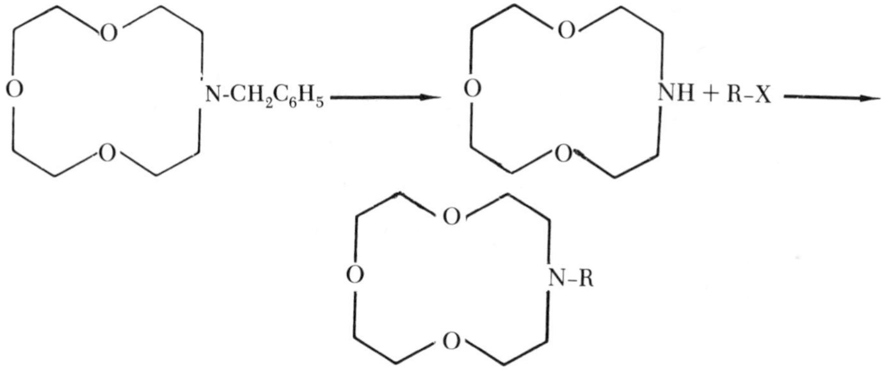

O O O $\text{N-CH}_2\text{C}_6\text{H}_5 \longrightarrow$ O O O $\text{NH} + \text{R–X} \longrightarrow$

O O O N–R

Other monoaza-12-crown-4 derivatives were prepared by using the cyclization method developed by Calverley and Dale (*14, 15*). This method involves reaction between a primary amine and tetraethylene glycol diiodide.

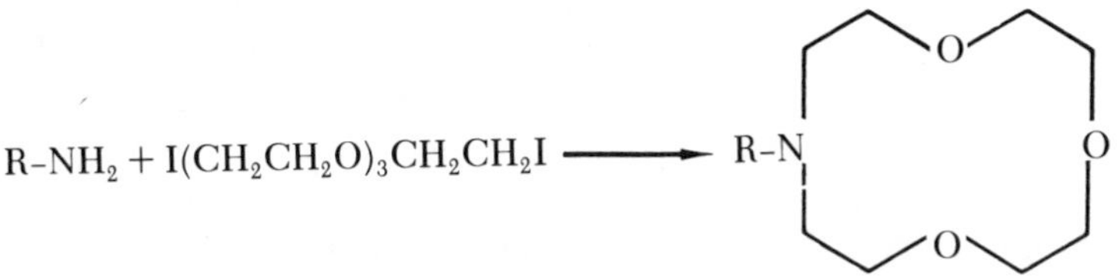

$$\text{R-NH}_2 + \text{I(CH}_2\text{CH}_2\text{O)}_3\text{CH}_2\text{CH}_2\text{I} \longrightarrow \text{R-N}$$

O O O

Cation Binding by Crown Ethers: The "Hole-Size Relationship"

Before proceeding with a discussion of how lariat ethers bind various cations, we will consider how simple monocyclic crown ethers (12-crown-4, 15-crown-5, 18-crown-6, etc.) interact with Na^+, K^+, Ca^{2+}, and NH_4^+. The binding of many other ions would doubtless be of interest, but those ions noted here are biologically relevant and of special interest to us.

The "hole-size" relationship between cations and crown ethers has been a part of the lore in the cation binding area for nearly two decades. Although, to our knowledge, no formal definition of this principle has ever been offered, the general concept seems to be that cation binding will be optimized when the cation diameter and macrocycle cavity size are identical. A simple consequence of this concept is the notion that 15-crown-5 is selective (binds more strongly) for Na^+ over K^+. We have measured the homogeneous (equilibrium) stability constants for the reaction

$$\text{ligand} + M^+ \rightleftarrows (\text{ligand}\cdot M)^+$$

with several macrocycles and cations (*16*). The Na^+-, K^+-, and NH_4^+-binding constants (determined in anhydrous methanol at 25 °C) are listed in Table I for the simple crown ethers.

Table I. Cation Stability Constants for Simple Crowns

Crown	Na^+	K^+	NH_4^+
12-crown-4	1.7	1.7	1.3
15-crown-5	3.24	3.43	3.03
18-crown-6	4.35	6.08	4.14
21-crown-7	2.54	4.35	3.27
24-crown-8	2.35	3.53	2.63

If Na^+ is selected by 15-crown-5 rather than 18-crown-6, then K_S for the formation of $(Na\cdot 15\text{-crown-}5)^+$ should exceed the binding constant for $(Na\cdot 18\text{-crown-}6)^+$. In fact, the reverse is true. An alternate interpretation of the hole-size rule is that 15-crown-5 binds Na^+ more strongly than 18-crown-6 binds Na^+. In fact, the log K_S values are 4.35 versus 3.24. This difference is more than 1 order of magnitude in *the wrong direction*. In fact, if any "rules" operate for these simple systems, these rules are as follows: (1) for this series of macrocycles, K^+ binding is greater for all ligands; and (2) all cations in this series are bound more strongly by 18-crown-6 than by any other ligand. These results are for homogeneous systems where a single equilibrium constant describes the reaction. "Extraction constants" give data of a different sort depending on how the experiments are conducted.

Several factors affect this selectivity, and they are unlikely to apply equally well to more rigid macrocycles than those noted so far. A major factor is the solvation enthalpy of each cation (*17*). Potassium cation is less charge dense (z/e is greater) than Li^+, Na^+, or Ca^{2+} (*see* Table II) (*18*). Geometry, conformation, and ridigity all play roles in determining binding strength and selectivity, but charge density is of paramount importance for flexible systems. The special stability of 18-crown-6 complexes is explained, at least in part, by symmetry and a lack of strain in the complexed (D_{3d}) conformation.

Table II. Ionic Radii and Charge Densities of Alkali and Alkaline Earth Cations

Ion	*r (Å)*	*Charge Densities*[a]
Li^+	0.68	12.1
Na^+	0.95	4.46
K^+	1.33	1.62
Rb^+	1.48	1.18
Mg^{2+}	0.66	26.8
Ca^{2+}	0.99	7.90
Sr^{2+}	1.12	5.46
Ba^{2+}	1.34	3.17

Source: Adapted from reference 18.
[a] In 10^{20} coulombs/Å^3.

Cation Binding by Monoaza Lariat Ethers

The series of *N*-substituted monoazacrown compounds having 12-, 15-, and 18-membered rings and short to relatively long $(CH_2CH_2O)_nCH_3$ side chains was prepared as described previously. Sodium cation binding was measured by ion-selective electrode methods (*19*) and is shown graphically in Figure 1.

If a simple hole-size relationship operated for these systems, a binding maximum might be expected for the 15-membered rings. This situation is not the case at all. The binding maximum in each case (filled circles, squares, and triangles) occurs when six oxygen and one nitrogen atoms are present in the donor group array. For the case of a hard ion like Na^+, it seems reasonable to focus on the number of oxygen atoms (*20*) in the donor array, as done in Figure 1. Peak Na^+ binding occurs when six oxygen atoms are present irrespective of how the donor atoms are arranged: The macroring may be 12-, 15-, or 18-membered and the side arm that complements it may contain one, two, or three oxygen atoms. Proof that the binding array is indeed three dimensional is found in the K^+ complexes of *N*-2-[2-(2-methoxyethoxy)ethoxy]ethylmonoaza-12-crown-4, **IV**, and *N*-(2-methoxyethyl)monoaza-18-crown-6, **V** (*21*). Complex **IV**·K^+ has the unique "calabash" structure we have previously reported (22) (Figure 2).

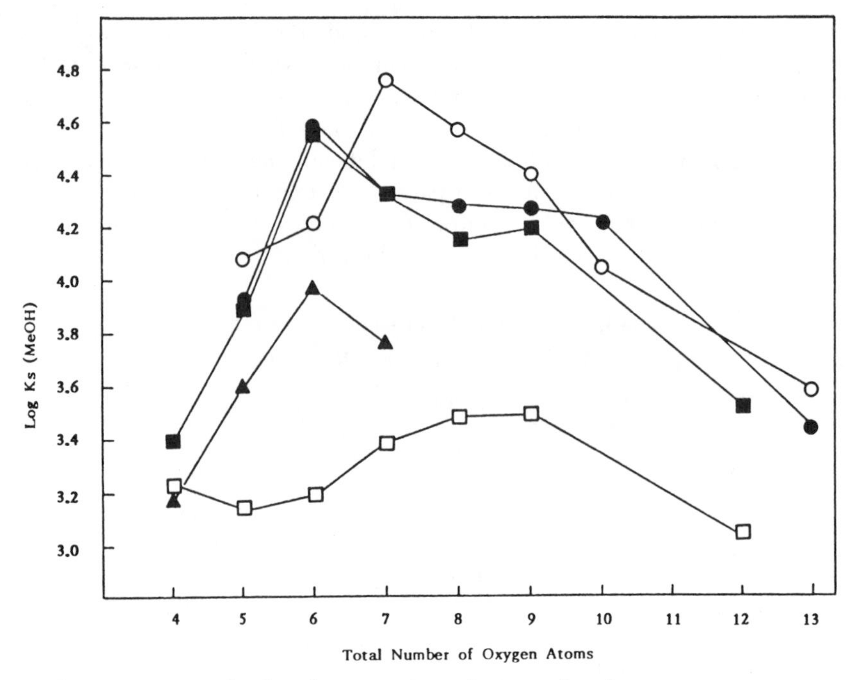

Figure 1. Cation binding by 12-, 15-, and 18-membered ring, nitrogen-pivot lariat ethers. Ring sizes: triangles, 12; squares, 15; circles, 18. Cations: filled symbols, Na^+*; open symbols,* NH_4^+.

Ammonium Cation Binding by Lariat Ethers

Another proof of intramolecular cation complexation is found in the NH_4^+ ion binding data (*23*). Ammonium ion differs from Na^+, K^+, or Ca^{2+} because it is tetrahedral rather than spherical: its N–H bonds exhibit directionality. Because the distance appropriate to form three O–H–N hydrogen bonds favors an 18-membered ring, NH_4^+ ion binding is predicted to be stronger for 18-membered rings than for either of the smaller macrocycles. In addition, an examination of Corey–Pauling–Koltun molecular models shows that the second oxygen atom in a $>NCH_2CH_2OCH_2CH_2OCH_3$ side chain is ideally placed to hydrogen bond the apical NH_4^+ hydrogen. The binding constant data (open circles and squares) shown in Figure 1 corroborate this appraisal.

Another interesting observation emerges from the ammonium cation binding data (*19*). Presumably, all four NH bonds are associated with an oxygen when optimal binding occurs. In methanol solution, peak binding occurs at 4.8 log units or 1.2 "binding" units per hydrogen bond. The 15-membered rings exhibit more modest binding and presumably never coordinate more than three NH groups at a time. Surely, that peak binding occurs for the 15-membered ring systems in methanol solution at 3.6 log units is no coincidence.

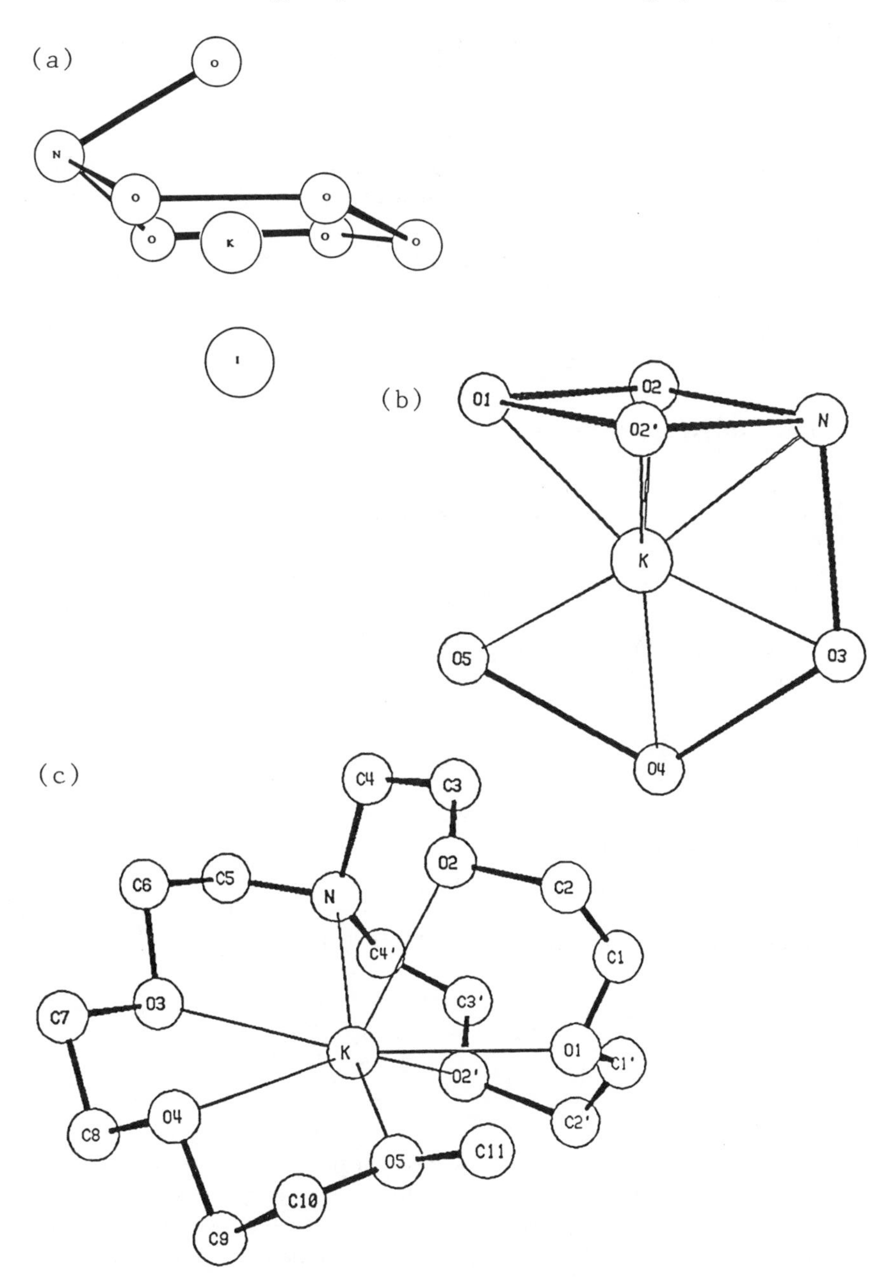

Figure 2. (a) Framework drawing of N-*(2-methoxyethyl)monoaza-18-crown-6 complexed by KI. Framework drawing (b) and molecular structure* (**c**) *of the complex between* K^+ *and* $CH_3OCH_2CH_2OCH_2CH_2OCH_2CH_2N(CH_2\ CH_2O)_3$*, the so-called "calabash" complex.*

Bibracchial Lariat Ethers, BiBLEs

Clearly a single arm augments the cation binding of monoaza lariat ethers. Will two arms increase or diminish cation binding strength and by how much? Will two-armed compounds exhibit similar or altered cation-binding selectivity? Will a higher level of cation involvement afford enhanced nucleophilicity of associated anions even though the lariat ethers and BiBLEs are "dynamic" cation binders? To answer these questions, we have constructed a series of N,N′-disubstituted diaza-15-crown-5 and -18-crown-6 compounds. When two arms are present, we use the Latin *bracchium* for arm and refer to the compounds as bibracchial lariat ethers, BiBLEs.

Synthesis of Diaza-BiBLEs

4,13-Diaza-18-crown-6 derivatives (*24–42*) can be prepared by using a unique, single-step reaction of primary amines with triethylene glycol diiodide (*42*):

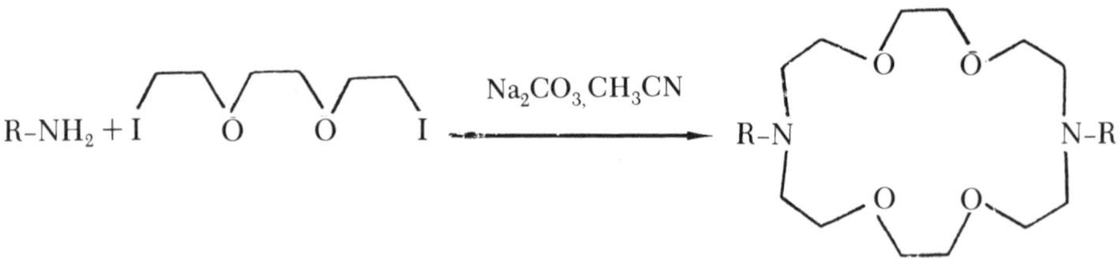

$$\text{R-NH}_2 + \text{I(CH}_2\text{CH}_2\text{O)}_2\text{CH}_2\text{CH}_2\text{I} \xrightarrow{\text{Na}_2\text{CO}_3,\ \text{CH}_3\text{CN}} \text{R-N}\!<\!\text{(18-crown-6 ring, O O / O O)}\!>\!\text{N-R}$$

The synthesis of the 15-membered rings is accomplished by a more conventional, two-step approach:

$$\text{R–NH}_2 + \text{Cl–CO–CH}_2\text{–O–CH}_2\text{–CO–Cl} \rightarrow (\text{R–NH–CO–CH}_2\text{–})_2 \rightarrow$$
$$\text{R–NH–CH}_2\text{–CH}_2\text{–O–CH}_2\text{–CH}_2\text{–NH–R} + \text{ICH}_2(\text{CH}_2\text{OCH}_2)_2\text{CH}_2\text{I} \rightarrow$$
$$\text{R-N}\!<\!\text{(15-crown-5 ring, O / O O)}\!>\!\text{N-R}$$

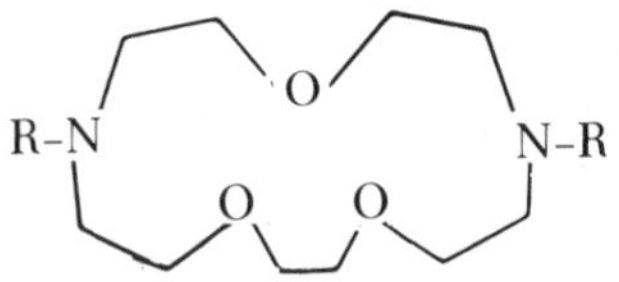

Although the latter method appears both more cumbersome and less imaginative when applied to the 18-membered ring systems, it has the advantages of high yield and ease of operation. Using this approach, we have been able to obtain both 15- (*43–47*) and 18-membered ring systems in yields of approximately 70% overall. Using this synthetic approach, we have been able to compile the cation-binding data shown in Table III.

Several conclusions can be made from the data tabulated in Table III. First, more polar donor groups in the side arms afford selectivity for the more charge-dense cations. This trend is clear from the binding constants for compounds **XIa,b** and **XVIa,b.** Second, there is a temptation to invoke a

Table III. Na^+, K^+, and Ca^{2+} Binding for 15- and 18-Membered Ring Bibracchial Lariat Ethers

Compd No.	Sidearm	*Stability Constant (log K_s in Methanol at 25 °C*					
		15-Membered Rings (**a**)			*18-Membered Rings* (**b**)		
		Na^+	K^+	Ca^{2+}	Na^+	K^+	Ca^{2+}
VI	H	<1.5	<1.5	nd[a]	1.5	1.8	nd
VII	*n*-butyl	—[b]	—	—	2.84	3.82	nd
VIII	*n*-hexyl	—	—	—	2.89	3.78	nd
IX	*n*-nonyl	—	—	—	2.95	3.70	nd
X	$CH_2C_6H_5$	2.59	2.12	2.34	2.72	3.38	2.79
XI	CH_2CH_2OH	—	—	—	4.87	5.08	6.02
XII	$CH_2CH_2OCH_3$	5.09	4.86	4.97	4.75	5.46	4.48
XIII	CH_2-2-furyl	3.99	3.87	3.45	3.77	4.98	nd
XIV	$CH_2C_6H_4$-2-OCH_3	3.59	3.13	3.04	3.65	4.94	3.27
XV	$CH_2C_6H_4$-2-OH	—	—	—	2.40	2.59	2.95
XVI	$CH_2COOCH_2CH_3$	5.34	4.65	6.04	5.51	5.78	6.78
XVII	CH_2COOH	—	—	—	nd	~1.8	nd

[a] nd is not determined;
[b] — is not prepared.

hole-size effect to account for the K^+ selectivity of **XIIb** or **XIVb** whereas the corresponding 15-membered rings (**XIIa** and **XIVa**) are Na^+ selective. Of course, neither the simple bis(benzyl) compounds **Xa** and **Xb** nor the glycine derivatives **XVI** exhibit this trend. Although hole size, at least as generalized to donor-group geometry, is not without its significance, charge density continues to play a major role. Finally, differences in binding strengths for ligands such as **XIIa, XIIIa,** and **XIVa** reflect not only the presence of sp^2- versus sp^3-hybridized oxygen donor groups in the sidearms but also quite different steric considerations for the sidearms (*48*). Studies to ascertain the influence of dynamic cation envelopment on associated anions are under way.

Cation-Binding Enhancement by Reduced Lariat Ethers and Cryptands

A variety of "switchable" lariat ethers containing ionizable functional groups like carboxylic acids and phenols have been shown to exhibit enhanced cation binding upon deprotonation. Our approach to cation-binding enhancement and potentially to "molecular switching" involves the incorporation of an electron-deficient side arm that can be reversibly reduced to its corresponding anion radical (*49, 50*). Upon reduction, the anionic side arm, if correctly disposed geometrically, is able to interact strongly with the macroring-bound cation (Scheme I). The result is an overall binding enhancement ($K_{SE}/K_S > 1$).

Of the componds tested so far, optimum enhancement factors have been observed for *N*-(2-nitrobenzyl)monoaza-15-crown-5·Na^+ (*see* the table in

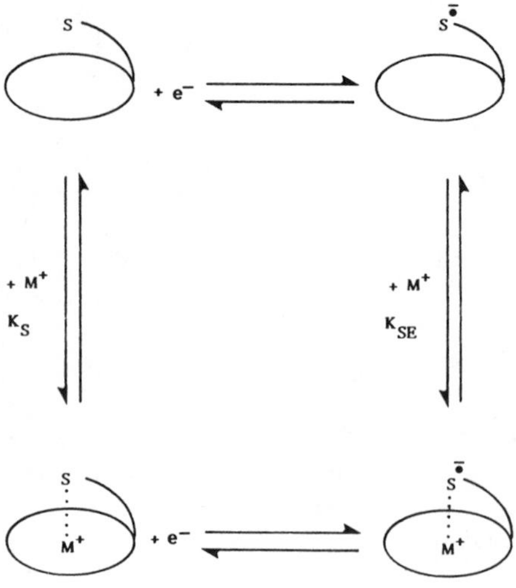

S

+ e^-

$S^{\overline{\bullet}}$

+ M^+

K_S

+ M^+

K_{SE}

S

M^+

+ e^-

$S^{\overline{\bullet}}$

M^+

Scheme I

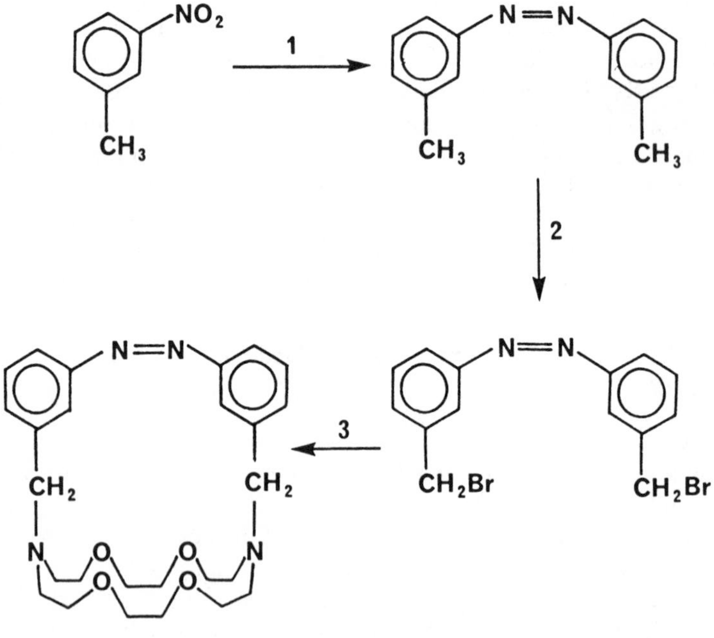

NO_2

CH_3

1

N=N

CH_3

CH_3

2

N=N

CH_2Br

CH_2Br

3

N=N

CH_2

CH_2

N O O N

O O

Scheme II. Synthesis of 4,13-[3,3'-bis(methyleneazobenzene)]diaza-18-crown-6. Synthetic steps: 1, $LiAlH_4$ in THF; 2, N-bromosuccinimide in CCl_4; 3, 4,13-diaza-18-crown-6 and Na_2CO_3 in CH_3CN.

reference 51). Enhancement values as high as 10^7 have been determined electrochemically, which result in appropriate overall binding near the range of some cryptand complexes. We noted (*49*) that enhancement factors are always largest for Li^+, than Na^+, and finally K^+, as expected on an electrostatic basis. Because binding by the neutral ligand (K_S) follows the reverse order (see previously), the enhanced constants (K_{SE}) are more or less constant. This finding translates into a lack of selectivity.

An attempt to confer selectivity for K^+ upon reduction succeeded by using the azocryptand pictured in Scheme II. Upon electrochemical reduction of the azocryptand portion of the cryptand, enhanced binding by more than a hundredfold was measurable for K^+ although no enhancement with Na^+ was observed (*52*). Fine tuning of molecular parameters should afford both binding enhancement and selectivity.

Acknowledgments

We thank W. R. Grace & Co. and the National Institutes of Health (NIGMS) for grants that supported portions of this work.

Literature Cited

1. Weber, W. P.; Gokel, G. W. *Phase Transfer Catalysis in Organic Synthesis:* Springer-Verlag, Berlin, 1977.
2. Starks, C. M.; Liotta, C. L. *Phase Transfer Catalysis;* Academic: New York, 1978.
3. Dehmlow, E. V.; Dehmlow, S. S. *Phase Transfer Catalysis;* Verlag Chemie: Deerfield Beach, 1980.
4. Pedersen, C. J. *J. Am. Chem. Soc.* **1967,** *89,* 7017.
5. Lehn, J. -M.; *Acc. Chem. Res.* **1978,** *11,* 49.
6. Lehn, J.-M.; *Science (Washington, D.C.)* **1985,** *227,* 849.
7. Dishong, D. M.; Diamond, C. J.; Cinoman, M. I.; Gokel, G. W.; *J. Am. Chem. Soc.* **1983,** *105,* 586.
8. Cook, F. L.; Caruso, T. C.; Byrne, M. P.; Bowers, C. W.; Speck, D. H.; Liotta, C. L.; *Tetrahedron Lett.* **1974,** 4029.
9. Gokel, G. W.; Dishong, D. M.; Diamond, C. J.; *J. Chem. Soc., Chem. Commun.* **1980,** 1053.
10. Izatt, R. M.; Bradshaw, J. S.; Nielsen, S. A.; Lamb, J. D.; Christensen, J. J.; Sen, D. *Chem. Rev.* **1985,** *85,* 271.
11. Nakatsuji, Y.; Nakamura, T.; Okahara, M.; Dishong, D. M.; Gokel, G. W. *Tetrahedron Lett.* **1982,** 1351.
12. Nakatsuji, Y.; Nakamura. T.; Okahara, M.; Dishong, D. M.; Gokel, G. W. *J. Org. Chem.* **1983,** *48,* 1237.
13. White, B. D.; Dishong, D. M.; Minganti, C. M.; Arnold, K. A.; Goli, D. M.; Gokel, G. W.; *Tetrahedron Lett.* **1985,** 151–154.
14. Calverley, M. J.; Dale, J. *J. Chem. Soc., Chem. Commun.* **1981,** 684.
15. Calverley, M. J.; Dale, J. *Acta Chem. Scand. B* **1982** *36,* 241.
16. Gokel, G. W.; Goli, D. M.; Minganti, C.; Echegoyen, L. *J. Am. Chem. Soc.* **1983,** *105,* 6786.
17. Michaux, G.; Reisse, J. *J. Am. Chem. Soc.* **1982,** *104,* 6895.
18. Hilgenfeld, R.; Saenger, W.; In *Host Guest Complex Chemistry Macrocycles;* Voegtle, F.; Weber, E., Eds.; Springer-Verlag, Berlin, 1985; p 50.

19. Schultz, R. A.; White, B. D.; Dishong, D. M.; Arnold, K. A.; Gokel, G. W. *J. Am. Chem. Soc.* **1985** *107*, 6559.
20. Day, M. C.; Medley, J. H.; Ahmad, N. *Can. J. Chem.* **1983,** *61*, 1719.
21. Fronczek, F. R.; Gatto, V. J.; Schultz, R. A.; Jungk, S. J.; Colucci, W. J.; Gandour, R. D.; Gokel, G. W. *J. Am. Chem. Soc.* **1983,** *105*, 6717.
22. White, B. D.; Arnold, K. A.; Fronczek, F. R.; Gandour, R. D.; Gokel, G. W. *Tetrahedron Lett.* **1985,** 4038.
23. Schultz, R. A.; Schlegel, E.; Dishong, D. M.; Gokel, G. W.; *J. Chem. Soc., Chem. Commun.* **1982,** 242.
24. Ricard, A.; Cappillon, J.; Quivoeron, C. *Polymer* **1985,** *25*, 1136.
25. Tsukube, H. *Bull. Chem. Soc. Jpn.* **1984,** *57*, 2685.
26. Tsukube, H. *J. Chem. Soc., Chem. Commun.* **1984,** 315.
27. Tsukube, H. *J. Chem. Soc., Chem. Commun.* **1983,** 970.
28. Keana, J. F. W.; Cuomo, J.; Lex, L.; Seyedrezai, S. E. *J. Org. Chem.* **1983,** *48*, 2647.
29. DeJong, F. A.; Van Zon, A.; Reinhoudt, D. H.; Torny, G. J.; Tomassen, H. P. M. *Recl. J. R. Neth. Chem. Soc.* **1983,** *102*, 164.
30. Shinkai, S.; Kinda, H.; Araragi, Y.; Manabe, O. *Bull. Chem. Soc. Jpn.* **1983,** *56*, 559.
31. Kobayashi, H.; Okahara, M. *J. Chem. Soc., Chem. Commun.* **1983,** 800.
32. Bogatsky, A. V.; Lukyanenko, N. G.; Pastushok, V. N.; Kostyanovsky, R. G. *Synthesis* **1983,** 992.
33. Frere, Y.; Gramain, P. *Makromol. Chem.* **1982,** *183*, 2163.
34. Tazaki, M.; Nita, K.; Takagi, M.; Ueno, K. *Chem. Lett.* **1982,** 571.
35. Cho, I.; Chang, S.-K. *Bull. Korean Chem. Soc.* **1980.** 145.
36. Gramain, P.; Kleiber, M.; Frere, Y. *Polymer* **1980,** *21*, 915.
37. Kulstad, S.; Malmsten, L. A.; *Acta Chem. Scand., Ser. B.* **1979,** *B33*, 469.
38. Webster, N.; Voegtle, F. *J. Chem. Res. (S)* **1978,** 400.
39. Takagi, M.; Tazaki, M.; Ueno, K.; *Chem. Lett.* **1978,** 1179.
40. Graf, E.; Lehn, J.-M. *J. Am. Chem. Soc.* **1975,** *97*, 5022.
41. Lehn, J.-M. U.S. Patent 3,888,887, June 10, 1975.
42. Gatto, V. J.; Gokel, G. W. *J. Am. Chem. Soc.* **1984,** *106*, 8240.
43. Maeda, H.; Kikui, T.; Nakatsuji, Y.; Okahara, M. *Synthesis* **1983,** 185.
44. Okahara, M.; Nakatsuji, Y. Japan Kokai SHO 58-154576, Sept 14, 1983.
45. Maeda, Y.; Nakatsuji, Y.; Okahara, M. Japan Kokai SHO 58-154565, Sept 14, 1983.
46. Okahara, M.; Maeda, Y.; Nakatsuji, Y. Japan Kokai SHO 58-154566, Sept 14, 1983.
47. Maeda, H.; Furuyoshi, S.; Nakatsuji, Y.; Okahara, M. *Bull. Chem. Soc. Jpn.* **1983,** *56*, 3073.
48. Gatto, V. J.; Arnold, K. A.; Viscariello, A. M.; Miller, S. R.; Gokel, G. W. *Tetrahedron Lett.,* **1986,** 327.
49. Kaifer, A.; Echegoyen, L.; Gustowski, D.; Goli, D. M.; Gokel, G. W. *J. Am. Chem. Soc.* **1983,** *105*, 7168.
50. Gustowski, D. A.; Echegoyen, L.; Goli, D. M.; Kaifer, A.; Schultz, R. A.; Gokel, G. W. *J. Am. Chem. Soc.* **1984,** *106*, 1633.
51. Kaifer, A.; Gustowski, D. A.; Echegoyen, L.; Gatto, V. J.; Schultz, R. A.; Cleary, T. P.; Morgan, C. R.; Goli, D. M.; Rios, A. M.; Gokel, G. W. *J. Am. Chem. Soc.* **1985,** *107*, 1958.
52. Gustowski, D. A.; Gatto, V. J.; Kaifer, A.; Echegoyen, L.; Godt, R. E.; Gokel, G. W. *J. Chem. Soc., Chem. Commun.* **1984,** 923.

RECEIVED for review October 21, 1985. ACCEPTED January 27, 1986.

31

Manipulation of Nucleophilic Displacement Reactions by Host–Guest Complexes

Models for Enzyme Analogue Catalysis and Inhibition

Hans-Jörg Schneider, Rainer Busch, Rüdiger Kramer, Ulrich Schneider, and Isolde Theis

Fachrichtung Organische Chemie der Universität des Saarlandes, D-6600 Saarbrücken 11, West Germany

Most of the examples in this review involve encapsulation of an organic substrate in the cavity of a macrocyclic ammonium ion in aqueous solution. Methods for the preparation of the macrocycles and for the characterization of the complexes are briefly discussed. This chapter describes how S_N2-type reactions are catalyzed by positively charged host compounds and S_N1-type reactions, apart from salt effects, by a negative environment. The shape of the cavity largely determines the substrate selectivity; a discrimination between S_N1 and S_N2-type reactions on the basis of different transition-state stabilization leads to a drastic regioselectivity change with nitrite as the ambident nucleophile. Selective inhibition is observed by competitive binding either of unreactive organic compounds or of inorganic nucleophiles in the case of smaller cavities.

NUCLEOPHILIC SUBSTITUTIONS BELONG TO THE MOST IMPORTANT and most thoroughly studied reactions of organic chemistry; these substitutions play, however, only a minor role in biological systems. Probably for this reason simple aliphatic or aromatic nucleophilic displacement reactions have only recently been studied in the context of enzyme–analogue catalysis. Most of the efforts in the application of host–guest complexes for catalysis have been directed toward the simulation of specific enzyme reactions, such as acyl transfer or transamination processes. Several reviews (*1–5*) are available on the exciting progress in biomimetic catalysis, which allows us to

0065-2393/87/0215-0457$06.50/0

restrict ourselves to some basic principles and prerequisites for artificial enzyme analogues and then to proceed to their use in the catalysis of nucleophilic substitutions.

Crown Ethers, Cryptates, and Other Chelating Reagents

Heteromacrocycles containing vicinal oxygen, nitrogen, or sulfur linkages are thus far the most prominent representatives both for specific encapsulation of guest molecules in macrocyclic cavities and for the use of such complexation capability for the catalysis of organic reactions, including nucleophilic substitutions. Synthesis, structure, dynamic behavior, and applications of these compounds, which are already of considerable commercial importance, have been aptly reviewed *(6–10)*. The essential features of the crown ethers and related macrocycles are their strong binding capacity mostly for metal ions and their ability to extract the corresponding salts into organic solvents by virtue of the lipophilic nature of the exterior walls of the cavities.

Chart I shows dicyclohexyl-18-crown-6, an often used compound, as well as a cryptate, which is an even stronger binder. Not only do salts such as alkali halides become soluble in unpolar organic solvents, but also the reactivity of the corresponding counteranions can be increased dramatically in the presence of these agents. This result is ascribed to their increased

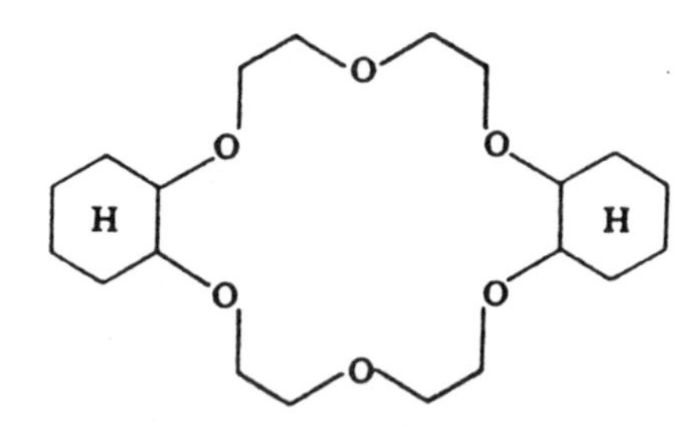

Crown Ethers

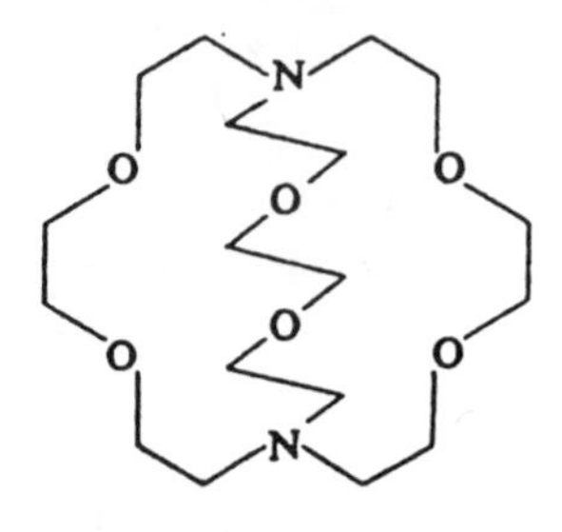

Cryptates

Essential Features:

1. Mask metal ions—make salts lipophilic
2. Extract salts into organic solvents
3. Activate anions for S_N reactions
4. Open chain polyethers often almost as effective

Chart I. Crown ethers, cryptates: essential features.

desolvation as compared to protic solutions. Moreover, encapsulation of the metal ions by the macrocycle prevents formation of less reactive internal ion pairs, at least in polar aprotic solvents. As a consequence, many nucleophilic displacements proceed much faster, in particular those involving usually weak nucleophiles such as carboxylate anions. In this way, crown ethers and related systems serve essentially the same purpose as phase-transfer catalysts (*11–14*), which most often are onium ions with lipophilic alkyl substituents. The role of the phase-transfer catalyst, be it onium salt or metal chelating agent, is essentially an auxiliary one: this catalyst provides for the extraction of the salt into the organic solvent and therefore also for anion activation. In nucleophilic substitutions of enolates, which, in this context, may be better characterized as anion alkylations (*13, 14*), often synthetically useful changes of regioselectivity are also obtained. For a more detailed discussion of catalytic effects in phase-transfer reactions including macrocyclic ion binders, see recent extensive monographs and reviews on this subject (*11–14*). Phase-transfer reagents are still expected to dominate many practical applications with respect to nucleophilic substitutions in the future, although similar results can often be obtained in aprotic polar solvents. Macrocyclic chelating agents, on the other hand, can often be substituted by structurally related so-called podands (*15*), which bear the chelating groups in open-chain sidearm tentacles, or simply by inexpensive polyethylene glycols (*16*).

The common feature of macrocyclic chelating agents and enzyme analogues with lipophilic cavities of course must be seen in the encapsulation of either ions or uncharged organic molecules by virtue of so-called nonbonded interactions. The fit and the association constants between host and guest molecules can not only be tailored by the ring geometry and the nature of the groups inside the cavity but also be altered by additional sidearms flanking those parts of the guest molecule that are not immersed in the cavity.

The careful engineering of suitable host structures has already led to new methods for separations of, for example, metal cations (*17*) or of chiral ammonium salts (*2*).

Properties of Enzyme Analogous Host Catalysts

Essential characteristics of synthetic enzyme analogues are summarized in the Box on page 460. Many of these features are closely interrelated, such as complexation, saturation kinetics, and substrate selectivity, as well as competitive inhibition by stronger binding substrates, which are less reactive or completely unreactive. Regio- or stereoselectivity, if applicable, can change in comparison to the uncatalyzed process by a variation in structure or by disposition of the transition state, for example, from early to late or from an S_N1 to an S_N2 mechanism.

The effectiveness of a catalyst can be described in terms of the rate enhancement achieved by stabilization of the transition state ($\triangleq k_{cat}$) and of

Enzyme Analogues—Essential Characteristics

Complexation in molecular cavity
Rate enhancements—saturation kinetics
Substrate selectivity
Reaction selectivity
Competitive inhibition

Important Contributions to Catalysis

Selective binding
Proximity effects
Entropy effects
Electrostatic interactions
Reaction field changes

the association constant between substrate and catalyst ($\hat{=} K_a$) (*18, 19*). Thus, an increase in effectiveness by a factor of 100 can be obtained in principle by an association constant of 100 alone, provided the rate of product formation is not lowered. This picture is oversimplified as it involves, for example, the rate constant comparison between the unimolecular reaction from the complex to the product with the uncatalyzed reaction of higher order. Also, too strong an association with the substrates or the related products can lead to inhibition or rate retardation. A thorough discussion of these factors and the factors affecting the performance of an enzymatic catalyst can be found in suitable reviews and monographs (*18, 19*). The Box lists only those elements that are essential for the understanding of the systems to be discussed.

Several principal differences exist between most host–guest catalysts and enzymes that are the result of the optimization of the enzymes, which has been accomplished by nature over many millions of years. In constructing synthetic enzyme analogues (*1–5*), strong—although not necessarily selective—binding to different substrates is less difficult to obtain than is an optimal stabilization of transition states, which usually requires not only a fitting of van der Waals contacts but also the optimal alignment of reacting groups, dipoles, and so on. The first step in developing host enzyme analogues should therefore be the characterization of a complexation, preferably by spectroscopic techniques, complemented by molecular modeling, which can be aided by interactive computer graphics.

Molecular Structures with Large Lipophilic Cavities

The most well-known compounds of this type are cycloamyloses or cyclodextrins (*20–22*), which form nearly cylindrical cavities of 5–8-Å inner diameter, depending on the chosen natural product. Chemical modification at the rim of the cavity, as well as improved binding by adding caps on the top, has already led to many promising biomimetic catalysts. Aside from their limited stability, the limitation of the cyclodextrins lies in the fact that as a gift of nature they come with a basically invariable cavity structure. Although in a preliminary study cyclodextrins did show moderate enhancements of hydrolysis rates with a tertiary alkyl chloride (*23*), cyclodextrins have so far not been used for the catalysis of nucleophilic displacement reactions. In view of the electroneutral nature of the cycloamyloses, this usage would probably require the introduction of polar groups or charges in the host structure.

In contrast to cyclodextrins, azamacrocycles (*24*) not only are accessible in variable geometries by synthesis but also can be converted to charged ammonium ions. The positively charged cavities make these systems promising candidates for the stabilization of negatively charged S_N2 transition states, whereas S_N1 reactions should be accelerated by negatively charged surroundings, if it is not the leaving group, but rather the organic substrate that is complexed. Moreover, multiple charges on the rings provide for water solubility even of macrocyclic structures containing up to 70 nonhydrogen atoms and approaching molecular weights of 1000. Solubility in aqueous solutions is a prerequisite for the application of many receptor and enzyme analogues, because an important driving force for the encapsulation of organic substrates derives from hydrophobic interactions. Complexations of anions instead of, or in addition to, lipophilic moieties also were observed in several macrocyclic ammonium ions (*25*, *26*); their impact on nucleophilic substitution reactions will be discussed.

In the last few years, Tabushi et al. (*27*), Koga and co-workers (*24*, *28*), Jarvi and Whitlock (*29*), Breslow and co-workers (*30*), Diederich and co-workers (*31*, *32*), and Vögtle and co-workers (*33*, *34*) as well as our group (*35*) have shown that azacyclophanes in the form of their ammonium salts can very effectively complex lipophilic substrates. Even hydrocarbons bind with association constants of 1000 or more in aqueous solutions. Our own studies of such host–guest systems were initiated by the desire to use complexations of alkanes for the selective functionalization of paraffins (*23*, *36*) [in addition to our interest (*37*, *38*) in conformations of complex ring systems and their study by NMR methods as well as by force field model calculations].

Methods for the Characterization of Host–Guest Catalysts

Methods for the study of host–guest complexes in solution have been reviewed (*39*) with emphasis on crown ethers or cryptates and on the bonding of ions; measurements of the often weaker complexes with lipophilic substrates are preferably done by NMR shift titration. Because K_a values of 10^3 require measurements in a concentration range of 10^{-3} M in order to see uncomplexed as well as complexed material, NMR is a more convenient method for such complexes as, for example, UV or fluorescence spectroscopy, which require more dilute solutions. UV spectra, moreover, often show only small changes upon complexation.

Older methods based on solubility changes upon complexation, or on partition coefficients between aqueous solutions and hydrophobic solvents, have been shown to lead to gross errors as compared to spectroscopic techniques (*40*) that are also less sensitive to the formation of emulsions, micelles, and so on. The traditional X-ray analysis of inclusion compounds is of limited significance for establishing complexation between lipophilic substrates and macrocyclic host, particularly in aqueous solution. The essential hydrophobic driving force for complexation, of course, is nonexistent in the crystal. The future development of NMR methods including shielding calculations and measurements of intermolecular nuclear Overhauser effects is expected to provide the most reliable information on intercavity inclusion complexes in solution as the basis for catalytic applications.

Considering the complex kinetic behavior of most enzymatically controlled reactions (*41*), the formal treatment of simple catalytic analogues should not pose additional problems. However, one consequence of the less perfect, but for most practical and mechanistical purposes sufficient performance of synthetic catalysts in comparison to enzymes is that in many kinetic studies, a large excess of substrate over the catalyst cannot be used, because then the uncatalyzed reaction will be too fast. Consequently, kinetic studies under catalyst saturation, or the steady-state methods that are most often used in the investigation of enzymes (*18*, *19*, *41*), are not suitable here. The formal treatment of the resulting, often quite complex, kinetics is greatly facilitated by computer-aided numerical simulations, which also help to design proper experimental conditions.

Synthesis of Heteromacrocycles (6, 8–10, 34)

As early as 1935, Lüttringhaus obtained large heterocyclic systems by reaction of, for example, 1,ω-dihalogen alkanes with bisphenols (*42*, *43*). The principal reaction sequence, starting from complementary bifunctional spacers A and B, is shown in Scheme I. The formation of smaller and larger rings, DR and TR, respectively, from two or four spacer units was also reported already by Lüttringhaus. The low yields often observed for cyclizations can indeed be also due to the easily overlooked simultaneous formation of different ring sizes (*44*). The condensation of 1,ω-ditosylamides with 1,ω-dihaloalkanes was introduced by Stetter and co-workers (*45–47*) and Fuson and House (*48*) and opened a widely used access to azamacrocyclic compounds. The high dilution, which is particularly necessary for the reaction with highly reactive educts, such as with dicarboxylic acid halides, to suppress excessive polymerization, however, means that sometimes only milligram quantities can be obtained over several days. This time factor is likely

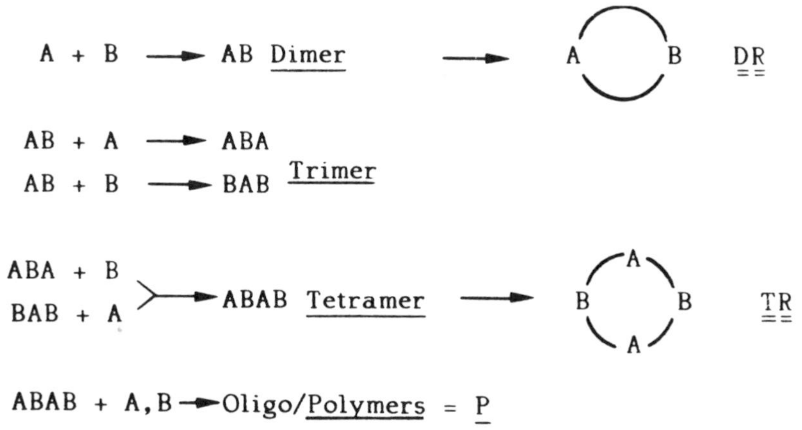

A + B ⟶ AB Dimer ⟶ A⌒B DR

AB + A ⟶ ABA
AB + B ⟶ BAB Trimer

ABA + B
BAB + A ⟶ ABAB Tetramer ⟶ cyclic (A, B, A, B) TR

ABAB + A,B ⟶ Oligo/Polymers = P

Scheme I. Reaction sequence for the formation of smaller and larger rings (DR and TR).

to be one of the reasons the majority of the many synthesized potential host compounds have until now rarely been subjected to further examination. By adjustment of the time scale and concentration of the cyclization on the basis of computer simulations (Figure 1), as well as by use of potassium carbonate as the base in heterogenous dimethylformamide (DMF) solutions (*46, 48*), both quantities as yields of the condensation reactions with tosylamides can be improved considerably (*44*) (Table I). The limited solubility of potassium carbonate in DMF leads to stationary concentration of the reactive deprotonated amide of $<10^{-5}$ M. Cesium carbonate, which, because of template effects (*49*), is often reported to enhance cyclization yields, is more soluble and has been found to be less satisfactory in the amide–alkyl halide condensation reaction in DMF (*44*).

The ratio of smaller to larger rings obtained can be further influenced by the applied concentrations or time periods, as well as by the reaction temperature (*44*) (Table I), in agreement with lower activation entropies for the cyclization of larger rings (*50*). After removal of the both protective and activating tosyl groups, the macrocycles are soluble enough in aqueous acids for many binding studies. Koga et al. (*24, 28*) have shown that the easily accessible protonated azamacrocycles—mostly containing the diphenylmethane spacer unit already used by Lüttringhaus for his preparations—are capable of complexing various lipophilic substrates. For most studies, however, the amines must be converted to neutral permethyl ammonium salts. A large number of macrocycles (*24, 51*) with cavities spanning approximately 5.0–10.0 Å between the most distant inner van der Waals surfaces are available by using the above-mentioned methods.

The investigation of catalytic applications usually requires larger quantities of the host compounds, which, for the reasons described previously,

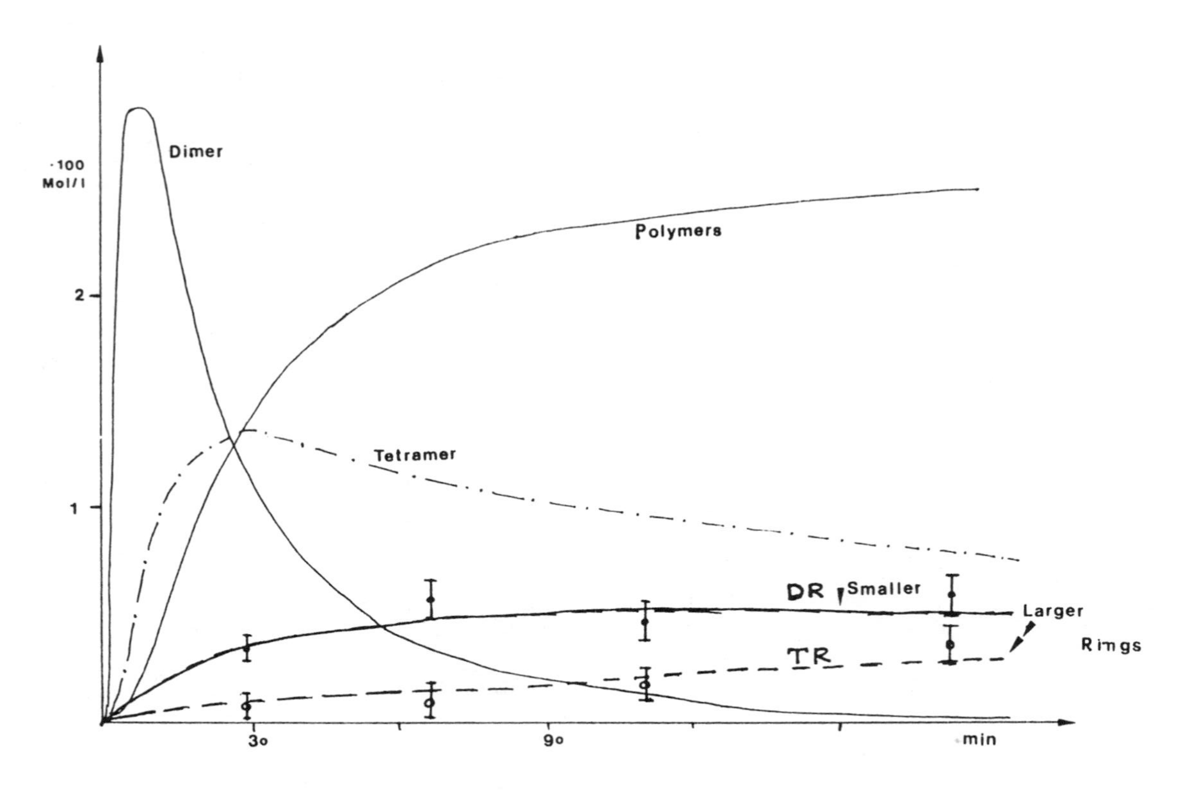

Figure 1. Computer simulations of the reaction sequence of Scheme I; the vertical bars on the curves for DR and TR correspond to experimentally observed concentrations of the smaller and larger rings.

Table I. Yields and Ratios of Smaller to Larger Rings (DR:TR) as a Function of Concentration and Temperature

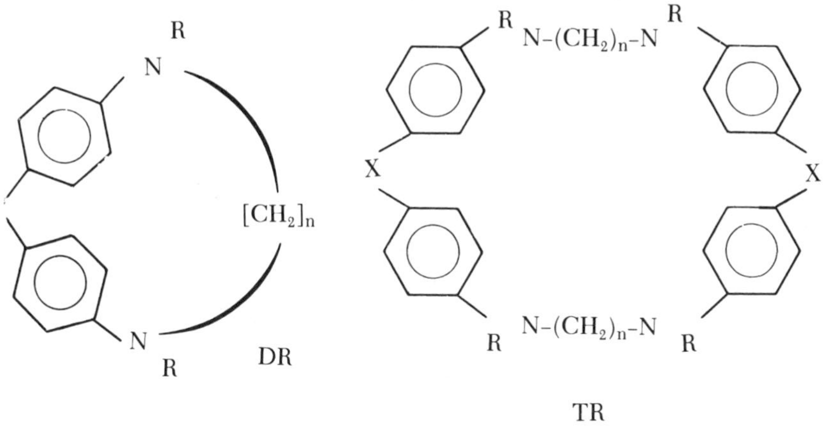

R N $[CH_2]_n$ N R

DR

R $N-(CH_2)_n-N$ R

X X

R $N-(CH_2)_n-N$ R

TR

Variable	*DR (%)*	*TR (%)*	*DR:TR*
0.01 M	59	3	20.0
0.10 M	36	24	1.5
25 °C	36	24	1.5
120 °C	37	10	3.7

NOTE: The values are for X = $-CH_2CH_2-$ and n = 10.
[a] Other temperature effects are, for example, from 2:1 to 30:1.

often have to be studied in excess over the substrates. Stronger binding constants can be expected in bi- or polycyclic cavities in which the substrate is exposed to the solution to a lesser degree. Furthermore, introduction of suitable functional groups that can stabilize transition states will also require synthetic efforts for the development of more optimized catalysts.

Catalysts for Nucleophilic Displacement Reactions

A Polycyclic Ammonium Salt as Host. Schmidtchen (52, 53) was the first to report examples of enzyme analogue catalysis of nucleophilic substitutions, mostly with aromatic compounds. He synthesized the two highly symmetrical cage compounds **I** and **II** (54) (Scheme II) and first examined their ability to form complexes with several anions (25). The smaller host with six methylene units as the spacer (cavity diameter of 4.5 Å) showed, for example, association constants of $\sim 10^2$ for chloride, $\sim 10^3$ for bromide, and $\sim 10^2$ for several biologically important anions such as adenosine triphosphate or nicotinamide dinucleotide diphosphate. Whereas the association of anions with **I** seems to be dominated by electrostatic interactions, the larger

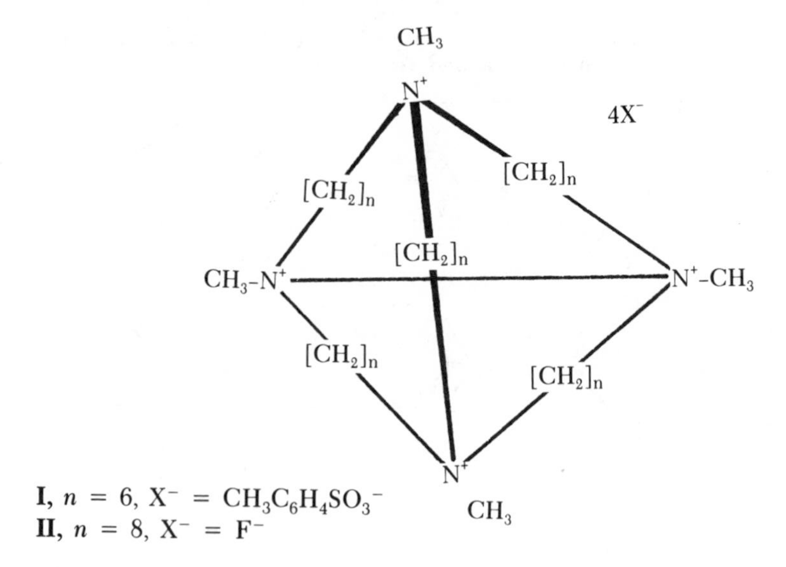

Schmidtchen:

A: III (X, NO_2, NO_2) $\xrightarrow{N_3^-}$ (N_3, X, NO_2, NO_2) $\longrightarrow$ (N_3, NO_2, NO_2) + X^- | IV (Cl, NO_2, NO_2, COO^-)

B: $CH_3I \longrightarrow N_3\cdots\bar{C}H_3\cdots I \longrightarrow CH_3N_3 \quad I^-$

Scheme II. A polycyclic ammonium host as catalyst for S_N2 reactions (references 52 and 53).

and more hydrophobic cavity of **II** (inner diameter of 7.5 Å) provides for the binding of bulkier and more lipophilic or polarizable ions, as evident from the enhanced constants with *p*-nitrophenolate, which is not complexed by **I** to an appreciable degree.

The strong association of **I** with small and hard anions leads only to a small and difficult-to-measure (*53*) rate decrease of nucleophilic aromatic substitution (*52*, *53*). Addition of the host **II**, however, to the S_N2 reactions A and B (Scheme II) produced remarkable rate enhancements (Table II), which showed a Michaelis–Menten-type dependence on ratio of nucleophile and catalyst concentrations. The catalytic effect of the host was ascribed to a desolvation of the azide anion, which was used in all cases. This situation is similar to the rationalization of corresponding micelle-catalyzed substitutions (*55*, *56*). Another important factor is the stabilization of the transition states,

Table II. Reactions of Nitrophenyl Halides with Azides

Compound	k_{un}	k_{cat}	K_A
III, X = F	4.0	190.0	140
III, X = Cl	0.0008	1.04	—[a]
III, X = Br	0.0012	1.52	—
III, X = I	0.0009	2.10	—
IV	0.0090	6.2	200

SOURCE: Data are extracted from reference 53.
NOTE: k_{un} (uncatalyzed) and k_{cat} (catalyzed): rate constants in 100 I mol^{-1} units, in 25% methanol at 25 °C. K_A association constant in L/mol. Compounds **III** and **IV**: *see* Scheme II.
[a] Not determined.

which bear an extendedly delocalized negative charge, by the positively charged environment. The significance of this contribution becomes nicely visible in the exceptionally high k_{cat}/k_{un} value for the negatively charged substrate **IV**, which, in comparison to the electroneutral phenyl halide **III** (X = F), does not show a significant alteration of the binding constant (Table II). When phenyl halides with different leaving groups are compared, a leveling off of the catalytic effect is found, reminiscent of the similar phase-transfer catalysis (*11*), which is particularly pronounced with systems notorious for their slow aromatic substitution. A study by NMR techniques of the degree to which the organic substrates can really enter the cavity of **II** would be interesting, particularly in view of the unknown rate dependence on the phenyl halide-to-catalyst ratio and the unknown sensitivity of the catalyst against variation of the organic substrate size.

An Azacyclophane Host: Conformational Fitting and Selective Catalysis Involving Changes between S_N1 and S_N2 Mechanisms. The monocyclic host compound **V** (Chart II), which was used for most of our studies (*57*, *58*), is similar to the systems first investigated by Koga and co-workers (*24*, *28*); **V** was, however, converted to a permethylamonium salt to allow for measurements under neutral conditions and has a longer alkane spacer containing six methylene groups. Inspection of the structure by Corey–Pauling–Koltun models as well as by interactive computer graphics shows that this number is the minimum chain length to accommodate guest molecules of the naphthalene shape fully immersed within the cavity. Molecular mechanics force-field calculations (*59*) indicate that the macrocycle maintains an all-trans conformation along the alkane chain both in the free and in the complexed state with naphthalene in the cavity (Figure 2).

Equilibrium measurements with **V** and different naphthalene derivatives in aqueous solution showed attractive free energies of up to 7 kcal/mol (Chart III) (*35*), if both lipophilic and electrostatic contributions are present. Comparison of all available association constants (*35*, *40*), however, clearly shows that the lipophilic part is always dominating [other than in the immo-

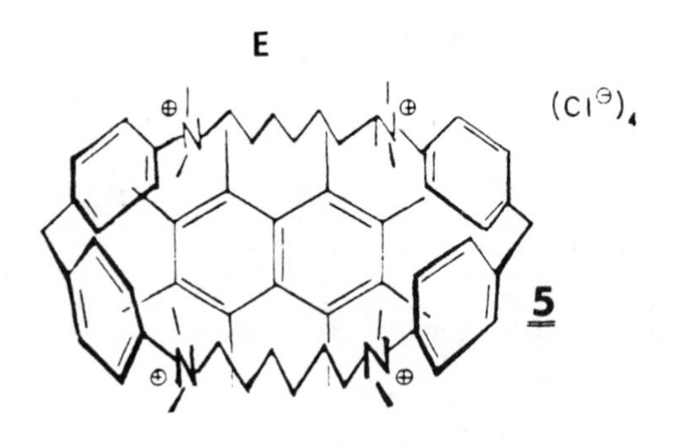

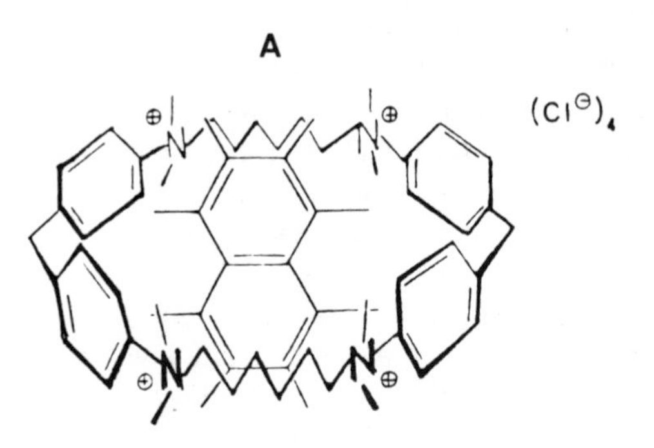

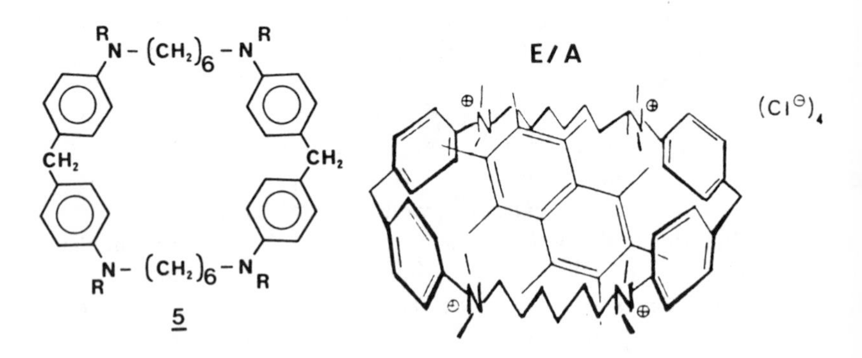

Chart II. The azacyclophane **V** *and conformations of its complexes with naphthalene.*

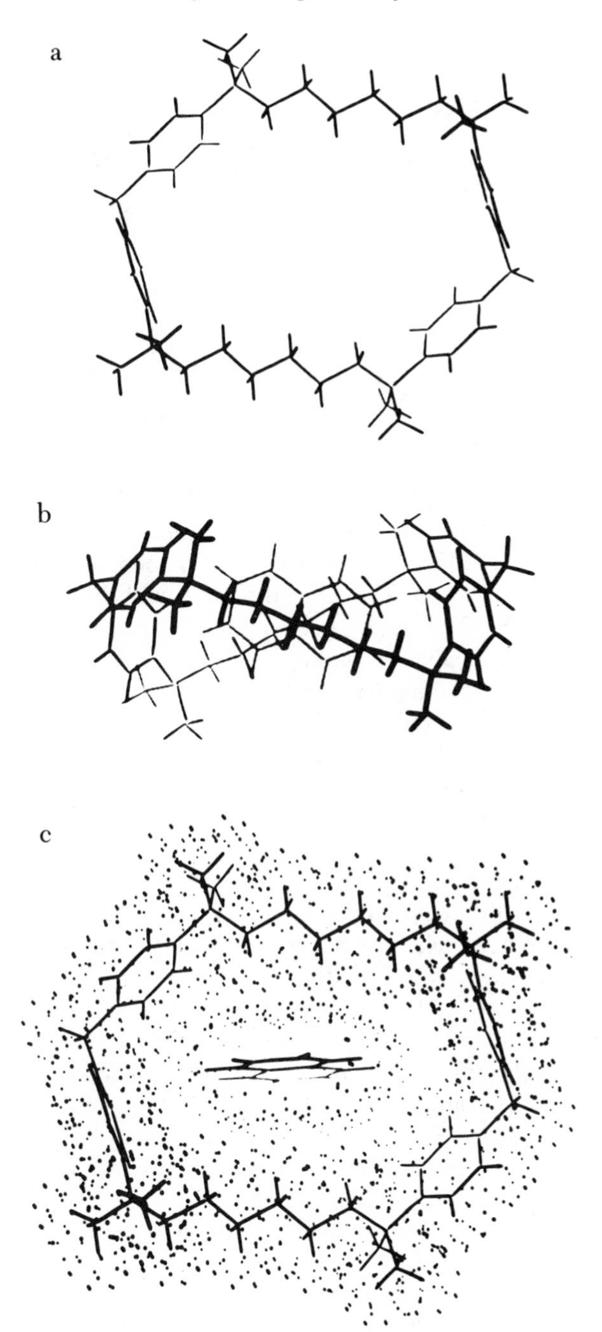

Figure 2. Energy-minimized conformations. (a) Macrocycle **V** *before complexation, top view. (b) Complex of* **V** *and naphthalene, side view. (c) Complex of* **V** *and naphthalene, top view.*

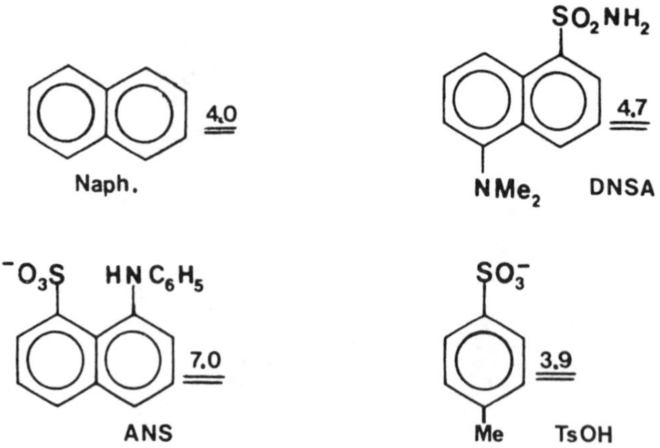

Naph. 4.0

DNSA 4.7

ANS 7.0

TsOH 3.9

Chart III. Complexation energies with **V**.

bilized state of the host, as evident from chromatographic investigations (*58*)]. Consequently, phenyl derivatives are not bound to an appreciable degree in absence of negatively charged groups, and their S_N reactions are not catalyzed. The host also discriminates tetralin or decalin from naphthalene but is capable of binding substrates as large as estradiol, because of the structural similarity to tetralin (*40*).

Solvent and salt effects on the equilibrium constants with **V** are very pronounced for negatively charged substrates (*35, 40*), which is of relevance for catalytic investigations. Thus, addition of buffer salts or alkali halides as nucleophiles at concentrations as low as 0.01 M decreases the complexation significantly. An opposite salting out effect on purely lipophilic substrates, for example, hydrocarbons, can only be expected at salt concentrations of ~1 M. If lipophilic solvents, such as methanol, are added, for example, to circumvent solubility problems, the association constants are again lowered to a degree that may prohibit catalytic applications.

The intercavity inclusion as well as the possible orientation of naphthalene derivatives in the macrocycle (Chart I) was studied by the NMR complexation shifts (*35*), which were obtained simultaneously with the association constants by computer fit of the shift titration values. The observed complexation shifts agree with either a pseudoequatorial orientation or a mixture of axial and equatorial geometries (Chart II and Table III), if, other than in earlier investigations (*28*), the electric field effects of the charged nitrogen atoms are taken into account in addition to the anisotropy effects of the aromatic cavity parts.

An energy minimization of the complex between **V** and naphthalene indeed showed a tight fit between the van der Waals surfaces in a nearly equatorial conformation (Figure 2). The force field minimized structure deviates from the idealized one shown in Chart II, which, however, represents a time-averaged picture of the ring. Although the computer models

Table III. Complex Geometry from NMR Shifts: Naphthalene in 50% MeOH

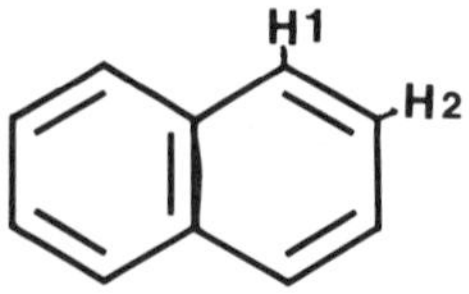

Geometry	*H-1*	*H-2*
axial	−2.9	−1.2
pseudo	−1.7	−1.5
equatorial	−0.4	−1.9
observed	−1.5	−1.5

NOTE: Ring-current (anisotropy) effects: 60–85% (orientation dependent). Linear electric field effects: 15–40% (orientation dependent); complexation-induced shifts in ppm.

represent only a first approximation, particularly in view of the largely neglected effects of solvent, charges, and polarizations, the models do exclude a completely equatorial inclusion. A more axial starting conformation was converted to the same more equatorial conformation, which was obtained from another starting geometry (Figure 2); this result shows the effective docking of the substrate in the cavity by the applied minimization procedure.

Attachment of a bromomethyl group to the β position of the almost fully immersed naphthalene brings a reactive center to the vicinity of both the charged ammonium ions and the accumulation of negatively charged nucleophiles around them. Substitution reactions at this center were therefore expected to be distinctively affected (Scheme III). Figure 3 shows the result of a kinetic study with nitrite as nucleophiles (57); numerical curve fitting of the observed pseudo-first-order rate constants for the different ratios of catalyst to substrate to a preequilibrium reaction sequence involving the complex formation with an apparent association constant K_a yielded k_{cat} = 0.02 s^{-1} and K_a = ~4 L/mol; the effectivity of the catalyst expressed as k_{cat}/k_{un} is ~30.

The most remarkable feature of the reaction with the ambident anion NO_2^- was the change in the product ratio R–ONO:R–NO_2, which corresponded exactly to the observed overall rate constant increase as a function of the catalyst-to-substrate concentration ratio (Figure 3). If the product ratio is

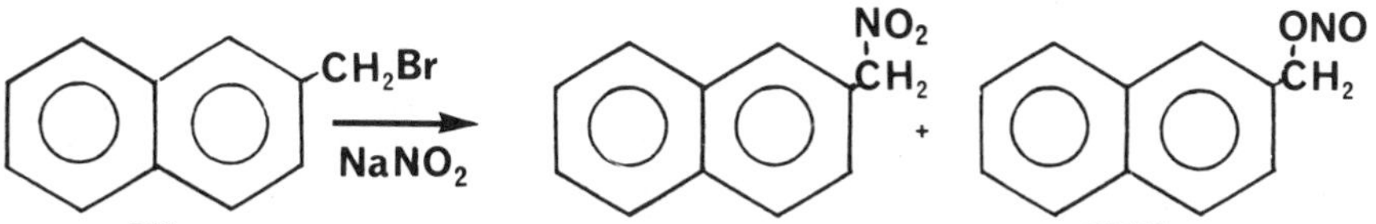

Scheme III. Catalytic effects of **V** *on* S_N2 *reactions.*

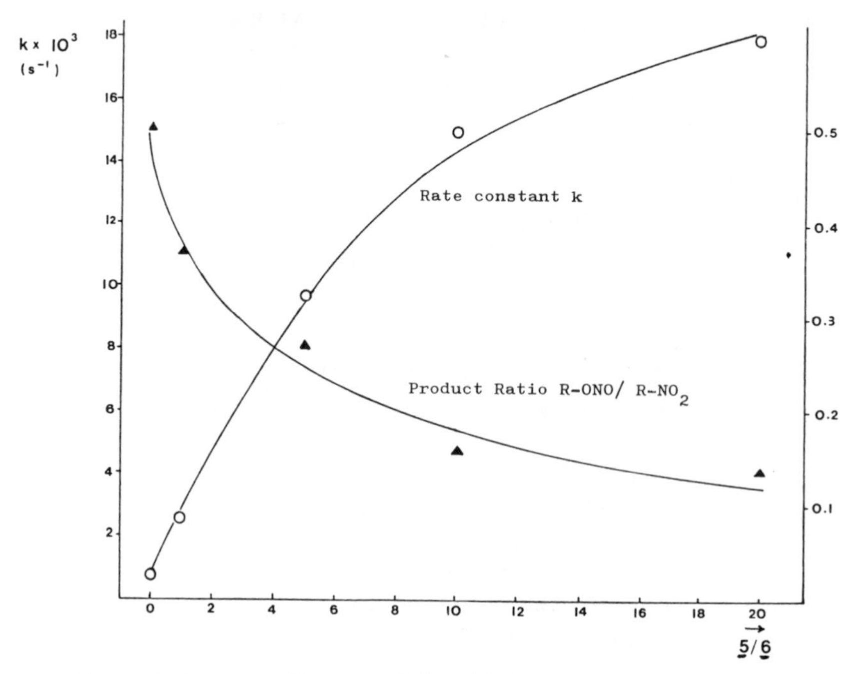

Figure 3. Reaction of bromomethylnaphthalene, **VI**, *with nitrite in the presence of host* **V**; *the rate constants and product ratio are a function of the catalyst-to-substrate ratio.*

corrected for the opposite effect observed upon addition of open-chain alkylammonium salts during these substitutions, which in contrast to the catalyst **V** leads to an increase of the R–ONO ester, the change in regioselectivity brought about by the host compound is almost complete. Reactions of **VI** with other ambident anions such as cyanide or isothiocyanate are also considerably accelerated by the host **V**; the products, however, are not changed as in the case with nitrite anions, which is due to the intrinsically much larger predominance for attack at C, or S, respectively, with the nucleophiles CN^- and SCN^- (*60*).

The observations presented in Scheme III and in references 51 and 57 document the functional similarity of the host catalyst **V** to an enzyme: the system discriminates not only between educts but also between transition states; the system also shows competitive inhibition by a stronger yet unreactive binder (Scheme III). The rationalization of these results is straightforward on the basis of the known principles ruling nucleophilic substitutions with ambident ions. The transition state with significant charge separation (S_N1 like), leading to the hard–hard combination product R–ONO, is suppressed by the positively charged environment, whereas the negatively charged transition state of the S_N2-type reaction is stabilized. Moreover, the anion accumulation around the ammonium ions also enhances the participation of the direct displacement reaction.

Complexation of Nucleophiles Instead of Organic Substrates: Inhibition

The diazacyclophane derivative **IX** (DR) is too small to accommodate the organic substrate **VI** within the cavity (Chart IV). Instead, the anion is complexed, as evident from preliminary NMR studies, for example, with cyanide (*26*). Here, we observe a ^{13}C-NMR shielding effect of 15 ppm upon complexation of CN^- with **IX**, as well as significant line broadening as a consequence of the dissociation of the complex being slow on the NMR time scale. In contrast, the interaction of the larger host **VI** with cyanide leads only to up to 3-ppm shielding, and the lines are not broadened.

These results are so far comparable to those of Schmidtchen (*25*), who also observed strong anion binding preferentially with the smaller host **I**. In contrast to the small kinetic effect reported for **I**, however, we find a large rate decrease in the reaction of **VI** with nitrite after adding the host **IX** (Figure 4), which levels off after the macrocycle has taken up to two molecules of anion. Preliminary NMR titrations support the view that the complex contains more than one anion. The use of smaller charged host structures thus opens up the possibility for a selective inhibition of nucleophilic substitution.

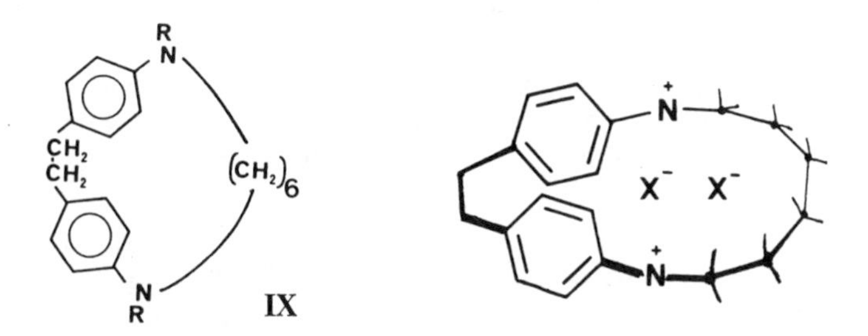

Chart IV. A smaller azacyclophane complexing anions.

An Outlook on S_N1-Type Reactions

Unimolecular nucleophilic substitutions, characterized by the buildup of a positive charge in the organic substrate, are not expected to undergo catalysis by macrocyclic onium ions unless the host stabilizes the leaving group by complexation. Nevertheless, preliminary measurements with **V** and **VI** in the presence of hard nucleophiles such as sodium tosylate or sulfate in water did show a moderate rate constant increase of up to 50% in comparison to experiments conducted in absence of the host, which can be accounted for by a salt effect due to the increased anion concentration around the reactive center containing the leaving group.

Significant stabilization of S_N1-like transition states requires the presence of a negatively charged cavity, which, however, possibly does not involve nucleophilic groups such as carboxylates that can undergo direct

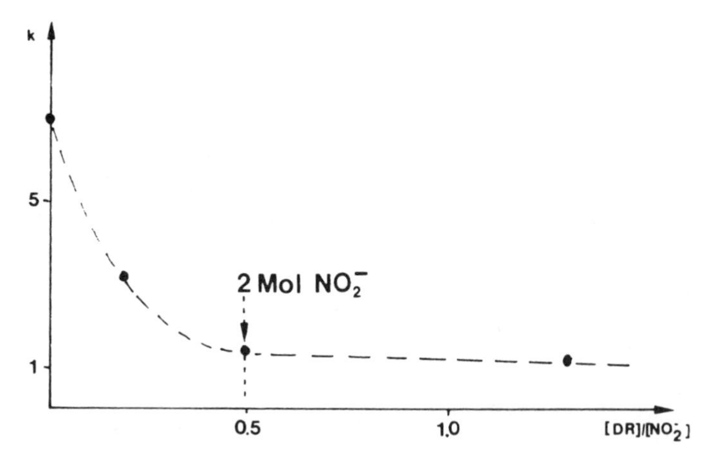

Figure 4. Reaction rate of **VI** *with nitrite in the presence of the smaller host* **IX** *as a function of the host-to-nitrite ratio.*

substitution by the electrophilic substrate. Host compounds with negative charges have been described in the literature (*61, 62*), but these contain carboxyl groups; others, like the so-called calixarenes (*63*) are not soluble in any suitable solvent. This finding may explain why calixarenes (*63*) have to our knowledge thus far not been studied with respect to their complexation abilities. We (*64*) recently found that phenol–aldehyde condensation products, similar to the calixarenes, are soluble in water and that they do complex organic substrates effectively, provided that these substrates bear a positive charge.

Macrocyclic products obtained from resorcinol and, for example, acetaldehyde were obtained a long time ago (*65–68*); their structures have been studied in the crystal as well as in solution, although invariably in form of substitution products of these phenols (*69, 70*). On the basis of the NMR observation of different isomers, we propose for one of the phenolic macrocycles the highly symmetric structure **X**, similar to an ester conformation described earlier (*69*) (Scheme IV). The octaphenol is easily deprotonated to the tetraanion **XI**, which then, however, resists any further deprotonation. We ascribe this unusual behavior to the very strong hydrogen bonds requiring the presence of at least four phenolic protons. Another consequence of the strong hydrogen bonds is that the phenols, even in basic solutions, are not attacked by the reacting substrates.

The macrocycle **XI** resembles an open cup with a delocalized negative charge near the rim. Because of the relatively open and polar cavity, organic substrates are bound only to a small degree, unless they bear a positive charge. The large association constants of tetraalkylammonium ion compounds decrease with increasing length of the alkyl residues, which reflects the decreasing electrostatic attraction resulting from the larger separation

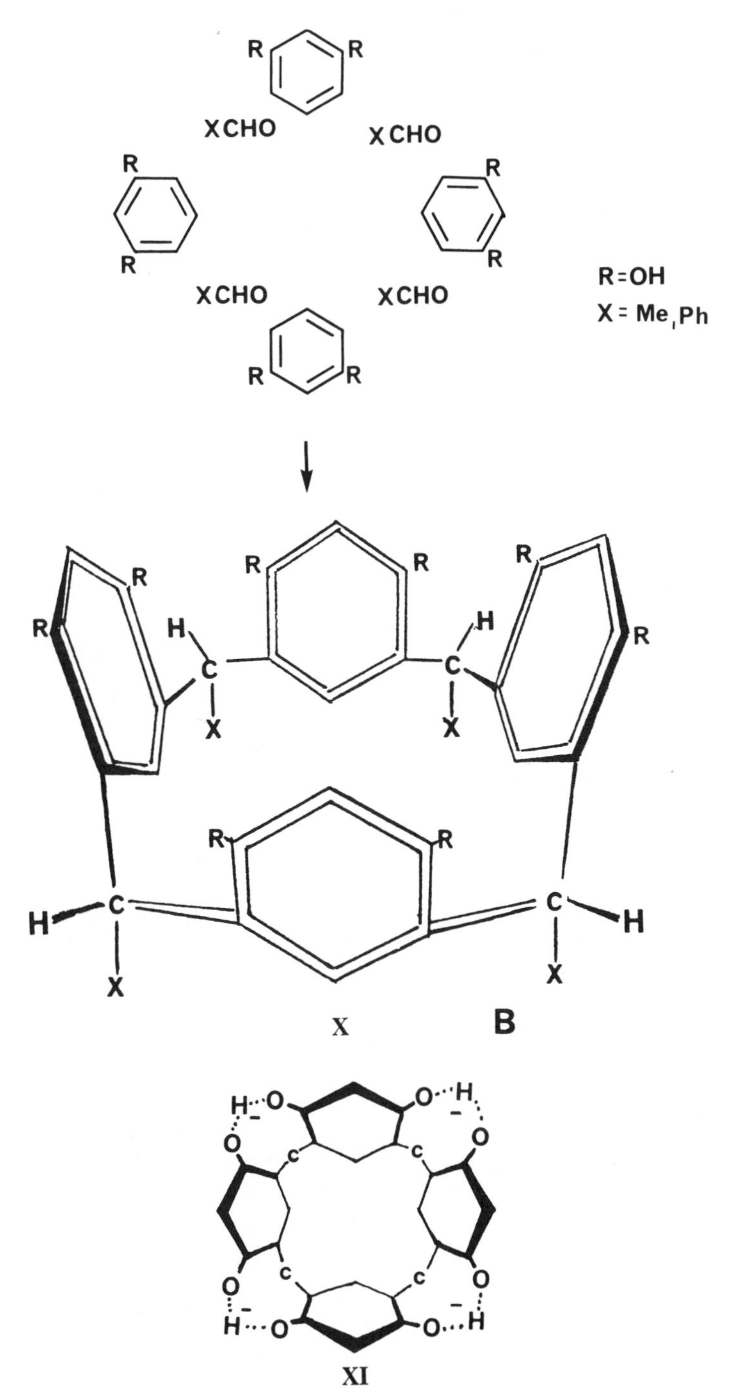

Scheme IV. Formation of the cyclic octaphenol **X** *and the tetraanion* **XI**.

between the charge centers. The increasing distance from the ring current in the cavity is simultaneously visible in the smaller complexation shifts (Chart V).

The presence of a positively charged center in a substrate leads to a low reactivity in nucleophilic substitution reactions, if the charge is not too isolated from the atom bearing the leaving group. Compounds such as **XII**

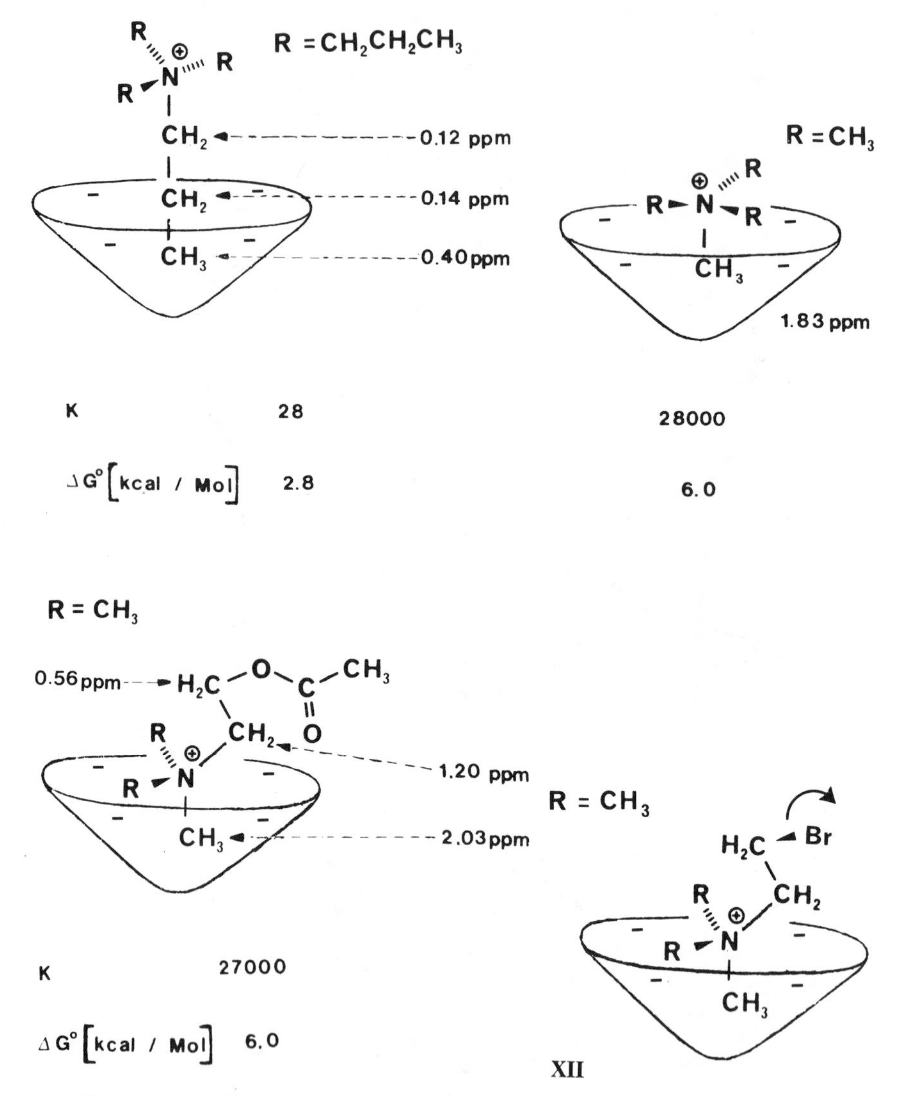

Chart V. Complexation free energies, ΔG, ^{1}H-NMR complexation shifts, and a possible catalytic mechanism with **XI**.

bind as strongly to **X** as acetylcholine; the hydrolysis of **XII**, however, is retarded because of a still predominating S_N2 mechanism. The use of charged macrocycles capable of complexing substrates with an opposite charge makes it possible to neutralize the intrinsic inductive or field effect of charges within a given substrate by through space-field effects of the molecular receptor.

Acknowledgments

This work was supported by the Deutsche Forschungsgemeinschaft, the Fonds der Chemischen Industrie, Frankfurt, and by an Akademie-Stipendium of the Volkswagen-Stiftung to H.-J.S. This review was written during a stay by H.-J.S. at the Department of Chemistry, University of California, San Diego. The generous help of R. C. Fahey, particularly with the manuscript preparation, is sincerely appreciated. Assistance with computer facilities by members of the Agouron Institute and of Biosym Technologies, San Diego, is gratefully acknowledged.

Literature Cited

1. Breslow, R. *Science* (Washington, D.C.) **1982,** *218*, 532.
2. Cram, D. J.; Trueblood, K. N. *Top. Curr. Chem.* **1981,** *98*, 43.
3. Tabushi, I. In *IUPAC—Frontiers of Chemistry*; Laidler, K. J. Ed.; Pergamon: Oxford, 1982; p 275.
4. Lehn, J. M. In *Biomimetic Chemistry*; Yoshida, Z.-I.; Ise, N., Eds.; Kodansha: Tokyo 1983, pp 1, 163.
5. Murakami, Y. *Top. Curr. Chem.* **1983,** *115*, 107.
6. Gokel, G. W.; Korzeniowski, S. H. *Macrocyclic Polyether Synthesis*; Springer Verlag: Berlin; 1982.
7. Hiraoka, M. *Crown Compounds;* Kodansha: Tokyo, Elsevier: Amsterdam; 1982.
8. Izatt, R. M.; Christensen, J. J., Eds. *Synthetic Multidentate Compounds*; Academic: New York; 1982.
9. Izatt, R. M.; Christensen, J. J., Eds. *Progress in Macrocyclic Chemistry*; Wiley: New York; 1979, 1981.
10. Vögtle, F.; Sieger, H.; Müller, W. H. *Top. Curr. Chem.* 1981, *98*, 107.
11. Dehmlov, E. V.; Dehmlov, S. S. *Phase Transfer Catalysis* Verlag Chemie: Weinheim; 2nd ed.; 1983.
12. Starks, C. M.; Liotta, C. *Phase Transfer Catalysis: Principles and Techniques* Academic Press: New York, 1978.
13. Makosza, M. *Pure Appl. Chem.* **1975,** *43*, 439.
14. Makosza, M. *Surv. Prog. Chem.* **1980,** *9*, 1.
15. Vögtle, F.; Weber, E. *Angew. Chem. Int. Ed. Engl.* **1979,** *18*, 753.
16. Stott, P. E.; Bradshaw, J. S.; Parish, W. J. *J. Am. Chem. Soc.* **1980,** *102*, 4810.
17. Blasius, E.; Janzen, K.-P. *Top. Curr. Chem.* **1981,** *98*, 163.
18. Fersht, A. *Enzyme Structure and Mechanism* Freeman: Reading, PA, 1977.
19. Jencks, W. P. *Catalysis in Chemistry and Enzymology* McGraw Hill Co.: New York; 1969.
20. Bender, M. L.; Komiyama, M., *Cyclodextrin Chemistry;* Springer Verlag: Berlin; 1977.

21. Szeijtli, J. *Cyclodextrins and their Inclusion Complexes* Akademiai Kiado: Budapest; 1982.
22. Saenger, W. *Angew. Chem. Int. Ed. Engl.* **1980,** *21,* 344.
23. Schneider, H.-J.; Philippi, K. *Chem. Ber.* **1984,** *117,* 3056.
24. Odashima, K.; Koga, K. In *Cyclophanes;* Academic: New York, 1970; Vol. 2, p. 629.
25. Schmidtchen, F. P. *Chem. Ber.* **1981,** *114,* 597.
26. Schneider, H.-J.; Busch, R.; Kramer, R., unpublished results.
27. Tabushi, I.; Sasaki, H.; Kuroda, Y. *J. Am. Chem. Soc.* **1976,** *98,* 5727.
28. Odashima, K.; Soga, T.; Koga, T. *Tetrahedron Lett.* **1981,** 5311, and earlier references cited therein.
29. Jarvi, E. T.; Whitlock, H. W. *J. Am. Chem. Soc.* **1982,** *104,* 1602, and earlier references cited therein.
30. Winkler, J.; Coutouli-Agryropulu, E.; Leppkes, R.; Breslow, R. *J. Am. Chem. Soc.* **1983,** *105,* 7198.
31. Diederich, F.; Dick, K. *Angew. Chem. Int. Ed. Engl.* **1984,** *23,* 810.
32. Diederich, F.; Griebel, D. *J. Am. Chem. Soc.* **1984,** *106,* 8037, and references cited therein.
33. Vögtle, F.; Merz, Th.; Wirtz, H. *Angew. Chem. Int. Ed. Engl.* **1985,** *24,* 221, and references cited therein.
34. Rossa, L.; Vögtle, F. *Top. Curr. Chem.* **1983,** *113,* 1.
35. Schneider, H.-J.; Philippi, K.; Pöhlmann, J. *Angew. Chem. Int. Ed. Engl.* **1984,** *23,* 908.
36. Schneider, H.-J.; Müller, W. *J. Org. Chem.* **1985,** *50,* 4609.
37. Schneider, H.-J.; Buchheit, U.; Becker, N.; Schmidt, G.; Siehl, U. *J. Am. Chem. Soc.* **1985,** *107,* 7027, and earlier references therein.
38. Schneider, H.-J.; Schmidt, G.; Thomas, F. *J. Am. Chem. Soc.* **1983,** *105,* 3556, and earlier publications cited therein.
39. Popov, A. I.; Lehn, J. M.: In *Coordination Chemistry of Macrocyclic Compounds;* Melson, G. A., Ed.; Plenum: New York; 1979.
40. Schneider, H.-J.; Philippi, K.; Pöhlmann, J.; Kramer, R., unpublished results.
41. Purdich, D. L., Ed. *Contemporary Enzume Kinetics and Mechanisms;* Academic: New York; 1983.
42. Lüttringhaus, A. *Liebigs Ann. Chem.* **1936,** 528, *211,* 223.
43. Lüttringhaus, A. *Ber. Dtsch. Chem. Ges.* **1939,** *72,* 887.
44. Schneider, H.-J.; Busch, R. *Chem. Ber.,* **1986,** *119,* 747.
45. Stetter, H. *Chem. Ber.* **1953,** *86,* 197.
46. Stetter, H.; Ross, E. E. *Chem. Ber.* **1985,** *88,* 1390.
47. Stetter, H.; Marx-Moll, L. *Chem. Ber.* **1953,** *91,* 677, 1775.
48. Fuson, R. R.; House, H. O. *J. Am. Chem. Soc.* **1953,** *75,* 1327, 5744.
49. Vriesama, B. K.; Buter, J.; Kellogg, R. M. *J. Org. Chem.* **1984,** *49,* 110, and references therein.
50. Mandolini, L.; Illuminati, G. *Acc. Chem. Res.* **1981,** *14,* 95, and references cited therein.
51. Busch, R. Ph.D. Dissertation, Universität des Saarlandes, Saarbrücken, 1984.
52. Schmidtchen, F. P. *Angew. Chem. Int. Ed. Engl.* **1981,** *20,* 466.
53. Schmidtchen, F. P. *Chem. Ber.* **1984,** *117,* 725.
54. Schmidtchen, F. P. *Chem. Ber.* **1980,** *113,* 864.
55. Broxton, T. J.; Jakovljevic, A. C. *Austr. J. Chem.* **1982,** *35,* 2557, and references cited therein.
56. Bunton, C. A.; Moffatt, J. R.; Rodenas, E. *J. Am. Chem. Soc.* **1982,** *104,* 2653.
57. Schneider, H.-J.; Busch, R. *Angew. Chem. Int. Ed. Engl.* **1984,** *23,* 912.
58. Schneider, H.-J.; Güttes, D. *Angew. Chem. Int. Ed. Engl.* **1984,** *23,* 910.

59. Burkert, U.; Allinger, N. L. *Molecular Mechanics;* ACS Monograph 177; American Chemical Society: Washington DC, 1981, and reviews cited in reference 38.
60. Friedrich, K.; Wallenfels, K. In *The Chemistry of the Cyanogroup* Rappoport, Z., Ed.; Interscience: London; 1970.
61. Dhaenens, M.; Lacombe, L.; Lehn, J.-M.; Vigneron, J.-P. *J. Chem. Soc. Chem. Comm. 1984,* 1097.
62. Andreetti, G. D.; Ungaro, R.; Pochini, A. J. *J. Chem. Soc. Chem. Comm.* **1979,** 1005.
63. Gutsche, C. D. *Acc. Chem. Res.* **1983,** *16,* 161.
64. Schneider, H.-J.; Güttes, D.; Schneider, U., *Angew. Chem., Intl. Ed. Engl.* **1986,** *25,* 647.
65. Bayer, A. v. *Ber. Dtsch. Chem. Ges.* **1872,** *5,* 25, 280, 1094.
66. Zinke, A.; Kretz, R.; Leggewie, E.; Hossinger, K. *Monatsh. Chem.* **1952,** *83,* 1213.
67. Niederl, J. B.; Vogel, H. J. *J. Am. Chem. Soc.* **1940,** *62,* 2512.
68. Erdtman, H.; Haglid, F.; Ryhage, R. *Acta Chem. Scand.* **1964,** *18,* 1249.
69. Högberg, A. G. S. *J. Am. Chem. Soc.* **1980,** *102,* 6046.
70. Palmer, K. J.; Wong, R. Y.; Jurd, L.; Stevens, K. *Acta Cryst. B* **1976,** *32,* 847.

RECEIVED for review October 21, 1985. ACCEPTED June 17, 1986.

AUTHOR INDEX

SUBJECT INDEX

A

B

D

N

O

P

Copy editing and indexing by Deborah H. Steiner
Production by Joan C. Cook
Jacket design by Pamela Lewis
Managing Editor: Janet S. Dodd

Recent ACS Books

Personal Computers for Scientists: A Byte at a Time
By Glenn I. Ouchi
288 pages; clothbound; ISBN 0-8412-1001-2

Writing the Laboratory Notebook
By Howard M. Kanare
145 pages; clothbound; ISBN 0-8412-0906-5

The ACS Style Guide: A Manual for Authors and Editors
Edited by Janet S. Dodd
264 pages; clothbound; ISBN 0-8412-0917-0

Chemical Demonstrations: A Sourcebook for Teachers
By Lee R. Summerlin and James L. Ealy, Jr.
192 pages; spiral bound; ISBN 0-8412-0923-5

Phosphorus Chemistry in Everyday Living, Second Edition
By Arthur D. F. Toy and Edward N. Walsh
342 pages; clothbound; ISBN 0-8412-1002-0

Pharmacokinetics: Processes and Mathematics
By Peter G. Welling
ACS Monograph 185; 290 pages; ISBN 0-8412-0967-7

Phase-Transfer Catalysis: New Chemistry, Catalysts, and Applications
Edited by Charles M. Starks
ACS Symposium Series 326; 195 pages; ISBN 0-8412-1007-1

Geochemical Processes at Mineral Surfaces
Edited by James A. Davis and Kim F. Hayes
ACS Symposium Series 323; 684 pages; ISBN 0-8412-1004-7

Polymeric Materials for Corrosion Control
Edited by Ray A. Dickie and F. Louis Floyd
ACS Symposium Series 322; 384 pages; ISBN 0-8412-0998-7

Porphyrins: Excited States and Dynamics
Edited by Martin Gouterman, Peter M. Rentzepis, and Karl D. Straub
ACS Symposium Series 321; 384 pages; ISBN 0-8412-0997-9

Water-Soluble Polymers: Beauty with Performance
Edited by J. E. Glass
Advances in Chemistry Series 213; 449 pages; ISBN 0-8412-0931-6

Historic Textile and Paper Materials: Conservation and Characterization
Edited by Howard L. Needles and S. Haig Zeronian
Advances in Chemistry Series 212; 464 pages; ISBN 0-8412-0900-6
